LONDON MATHEMATICAL SOCIETY STUDENT TEXTS

Managing editor: Professor D. J. Benson,
Department of Mathematics, University of Aberdeen, UK

London Mathematical Society Student Texts 72

Elements of the Representation Theory of Associative Algebras

Volume 3 Representation-Infinite Tilted Algebras

DANIEL SIMSON
Nicolaus Copernicus University

ANDRZEJ SKOWROŃSKI
Nicolaus Copernicus University

CAMBRIDGE
UNIVERSITY PRESS

CAMBRIDGE
UNIVERSITY PRESS

Shaftesbury Road, Cambridge CB2 8EA, United Kingdom

One Liberty Plaza, 20th Floor, New York, NY 10006, USA

477 Williamstown Road, Port Melbourne, VIC 3207, Australia

314–321, 3rd Floor, Plot 3, Splendor Forum, Jasola District Centre, New Delhi – 110025, India

103 Penang Road, #05–06/07, Visioncrest Commercial, Singapore 238467

Cambridge University Press is part of Cambridge University Press & Assessment, a department of the University of Cambridge.

We share the University's mission to contribute to society through the pursuit of education, learning and research at the highest international levels of excellence.

www.cambridge.org
Information on this title: www.cambridge.org/9780521708760

First published 2007

A catalogue record for this publication is available from the British Library

ISBN	978-0-521-88218-7	Hardback
ISBN	978-0-521-70876-0	Paperback

To our Wives

Sabina and Mirosława

Contents

Introduction

The first volume serves as a general introduction to some of the techniques most commonly used in representation theory. The quiver technique, the Auslander–Reiten theory and the tilting theory were presented with some application to finite dimensional algebras over a fixed algebraically closed field. In particular, a complete classification of those hereditary algebras that are representation-finite (that is, admit only finitely many isomorphism classes of indecomposable modules) is given. The result, known as Gabriel's theorem, asserts that a basic connected hereditary algebra A is representation-finite if and only if the quiver Q_A of A is a Dynkin quiver.

In Volume 2 we study in detail the indecomposable modules and the shape of the Auslander–Reiten quiver $\Gamma(\mathrm{mod}\,A)$ of the class of hereditary algebras A that are representation-infinite and minimal with respect to this property. They are just the hereditary algebras of Euclidean type, that is, the path algebras KQ, where Q is a connected acyclic quiver whose underlying non-oriented graph $\overline{Q}$ is one of the following Euclidean diagrams

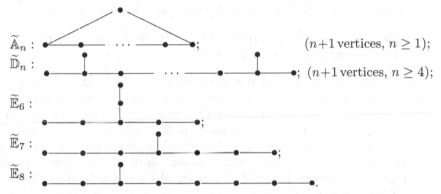

$\widetilde{\mathbb{A}}_n$: $\qquad\qquad\qquad\qquad\qquad\qquad$ $(n+1\,\text{vertices},\ n \geq 1)$;

$\widetilde{\mathbb{D}}_n$: $\qquad\qquad\qquad\qquad\qquad\qquad$ $(n+1\,\text{vertices},\ n \geq 4)$;

$\widetilde{\mathbb{E}}_6$:

$\widetilde{\mathbb{E}}_7$:

$\widetilde{\mathbb{E}}_8$:

In Volume 2, we also study in detail the indecomposable modules and the shape of the Auslander–Reiten quiver $\Gamma(\mathrm{mod}\,B)$ of concealed algebras of Euclidean type, that is, the tilted algebras B of the form

$$B = \mathrm{End}\,T_{KQ},$$

where KQ is a hereditary algebra of Euclidean type and T_{KQ} is a postprojective tilting KQ-module.

The main aim of the first part of Volume 3 is to study arbitrary representation-infinite tilted algebras $B = \mathrm{End}\,T_{KQ}$ of a Euclidean type Q, where T_{KQ} is a tilting T_{KQ}-module, and to give a fairly complete description of their indecomposable modules, their module categories $\mathrm{mod}\,B$, and the Auslander–Reiten quivers $\Gamma(\mathrm{mod}\,B)$.

For this purpose, we introduce in Chapters XV-XVII some concepts and tools that allow us to give in Chapter XVII a complete description of arbitrary representation-infinite tilted algebras B of Euclidean type and their module categories mod B, due to Ringel [525]. In particular, we show that:

- the Auslander–Reiten quiver $\Gamma(\text{mod}B)$ of any such an algebra B has a disjoint union decomposition

$$\Gamma(\text{mod}B) = \mathcal{P}(B) \cup \boldsymbol{\mathcal{T}}^B \cup \mathcal{Q}(B),$$

  where $\mathcal{P}(B)$ is a unique postprojective component, $\mathcal{Q}(B)$ is a unique preinjective component, and

$$\boldsymbol{\mathcal{T}}^B = \{\mathcal{T}_\lambda^B\}_{\lambda \in \mathbb{P}_1(K)}$$

  is a $\mathbb{P}_1(K)$-family of pairwise orthogonal standard ray or coray tubes $\mathcal{T}_\lambda^B$ separating $\mathcal{P}(B)$ from $\mathcal{Q}(B)$;

- the module category mod B of a tilted algebra B of Euclidean type is controlled by the Euler quadratic form $q_B : K_0(B) \longrightarrow \mathbb{Z}$ of B, and

- the number of the isomorphism classes of tilted algebras of Euclidean type of any fixed dimension is finite.

In Chapter XVIII, we turn our attention to the representation theory of wild hereditary algebras $A = KQ$, where Q is an acyclic quiver such that the underlying graph is neither a Dynkin nor a Euclidean diagram. The shape of the components of the regular part $\mathcal{R}(A)$ of $\Gamma(\text{mod }A)$ is described and, for any such an algebra A, a wild behaviour of the category mod A is established. Moreover, an important theorem on homomorphisms between the regular modules over a wild hereditary algebra, due to Baer [35] and Kerner [343], is proved.

An essential rôle in the investigation is played by the notion of a perpendicular category associated to a partial tilting module, introduced by Geigle and Lenzing [247] and Schofield [559].

We also exhibit some classes of tilted algebras B of wild type and we discuss the structure of their module categories mod B. In particular, we prove a theorem of Ringel [526] on the existence of a regular tilting module over a hereditary algebra, and we present an efficient procedure of Baer [35], [36] allowing us to construct regular tilting modules over any wild hereditary algebra A with at least three pairwise non-isomorphic simple modules.

In Chapter XIX, we introduce the concepts of tame representation type and of wild representation type for algebras, and we discuss the tame and the wild nature of module categories mod B. We prove that the concealed algebras of Euclidean type are of tame representation type, and the concealed algebras of wild type are of wild representation type.

In the final Chapter XX, we present (without proofs) selected results of the representation theory of finite dimensional algebras that are related to the material discussed in the previous chapters. This, together with a rather long list of complementary references, should provide the reader with the right tives for further study and interesting research directions.

Unfortunately, many important topics from the theory have been left out. Among the most notable omissions are covering techniques, the use of derived categories and partially ordered sets. Some other aspects of the theory presented here are discussed in the books [34], [53], [54], [242], [318], [276], [575], and especially [525].

We assume that the reader is familiar with Volumes 1 and 2, but otherwise the exposition is reasonably self-contained, making it suitable either for courses and seminars or for self-study. The text includes many illustrative examples and a large number of exercises at the end of each of the Chapters XV-XIX.

The book is addressed to graduate students, advanced undergraduates, and mathematicians and scientists working in representation theory, ring and module theory, commutative algebra, abelian group theory, and combinatorics. It should also, we hope, be of interest to mathematicians working in other fields.

Throughout this book we use freely the terminology and notation introduced in Volumes 1 and 2. We denote by K a fixed algebraically closed field. The symbols $\mathbb{N}$, $\mathbb{Z}$, $\mathbb{Q}$, $\mathbb{R}$, and $\mathbb{C}$ mean the sets of natural numbers, integers, rational, real, and complex numbers. The cardinality of a set X is denoted by $|X|$. Given an algebra A, the A-module means a finite dimensional right A-module. We denote by $\mathrm{Mod}\, A$ the category of all right A-modules, by $\mathrm{mod}\, A$ the category of finite dimensional right A-modules, and by $\Gamma(\mathrm{mod}\, A)$ the Auslander–Reiten translation quiver of A. The ordinary quiver of an algebra A is denoted by Q_A. Given a matrix $C = [c_{ij}]$, we denote by C^t the transpose of C.

A finite quiver $Q = (Q_0, Q_1)$ is called a **Euclidean quiver** if the underlying graph $\overline{Q}$ of Q is any of the Euclidean diagrams $\widetilde{\mathbb{A}}_m$, with $m \geq 1$, $\widetilde{\mathbb{D}}_m$, with $m \geq 4$, $\widetilde{\mathbb{E}}_6$, $\widetilde{\mathbb{E}}_7$, and $\widetilde{\mathbb{E}}_8$. Analogously, Q is called a **Dynkin quiver** if the underlying graph $\overline{Q}$ of Q is any of the Dynkin diagrams $\mathbb{A}_m$, with $m \geq 1$, $\mathbb{D}_m$, with $m \geq 4$, $\mathbb{E}_6$, $\mathbb{E}_7$, and $\mathbb{E}_8$.

We take pleasure in thanking all our colleagues and students who helped us with their useful comments and suggestions. We wish particularly to express our appreciation to Ibrahim Assem, Sheila Brenner, Otto Kerner, and Kunio Yamagata for their helpful discussions and suggestions. Particular thanks are due to Dr. Jerzy Białkowski and Dr. Rafał Bocian for their help in preparing a print-ready copy of the manuscript.

Chapter XV

Tubular extensions and tubular coextensions of algebras

In Volume 2, we study in detail the indecomposable modules and the shape of the Auslander–Reiten quiver $\Gamma(\mathrm{mod}\, B)$ of concealed algebras of Euclidean type, that is, the tilted algebras B of the form

$$B = \mathrm{End}\, T_{KQ},$$

where KQ is a hereditary algebra of Euclidean type and T_{KQ} is a post-projective tilting KQ-module. We recall that every concealed algebra B of Euclidean type is representation-infinite and the Auslander–Reiten quiver $\Gamma(\mathrm{mod}\, B)$ of B has the shape

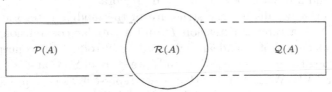

where $\mathrm{mod}\, B$ is the category of finite dimensional right B-modules, $\mathcal{P}(B)$ is the unique postprojective component of $\Gamma(\mathrm{mod}\, B)$ containing all the inde-composable projective B-modules, $\mathcal{Q}(B)$ is the unique preinjective compo-nent of $\Gamma(\mathrm{mod}\, B)$ containing all the indecomposable injective B-modules, and $\mathcal{R}(B)$ is the (non-empty) regular part consisting of the remaining com-ponents of $\Gamma(\mathrm{mod}\, B)$. We recall also that:

- the regular part $\mathcal{R}(B)$ of the Auslander–Reiten quiver $\Gamma(\mathrm{mod}\, B)$ is a disjoint union of the $\mathbb{P}_1(K)$-family

$$\boldsymbol{\mathcal{T}}^B = \{\mathcal{T}_\lambda^B\}_{\lambda \in \mathbb{P}_1(K)}$$

of pairwise orthogonal standard stable tubes $\mathcal{T}_\lambda^B$, where $\mathbb{P}_1(K)$ is the projective line over K,
- the family $\boldsymbol{\mathcal{T}}^B$ separates the postprojective component $\mathcal{P}(B)$ from the preinjective component $\mathcal{Q}(B)$,
- the module category $\mathrm{mod}\, B$ is controlled by the Euler quadratic form $q_B : K_0(B) \longrightarrow \mathbb{Z}$ of the algebra B.

In Volume 3, we study the representation-infinite tilted algebras $B =$ End T_{KQ} of a Euclidean type Q, where T_{KQ} is a tilting T_{KQ}-module. We give a fairly complete description of their indecomposable modules, their module categories mod B, and the Auslander–Reiten quivers $\Gamma(\text{mod } B)$.

The aim of the present chapter is to introduce concepts playing a fundamental rôle in the classification of arbitrary representation-infinite tilted algebras of Euclidean type, presented in Chapter XVII.

In Section 1, we introduce the concepts of a one-point extension and a one-point coextension of an algebra, and we discuss a behavior of almost split sequences under the one-point extension and the one-point coextension procedure.

In Section 2, we introduce the concepts of a tubular extension and a tubular coextension of an algebra, and the related concepts of ray tubes and coray tubes. As we shall see in Chapter XVII, the components of a representation-infinite tilted algebra of Euclidean type that are neither postprojective nor preinjective, are ray tubes or coray tubes.

In Section 3, we show that the concepts of the tubular extension and the tubular coextension of an algebra coincide with the concepts of a branch extension and a branch coextension of an algebra.

In Section 4, we discuss the structure of the module categories mod B and mod B' of a tubular extension B and a tubular coextension B' of a concealed algebra A of Euclidean type, and we introduce the concept of the tubular type of such algebras. The study we start in Section 4 is continued in Chapter XVII. We show there that every representation-infinite tilted algebra of Euclidean type is either a domestic tubular extension or a domestic tubular coextension of a concealed algebra of Euclidean type.

XV.1. One-point extensions and one-point coextensions of algebras

We start by explaining the idea of a one-point extension algebra. Assume that B is a K-algebra such that the quiver Q_B of B has a source vertex 0. We form a K-algebra A in such a way that the quiver Q_A of A is obtained by deleting from the quiver Q_B of B the source 0, as well as all the arrows passing through 0. We are interested in the relation between the representation theories of B and of the algebra A.

Let e_0 be the idempotent of B corresponding to the source vertex 0, and we set

$$A = (1 - e_0)B(1 - e_0).$$

Since the algebra A is isomorphic the quotient algebra B/Be_0B of B (modulo the two-sided ideal Be_0B generated by the idempotent e_0), then the canonical algebra epimorphism

$$B \longrightarrow A \cong B/Be_0B$$

induces an embedding of module categories $\mathrm{mod}\, A \hookrightarrow \mathrm{mod}\, B$, called the standard embedding.

On the other hand, because 0 is a source of Q_B, the injective right B-module $I(0)_B = D(Be_0)$ is simple and, consequently, there is a K-algebra isomorphism $e_0Be_0 \cong \mathrm{End}\,(D(Be_0)) \cong K$, and

$$(1 - e_0)Be_0 \cong \mathrm{Hom}_B(D(Be_0), D(B(1 - e_0))) = 0.$$

Moreover, the (right) K-vector space $X = e_0B(1 - e_0)$ has a canonical right A-module structure and a canonical left K-module structure induced by the right one (that is, defined by the formula $\lambda \cdot x = x\lambda$, for all $x \in X$ and $\lambda \in K$); they define a K-A-bimodule structure $_KX_A$ on X, see (I.2.10). It follows that we can view the algebra B in the matrix form

$$B = \begin{bmatrix} (1 - e_0)B(1 - e_0) & (1 - e_0)Be_0 \\ & \\ e_0B(1 - e_0) & e_0Be_0 \end{bmatrix} = \begin{bmatrix} A & 0 \\ _KX_A & K \end{bmatrix},$$

where $A = (1 - e_0)B(1 - e_0) \cong B/Be_0B$, $_KX_A = X = e_0B(1 - e_0)$, $K \cong e_0Be_0$, and the multiplication is induced from the K-A-bimodule structure of $_KX_A$, see (A.2.7) of Volume 1. Because the right ideal

$$e_0B = \begin{bmatrix} 0 & 0 \\ _KX_A & K \end{bmatrix} \quad \text{of} \quad B = \begin{bmatrix} A & 0 \\ _KX_A & K \end{bmatrix}$$

is an indecomposable projective B-module, then the A-module $_KX_A$ identified with the B-submodule $\begin{bmatrix} 0 & 0 \\ _KX_A & 0 \end{bmatrix}$ of e_0B equals the radical $\mathrm{rad}\, e_0B$ of e_0B, that is, we make the identification

$$\mathrm{rad}\, e_0B = \begin{bmatrix} 0 & 0 \\ _KX_A & 0 \end{bmatrix} \equiv {}_KX_A.$$

These considerations, already used implicitly, for instance in (VII.2.5) and (IX.4), and their duals, lead to the following definitions.

1.1. Definition. Let A be a K-algebra, and X be a right A-module.

(a) The **one-point extension** of A by X, which we denote by $A[X]$, is the 2×2-matrix algebra

$$A[X] = \begin{bmatrix} A & 0 \\ {}_K X_A & K \end{bmatrix}$$

with the ordinary addition of matrices, and the multiplication induced from the usual K-A-bimodule structure ${}_K X_A$ of X, see (A.2.7).

(b) The **one-point coextension** of A by X, which we denote by $[X]A$, is the 2×2-matrix algebra

$$[X]A = \begin{bmatrix} K & 0 \\ DX & A \end{bmatrix}$$

with the ordinary addition of matrices, and the multiplication induced from the A-K-bimodule structure of $DX = \mathrm{Hom}_K({}_K X_A, K)$ induced by the K-A-bimodule structure of ${}_K X_A$, see Section I.2.9 of Volume 1.

We recall that given two K-algebras A, C, and a finite dimensional C-A-bimodule ${}_C X_A$, the set

$$B = \begin{bmatrix} A & 0 \\ {}_C X_A & C \end{bmatrix}$$

of all matrices $\begin{bmatrix} a & 0 \\ x & c \end{bmatrix}$, where $a \in A$, $c \in C$, and $x \in X$, endowed with the usual matrix addition and the multiplication given by the formula

$$\begin{bmatrix} a & 0 \\ x & c \end{bmatrix} \cdot \begin{bmatrix} a' & 0 \\ x' & c' \end{bmatrix} = \begin{bmatrix} aa' & 0 \\ xa'+cx' & cc' \end{bmatrix},$$

is a finite dimensional K-algebra with identity element $1 = e_A + e_C$, where $e_A = \begin{bmatrix} 1 & 0 \\ 0 & 0 \end{bmatrix}$ and $e_C = \begin{bmatrix} 0 & 0 \\ 0 & 1 \end{bmatrix}$, see (A.2.7).

It is easy to see that the quiver $Q_{A[X]}$ of the one-point extension algebra $A[X]$ contains the quiver Q_A of A as a full convex subquiver, and there is a single additional point in $Q_{A[X]}$, which is a source vertex. One may thus visualise the quiver $Q_{A[X]}$ of $A[X]$ as follows

Conversely, the considerations preceding the definition show that any algebra B having a source in the quiver Q_B can be written as a one-point extension $A[X]$ of a quotient A of B by the two-sided ideal generated by the idempotent corresponding to this source.

Dually, the quiver $Q_{[X]A}$ of the one-point coextension K-algebra $[X]A$ contains the quiver Q_A of A, as a full convex subquiver, and there is a single additional point in $Q_{[X]A}$, which is a sink vertex. One may thus visualise the quiver $Q_{[X]A}$ of $[X]A$ as follows

In this chapter, and contrary to our custom in this book, but for the sake of brevity, we only state the results for one-point extensions, but not their duals (for one-point coextensions). We urge the reader to do the primal-dual translation work.

For our purposes, an equivalent description of the category mod $A[X]$ in terms of the representations of bimodules is needed. It is well-known that modules over a 2×2 triangular matrix algebra may be represented as triples, each consisting of a pair of modules and a homomorphism.

Now we illustrate the definition with two simple examples.

• Assume that $A = K$ and $X = K$. Then the one-point extension $A[X]$ of the algebra K by $X = K$ is the algebra

$$K[K] = \begin{bmatrix} K & 0 \\ K & K \end{bmatrix}$$

consisting of 2×2 lower triangular matrices with coefficients in K. In other words, the one-point extension $K[K]$ is the path K-algebra of the Dynkin quiver $1 \circ \longleftarrow \circ\, 2$.

• Assume that $A = K$ and $X = K^2$. Then the one-point extension $A[X]$ of the algebra K by $X = K^2$ is the Kronecker algebra

$$K[K^2] = \begin{bmatrix} K & 0 \\ K^2 & K \end{bmatrix},$$

see (I.2.5). Equivalently, $K[K^2]$ is the path K-algebra of the Kronecker quiver $1 \circ \rightrightarrows \circ\, 2$, of the Euclidean type $\widetilde{\mathbb{A}}_1$.

Given a K-algebra A and a K-A-bimodule $_KX_A$, we define the K-category

$$\operatorname{rep}(X) = \operatorname{rep}(_KX_A) \qquad\qquad (1.2)$$

of all K-linear **representations of the bimodule** $_KX_A$ as follows, see (A.2.7) in Volume 1.

   (i) An object $M = (M_0, M_1, \psi_M)$ in $\operatorname{rep}(_KX_A)$ consists of a K-vector space M_0, a right A-module M_1, and a homomorphism $\psi_M : M_0 \otimes_K X_A \longrightarrow M_1$ of right A-modules.
  (ii) A morphism from $M = (M_0, M_1, \psi_M)$ to $M' = (M_0', M_1', \psi_{M'})$ in $\operatorname{rep}(_KX_A)$ is a pair $f = (f_0, f_1)$, where $f_0 : M_0 \longrightarrow M_0'$ is a homomorphism of K-vector spaces and $f_1 : M_1 \longrightarrow M_1'$ is a homomorphism of A-modules, which are compatible with the structural homomorphisms ψ_M and $\psi_{M'}$, that is, the following square commutes

$$
\begin{array}{ccc}
M_0 \otimes_K X_A & \xrightarrow{\ \psi_M\ } & M_1 \\
{\scriptstyle f_0 \otimes 1_X}\big\downarrow & & \big\downarrow {\scriptstyle f_1} \\
M_0' \otimes_K X_A & \xrightarrow{\ \psi_{M'}\ } & M_1'.
\end{array}
$$

 (iii) The composition of morphisms in $\operatorname{rep}(_KX_A)$ is induced by the composition of homomorphisms in $\operatorname{mod} K$ and $\operatorname{mod} A$, respectively.
 (iv) The direct sum of two objects

$$M = (M_0, M_1, \psi_M) \quad\text{and}\quad M' = (M_0', M_1', \psi_{M'})$$

in $\operatorname{rep}(_KX_A)$ is the object in $\operatorname{rep}(_KX_A)$

$$M \oplus M' = (M_0 \oplus M_0', M_1 \oplus M_1', \psi_M \oplus \psi_M')$$

in $\operatorname{rep}(_KX_A)$.

It is easy to check that $\operatorname{rep}(_KX_A)$ is an additive K-category. Moreover, using the **adjunction isomorphism** (I.2.11)

$$\operatorname{Hom}_A(M_0 \otimes_K X, M_1) \cong \operatorname{Hom}_K(M_0, \operatorname{Hom}_A(X, M_1)),$$

we see that the category $\operatorname{rep}(_KX_A)$ is equivalent to the category $\overline{\operatorname{rep}}(_KX_A)$ defined as follows.

   (i) An object $M = (M_0, M_1, \varphi_M)$ in $\overline{\operatorname{rep}}(_KX_A)$ consists of a K-vector space M_0, a right A-module M_1 and a homomorphism of K-vector spaces $\varphi_M : M_0 \longrightarrow \operatorname{Hom}_A(_KX_A, M_1)$.

(ii) A morphism from $M = (M_0, M_1, \varphi_M)$ to $M' = (M'_0, M'_1, \varphi_{M'})$ in $\overline{\mathrm{rep}}(_K X_A)$ is a pair $f = (f_0, f_1)$, where $f_0 : M_0 \longrightarrow M'_0$ is a homomorphism of K-vector spaces, and $f_1 : M_1 \longrightarrow M'_1$ is a homomorphism of A-modules, which are compatible with the structural homomorphisms φ_M and $\varphi_{M'}$, that is, the following square commutes

$$
\begin{array}{ccc}
M_0 & \xrightarrow{\varphi_M} & \mathrm{Hom}_A(_K X_A, M_1) \\
{\scriptstyle f_0}\Big\downarrow & & \Big\downarrow{\scriptstyle \mathrm{Hom}_A(_K X_A, f_1)} \\
M'_0 & \xrightarrow{\varphi_{M'}} & \mathrm{Hom}_A(_K X_A, M'_1).
\end{array}
$$

(iii) The composition of morphisms in $\overline{\mathrm{rep}}(_K X_A)$ is induced from the composition of homomorphisms in $\mathrm{mod}\, K$ and $\mathrm{mod}\, A$, respectively. The direct sum is defined componentwise.

It is easy to check that $\overline{\mathrm{rep}}(_K X_A)$ is an additive K-category. A K-linear category equivalence

$$
H : \mathrm{rep}(_K X_A) \xrightarrow{\ \simeq\ } \overline{\mathrm{rep}}(_K X_A)
$$

is defined by assigning to any object $M = (M_0, M_1, \psi_M)$ of $\mathrm{rep}(_K X_A)$ the object $H(M) = (M_0, M_1, \overline{\psi_M})$ of $\overline{\mathrm{rep}}(_K X_A)$, where

$$
\overline{\psi_M} : M_0 \longrightarrow \mathrm{Hom}_A(_K X_A, M_1)
$$

is the K-linear map adjoint to the homomorphism $\psi_M : M_0 \otimes_K X_A \longrightarrow M_1$ of right A-modules, that is, $\overline{\psi_M}(m_0)(x) = \psi_M(m_0 \otimes x)$, for all $m_0 \in M_0$ and $x \in X$. We also set $H(f_0, f_1) = (f_0, f_1)$.

1.3. Lemma. *Under the notation introduced above, the following two statements hold.*

(a) *The additive K-categories $\mathrm{rep}(X)$ and $\overline{\mathrm{rep}}(X)$ are abelian, and there exist K-linear equivalences of categories*

$$
\mathrm{mod}\, A[X] \xrightarrow[\simeq]{F} \mathrm{rep}(_K X_A) \xrightarrow[\simeq]{H} \overline{\mathrm{rep}}(_K X_A).
$$

(b) *If M is a module in $\mathrm{mod}\, A[X]$ and $F(M) = (M_0, M_1, \psi_M)$, then the dimension vector $\mathbf{dim}\, M$ of M has the form*

$$
\mathbf{dim}\, M = (\mathbf{dim}\, M_1, \mathbf{dim}\, M_0),
$$

where $\mathbf{dim}\, M_0 = \dim_K M_0$ and $\mathbf{dim}\, M_1$ is the dimension vector of the A-module M_1, with $(\mathbf{dim}\, M)_a = (\mathbf{dim}\, M_1)_a = \dim_K M e_a$, for any vertex $a \neq 0$ of the quiver $Q_{A[X]}$ of $A[X]$. Here $0 \in (Q_{A[X]})_0$ is the source vertex defined by the one-point extension structure of the algebra $A[X]$.

Proof. (a) The second equivalence H is described above. To establish the first one, we define a K-linear functor

$$F : \operatorname{mod} A[X] \longrightarrow \operatorname{rep}({}_K X_A)$$

as follows. Let, as before, 0 denote the source which belongs to the quiver $Q_{A[X]}$ of the one-point extension algebra $A[X]$, but not to the quiver Q_A of A, and $e_0 \in A[X]$ denote the corresponding idempotent.

For a right $A[X]$-module M, we set $M_0 = Me_0$ and $M_1 = M(1-e_0)$, and we denote by $\psi_M : M_0 \otimes_K X_A \longrightarrow M_1$ the homomorphism induced by the multiplication map $me_0 \otimes x \mapsto me_0 x$ for $m \in M$ and $x \in X$, where we use that $X = e_0 A[X](1 - e_0)$. It is easy to see that $F(M) = (M_0, M_1, \psi_M)$ is an object of the category $\operatorname{rep}({}_K X_A)$. If $f : M \longrightarrow M'$ is a homomorphism of $A[X]$-modules, we define $f_0 : M_0 \longrightarrow M_0'$ and $f_1 : M_1 \longrightarrow M_1'$ to be the restrictions of f to M_0 and M_1, respectively. This is possible, because, for any $m \in M$, we have $f(me_0) = f(m)e_0 \in M_0'$ and $f(m(1 - e_0)) = f(m)(1-e_0) \in M_1'$. It is clear that $F(f) = (f_0, f_1)$ is a morphism from $F(M)$ to $F(M')$ in the category $\operatorname{rep}({}_K X_A)$. A routine calculation shows that we have defined an additive K-linear functor $F : \operatorname{mod} A[X] \longrightarrow \operatorname{rep}({}_K X_A)$.

We also define a functor

$$G : \operatorname{rep}({}_K X_A) \longrightarrow \operatorname{mod} A[X]$$

as follows. For an object $M = (M_0, M_1, \psi_M)$ in $\operatorname{rep}({}_K X_A)$, we let $G(M_0, M_1, \psi_M)$ be the right $A[X]$-module having

$$G(M) = M_1 \oplus M_0$$

as underlying vector space, with the multiplication

$$\cdot : G(M) \times A[X] \longrightarrow G(M)$$

defined by the formula

$$(m_1, m_0) \cdot \begin{bmatrix} a & 0 \\ x & \lambda \end{bmatrix} = (m_1 a + \psi_M(m_0 \otimes x), m_0 \lambda),$$

for $(m_1, m_0) \in G(M)$ and $\begin{bmatrix} a & 0 \\ x & \lambda \end{bmatrix} \in A[X]$. If

$$(f_0, f_1) : (M_0, M_1, \psi_M) \longrightarrow (M_0', M_1', \psi_{M'})$$

is a morphism in the category $\operatorname{rep}({}_K X_A)$, we define the K-linear map

$$G(f_0, f_1) : G(M_0, M_1, \psi_M) \longrightarrow G(M_0', M_1', \psi_{M'})$$

by the formula

$$G(f_0, f_1)(m_1, m_0) = (f_1(m_1), f_0(m_0)).$$

Then $G(f_0, f_1)$ is indeed a homomorphism of $A[X]$-modules, because the following equalities hold

$$
\begin{aligned}
G(f_0, f_1)\Big((m_1, m_0) \cdot \begin{bmatrix} a & 0 \\ x & \lambda \end{bmatrix}\Big) &= G(f_0, f_1)(m_1 a + \psi_M(m_0 \otimes x), m_0 \lambda) \\
&= (f_1(m_1 a) + f_1 \psi_M(m_0 \otimes x), f_0(m_0 \lambda)) \\
&= (f_1(m_1)a + \psi_{M'}(f_0 \otimes 1_X)(m_0 \otimes x), f_0(m_0)\lambda) \\
&= (f_1(m_1)a + \psi_{M'}(f_0(m_0) \otimes x), f_0(m_0)\lambda) \\
&= (f_1(m_1), f_0(m_0)) \cdot \begin{bmatrix} a & 0 \\ x & \lambda \end{bmatrix} \\
&= [G(f_0, f_1)(m_1, m_0)] \cdot \begin{bmatrix} a & 0 \\ x & \lambda \end{bmatrix},
\end{aligned}
$$

for all $(m_1, m_0) \in G(M)$ and $\begin{bmatrix} a & 0 \\ x & \lambda \end{bmatrix} \in A[X]$.

It is easily shown that F and G are additive K-linear functors, and quasi-inverse to each other.

(b) The description of the functors F and G shows that, if a triple $M = (M_0, M_1, \varphi_M)$ is viewed as a right $A[X]$-module via G, then the dimension vector $\mathbf{dim}\, M$ of M in mod $A[X]$ is computed as follows. If a is a point in the quiver $Q_{A[X]}$ of $A[X]$, then $(\mathbf{dim}\, M)_a = (\mathbf{dim}\, M_1)_a = \dim_K M e_a$, if $a \neq 0$, and $(\mathbf{dim}\, M)_0 = \dim_K M_0$. This finishes the proof.   □

The reader is referred to (I.2.4) and (I.2.5) for simple examples explaining the functors F and H of Lemma (1.3). In the sequel, the equivalences F and H of (1.3) are treated as identifications.

The category $\overline{\mathrm{rep}}(_K X_A)$ being more suited for our purposes, we consider in fact $A[X]$-modules as being objects in $\overline{\mathrm{rep}}(_K X_A)$.

Another consequence of (1.3) is the following useful fact.

1.4. Corollary. *Let $A[X]$ be a one-point extension algebra as above. Then there exist two essentially distinct full and faithful embeddings of mod A inside mod $A[X]$ preserving the indecomposablity:*

   (a) *the standard embedding of mod A inside mod $A[X]$, that associates to an A-module M the triple $(0, M, 0)$ (we simply identify $(0, M, 0)$ with M),*

   (b) *the functor associating to an A-module M the triple*

$$\overline{M} = (\mathrm{Hom}_A(X, M), M, 1_{\mathrm{Hom}_A(X,M)}).$$

                                                         □

We now use the embeddings of (1.4) to describe those almost split sequences in the category $\operatorname{mod} A[X]$ whose right end term is an A-module. We need a technical lemma.

1.5. Lemma. *Let A be an algebra, and X be an A-module.*

(a) *If $f : L \longrightarrow M$ is left minimal almost split in $\operatorname{mod} A$, then*

$$(1, f) : \overline{L} \longrightarrow (\operatorname{Hom}_A(X, L), M, \operatorname{Hom}_A(X, f))$$

is left minimal almost split in $\operatorname{mod} A[X]$.

(b) *If $g : M \longrightarrow N$ is right minimal almost split in $\operatorname{mod} A$, and*

$$j : \operatorname{Ker} \operatorname{Hom}_A(X, g) \longrightarrow \operatorname{Hom}_A(X, M)$$

denotes the inclusion, then

$$(0, g) : (\operatorname{Ker} \operatorname{Hom}_A(X, g), M, j) \longrightarrow N$$

is right minimal almost split in $\operatorname{mod} A[X]$.

Proof. (a) Clearly, $(1, f)$ is a morphism

$$\overline{L} = (\operatorname{Hom}_A(X, L), L, 1) \longrightarrow (\operatorname{Hom}_A(X, L), M, \operatorname{Hom}_A(X, f)).$$

If it were a section, then so would be f, a contradiction. Thus $(1, f)$ is not a section.

Let $u = (u_0, u_1) : \overline{L} \longrightarrow (U_0, U_1, \varphi_U) = U$ be a morphism which is not a section. We claim that $u_1 : L \longrightarrow U_1$ is not a section in $\operatorname{mod} A$. For, assume to the contrary that this is the case. Then there exists $u_1' : U_1 \longrightarrow L$ such that $u_1' \circ u_1 = 1_L$. Hence the pair of maps $u' = (\operatorname{Hom}_A(X, u_1') \circ \varphi_U, u_1')$ is a morphism from (U_0, U_1, φ_U) to $\overline{L}$, which satisfies $u' \circ u = 1_{\overline{L}}$, because

$$\operatorname{Hom}_A(X, u_1') \circ \varphi_U \circ u_0 = \operatorname{Hom}_A(X, u_1') \circ \operatorname{Hom}_A(X, u_1) = 1,$$

contrary to our hypothesis that u is not a section. This establishes our claim.

Because, by hypothesis, f is left almost split, there exists a homomorphism $\widetilde{u}_1 : M \longrightarrow U_1$ in $\operatorname{mod} A$ such that $\widetilde{u}_1 \circ f = u_1$. Then the pair of homomorphisms $\widetilde{u} = (u_0, \widetilde{u}_1)$ is a morphism

$$(\operatorname{Hom}_A(X, L), M, \operatorname{Hom}_A(X, f)) \longrightarrow (U_0, U_1, \varphi_U).$$

Indeed, we have

$$\operatorname{Hom}_A(X, \widetilde{u}_1) \circ \operatorname{Hom}_A(X, f) = \operatorname{Hom}_A(X, u_1) = \varphi_U u_0,$$

that is, the diagram

$$
\begin{array}{ccc}
\operatorname{Hom}_A(X,L) & \xrightarrow{\;\operatorname{Hom}_A(X,f)\;} & \operatorname{Hom}_A(X,M) \\
{\scriptstyle u_0}\downarrow & & \downarrow{\scriptstyle \operatorname{Hom}_A(X,\tilde{u}_1)} \\
U_0 & \xrightarrow{\quad \varphi_U \quad} & \operatorname{Hom}_A(X,U_1)
\end{array}
$$

is commutative. Further, we have $\tilde{u} \circ (1, f) = u$. We have shown that $(1, f)$ is left almost split in $\operatorname{mod} A[X]$.

To show that $(1, f)$ is left minimal, assume that $g = (g_0, g_1)$ is an endomorphism of $(\operatorname{Hom}_A(X,L), M, \operatorname{Hom}_A(X,f))$ such that $g \circ (1, f) = (1, f)$. Then $g_1 \circ f = f$, which implies that g_1 is an automorphism of M, while $g_0 = 1$. This shows that g is an automorphism.

(b) That $(0, g)$ is a morphism from $(\operatorname{Ker} \operatorname{Hom}_A(X, g), M, j)$ to $N = (0, N, 0)$ follows from the commutativity of the square

$$
\begin{array}{ccc}
\operatorname{Ker} \operatorname{Hom}_A(X,g) & \xrightarrow{\;\;j\;\;} & \operatorname{Hom}_A(X,M) \\
\downarrow & & \downarrow{\scriptstyle \operatorname{Hom}_A(X,g)} \\
0 & \xrightarrow{\hspace{2cm}} & \operatorname{Hom}_A(X,N).
\end{array}
$$

On the other hand, $(0, g)$ is clearly not a retraction in $\operatorname{mod} A[X]$, because the homomorphism $g : M \longrightarrow N$ is not a retraction in $\operatorname{mod} A$. Let then

$$ v = (v_0, v_1) : (V_0, V_1, \varphi_V) = V \longrightarrow (0, N, 0) = N $$

be a morphism, which is not a retraction. Clearly, $v_0 = 0$, while $v_1 : V_1 \longrightarrow N$ is not a retraction in $\operatorname{mod} A$. Because g is right almost split, there exists $\tilde{v}_1 : V_1 \longrightarrow M$ in $\operatorname{mod} A$ such that $v_1 = g \circ \tilde{v}_1$. Now

$$ \operatorname{Hom}_A(X,g) \circ \operatorname{Hom}_A(X, \tilde{v}_1) \circ \varphi_V = \operatorname{Hom}_A(X, v_1) \circ \varphi_V = 0. $$

Hence the map $\operatorname{Hom}_A(X, \tilde{v}_1)\varphi_V$ factors through j, that is, there exists a K-linear map

$$ \tilde{v}_0 : V_0 \longrightarrow \operatorname{Ker} \operatorname{Hom}_A(X, g) $$

such that $j \circ \tilde{v}_0 = \operatorname{Hom}_A(X, \tilde{v}_1) \circ \varphi_V$. The latter equality expresses the fact that $\tilde{v} = (\tilde{v}_0, \tilde{v}_1)$ is a morphism from (V_0, V_1, φ_V) to $(\operatorname{Ker} \operatorname{Hom}_A(X, g), M, j)$. Further, we have $(0, g)\tilde{v} = v$. This shows that $(0, g)$ is right almost split.

To show that $(0, g)$ is right minimal, let $h = (h_0, h_1)$ be an endomorphism of $(\operatorname{Ker} \operatorname{Hom}_A(X, g), M, j)$ such that $(0, g)\circ h = (0, g)$. Thus we have $g \circ h_1 = g$, showing that h_1 is an automorphism of M. On the other hand, the condition for h to be a morphism reads

$$ j \circ h_0 = \operatorname{Hom}_A(X, h_1) \circ j, $$

and this says that h_0 is the restriction to $\operatorname{Ker}\operatorname{Hom}_A(X, g)$ of the automorphism $\operatorname{Hom}_A(X, h_1)$. Consequently h_0 is itself an automorphism of $\operatorname{Ker}\operatorname{Hom}_A(X, g)$. Hence h is an automorphism.   □

1.6. Theorem. *Let A be an algebra, and X be an A-module. If*

$$0 \longrightarrow L \xrightarrow{\ f\ } M \xrightarrow{\ g\ } N \longrightarrow 0$$

is an almost split sequence in $\operatorname{mod} A$, *then*

$$0 \longrightarrow \overline{L} \xrightarrow{\ (1,f)\ } (\operatorname{Hom}_A(X, L), M, \operatorname{Hom}_A(X, f)) \xrightarrow{\ (0,g)\ } N \longrightarrow 0$$

is an almost slit sequence in $\operatorname{mod} A[X]$.

Proof. The left exactness of the given sequence follows from the fact that there is a commutative diagram with exact rows

$$
\begin{array}{ccccccccc}
0 & \longrightarrow & \operatorname{Hom}_A(X, L) & \xrightarrow{\ 1\ } & \operatorname{Hom}_A(X, L) & \longrightarrow & 0 \\
& & {\scriptstyle 1}\big\downarrow & & \big\downarrow{\scriptstyle \operatorname{Hom}_A(X,f)} & & \big\downarrow \\
0 & \longrightarrow & \operatorname{Hom}_A(X, L) & \xrightarrow{\operatorname{Hom}_A(X,f)} & \operatorname{Hom}_A(X, M) & \xrightarrow{\operatorname{Hom}_A(X,g)} & \operatorname{Hom}_A(X, N).
\end{array}
$$

On the other hand, the surjectivity of $(0, g)$ follows from the computation of the dimension of the cokernel of $(1, f)$:

$$\dim_K(\operatorname{Hom}_A(X, L), M, \operatorname{Hom}_A(X, f)) - \dim_K \overline{L}$$
$$= \dim_K \operatorname{Hom}_A(X, L) + \dim_K M - \dim_K \operatorname{Hom}_A(X, L) - \dim_K L$$
$$= \dim_K M - \dim_K L$$
$$= \dim_K N.$$

Applying (1.5)(a), we see that $(1, f)$ is left minimal almost split. Because the functor $\operatorname{Hom}_A(X, -)$ is left exact, then $\operatorname{Hom}_A(X, L)$ is a kernel of $\operatorname{Hom}_A(X, g)$. Therefore (1.5)(b) yields that $(0, g)$ is right minimal almost split.   □

1.7. Corollary. *Let A be an algebra, and X be an A-module. If*

$$0 \longrightarrow L \xrightarrow{\ f\ } M \xrightarrow{\ g\ } N \longrightarrow 0$$

is an almost split sequence in $\operatorname{mod} A$ *such that* $\operatorname{Hom}_A(X, L) = 0$, *then this sequence remains almost split in* $\operatorname{mod} A[X]$ *under the standard embedding.*

Proof. Apply (1.6).   □

We note also the following useful fact.

1.8. Lemma. *Let A be an algebra, X an A-module, and $V = (V_0, V_1, \varphi_V)$ be an indecomposable $A[X]$-module with $\varphi_V \neq 0$. Then $\operatorname{Hom}_A(X, N) \neq 0$, for any indecomposable direct summand N of V_1 in $\operatorname{mod} A$.*

Proof. Assume, to the contrary, that N is an indecomposable direct summand of V_1 in $\operatorname{mod} A$ such that $\operatorname{Hom}_A(X, N) = 0$. Let $p : V_1 \longrightarrow N$ be the canonical retraction and $u : N \longrightarrow V_1$ the canonical section in $\operatorname{mod} A$. It follows that

$$(0, p) : V \longrightarrow (0, N, 0) = N$$

is a retraction in $\operatorname{mod} A[X]$ with section

$$(0, u) : N = (0, N, 0) \longrightarrow V.$$

Hence, $(0, p)$ is an isomorphism $V \cong N$ in $\operatorname{mod} A[X]$, because V is assumed to be indecomposable. Hence we get a contradiction, because $\varphi_V \neq 0$ implies $V_0 \neq 0$. $\qquad\square$

1.9. Example. Consider the algebra A given by the quiver

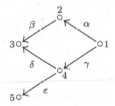

bound by two relations $\alpha\beta = \gamma\delta$ and $\gamma\varepsilon = 0$. Then the Auslander–Reiten quiver $\Gamma(\operatorname{mod} A)$ is of the form

$$
\begin{array}{ccccccccc}
& 1{}^{1}_{0}0{}_{0} & & & & 0{}^{0}_{1}0 & & & \\[2pt]
\nearrow & & \searrow & & \nearrow & & \searrow & & \\[2pt]
1{}^{0}_{0}0{}_{0} & & 1{}^{1}_{1}0{}_{1} & & 0{}^{0}_{0}0{}_{1} & & 0{}^{1}_{0}1{}_{0} & & \\[2pt]
& \searrow & & \nearrow\;\searrow & & \nearrow\;\searrow & & \nearrow & \\[2pt]
1{}^{0}_{1}0{}_{1} & & 1{}^{1}_{0}0{}_{1} \to 1{}^{1}_{1}1{}_{0} \to 0{}^{1}_{0}1{}_{0} & & & & 0{}^{0}_{0}1{}_{0} & & \\[2pt]
\nearrow & & \searrow\;\nearrow & & \searrow\;\nearrow & & \searrow\;\nearrow & & \\[2pt]
0{}^{0}_{0}0{}_{1} & & 1{}^{0}_{1}0 & & 0{}^{1}_{0}0 & & 0{}^{0}_{1}1 & &
\end{array}
$$

where the indecomposable modules are represented by their dimension vectors. Let $X = I(5)_A$. Then $X = 0\,{}^{0}_{1}\,1$ $_0$ and the quiver $Q_{A[X]}$ of the one-point extension algebra $A[X]$ is of the form

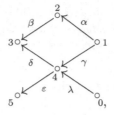

bound by three relations $\alpha\beta = \gamma\delta$, $\gamma\varepsilon = 0$ and $\lambda\delta = 0$. The Auslander–Reiten quiver $\Gamma(\operatorname{mod} A[X])$ is of the form

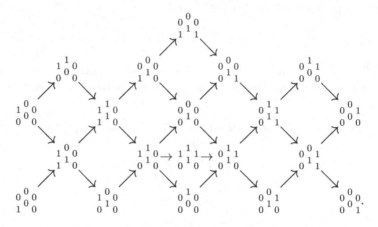

The image of an indecomposable A-module under the standard embedding is easily recognised: its dimension vector as an $A[X]$-module has a zero coordinate at the point 0. The other embedding is different: both coincide for those indecomposable modules M such that $\operatorname{Hom}_A(X, M) = 0$. For the other, using dimension vectors, we have

$$\overline{\begin{smallmatrix} & 0 & \\ 0 & & 0 \\ & 1 & \\ 1 & & \end{smallmatrix}} = \begin{smallmatrix} & 0 & \\ 0 & & 0 \\ & 1 & \\ 1 & & 1 \end{smallmatrix}, \quad \overline{\begin{smallmatrix} & 0 & \\ 0 & & 0 \\ & 1 & \\ 0 & & \end{smallmatrix}} = \begin{smallmatrix} & 0 & \\ 0 & & 0 \\ & 1 & \\ 0 & & 1 \end{smallmatrix}, \quad \overline{\begin{smallmatrix} & 1 & \\ 0 & & 1 \\ & 1 & \\ 0 & & \end{smallmatrix}} = \begin{smallmatrix} & 1 & \\ 0 & & 1 \\ & 1 & \\ 0 & & 1 \end{smallmatrix}, \quad \overline{\begin{smallmatrix} & 0 & \\ 0 & & 1 \\ & 1 & \\ 0 & & \end{smallmatrix}} = \begin{smallmatrix} & 0 & \\ 0 & & 1 \\ & 1 & \\ 0 & & 1 \end{smallmatrix}.$$

We now consider the almost split sequence in $\operatorname{mod} A$

$$0 \longrightarrow \begin{smallmatrix} & 1 & \\ 0 & & 1 \\ & 1 & \\ & 0 & \end{smallmatrix} \overset{f}{\longrightarrow} \begin{smallmatrix} & 1 & \\ 0 & & 1 \\ & 0 & \\ & 0 & \end{smallmatrix} \oplus \begin{smallmatrix} & 0 & \\ 0 & & 1 \\ & 1 & \\ & 0 & \end{smallmatrix} \overset{g}{\longrightarrow} \begin{smallmatrix} & 0 & \\ 0 & & 1 \\ & 0 & \\ & 0 & \end{smallmatrix} \longrightarrow 0.$$

Because $\dim_K \operatorname{Hom}_A\left(X, \begin{smallmatrix} & 1 & \\ 0 & & 1 \\ & 1 & \\ & 0 & \end{smallmatrix}\right) = 1$, then

$$\left(\operatorname{Hom}_A\left(X, \begin{smallmatrix} & 1 & \\ 0 & & 1 \\ & 1 & \\ & 0 & \end{smallmatrix}\right), \begin{smallmatrix} & 1 & \\ 0 & & 1 \\ & 0 & \\ & 0 & \end{smallmatrix} \oplus \begin{smallmatrix} & 0 & \\ 0 & & 1 \\ & 1 & \\ & 0 & \end{smallmatrix}, \operatorname{Hom}_A(X, f)\right) = \begin{smallmatrix} & 1 & \\ 0 & & 1 \\ & 0 & \\ & 0 & \end{smallmatrix} \oplus \begin{smallmatrix} & 0 & \\ 0 & & 1 \\ & 1 & \\ & 0 & \end{smallmatrix},$$

so that there is an almost split sequence in mod $A[X]$

$$0 \longrightarrow 0 \begin{smallmatrix} & 1 & \\ & & 1 \\ 0 & & 1 \\ & 1 & \end{smallmatrix} \longrightarrow 0 \begin{smallmatrix} & 1 & \\ & & 1 \\ 0 & & 0 \\ & 0 & \end{smallmatrix} \oplus 0 \begin{smallmatrix} & 0 & \\ & & 1 \\ 0 & & 1 \\ & 1 & \end{smallmatrix} \longrightarrow 0 \begin{smallmatrix} & 0 & \\ & & 1 \\ 0 & & 0 \\ & 0 & \end{smallmatrix} \longrightarrow 0.$$

We say that the tubes $\mathcal{T}_1$ and $\mathcal{T}_2$ of the Auslander–Reiten quiver $\Gamma(\operatorname{mod} A)$ of an algebra A are **orthogonal** if $\operatorname{Hom}_A(\mathcal{T}_1, \mathcal{T}_2) = 0$ and $\operatorname{Hom}_A(\mathcal{T}_2, \mathcal{T}_1) = 0$, that is, $\operatorname{Hom}_A(X_1, X_2) = 0$ and $\operatorname{Hom}_A(X_2, X_1) = 0$, for any module X_1 in $\mathcal{T}_1$ and any module X_2 in $\mathcal{T}_2$.

1.10. Example. Let A be a K-algebra given by the quiver

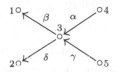

We use the notation of (X.2.12). Let $X = S = S(3)$. The bound quiver of the one-point extension algebra $A[S]$ is

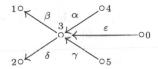

bound by two zero relations $\varepsilon\beta = 0$ and $\varepsilon\delta = 0$. We now compute the module category $\operatorname{mod} A[S]$. Because S belongs to a stable tube $\mathcal{T}_0$ of rank 2 in $\Gamma(\operatorname{mod} A)$, by (X.2.12), we have $\operatorname{Hom}_A(S, L) = 0$, for any indecomposable A module L lying in the postprojective component $\mathcal{P}$ of $\Gamma(\operatorname{mod} A)$, see (VIII.2.6). Then, in view of (1.7), we conclude that $\mathcal{P}$ becomes a component $\overline{\mathcal{P}}$ of $\Gamma(\operatorname{mod} A[S])$ under the standard embedding, and $\overline{\mathcal{P}}$ is clearly a postprojective component of $\Gamma(\operatorname{mod} A[S])$.

Similarly, the orthogonality of the tubes in $\Gamma(\operatorname{mod} A)$ (see (XI.2.8)) implies that all the tubes except $\mathcal{T}_0$, the one containing S, become components of $\Gamma(\operatorname{mod} A[S])$ under the standard embedding, and these components are clearly stable tubes. Thus, only $\mathcal{T}_0$ and the preinjective component may change.

We first look at $\mathcal{T}_0$. Because the tube $\mathcal{T}_0$ is standard, then we have $\operatorname{Hom}_A(S, M) \neq 0$, for an indecomposable module M in $\mathcal{T}_0$, if and only if M belongs to the ray

$$S = S[1] \xrightarrow{u_2} S[2] \xrightarrow{u_3} S[3] \xrightarrow{u_4} \cdots$$

starting with S, and moreover, for any $j \geq 1$, we have $\dim_K \operatorname{Hom}_A(S, S[j]) = 1$. In particular, for any module $E[j]$ lying on the ray

$$E = E[1] \xrightarrow{\;u_2'\;} E[2] \xrightarrow{\;u_3'\;} E[3] \xrightarrow{\;u_4'\;} \cdots$$

given by the ray module $E = \tau S$, we have $\operatorname{Hom}_A(S, E[j]) = 0$. We denote the canonical epimorphisms in $\mathcal{T}_0$ by

$$p_j : S[j] \longrightarrow E[j-1] \quad \text{and} \quad p_j' : E[j] \longrightarrow S[j-1],$$

for $j \geq 2$, respectively. Theorem (1.6) thus gives, for each $j \geq 1$, an almost split sequence

$$0 \longrightarrow (K, S[j], 1) \xrightarrow{\left[\begin{smallmatrix}(1, u_{j+1})\\(0, p_j)\end{smallmatrix}\right]} (K, S[j+1], 0) \oplus E[j-1] \xrightarrow{[(0, p_{j+1}), (0, u_j')]} E[j] \longrightarrow 0.$$

Finally, for each $j \geq 1$, the morphism

$$(0, 1) : S[j] \longrightarrow (K, S[j], 1) = \overline{S[j]}$$

is irreducible. Indeed, it is clearly neither a section nor a retraction (because $S[j]$ and $\overline{S[j]}$ are non-isomorphic indecomposable modules). If we have $(0, 1) = (f_0, f_1)(g_0, g_1)$ for some morphisms

$$(g_0, g_1) : S[j] \longrightarrow U = (U_0, U_1, \varphi_U) \quad \text{and} \quad (f_0, f_1) : \; U \longrightarrow \overline{S[j]},$$

then $g_0 = 0$ and $f_1 g_1 = 1$. Thus we have a direct sum decomposition $U_1 = U_1' \oplus U_1''$ such that g_1 and f_1 induce mutually inverse isomorphisms between $S[j]$ and U_1'. Set

$$\varphi_U = \begin{bmatrix}\varphi'\\\varphi''\end{bmatrix} : U_0 \longrightarrow \operatorname{Hom}_A(S, U_1') \oplus \operatorname{Hom}_A(S, U_1''),$$

where φ' (respectively, φ'') is the composition of φ_U with the canonical projection of $\operatorname{Hom}_A(S, U_1)$ onto $\operatorname{Hom}_A(S, U_1')$ (respectively, onto $\operatorname{Hom}_A(S, U_1'')$). Now, notice that there are isomorphisms

$$\operatorname{Hom}_A(S, U_1') \cong \operatorname{Hom}_A(S, S[j]) \cong K.$$

We thus have two cases. If $\varphi' = 0$, then clearly $(g_0, g_1) = (0, g_1)$ is a section. If $\varphi' \neq 0$, then φ' is an epimorphism, hence a retraction. Therefore, it follows from the commutative diagram

$$
\begin{array}{ccc}
0 & \longrightarrow & \operatorname{Hom}_A(S, S[j]) \\
\downarrow & & \downarrow{\scriptstyle \operatorname{Hom}_A(S, g_1)} \\
U_0 & \xrightarrow{\left[\begin{smallmatrix}\varphi'\\\varphi''\end{smallmatrix}\right]} & \operatorname{Hom}_A(S, U_1') \bigoplus \operatorname{Hom}_A(S, U_1'') \\
{\scriptstyle f_0}\downarrow & & \downarrow{\scriptstyle \operatorname{Hom}_A(S, f_1)} \\
K & \xrightarrow{\;\;1\;\;} & \operatorname{Hom}_A(S, S[j])
\end{array}
$$

that f_0 is itself a retraction. Hence (f_0, f_1) is a retraction and, consequently, the map $(0,1) : S[j] \longrightarrow \overline{S[j]}$ is an irreducible monomorphism.

We now consider, for each $j \geq 1$, the short exact sequence in mod $A[S]$

$$0 \longrightarrow S[j] \xrightarrow{\left[\begin{smallmatrix}(0,u_{j+1})\\(0,1)\end{smallmatrix}\right]} S[j+1] \oplus \overline{S[j]} \xrightarrow{[(0,1)\,(1,u_{j+1})]} \overline{S[j+1]} \longrightarrow 0.$$

Each of the morphisms shown is irreducible, and both end terms are indecomposable. Hence it is almost split. Thus the tube $\mathcal{T}_0$ yields in mod $A[S]$ the following component

$$
\begin{array}{ccccccccc}
 & & & & \overline{S[1]} & & & & E[1] \\
 & & & {}^{(0,1)}\nearrow & & \searrow{}^{(1,u_2)} & {}^{(0,p_2)}\nearrow & & | \\
E[1] & & S[1] & & & & \overline{S[2]} & & | \\
 {}^{(0,u_2')}\searrow & {}^{(0,p_2')}\nearrow & & \searrow{}^{(0,u_2)} & & {}^{(0,1)}\nearrow & & \searrow{}^{(1,u_3)} & | \\
 & E[2] & & & S[2] & & & & \overline{S[3]} \\
 {}^{(0,p_3)}\nearrow & & \searrow{}^{(0,u_3')} & {}^{(0,p_3')}\nearrow & & \searrow{}^{(0,u_3)} & & {}^{(0,1)}\nearrow & | \\
\overline{S[3]} & & & E[3] & & & S[3] & & | \\
 {}^{(1,u_4)}\searrow & {}^{(0,p_4)}\nearrow & & \searrow{}^{(0,u_4')} & {}^{(0,p_4')}\nearrow & & \searrow{}^{(0,u_4)} & & | \\
 & \overline{S[4]} & & & E[4] & & & & S[4] \\
 & \nearrow & \searrow & & \nearrow & \searrow & & \nearrow & | \\
 & & \vdots & & & \vdots & & &
\end{array}
$$

where as usual we identify along the vertical dotted lines. This component $\overline{\mathcal{T}_0}$ is not a stable tube, actually all its indecomposable modules belong to the τ-orbit of the projective $A[S]$-module

$$P(0) = \overline{S[1]}.$$

In the next section, such a component is called a ray tube.

We now turn to the preinjective component. Because the quiver of $A[S]$ is a tree, it satisfies the coseparation condition (by the dual of (IX.4.3)). Hence $\Gamma(\mathrm{mod}\,A[S])$ has a preinjective component, which is easily computed as follows

$$
\begin{array}{ccccccccc}
 & & {\begin{smallmatrix}0&1\\1&0\\0&0\end{smallmatrix}} & & {\begin{smallmatrix}1&1\\1&0\\0&1\end{smallmatrix}} & & {\begin{smallmatrix}0&0\\1&1\\0&1\end{smallmatrix}} & & {\begin{smallmatrix}0&1\\0&0\\0&0\end{smallmatrix}} \\
 & \nearrow & & \searrownearrow & & \searrownearrow & & \searrow & \nearrow \\
\cdots \longrightarrow & {\begin{smallmatrix}1&2\\3&1\\1&2\end{smallmatrix}} \longrightarrow & {\begin{smallmatrix}0&1\\2&1\\0&1\end{smallmatrix}} \longrightarrow & {\begin{smallmatrix}0&1\\1&0\\0&1\end{smallmatrix}} \longrightarrow & {\begin{smallmatrix}0&1\\1&1\\0&1\end{smallmatrix}} \longrightarrow & {\begin{smallmatrix}0&0\\0&1\\0&0\end{smallmatrix}} & & & \\
 & \searrow & & \nearrowsearrow & & \nearrowsearrow & & \nearrow & \searrow \\
 & & {\begin{smallmatrix}0&0\\1&0\\0&1\end{smallmatrix}} & & {\begin{smallmatrix}0&1\\1&0\\1&1\end{smallmatrix}} & & {\begin{smallmatrix}0&1\\1&1\\0&0\end{smallmatrix}} & & {\begin{smallmatrix}0&0\\0&0\\0&1\end{smallmatrix}}.
\end{array}
$$

We have represented here the indecomposable modules by their dimension vectors.

We now claim that we have obtained in this way all the indecomposable $A[S]$-modules. Notice first that $A[S]$ is a tilted algebra: indeed, applying (VIII.5.6) to the faithful section consisting of the indecomposable preinjective $A[S]$-modules

$$\left\{ \begin{array}{cccccc} {\begin{smallmatrix} 1 & 1 \\ 1 & 0 \\ 0 & 1 \end{smallmatrix}}, & {\begin{smallmatrix} 0 & 1 \\ 1 & 0 \\ 1 & 1 \end{smallmatrix}}, & {\begin{smallmatrix} 0 & 1 \\ 1 & 0 \\ 0 & 1 \end{smallmatrix}}, & {\begin{smallmatrix} 0 & 1 \\ 1 & 1 \\ 0 & 1 \end{smallmatrix}}, & {\begin{smallmatrix} 0 & 1 \\ 0 & 0 \\ 0 & 0 \end{smallmatrix}}, & {\begin{smallmatrix} 0 & 0 \\ 0 & 0 \\ 0 & 1 \end{smallmatrix}} \end{array} \right\},$$

it follows directly that $A[S]$ is the endomorphism algebra of a tilting B-module T, where B is the path algebra of the quiver

$$\Delta' : \qquad \begin{array}{c} \overset{1}{\circ} \qquad\qquad \overset{5}{\circ} \\ \nwarrow \qquad\qquad \swarrow \\ \overset{3}{\circ} \longleftarrow \overset{4}{\circ} \\ \swarrow \qquad\qquad \searrow \\ \underset{2}{\circ} \qquad\qquad \underset{6}{\circ} \end{array}$$

We now compute the module T_B, using the technique shown in (VIII.5.7). We have

$$\mathrm{Hom}_B(T, I(1)) = {\begin{smallmatrix} 1 & 1 \\ 1 & 0 \\ 0 & 1 \end{smallmatrix}}, \qquad\qquad \mathrm{Hom}_B(T, I(2)) = {\begin{smallmatrix} 0 & 1 \\ 1 & 0 \\ 1 & 1 \end{smallmatrix}},$$

$$\mathrm{Hom}_B(T, I(3)) = {\begin{smallmatrix} 0 & 1 \\ 1 & 0 \\ 0 & 1 \end{smallmatrix}}, \qquad\qquad \mathrm{Hom}_B(T, I(4)) = {\begin{smallmatrix} 0 & 1 \\ 1 & 1 \\ 0 & 1 \end{smallmatrix}},$$

$$\mathrm{Hom}_B(T, I(5)) = {\begin{smallmatrix} 0 & 1 \\ 0 & 0 \\ 0 & 0 \end{smallmatrix}}, \qquad\qquad \mathrm{Hom}_B(T, I(6)) = {\begin{smallmatrix} 0 & 0 \\ 0 & 0 \\ 0 & 1 \end{smallmatrix}}.$$

Thus

$$T_B = \begin{bmatrix} 1 \\ 0 \end{bmatrix} 0\ 0\begin{bmatrix} 0 \\ 0 \end{bmatrix} \oplus \begin{bmatrix} 0 \\ 1 \end{bmatrix} 0\ 0 \begin{bmatrix} 0 \\ 0 \end{bmatrix} \oplus \begin{bmatrix} 1 \\ 1 \end{bmatrix} 1\ 1 \begin{bmatrix} 0 \\ 0 \end{bmatrix} \oplus \begin{bmatrix} 1 \\ 1 \end{bmatrix} 1\ 1 \begin{bmatrix} 1 \\ 1 \end{bmatrix} \oplus \begin{bmatrix} 1 \\ 1 \end{bmatrix} 1\ 1 \begin{bmatrix} 0 \\ 1 \end{bmatrix} \oplus \begin{bmatrix} 0 \\ 0 \end{bmatrix} 0\ 1 \begin{bmatrix} 0 \\ 0 \end{bmatrix}.$$

Observe that the first five direct summands of T_B are the indecomposable projective B-modules, and hence they lie in the postprojective component $\mathcal{P}(B)$ of B. On the other hand, the final direct summand of T_B is a simple regular module lying on the mouth of the unique stable tube of rank 3 of $\Gamma(\mathrm{mod}\,B)$, whose remaining two simple regular modules have the dimension vectors

$$\begin{array}{c} 1 \\ 1\ 1\ 1 \\ 1\ \ 1 \end{array} \quad \text{and} \quad \begin{array}{c} 0 \\ 0\ 1\ 0 \\ 0\ \ 0 \end{array},$$

see (XIII.2.6)(c) and (XIII.2.9). The straightforward computation of the subcategories $\mathcal{T}(T_B)$ and $\mathcal{F}(T_B)$ shows that we have indeed obtained above all the indecomposable $A[S]$-modules. In particular, the preinjective component $\mathcal{Q}(A)$ of $\Gamma(\mathrm{mod}\,A)$ presented above is the connecting component $\mathcal{C}_T$ determined by the tilting B-module T. The details are left to the reader as an exercise.

XV.2. Tubular extensions and tubular coextensions of algebras

As seen in the above example (1.10) the effect of a one-point extension by a mouth module in a stable tube yields a new component, which resembles a tube, but is certainly not stable (indeed, each of its points lies in the τ-orbit of a projective, thus τ does not act as an automorphism) or, equivalently, contains no τ-periodic indecomposable module. Our objective in this section is to generalise and to iterate this procedure. For this purpose, we need a definition.

2.1. Definition. Let $\mathcal{C}$ be a standard component of the Auslander–Reiten quiver of an algebra A.

(a) A ray module X in $\mathcal{C}$ is said to be **admissible** if the ray

$$X = X_0 \xrightarrow{u_1} X_1 \xrightarrow{u_2} X_2 \longrightarrow \ldots \xrightarrow{u_i} X_i \xrightarrow{u_{i+1}} \ldots$$

starting at X satisfies the following three conditions:

(i) If M is an indecomposable module in $\mathcal{C}$ such that $\operatorname{Hom}_A(X, M) \neq 0$, then there exists $i \geq 0$ such that $M \cong X_i$.

(ii) If $f : M \longrightarrow N$ is a homomorphism between indecomposable modules in $\mathcal{C}$ such that $\operatorname{Hom}_A(X, f) \neq 0$, then there exist i, j with $i \leq j$ such that $M \cong X_i$, $N \cong X_j$ and f is a scalar multiple of the composite homomorphism $u_j \ldots u_{i+1} : X_i \longrightarrow X_j$, if $i < j$, and f is a scalar multiple of the identity, if $i = j$.

(iii) None of the modules X_i is injective.

(b) A coray module X in $\mathcal{C}$ is said to be **admissible** if the coray ending with X

$$\ldots \xrightarrow{p_{i+1}} X_i \xrightarrow{p_i} \ldots \longrightarrow X_2 \xrightarrow{p_2} X_1 \xrightarrow{p_1} X_0 = X$$

satisfies the following three conditions:

(i) If N is an indecomposable module in $\mathcal{C}$ such that $\operatorname{Hom}_A(N, X) \neq 0$, then there exists $i \geq 0$ such that $N \cong X_i$.

(ii) If $g : N \longrightarrow M$ is a homomorphism between indecomposable modules in $\mathcal{C}$ such that $\operatorname{Hom}_A(g, X) \neq 0$, then there exist i, j with $i \leq j$ such that $M \cong X_i$, $N \cong X_j$ and g is a scalar multiple of the composite homomorphism $p_{i+1} \ldots p_j : X_j \longrightarrow X_i$, if $i < j$, and g is a scalar multiple of the identity, if $i = j$.

(iii) None of the modules X_i is projective.

2.2. Lemma. *Let C be a standard stable tube in $\Gamma(\operatorname{mod} A)$, and X be a module in C.*

(a) *X is an admissible ray module if and only if X is a mouth module.*
(b) *X is an admissible coray if and only if X is a mouth module.*

Proof. We only show (a), because the proof of (b) is similar. Because X is a ray module if and only if it is a mouth module, it suffices to prove that every mouth module X in C is admissible. It follows from the standardness of C and the fact that almost split sequences in C have at most two middle terms that, if M is an indecomposable module in C such that $\operatorname{Hom}_A(X, M) \neq 0$, then M belongs to the ray

$$X = X_0 \xrightarrow{u_1} X_1 \xrightarrow{u_2} X_2 \longrightarrow \ldots \xrightarrow{u_i} X_i \longrightarrow \ldots$$

starting with the module X. Furthermore, if $M \cong X_i$, then the vector space $\operatorname{Hom}_A(X, M)$ is one dimensional and generated by the homomorphism $u_i \ldots u_1 : X \longrightarrow M$ which is the composition of the monomorphisms on this sectional path, if $i \geq 1$, or by the identity map 1_X, if $i = 0$. This implies the statement.   □

The following definition is very useful.

2.3. Definition. Given an integer $t \geq 1$, we denote by

$$H = H_t = \begin{bmatrix} K & 0 & \ldots & 0 \\ K & K & \ldots & 0 \\ \vdots & \vdots & \ddots & \vdots \\ K & K & \ldots & K \end{bmatrix}$$

the subalgebra of $\mathbb{M}_t(K)$ consisting of all $t \times t$-lower triangular matrices. We identify H_t with the path algebra of the linear quiver

$$\overset{1}{\circ} \longleftarrow \overset{2}{\circ} \longleftarrow \overset{3}{\circ} \longleftarrow \ldots \longleftarrow \overset{t-1}{\circ} \longleftarrow \overset{t}{\circ}.$$

We denote by

$$P(t) = P(t)_H = I(1)_H$$

the unique indecomposable projective-injective H-module $e_t H_t$. If $t = 0$, we agree to denote by $H = H_0$ the zero algebra, and by $P(t)$ the zero module.

Let A be a K-algebra and assume that X is a module in a standard component C of $\Gamma(\operatorname{mod} A)$.

(a) If X is an admissible ray module, then the one-point extension algebra

$$A(X, t) = (A \times H_t)[X \oplus P(t)] = \begin{bmatrix} A \times H_t & 0 \\ X \oplus P(t) & K \end{bmatrix}$$

is called the t-**linear extension** of A at X.

An algebra A' is said to be a **line extension** of A if there exists a standard component $\mathcal{C}$ of $\Gamma(\mathrm{mod}\,A)$, an admissible ray module X in $\mathcal{C}$, and $t \geq 0$ such that $A' \cong A(X,t)$.

(b) If X is an admissible coray module, then the one-point coextension algebra

$$(X,t)A = [X \oplus P(t)](A \times H_t) = \begin{bmatrix} K & 0 \\ D(X \oplus P(t)) & A \times H_t \end{bmatrix}$$

is called the t-**linear coextension** of A at X.

An algebra A'' is said to be a **line coextension** of A if there exists a standard component $\mathcal{C}$ of $\Gamma(\mathrm{mod}\,A)$, an admissible coray module X in $\mathcal{C}$, and $t \geq 0$ such that $A'' \cong (X,t)A$.

Thus, whenever $t = 0$, the t-linear extension (or coextension) of A at X reduces to the one-point extension (or coextension) of A by X. Hence the above concept is a generalisation of that of one-point extension (or coextension) by an admissible ray (or coray, respectively) module, thus, for instance, by a mouth module in a stable tube.

Note that the bound quiver of a t-linear extension $A(X,t)$ is of the form

where 0 denotes the extension point, the shaded part denotes the bound quiver of A, and there are possibly additional relations from 0 to Q_A so that X equals the summand of $\mathrm{rad}\,P(0)_{A(X,t)}$ which is an A-module.

Similarly, the bound quiver of a t-linear coextension $(X,t)A$ is of the form

where 0 denotes the coextension point, the shaded part denotes the bound quiver of A, and there are possibly additional relations from Q_A to 0 so that X is a summand of

$$(I(0)/S(0))_{(X,t)A}$$

which is an A-module.

Now we illustrate this bound quiver description with the following simple example.

2.4. Example. Assume that A is a K-algebra given by the quiver

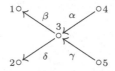

and let $X = S(3)$ (see examples (X.2.12) and (1.10)). Note that X is an admissible ray module, because X is a mouth module of a stable tube of $\Gamma(\mathrm{mod}\,A)$. While the algebra $A(X,0)$ is the one-point extension of A by X, as computed in (1.10), then taking $t = 2$, we get the algebra $A(X,2)$ given by the quiver

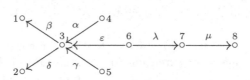

bound by two zero relations $\varepsilon\beta = 0$ and $\varepsilon\delta = 0$.

2.5. A rectangle insertion. Assume that $t \geq 0$ is an integer, A is an algebra, $\mathcal{C}$ a standard component of the Auslander–Reiten quiver $\Gamma(\mathrm{mod}\,A)$ of A, and X is an indecomposable admissible ray module lying in the component $\mathcal{C}$. We view $\mathcal{C}$ as a translation quiver. Let

$$A' = A(X,t)$$

be the t-linear extension of the algebra A by the admissible ray module X of $\mathcal{C}$.

Our purpose is to describe the component of $\Gamma(\mathrm{mod}\,A')$ which contains the module X, viewed as an A'-module. To do it, we first construct a translation quiver

$$(\mathcal{C}', \tau') \subseteq \Gamma(\mathrm{mod}\,A'),$$

which we later show to equal this particular component in $\Gamma(\mathrm{mod}\,A')$.

We recall that, given $t \geq 1$, $H = H_t$ is the path algebra of the linear quiver

$$\underset{1}{\circ} \longleftarrow \underset{2}{\circ} \longleftarrow \underset{3}{\circ} \longleftarrow \cdots \longleftarrow \underset{t-1}{\circ} \longleftarrow \underset{t}{\circ}.$$

and the Auslander–Reiten quiver $\Gamma(\operatorname{mod} H)$ of $\operatorname{mod} H$ has the form

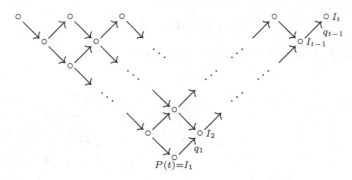

where we denote by

$$I_1 = I(1)_H, I_2 = I(2)_H, \ldots, I_t = I(t)_H$$

the indecomposable injective H-modules, viewed as A'-modules; that is, we set

$$I_j = I(j) = D(e_j H),$$

for $j = 1, \ldots, t$. In particular, $I_1 = I(1)_H = D(e_1 H) \cong e_t H = P(t)$ is the unique indecomposable projective-injective right H-module, and $I_t = I(t)_H = D(e_t H)$ is the unique simple injective right H-module.

On the other hand, the component $\mathcal{C}$ of $\Gamma(\operatorname{mod} A)$ being standard, the finite ray admissibility conditions on X imply that $\mathcal{C}$ looks as follows

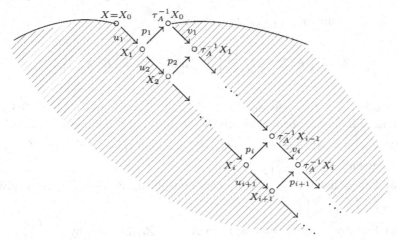

where, to avoid ambiguity, we denote by τ_A the Auslander–Reiten translation in $\operatorname{mod} A$.

We associate to the component $\mathcal{C}$ of $\Gamma(\mathrm{mod}\,A)$ the following translation quiver

$$(\mathcal{C}',\tau') \subseteq \Gamma(\mathrm{mod}\,A')$$

consisting of A'-modules, and homomorphisms of A'-modules

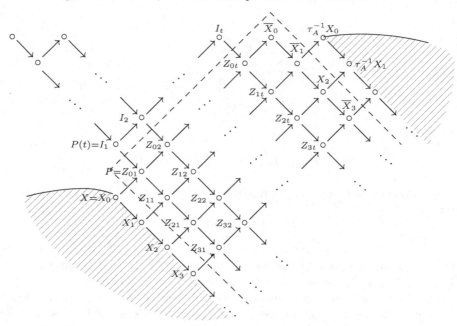

Here, for $i \geq 0$ and $j \in \{1,\dots,t\}$, we set

$$Z_{ij} = \left(K, X_i \oplus I_j, \begin{bmatrix} 1 \\ 1 \end{bmatrix}\right) \quad \text{and} \quad \overline{X}_i = (K, X_i, 1).$$

It is easily seen that the modules Z_{ij} and $\overline{X}_i$ are indecomposable A'-modules. The homomorphisms are the obvious ones, defined as follows.

(i) For a fixed $j \in \{1,\dots,t\}$, the homomorphism $Z_{ij} \longrightarrow Z_{i+1,j}$ is given by $\left(1, \begin{bmatrix} u_i & 0 \\ 0 & 1 \end{bmatrix}\right)$.

(ii) For a fixed $i \geq 0$, the homomorphism $Z_{ij} \longrightarrow Z_{i,j+1}$ is given by $\left(1, \begin{bmatrix} 1 & 0 \\ 0 & q_j \end{bmatrix}\right)$.

(iii) For any i, the homomorphism $X_i \longrightarrow Z_{i1}$ is given by $\left(0, \begin{bmatrix} 1 \\ 0 \end{bmatrix}\right)$.

(iv) For any $i \geq 0$, the homomorphism $Z_{it} \longrightarrow \overline{X}_i$ is given by $(1, [1\ 0])$.

(v) For any $i \geq 0$, the homomorphism $\overline{X}_i \longrightarrow \overline{X}_{i+1}$ is given by $(1, u_i)$.

(vi) For any $i \geq 0$, the homomorphism $\overline{X}_i \longrightarrow \tau_A^{-1} X_{i-1}$ is given by $(0, p_i)$.

(vii) For any $j \in \{1, \dots, t\}$, the homomorphism $I_j \longrightarrow Z_{0j}$ is given by $\left(0, \begin{bmatrix} 0 \\ 1 \end{bmatrix}\right)$.

The translation τ' of $\mathcal{C}'$ is defined as follows.

(i) If $i \geq 1$ and $j \geq 2$, then $\tau' Z_{ij} = Z_{i-1,j-1}$.

(ii) If $i \geq 1$, then $\tau' Z_{i1} = X_{i-1}$.

(iii) If $j \geq 2$, then $\tau' Z_{0j} = I_{j-1}$.

(iv) The module $P = Z_{01}$ is projective.

(v) $\tau' \overline{X}_0 = I_t$.

(vi) If $i \geq 1$, then $\tau' \overline{X}_i = Z_{i-1,t}$.

(vii) If $i \geq 0$, then $\tau'(\tau_A^{-1} X_i) = \overline{X}_i$.

For the remaining points of $\mathcal{C}$ (or of $\Gamma(\bmod H)$), the translation τ' coincides with τ_A (or with the translation τ_H in $\Gamma(\bmod H)$, respectively). This finishes the construction of the translation quiver $(\mathcal{C}', \tau')$.

The procedure presented above is called the **rectangle insertion**.

Intuitively, the construction of the translation quiver $(\mathcal{C}', \tau')$ may be understood as the following four step procedure:

(1°) take the standard component $\mathcal{C}$ of $\Gamma(\bmod A)$ and an admissible ray module X in $\mathcal{C}$,

(2°) 'cut it' along the arrows $p_1, p_2, \dots, p_i, p_{i+1}, \dots$,

(3°) 'insert in it' an infinite rectangle of width $t + 1$ consisting of the modules Z_{ij} and the modules $\overline{X}_i$, and finally

(4°) 'glue' the quiver $\Gamma(\bmod H)$ to its upper end.

It is useful to illustrate this construction with an example.

2.6. Example. Assume, as in (2.4), that A is a K-algebra given by the quiver

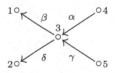

Let $X = S(3)$ and $t = 2$. We have seen in (2.4) that the algebra $A' = A(X, 2)$ is given by the quiver

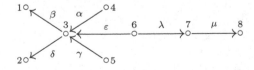

bound by two zero relations $\varepsilon\beta = 0$ and $\varepsilon\delta = 0$. In this case,

$$H = \begin{bmatrix} K & 0 \\ K & K \end{bmatrix}$$

and the Auslander–Reiten quiver $\Gamma(\mathrm{mod}\,H)$ is of the form

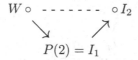

while, as seen in (X.2.12), the component $\mathcal{C}$ is a stable tube of rank two given by

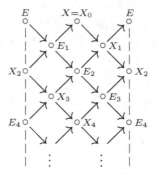

where we identify along the vertical dotted lines. Here, the translation quiver $(\mathcal{C}', \tau')$ is given by

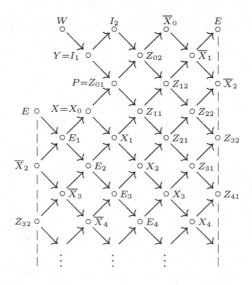

where again, we identify along the vertical dotted lines. We summarise by collecting the main properties of the translation quiver $(\mathcal{C}', \tau')$.

(a) The ray modules in $\mathcal{C}'$ are precisely the simple H-modules W and I_2, the module $\overline{X}_0 = (K, X_0, 1)$ and the mouth module E. We have exactly five rays in $\mathcal{C}'$ (three more than in $\mathcal{C}$), and every module in $\mathcal{C}'$ lies on a ray.

(b) $\mathcal{C}'$ contains three projectives, namely W, Y and $P = Z_{01}$, and any module in $\mathcal{C}'$ belongs to the τ-orbit of one of these modules.

(c) $\mathcal{C}'$ is right stable, that is, no module in $\mathcal{C}'$ is injective.

Now we can establish one of the main properties of the translation quiver $(\mathcal{C}', \tau')$.

2.7. Proposition. *Let $\mathcal{C}$ be a standard component in $\Gamma(\mathrm{mod}\,A)$, X be an admissible ray module in $\mathcal{C}$, and $t \geq 0$. Let*

$$A' = A(X, t) = \begin{bmatrix} A \times H_t & 0 \\ X \oplus P(t) & K \end{bmatrix}$$

and $(\mathcal{C}', \tau')$ be the translation quiver associated to $\mathcal{C}$ in the rectangle insertion (2.5).

(a) *$\mathcal{C}'$ is the Auslander–Reiten component of $\Gamma(\mathrm{mod}\,A')$ containing X, when viewed as an A'-module.*

(b) *An indecomposable A'-module $U = (U_0, U_1, \varphi_U)$ belongs to $\mathcal{C}'$ if and only if U is an H_t-module or the restriction $\mathrm{res}_A U = U_1$ of U to A has an indecomposable direct summand from $\mathcal{C}$.*

Proof. (a) We use freely the notation introduced above. By construction, P is the only indecomposable projective A'-module which is neither an indecomposable projective A-module, nor an indecomposable projective H-module. Also, there are inclusion homomorphisms of X and $P(t)$ as summands of $\mathrm{rad}\,P$, which are therefore irreducible in $\mathrm{mod}\,A'$. We also recall from (1.5) that, if $g : M \longrightarrow N$ is a right minimal almost split morphism in $\mathrm{mod}\,A$, then

$$(0, g) : (\mathrm{Ker}\,\mathrm{Hom}_{A \times H}(X \oplus P(t), g), M, u) \longrightarrow (0, N, 0)$$

is right minimal almost split in $\mathrm{mod}\,A'$, where u denotes the canonical inclusion.

Let $i \geq 0$. The right minimal almost split morphism in $\mathrm{mod}\,A'$ ending in $(0, X_i, 0)$ is given by $(0, g_i)$, with $g_i : M \longrightarrow X_i$ right minimal almost split in $\mathrm{mod}\,A$. Clearly, $M = X_{i-1} \oplus M'$ (where we agree that $X_{-1} = 0$) and, for

$i \geq 1$, the module X_{i-1} does not lie in add M'. Because M is an A-module, $\mathrm{Hom}_{A \times H}(P(t), M) = 0$, so

$$\mathrm{Ker}\,\mathrm{Hom}_{A \times H}(X \oplus P(t), g_i) = \mathrm{Ker}\,\mathrm{Hom}_A(X, g_i),$$

where $\mathrm{Hom}_A(X, g_i) : \mathrm{Hom}_A(X, X_{i-1} \oplus M') \longrightarrow \mathrm{Hom}_A(X, X_i)$. By our assumption that X is an admissible ray module in $\mathcal{C}$, we have $\mathrm{Hom}_A(X, M') = 0$, while

$$\mathrm{Hom}_A(X, g_i) : \mathrm{Hom}_A(X, X_{i-1}) \longrightarrow \mathrm{Hom}_A(X, X_i)$$

is a monomorphism. It follows that $(0, g_i) : (0, M, 0) \longrightarrow (0, X_i, 0)$ is a right minimal almost split morphism in mod A'. In particular, the irreducible morphisms $X_i \longrightarrow X_{i+1}$ remain irreducible in mod A'. Moreover, for each $i \geq 0$, there exists an almost split sequence of the form

$$0 \longrightarrow X_i \longrightarrow X_{i+1} \oplus Z'_{i,1} \longrightarrow Z'_{i+1,1} \longrightarrow 0,$$

in mod A', where $Z'_{0,1} = Z_{0,1}$ is a projective module and $Z'_{j,1}$ is indecomposable non-projective such that $Z'_{j,1} \not\cong X_{j+1}$, for any $j \geq 1$. To see this, we note that if $X_i \longrightarrow N$ is an irreducible morphism in mod A', with N indecomposable non-projective, then there is an irreducible morphism $\tau_{A'} N \longrightarrow X_i$ in mod A', and the claim follows from the preceding description of the right almost split morphisms in mod A' ending at X_i. Because $Z'_{0,1} = Z_{0,1}$, we deduce inductively that $Z'_{i,1} = Z_{i,1}$, for any $i \geq 0$.

Applying again (1.5), we see that all the irreducible morphisms in mod H remain irreducible in mod A'. Then, as above, we conclude that there exist almost split sequences

$$0 \longrightarrow I_t \longrightarrow Z_{0,t} \longrightarrow \overline{X}_0 \longrightarrow 0, \text{ and}$$

$$0 \longrightarrow I_j \longrightarrow I_{j+1} \oplus Z_{0,j} \longrightarrow Z_{0,j+1} \longrightarrow 0,$$

in mod A', for any $j \in \{1, \ldots, t-1\}$. They connect the Auslander–Reiten quiver $\Gamma(\mathrm{mod}\,H)$ of $H = H_t$ with the component of $\Gamma(\mathrm{mod}\,A')$ containing the projective module $Z'_{0,1} = Z_{0,1}$.

A straightforward induction on the construction of the cokernel term in the respective sequences shows that we have indeed the almost split sequences starting at all the Z_{ij}, with $i \geq 1$ and $2 \leq j \leq t$.

There remains to compute the almost split sequences starting at the $\overline{X}_i$. Assume that there exists an irreducible morphism $X_i \longrightarrow L$ in mod A, with L indecomposable. By our assumption that X is an admissible ray module, we have

$$L \cong X_{i+1} \text{ or } L \cong \tau_A^{-1} X_{i-1}.$$

The left minimal almost split morphism starting at X_i in $\operatorname{mod} A$ is thus

$$f : X_i \longrightarrow X_{i+1} \oplus \tau_A^{-1} X_{i-1}.$$

Hence we deduce that the homomorphism

$$\overline{X}_i \xrightarrow{(1,f)} \left(\operatorname{Hom}_{A \times H}(X \oplus P(t), X_i), \, X_{i+1} \oplus \tau_A^{-1} X_{i-1}, \, \operatorname{Hom}_{A \times H}(X \oplus P(t), f)\right)$$
$$\cong (K, X_{i+1}, 1) \bigoplus (0, \tau_A^{-1} X_{i-1}, 0)$$

is a left minimal almost split morphism in $\operatorname{mod} A'$, again by (1.5).

(b) It follows from (a) and (2.5) that if V is an indecomposable module in $\mathcal{C}'$ then either V is an H_t-module or the restriction $\operatorname{res}_A V = V_1$ of V to A is an indecomposable A-module from $\mathcal{C}$.

Let $U = (U_0, U_1, \varphi_U)$ be an indecomposable A'-module such that the restriction $\operatorname{res}_A U = U_1$ of U to A has an indecomposable direct summand Z from $\mathcal{C}$. Then Z is an A-module and, hence,

$$\operatorname{Hom}_{A \times H}(X \oplus P(t), Z) = \operatorname{Hom}_{A \times H}(X, Z) = \operatorname{Hom}_A(X, Z).$$

First we assume that the K-linear map

$$\varphi_U : U_0 \longrightarrow \operatorname{Hom}_{A \times H}(X \oplus P(t), U_1)$$

is zero. Then $U_0 = 0$ and $U = (0, U_1, 0) = (0, Z, 0) = Z$, because the module Z is indecomposable. It follows that the module $U = Z$ belongs to $\mathcal{C}'$, because all modules from $\mathcal{C}$ belong to $\mathcal{C}'$.

Next we assume that the K-linear map $\varphi_U \neq 0$. Let $p : U_1 \longrightarrow Z$ be the retraction on Z and $q : Z \longrightarrow U_1$ its section. We show that the composite K-linear map

$$U_0 \xrightarrow{\varphi_U} \operatorname{Hom}_{A \times H}(X \oplus P(t), U_1) \xrightarrow{\operatorname{Hom}_{A \times H}(X \oplus P(t), p)} \operatorname{Hom}_{A \times H}(X \oplus P(t), Z)$$

is non-zero. Assume, to the contrary, that the composite map is zero. Then $(0, p) : U \xrightarrow{\varphi_U} Z = (0, Z, 0)$ is a retraction, $(0, q) : Z \xrightarrow{\varphi_U} U$ is its section, and the indecomposability of U yields $U \cong Z$. On the other hand, the assumption $\varphi_U \neq 0$ yields $U \not\cong Z$, and we get a contradiction. This proves our claim.

Hence, we conclude that $\operatorname{Hom}_A(X, Z) = \operatorname{Hom}_{A \times H}(X \oplus P(t), Z) \neq 0$. Because Z is an indecomposable A-module from $\mathcal{C}$ and $X = X_0$ is an admissible ray module then there is an isomorphism $Z \cong X_i$ of A-modules, for some $i \geq 0$. It is easy to see that $\operatorname{Hom}_{A \times H}(X \oplus P(t), p) \circ \varphi_U \neq 0$ yields the existence of a commutative diagram of K-linear maps

$$
\begin{array}{ccc}
U_0 & \xrightarrow{\varphi_U} & \operatorname{Hom}_{A \times H}(X \oplus P(t), U_1) \\
\Big\downarrow{f_0} & & \Big\downarrow{\operatorname{Hom}_{A \times H}(X \oplus P(t), p)} \\
K & \xrightarrow{\quad 1 \quad} & \operatorname{Hom}_{A \times H}(X \oplus P(t), Z) \cong K.
\end{array}
$$

This shows that $(f_0, f_1) : U \longrightarrow \overline{X}_i = (K, X_i, 1)$, with $f_1 = p$, is a homomorphism in mod A'.

Choose the index i to be minimal with respect to this property. If f is an isomorphism then U belongs to $\mathcal{C}'$ and (b) follows. Assume that f is not an isomorphism. Then f has a factorisation through the minimal right almost split morphism ending at $\overline{X}_i$ and, by the minimality of i, there is a homomorphism $g = (g_0, g_1) : U \longrightarrow Z_{i,t}$, with $g_0 \neq 0$. Choose a homomorphism $h = (h_0, h_1) : U \longrightarrow Z_{i,j}$ such that $h_0 \neq 0$, $i \geq 0$, $j \in \{1, \ldots, t\}$, and $i + j$ is minimal. We consider four cases.

<u>Case 1°</u>. Assume that $i = 0$ and $j = 1$. Then the minimal right almost split sequence ending at the projective module $P = Z_{0,1}$ is of the form

$$
X \oplus P(t) = \operatorname{rad} P \longrightarrow P = Z_{0,1}.
$$

Because $h_0 \neq 0$ then the induced homomorphism $\overline{h} : U/\operatorname{rad} U \longrightarrow P/\operatorname{rad} P$ is surjective and, consequently, $h : U \longrightarrow P = Z_{0,1}$ is an epimorphism. It follows that $U \cong Z_{0,1}$, because U is indecomposable and $Z_{0,1}$ is projective. Consequently, the module U belongs to $\mathcal{C}'$, and (b) follows in this case.

<u>Case 2°</u>. Assume that $i = 0$ and $j \geq 2$. Then there exists an almost split sequence

$$
0 \longrightarrow I_{j-1} \longrightarrow I_j \oplus Z_{0,j-1} \longrightarrow Z_{0,j} \longrightarrow 0
$$

in mod A'. It follows that the homomorphism $h : U \longrightarrow Z_{0,j}$ is an isomorphism and, hence, the module U belongs to $\mathcal{C}'$. To prove this, assume to the contrary, that h is not an isomorphism. Then h admits a factorisation through the module $I_j \oplus Z_{0,j-1}$ and $h_0 \neq 0$ yields the existence of a homomorphism $h' = (h'_0, h'_1) : U \longrightarrow Z_{0,j}$, with $h'_0 \neq 0$. This contradicts the minimality of $i + j = 0 + j = j$.

<u>Case 3°</u>. Assume that $i \geq 1$ and $j = 1$. Then there exists an almost split sequence

$$
0 \longrightarrow X_{i-1} \longrightarrow X_i \oplus Z_{i-1,1} \longrightarrow Z_{i,1} \longrightarrow 0
$$

in mod A'. It follows that the homomorphism $h : U \longrightarrow Z_{i,1}$ is an isomorphism and, hence, the module U belongs to $\mathcal{C}'$. To prove this, assume to the contrary, that h is not an isomorphism. Then h admits a factorisation through the module $X_i \oplus Z_{i-1,1}$ and $h_0 \neq 0$ yields the existence of a homomorphism $h'' = (h_0'', h_1'') : U \longrightarrow Z_{i-1,1}$, with $h_0'' \neq 0$. This contradicts the minimality of $i + j = i + 1$.

Case 4°. Assume that $i \geq 1$ and $j \geq 2$. Then there exists an almost split sequence

$$0 \longrightarrow Z_{i-1,j-1} \longrightarrow Z_{i,j-1} \oplus Z_{i-1,j} \longrightarrow Z_{i,j} \longrightarrow 0$$

in $\operatorname{mod} A'$. It follows that the homomorphism $h : U \longrightarrow Z_{i,j}$ is an isomorphism and, hence, the module U belongs to $\mathcal{C}'$. To prove this, assume to the contrary, that h is not an isomorphism. Then h admits a factorisation through the module $Z_{i,j-1} \oplus Z_{i-1,j}$, and $h_0 \neq 0$ yields the existence of a homomorphism $h' = (h_0', h_1') : U \longrightarrow Z_{i,j-1}$, with $h_0' \neq 0$, or of a homomorphism $h'' = (h_0'', h_1'') : U \longrightarrow Z_{i-1,j}$, with $h_0'' \neq 0$. This contradicts again the minimality of $i + j$.

We have proved in Cases 1°–4° that the homomorphism $h : U \longrightarrow Z_{i,j}$ is an isomorphism, that is, the module $U \cong Z_{i,j}$ belongs to $\mathcal{C}'$. This finishes the proof of (b) and completes the proof of the theorem.   □

2.8. Proposition. *Under the hypothesis and notation of the rectangle insertion (2.5), the component $\mathcal{C}'$ is standard.*

Proof. Let $L : K(\mathcal{C}) \longrightarrow \operatorname{ind} \mathcal{C}$ and $L' : K(\mathcal{C}') \longrightarrow \operatorname{ind} \mathcal{C}'$ be the obvious functors (as defined in the proof of (1.6)), where $K(\mathcal{C})$ and $K(\mathcal{C}')$ are the mesh K-categories of $\mathcal{C}$ and $\mathcal{C}'$. By hypothesis, L is an equivalence, and we want to show that so is L'. Because the functor L' is clearly dense, it remains to prove that L is full and faithful, that is, for all M, N in $\mathcal{C}'$, the functor L' induces an isomorphism $\operatorname{Hom}_{K(\mathcal{C}')}(M, N) \cong \operatorname{Hom}_{A'}(M, N)$.

Let $J : K(\mathcal{C}) \longrightarrow K(\mathcal{C}')$ be the K-linear embedding which is the identity on all objects and all arrows except those of the form $X_i \longrightarrow \tau_A^{-1} X_{i-1}$, the image of which is the corresponding sectional path in $\mathcal{C}'$.

Let $J' : \operatorname{ind} \mathcal{C} \longrightarrow \operatorname{ind} \mathcal{C}'$ be the functor induced by J. We clearly have a commutative square

$$
\begin{array}{ccc}
K(\mathcal{C}) & \xrightarrow{\;\;J\;\;} & K(\mathcal{C}') \\
{\scriptstyle L}\downarrow & & \downarrow{\scriptstyle L'} \\
\operatorname{ind} \mathcal{C} & \xrightarrow{\;\;J'\;\;} & \operatorname{ind} \mathcal{C}'.
\end{array}
$$

In particular, if $M, N \in \operatorname{ind} \mathcal{C}$, then $\operatorname{Hom}_{K(\mathcal{C}')}(M, N) \cong \operatorname{Hom}_{A'}(M, N)$.

If M is an H-module, and $\operatorname{Hom}_{A'}(M, N) \neq 0$, then either N is an H-module or N is of the form Z_{ij}. Similarly, if N is an H-module and

$\operatorname{Hom}_{A'}(M, N) \neq 0$, then M is an H-module. Hence, if M or N is an H-module, then L' induces the required isomorphism

$$\operatorname{Hom}_{K(\mathcal{C}')}(M, N) \cong \operatorname{Hom}_{A'}(M, N).$$

We may thus assume that neither M nor N is an H-module.

We note that the homomorphisms $Z_{ij} \longrightarrow \overline{X}_i$ in $\operatorname{mod} A'$ induced by the corresponding sectional paths in $\mathcal{C}'$ are surjective. Moreover, the irreducible morphisms

$$\overline{X}_i \longrightarrow \tau_A^{-1} X_{i-1}$$

are surjective. Let thus $N \in \operatorname{ind} \mathcal{C}$ and $M \notin \operatorname{ind} \mathcal{C}$. Then $M = Z_{ij}$ or $\overline{X}_i$ for some i, j. A non-zero homomorphism $f : M \longrightarrow N$ in $\operatorname{mod} A$ can always be written as $f = gv$, where

$$v : M \longrightarrow \tau_A^{-1} X_{i-1}$$

is induced by the corresponding sectional path in $\mathcal{C}'$. Because v lies in the image of L', and so does g (by the commutativity of the above square), L' induces a surjection

$$\operatorname{Hom}_{K(\mathcal{C}')}(M, N) \longrightarrow \operatorname{Hom}_{A'}(M, N).$$

On the other hand, as we have observed, v is an epimorphism in $\operatorname{mod} A'$, and J' is faithful. Hence, the previous surjection is an isomorphism.

Similarly, if $f : M \longrightarrow N$ is a non-zero homomorphism in $\operatorname{mod} A'$ with $M \in \operatorname{ind} \mathcal{C}$, $N \notin \operatorname{ind} \mathcal{C}$, then f can be written as $f = uh$, for some $h : M \longrightarrow X_i$ and $u : X_i \longrightarrow N$ induced by the corresponding sectional path. Because u is a monomorphism, it follows from the commutativity of the above square that L' induces the required isomorphism $\operatorname{Hom}_{K(\mathcal{C}')}(M, N) \cong \operatorname{Hom}_{A'}(M, N)$.

There remains to consider the case when both M and N are of the form Z_{ij}. In this case, a non-zero homomorphism $f : M \longrightarrow N$ in $\operatorname{mod} A'$ can be written as $f = ugv + h$, where

$$u : X_r \longrightarrow N \quad \text{and} \quad v : M \longrightarrow \tau_A^{-1} X_{s-1}$$

are induced by the corresponding sectional paths,

$$g : \tau^{-1} X_{s-1} \longrightarrow X_r,$$

and h is zero or is a composition of irreducible morphisms corresponding to arrows between modules of the form Z_{ij}. Because the homomorphisms h, u,

v belong to the image of L', and so does g (by the previous considerations), L' induces a surjection

$$\text{Hom}_{K(\mathcal{C}')}(M, N) \longrightarrow \text{Hom}_{A'}(M, N).$$

Now h is non-zero in mod A' if and only if it is non-zero in $K(\mathcal{C}')$. Similarly, because u is injective and v surjective in mod A', and moreover J is faithful, then ugv is non-zero in mod A' if and only if it is non-zero in $K(\mathcal{C}')$. Now, any non-zero morphisms $f : M \longrightarrow N$ in $K(\mathcal{C}')$ can be written as $f = ugv+h$ with u, g, v as above. Then $L'(f) = 0$ implies $0 \neq L'(h) = -L'(ugv)$. But h does not factor through modules in $\mathcal{C}$, while g does. This shows that L' induces an isomorphism $\text{Hom}_{K(\mathcal{C}')}(M, N) \cong \text{Hom}_{A'}(M, N)$. □

We can iterate the procedures of a line extension and a line coextension starting from an arbitrary K-algebra. This leads to the following definition.

2.9. Definition. Assume that C is an arbitrary K-algebra and

$$\boldsymbol{\mathcal{T}} = \{\mathcal{T}_\lambda\}_{\lambda \in \Lambda}$$

is a family of pairwise orthogonal standard stable tubes of $\Gamma(\text{mod}\, C)$.

(a) An algebra A is said to be a $\boldsymbol{\mathcal{T}}$-**tubular extension** of C if there exists a sequence of algebras

$$A_0 = C, A_1, \ldots, A_m = A$$

such that, for each $i \in \{1, \ldots, m\}$, the algebra A_i is a line extension of A_{i-1}, with respect to an admissible ray module X_i lying in a standard stable tube of $\boldsymbol{\mathcal{T}}$ or in a component of $\Gamma(\text{mod}\, A_{i-1})$, obtained from a standard stable tube of $\boldsymbol{\mathcal{T}}$ by the rectangle insertions created by the line extensions done so far.

(b) An algebra A is said to be a $\boldsymbol{\mathcal{T}}$–**tubular coextension** of C if there exists a sequence of algebras

$$A_0 - C, A_1, \ldots, A_m = A$$

such that, for each $i \in \{1, \ldots, m\}$, the algebra A_i is a line coextension of A_{i-1}, with respect to an admissible coray module X_i lying in a standard stable tube of $\boldsymbol{\mathcal{T}}$ or in a component of $\Gamma(\text{mod}\, A_{i-1})$, obtained from a standard stable tube of $\boldsymbol{\mathcal{T}}$ by the rectangle coinsertions created by the line coextensions done so far.

It follows that, if a K-algebra A is a $\boldsymbol{T}$-tubular extension of the K-algebra C then there exist:

- a sequence $\mathcal{C}_0, \mathcal{C}_1, \ldots, \mathcal{C}_m$ of standard components of the Auslander–Reiten quivers $\Gamma(\operatorname{mod} A_0), \Gamma(\operatorname{mod} A_1), \ldots, \Gamma(\operatorname{mod} A_m)$, respectively,
- a sequence of admissible ray modules $X_0, X_1, \ldots, X_{m-1}$, with X_j in $\mathcal{C}_j$, and
- a sequence of non-negative integers $t_0, t_1, \ldots, t_{m-1}$

such that

$$A_1 = A_0(X_0, t_0), \quad A_2 = A_1(X_1, t_1), \quad \ldots, \quad A_m = A_{m-1}(X_{m-1}, t_{m-1}).$$

We denote by $\mathcal{C}_i'$ the component of $\Gamma(\operatorname{mod} A_{i+1})$ obtained from $\mathcal{C}_i$ as in (2.3), for $i = 0, 1, \ldots, m-1$.

Because of the statement (c) in (2.11), we introduce the following definition.

2.10. Definition. (i) The component $\mathcal{C}_i'$ of $\Gamma(\operatorname{mod} A_{i+1})$ obtained from $\mathcal{C}_i$ as in (2.3), for $i = 0, 1, \ldots, m-1$, is called a **ray tube**. The **rank** $\operatorname{rk} \mathcal{C}_i'$ of a ray tube $\mathcal{C}_i'$ is defined to be the number of rays the tube contains.

(ii) A **coray tube**, and its **rank**, are defined dually.

2.11. Corollary. *Under the hypothesis and notation made in (2.9), for any $i \in \{0, 1, \ldots, m-1\}$, the following statements hold.*

(a) *The component $\mathcal{C}_i'$ is standard.*

(b) *The admissible ray modules in $\mathcal{C}_i'$ are the admissible ray modules of $\mathcal{C}_i$, except X_i, the simple H_{t_i}-modules, and the module*

$$\overline{X}_i = (K, X_i, 1).$$

(c) *Each module in $\mathcal{C}_i'$ belongs to a ray, and the number of rays in $\mathcal{C}_i'$ equals the number of rays in $\mathcal{C}_i$, plus $t_i + 1$.*

(d) *The component $\mathcal{C}_i'$ is right stable, that is, it contains no injectives.*

Proof. The statements (a), (b) and (c) follow from (2.3), (2.7) and induction, while (d) follows from the construction of the component $\mathcal{C}_i'$ starting from $\mathcal{C}_i$. $\qquad\square$

2.12. Corollary. *Let C be a K-algebra and*

$$\boldsymbol{T} = \{\mathcal{T}_\lambda\}_{\lambda \in \Lambda}$$

a family of pairwise orthogonal standard stable tubes of $\Gamma(\operatorname{mod} C)$.

(a) *If A is a $\boldsymbol{T}$-tubular extension of C then $\Gamma(\operatorname{mod} A)$ admits a family*

$$\boldsymbol{C} = \{\mathcal{C}_\lambda\}_{\lambda \in \Lambda}$$

of pairwise orthogonal standard ray tubes $\mathcal{C}_\lambda$ obtained from the family
$\mathcal{T}$ by rectangle insertions.

(b) If A is a $\mathcal{T}$-tubular coextension of C then $\Gamma(\operatorname{mod} A)$ admits a family

$$\mathcal{C} = \{\mathcal{C}_\lambda\}_{\lambda \in \Lambda}$$

of pairwise orthogonal standard coray tubes $\mathcal{C}_\lambda$ obtained from the family $\mathcal{T}$ by rectangle coinsertions.

Proof. Apply (2.11) and its dual. □

2.13. Example. Let A be the hereditary algebra given by the quiver

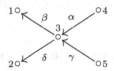

We take $X = S(3)$ and $t = 2$. Then the algebra

$$A_1 = A(X, 2)$$

is given by the quiver

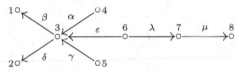

bound by two zero relations $\varepsilon\beta = 0$ and $\varepsilon\delta = 0$, see (2.4).

Letting $\mathcal{C}_1$ denote the stable tube of $\Gamma(\operatorname{mod} A)$ containing X, it follows from (1.10) and (2.6) that the component $\mathcal{C}_1'$ has the form

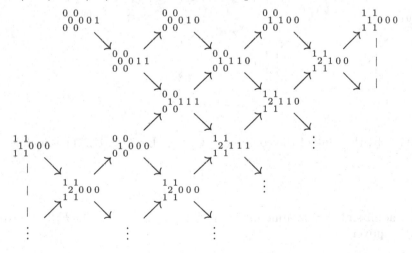

where we identify along the vertical dotted lines. Note also that the rank of the component C_1' is 5.

Let $C_2 = C_1'$. The simple module $X' = S(7)$ is an admissible ray module in C_2. The K-algebra

$$A_2 = A_1(X', 0)$$

is given by the quiver

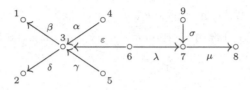

bound by three zero relations $\varepsilon\beta = 0$, $\varepsilon\delta = 0$ and $\sigma\mu = 0$. Then the component C_2' is given by

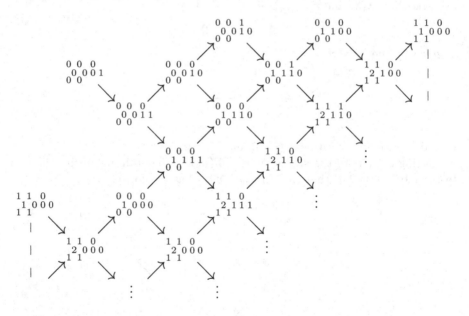

Obviously, the rank of the component C_2' is 6. Let $C_3 = C_2'$. The module

$$E = \begin{smallmatrix} & 1\ 1\ 0 \\ 1 & 0\ 0\ 0 \\ & 1\ 1 \end{smallmatrix}$$

is an admissible ray module in C_3. The K-algebra $A_3 = A_2(E, 0)$ is given by the quiver

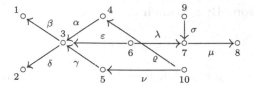

bound by the relations

$$\varepsilon\beta = 0, \quad \varepsilon\delta = 0, \quad \sigma\mu = 0, \quad \text{and } \varrho\alpha = \nu\gamma.$$

The component $\mathcal{C}'_3$ is given by

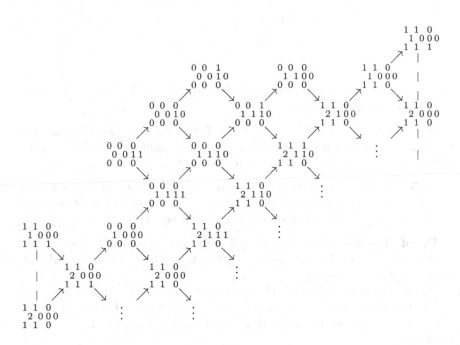

Obviously, the rank of the component $\mathcal{C}'_3$ is 7.

Our final objective in this section is to determine how the bound quiver of the algebra C, the tubular extension we start from, is modified.

We know that the quivers of one-point extensions have the effect of adding a source, and the quivers of line extensions have the effect of adding an oriented line. We now see the effect of the quiver Q_A of a tubular extension A of an algebra C. For this purpose, we need a definition.

2.14. Definition. By a **branch**

$$\mathcal{L} = (L, I)$$

we mean a finite connected full bound subquiver, containing the lowest vertex 0, of the following infinite tree

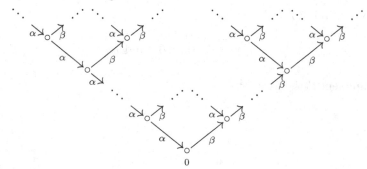

bound by all possible zero relations of the form $\alpha\beta = 0$. The lowest vertex 0 of $\mathcal{L}$ is called a **germ** of $\mathcal{L}$.

We observe that, if a K-algebra A is a $\mathcal{T}$-tubular extension of an algebra C then it follows from the definition that the bound quiver of C is a full bound convex subquiver of the bound quiver of A.

2.15. Lemma. *Let A be a $\mathcal{T}$-tubular extension of an arbitrary algebra C. Then the full bound subquiver of the bound quiver of A, consisting of all points not lying in C, is a disjoint union of branches.*

Proof. Because the algebra A is a $\mathcal{T}$-tubular extension of C, there exists a sequence $A_0 = C, A_1, \ldots, A_m = A$, with A_{i+1} a line extension of A_i, for each $i \in \{0, 1, \ldots, m-1\}$.

We use the induction on i. The statement holds for $i = 0$ (trivially) and $i = 1$ (by definition). Let $i \geq 1$ and $A_{i+1} = A_i(X_i, t_i)$. Denote by $\mathcal{C}_i$ the component of $\Gamma(\mathrm{mod}\, A_i)$ containing X_i. The bound quiver of A_i is of the form

By (2.11), the admissible ray modules in $\mathcal{C}'_{i-1}$ are the simple $H_{t_{i-1}}$-modules, the module $\overline{X}_{i-1}$, and the admissible ray modules of $\mathcal{C}_{i-1}$ except X_{i-1}. If $\mathcal{C}'_{i-1} \neq \mathcal{C}_i$ or if $\mathcal{C}'_{i-1} = \mathcal{C}_i$ and X_i is a ray module of $\mathcal{C}_{i-1}$ distinct from $\overline{X}_{i-1}$, then the bound quiver of A_{i+1} is of the form

$$
Q_{A_{i-1}} \quad
\begin{array}{c}
\overset{0'}{\circ} \longrightarrow \overset{t_i}{\circ} \longrightarrow \cdots \longrightarrow \overset{1'}{\circ} \\
\overset{0}{\circ} \longrightarrow \overset{t_{i-1}}{\circ} \longrightarrow \cdots \longrightarrow \overset{1}{\circ}
\end{array}
$$

If $\mathcal{C}'_{i-1} = \mathcal{C}_i$ and X_i is a simple H_{t_i}-module, then the bound quiver of A_{i+1} is of the form

$$
Q_{A_{i-1}} \quad
\begin{array}{c}
\overset{0'}{\circ} \overset{t_i}{\longrightarrow} \circ \longrightarrow \cdots \longrightarrow \overset{1'}{\circ} \\
\Big\downarrow \alpha \\
\overset{0}{\circ} \longrightarrow \overset{t_{i-1}}{\circ} \longrightarrow \cdots \longrightarrow \underset{\beta}{\circ} \longrightarrow \cdots \longrightarrow \overset{1}{\circ}
\end{array}
$$

with $\alpha\beta = 0$. If $\mathcal{C}'_{i-1} = \mathcal{C}_i$ and $X_i = \overline{X}_{i-1}$, then the bound quiver of A_{i+1} is of the form

$$
Q_{A_{i-1}} \quad
\begin{array}{c}
\overset{0'}{\circ} \longrightarrow \overset{t_i}{\circ} \longrightarrow \cdots \longrightarrow \overset{1'}{\circ} \\
\Big\downarrow \alpha \\
\overset{0}{\circ} \overset{\beta}{\longrightarrow} \overset{t_{i-1}}{\circ} \longrightarrow \cdots \longrightarrow \overset{1}{\circ}
\end{array}
$$

with $\alpha\beta = 0$. This completes the proof. □

For instance, in Example (2.13), we have two branches, namely

$$
\begin{array}{ccc}
& \overset{\circ}{\Big\downarrow \sigma} & \\
\underset{6}{\circ} \overset{\lambda}{\longrightarrow} \underset{7}{\circ} \overset{\mu}{\longrightarrow} \underset{8}{\circ} & \quad \text{and} \quad & \circ\ 10
\end{array}
$$

bound by the zero relation $\sigma\mu = 0$.

XV.3. Branch extensions and branch coextensions of algebras

One of the main aims of this section is to show that the concepts of the tubular extension and the tubular coextension discussed in the previous section coincide with the concepts of the tubular extension and the tubular coextension defined by Ringel in [525, Section 4.7].

First we introduce the concepts of a branch extension and a branch coextension.

3.1. Definition. Assume that C is an arbitrary K-algebra and

$$\mathcal{T} = \{\mathcal{T}_\lambda\}_{\lambda \in \Lambda}$$

is a family of pairwise orthogonal standard stable tubes of $\Gamma(\mathrm{mod}\, C)$. Let $E_1, \ldots, E_s$ be a set of pairwise different modules lying on the mouth of the tubes of $\mathcal{T}$.

(a) We construct inductively the sequence $C_1, \ldots, C_s$ of iterated one-point extension algebras, where $C_0 = C$ and $C_j = C_{j-1}[E_j]$, for each $j \in \{1, \ldots, s\}$. The algebra

$$C[E_1, \ldots, E_s] = C_{s-1}[E_s] \tag{3.2}$$

is called a **multiple one-point $\mathcal{T}$-extension** of C, or simply a **multiple one-point extension** of C.

(b) Dually, we construct inductively the sequence $C'_1, \ldots, C'_s$ of iterated one-point coextension algebras, where $C'_0 = C$ and $C'_j = C'_{j-1}[E_j]$, for each $j \in \{1, \ldots, s\}$. The algebra

$$[E_1, \ldots, E_s]C = [E_s]C'_{s-1} \tag{3.3}$$

is called a **multiple one-point $\mathcal{T}$-coextension** of C, or simply a **multiple one-point coextension** of C.

It is clear that the multiple one-point $\mathcal{T}$-extension algebra $C[E_1, \ldots, E_s]$ and the multiple one-point $\mathcal{T}$-coextension algebra $[E_1, \ldots, E_s]C$ have the following lower triangular matrix forms

$$C[E_1, \ldots, E_s] = \begin{bmatrix} C & 0 \\ E_1 \oplus \ldots \oplus E_s & K_1 \times \ldots \times K_s \end{bmatrix}, \tag{3.4}$$

$$[E_1, \ldots, E_s]C = \begin{bmatrix} K_1 \times \ldots \times K_s & 0 \\ D(E_1 \oplus \ldots \oplus E_s) & C \end{bmatrix}, \tag{3.5}$$

where $K_1 = \ldots = K_s = K$, and the left module structure of $E_1 \oplus \ldots \oplus E_s$ over the product $K_1 \times \ldots \times K_s$ of s copies of the field K is given by the formula $(\mu_1, \ldots, \mu_s) \cdot (u_1, \ldots, u_s) = (\mu_1 u_1, \ldots, \mu_s u_s)$.

For each $j \in \{1, \ldots, s\}$, we denote by O_j the extension vertex of the algebra

$$C_j = C_{j-1}[E_j]$$

in the ordinary quiver Q_{C_j} of C_j. Then the ordinary quiver $Q_{C[E_1, \ldots, E_s]}$ of the algebra $C[E_1, \ldots, E_s]$ is of the form

where Q_C is the quiver of C. Obviously, the ordinary quiver $Q_{[E_1,\ldots,E_s]C}$ of the algebra $[E_1,\ldots,E_s]C$ has the form dual to the above one.

3.6. Definition. Assume that C is an arbitrary K-algebra and
$$\mathcal{T} = \{\mathcal{T}_\lambda\}_{\lambda\in\Lambda}$$
is a family of pairwise orthogonal standard stable tubes of $\Gamma(\operatorname{mod}C)$. Let $E_1,\ldots,E_s$ be a set of pairwise different modules lying on the mouth of the tubes of $\mathcal{T}$.

(a) Let $C[E_1,\ldots,E_s]$ be the multiple one-point $\mathcal{T}$-extension algebra (3.4) and let $O_1,\ldots,O_s$ be the extension vertices of the algebras
$$C_1 = C_0[E_1],\ldots,C_s = C_{s-1}[E_s],$$
respectively, where $C_0 = C$. For each $j \subset \{1,\ldots,s\}$, we choose a branch $\mathcal{L}^{(j)}$, with the germ O_j^*. Further, we fix a bound quiver
$$(Q_{C[E_1,\ldots,E_s]}, I_{C[E_1,\ldots,E_s]})$$
defining the multiple one-point $\mathcal{T}$-extension algebra $C[E_1,\ldots,E_s]$, that is, there is an algebra isomorphism
$$C[E_1,\ldots,E_s] \cong KQ_{C[E_1,\ldots,E_s]}/I_{C[E_1,\ldots,E_s]}.$$
The **branch $\mathcal{T}$-extension** of the algebra C by means of the branches $\mathcal{L}^{(1)},\ldots,\mathcal{L}^{(s)}$ is defined to be the bound quiver algebra
$$C[E_1,\mathcal{L}^{(1)},\ldots,E_s,\mathcal{L}^{(s)}] = C[E_j,\mathcal{L}^{(j)}]_{j=1}^s = KQ/I \qquad (3.7)$$
of the bound quiver (Q,I) obtained from $(Q_{C[E_1,\ldots,E_s]}, I_{C[E_1,\ldots,E_s]})$ by adding the bound quivers of the branches $\mathcal{L}^{(1)},\ldots,\mathcal{L}^{(s)}$ and making the identification of the vertices $O_1,\ldots,O_s$ with the vertices $O_1^*,\ldots,O_s^*$, respectively (compare with [525, Section 4.7]).

(b) Let $[E_1,\ldots,E_s]C$ be the multiple one-point coextension algebra (3.5) and let $O_1',\ldots,O_s'$ be the coextension vertices of the algebras
$$C_1' = [E_1]C_0',\ldots,C_s' = [E_s]C_{s-1}',$$
respectively, where $C_0' = C$. For each $j \in \{1,\ldots,s\}$, we choose a branch $\mathcal{L}^{(j)}$, with the germ O_j^*. Further, we fix a bound quiver $(Q_{[E_1,\ldots,E_s]C}, I_{[E_1,\ldots,E_s]C}')$ such that there is an algebra isomorphism
$$[E_1,\ldots,E_s]C \cong KQ_{[E_1,\ldots,E_s]C}/I_{[E_1,\ldots,E_s]C}'.$$

The **branch $\mathcal{T}$-coextension** of the algebra C by means of the branches

$$\mathcal{L}^{(1)} = (L^{(1)}, I^{(1)}), \mathcal{L}^{(2)} = (L^{(2)}, I^{(2)}), \ldots, \mathcal{L}^{(s)} = (L^{(s)}, I^{(s)})$$

is defined to be the bound quiver algebra

$$[E_1, \mathcal{L}^{(1)}, \ldots, E_s, \mathcal{L}^{(s)}]C = {}^s_{j=1}[E_j, \mathcal{L}^{(j)}]C = KQ'/I' \qquad \textbf{(3.8)}$$

of the bound quiver (Q', I') obtained from $(Q_{[E_1,\ldots,E_s]C}, I'_{[E_1,\ldots,E_s]C})$ by adding the bound quivers of the branches $\mathcal{L}^{(1)}, \ldots, \mathcal{L}^{(s)}$ and making the identification of the vertices $O'_1, \ldots, O'_s$ with the vertices $O^*_1, \ldots, O^*_s$, respectively.

It is easy to see that the ordinary quiver $Q_{C[E_1, \mathcal{L}^{(1)}, \ldots, E_s, \mathcal{L}^{(s)}]}$ of the branch $\mathcal{T}$-extension algebra $C[E_1, \mathcal{L}^{(1)}, \ldots, E_s, \mathcal{L}^{(s)}]$ is of the form

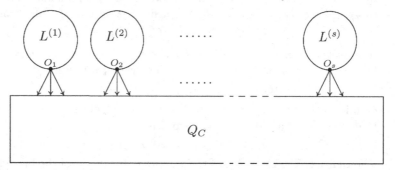

It is easy to see that the ordinary quiver $Q_{[E_1, \mathcal{L}^{(1)}, \ldots, E_s, \mathcal{L}^{(s)}]C}$ of the branch $\mathcal{T}$-coextension algebra $[E_1, \mathcal{L}^{(1)}, \ldots, E_s, \mathcal{L}^{(s)}]C$ has the form dual to the above one.

The following theorem shows that the concepts of the tubular extension and the tubular coextension discussed in the previous section coincide with the concepts of the tubular extension and the tubular coextension defined by Ringel in [525, Section 4.7].

3.9. Theorem. *Let A and C be algebras, and let $\mathcal{T} = \{\mathcal{T}_\lambda\}_{\lambda \in \Lambda}$ be a family of pairwise orthogonal standard stable tubes of the Auslander–Reiten quiver $\Gamma(\operatorname{mod} C)$.*

(a) *The algebra A is a $\mathcal{T}$-tubular extension of C if and only if A is a branch $\mathcal{T}$-extension of C.*

(b) *The algebra A is a $\mathcal{T}$-tubular coextension of C if and only if A is a branch $\mathcal{T}$-coextension of C.*

Proof. We only prove the statement (a), because the proof of (b) is dual.

Let $\boldsymbol{\mathcal{T}} = \{\mathcal{T}_\lambda\}_{\lambda \in \Lambda}$ be a family of pairwise orthogonal standard stable tubes of $\Gamma(\mathrm{mod}\, C)$.

Assume that A is a $\boldsymbol{\mathcal{T}}$-tubular extension of C. Then there exists a sequence of algebras

$$A_0 = C, A_1, \ldots, A_m = A$$

such that, for each $j \in \{1, \ldots, m\}$, the algebra A_j is a line extension of A_{j-1} with respect to an admissible ray module X_j lying in a standard stable tube of $\boldsymbol{\mathcal{T}}$ or in a ray tube of $\Gamma(\mathrm{mod}\, A_{j-1})$ obtained from a stable tube of $\boldsymbol{\mathcal{T}}$ by rectangle insertions created by the line extensions done so far. Let $E_1, \ldots, E_s$ be the modules in the family $\{X_1, \ldots, X_m\}$ of admissible ray modules that belong to the stable tubes of $\boldsymbol{\mathcal{T}}$. It follows from (2.2) that, for each $i \in \{1, \ldots, s\}$, the module E_i lies on the mouth of a standard stable tube $\mathcal{T}_{\lambda_i}$ of $\boldsymbol{\mathcal{T}}$. Hence, by applying (2.15), we conclude that the algebra A is a branch $\boldsymbol{\mathcal{T}}$-extension $C[E_1, \mathcal{L}^{(1)}, \ldots, E_s, \mathcal{L}^{(s)}]$ of C.

Conversely, assume that

$$A = C[E_1, \mathcal{L}^{(1)}, \ldots, E_s, \mathcal{L}^{(s)}]$$

is a branch $\boldsymbol{\mathcal{T}}$-extension of C. We show that A is a $\boldsymbol{\mathcal{T}}$-tubular extension of C by induction on the cardinality

$$\ell(A) = |\mathcal{L}^{(1)}| + \ldots + |\mathcal{L}^{(s)}|,$$

where $|\mathcal{L}^{(j)}|$ is the cardinality of the set of vertices of the branch $\mathcal{L}^{(j)}$, for $j \in \{1, \ldots, s\}$.

If $\ell(A) = 0$, then $A = C$ and there is nothing to show.

Assume that $\ell(A) \geq 1$, and fix $j \in \{1, \ldots, s\}$ such that $|\mathcal{L}^{(j)}| \geq 1$. Because $\mathcal{L}^{(j)}$ is a full connected bound subquiver of the infinite bound tree presented in (2.14) containing the lowest vertex 0, we may choose a source ω_j of $Q_{\mathcal{L}^{(j)}}$ such that $Q_{\mathcal{L}^{(j)}}$ admits a subquiver

$$\Delta: \quad \overset{1}{\circ} \longleftarrow \overset{2}{\circ} \longleftarrow \overset{3}{\circ} \longleftarrow \cdots \longleftarrow \overset{t-1}{\circ} \longleftarrow \overset{t}{\circ} \longleftarrow \overset{t+1 = a}{\circ},$$

with $t \geq 0$, such that,

- each of the vertices $1, 2, \ldots, t$ of Δ is the sink of precisely one arrow in $Q_{\mathcal{L}^{(j)}}$,
- each of the vertices $2, \ldots, t$ of Δ is the source of precisely one arrow in $Q_{\mathcal{L}^{(j)}}$, and
- the vertex 1 is a sink of $Q_{\mathcal{L}^{(j)}}$ (that is, there is no arrow in $Q_{\mathcal{L}^{(j)}}$ with 1 as a source).

Let $\mathcal{L}_\vee^{(j)}$ be the branch obtained from $\mathcal{L}^{(j)}$ by deleting the vertices $1, 2, \ldots, t$, and $t + 1 = a$ of Δ. Given $i \in \{1, 2, \ldots, s\} \setminus \{j\}$, we set

$$\mathcal{L}_\vee^{(i)} = \mathcal{L}^{(i)}.$$

Consider the $\boldsymbol{\mathcal{T}}$-tubular extension $A^\vee$ of C defined by the formula

$$A^\vee = \begin{cases} C[E_1, \mathcal{L}_\vee^{(1)}, \ldots, E_j, \mathcal{L}_\vee^{(j)}, \ldots, E_s, \mathcal{L}_\vee^{(s)}], & \text{if } \mathcal{L}^{(j)} \neq \mathcal{L}_\vee^{(j)}, \\ C[E_1, \mathcal{L}_\vee^{(1)}, \ldots, E_{j-1}, \mathcal{L}_\vee^{(j-1)}, E_{j+1}, \mathcal{L}_\vee^{(j+1)}, \ldots, E_s, \mathcal{L}_\vee^{(s)}], \\ & \text{if } \mathcal{L}^{(j)} = \mathcal{L}_\vee^{(j)}. \end{cases}$$

It is clear that

$$\ell(A) = |\mathcal{L}^{(1)}| + \ldots + |\mathcal{L}^{(s)}| > |\mathcal{L}_\vee^{(1)}| + \ldots + |\mathcal{L}_\vee^{(s)}| = \ell(A^\vee).$$

Then, by the inductive hypothesis, the algebra $A^\vee$ is a $\boldsymbol{\mathcal{T}}$-tubular extension of C. Moreover, the quiver $\Gamma(\operatorname{mod} A^\vee)$ admits a family

$$\boldsymbol{\mathcal{C}}^\vee = \{\mathcal{C}_\lambda^\vee\}_{\lambda \in \Lambda}$$

of pairwise orthogonal standard ray tubes obtained from the family $\boldsymbol{\mathcal{T}} = \{\mathcal{T}_\lambda\}_{\lambda \in \Lambda}$ of pairwise orthogonal standard stable tubes of $\Gamma(\operatorname{mod} C)$ by a sequence of iterated rectangle insertions corresponding to the sequence of line extensions leading from C to $A^\vee$, see (2.7), (2.8), and (2.11).

Note that, if $P(a) = e_a A$ is the indecomposable projective A-module of the vertex a then the radical $\operatorname{rad} P(a)$ of $P(a)$ admits the decomposition

$$\operatorname{rad} P(a) = X \oplus P(t),$$

where X is an A-module, and

- $P(t) = 0$, if $t = 0$, and
- $P(t) = e_t H_t$ is the unique indecomposable projective-injective module over the path algebra

$$H_t = \begin{bmatrix} K & 0 & \cdots & 0 \\ K & K & \cdots & 0 \\ \vdots & \vdots & \ddots & \vdots \\ K & K & \cdots & K \end{bmatrix}$$

of the quiver $\overset{1}{\circ} \longleftarrow \overset{2}{\circ} \longleftarrow \overset{3}{\circ} \longleftarrow \cdots \longleftarrow \overset{t-1}{\circ} \longleftarrow \overset{t}{\circ}$, if $t \geq 1$.

Now we show that X is an admissible ray module of a ray tube of $\mathcal{C}^\vee$. We consider two cases.

Case 1°. Assume that X is a C-module. Then X lies on the mouth of a stable tube $\mathcal{T}_\lambda$ of the family $\boldsymbol{\mathcal{T}}$, and hence, X is an admissible ray module of the ray tube of $\mathcal{C}_\lambda^\vee$ obtained from the stable tube $\mathcal{T}_\lambda$ by iterated rectangle insertions, with the rays parallel to the ray starting from X.

<u>Case 2°</u>. Assume that X is not a C-module. By (2.11) and (2.15), there is a unique arrow $a \longrightarrow b$ in Q_A with source a such that

- $b \neq a$ and b is a vertex of the branch $\mathcal{L}_\vee^{(j)}$,
- X is an admissible ray module of the ray tube of $\mathcal{C}_\lambda^\vee$ containing the indecomposable projective $A^\vee$-module $P^\vee(b) = e_b A^\vee$ of the vertex b, and
- the module X is an epimorphic image of $P^\vee(b)$.

It follows that the algebra A is a line extension of $A^\vee$ and has the form

$$A = A^\vee[X, t] = (A^\vee \times H_t)[X \oplus P(t)]$$

and consequently, A is a $\mathcal{T}$-tubular extension of C. This finishes the proof.$\square$

XV.4. Tubular extensions and tubular coextensions of concealed algebras of Euclidean type

In this section we study the structure of the module category $\bmod B$ over any algebra B that is a tubular extension or tubular coextension of a concealed algebra A of Euclidean type. Moreover, we study homological properties of connected components of the Auslander Reiten quiver $\Gamma(\bmod B)$ of such an algebra B. In particular, we show that $\mathrm{gl.dim}\, B \leq 2$, and, for each indecomposable B-module X in $\bmod B$, one of the following inequalities holds $\mathrm{pd}\, X \leq 1$ or $\mathrm{id}\, X \leq 1$.

We recall from the structure theorem (XII.3.4) that, given a concealed algebra A of Euclidean type, the Auslander–Reiten quiver $\Gamma(\bmod A)$ of A has a disjoint union decomposition

$$\Gamma(\bmod A) = \mathcal{P}(A) \cup \mathcal{T}^A \cup \mathcal{Q}(A)$$

where $\mathcal{P}(A)$ is the unique postprojective component of $\Gamma(\bmod A)$ containing all the indecomposable projective A-modules, $\mathcal{Q}(A)$ is the unique preinjective component containing all the indecomposable injective A-modules, and $\mathcal{T}^A = \{\mathcal{T}_\lambda^A\}_{\lambda \in \mathbb{P}_1(K)}$ is a $\mathbb{P}_1(K)$-family of pairwise orthogonal hereditary standard stable tubes $\mathcal{T}_\lambda^A$. The family $\mathcal{T}^A$ separates the component $\mathcal{P}(A)$ from $\mathcal{Q}(A)$ in the sense of (XII.3.3), and at most 3 of the tubes $\mathcal{T}_\lambda^A$ are of rank greater than or equal to 2.

Throughout we use the following definition.

4.1. Definition. Assume that A is a concealed algebra of Euclidean type and

$$\boldsymbol{\mathcal{T}}^A = \{\mathcal{T}_\lambda^A\}_{\lambda \in \mathbb{P}_1(K)}$$

is the complete $\mathbb{P}_1(K)$-family of all standard stable tubes of $\Gamma(\operatorname{mod} A)$.

(i) An algebra B is defined to be a **tubular extension** of A, if B is a $\boldsymbol{\mathcal{T}}^A$-tubular extension of A in the sense of (2.9)(a).

(ii) An algebra B is defined to be a **tubular coextension** of A, if B is a $\boldsymbol{\mathcal{T}}^A$-tubular coextension of A in the sense of (2.9)(b).

It follows from (3.9) that an algebra B is a tubular extension of a concealed algebra A of Euclidean type if and only if B is a branch $\boldsymbol{\mathcal{T}}^A$-extension of A in the sense of (3.5), where $\boldsymbol{\mathcal{T}}^A$ is the complete $\mathbb{P}_1(K)$-family of all standard stable tubes of $\Gamma(\operatorname{mod} A)$. Analogously, an algebra B is a tubular coextension of A if and only if B is a branch $\boldsymbol{\mathcal{T}}^A$-coextension of A in the sense of (3.5).

It follows that a tubular extension algebra B of an algebra A has the lower triangular matrix form

$$B \cong \begin{bmatrix} A & 0 \\ {}_C N_A & C \end{bmatrix},$$

where C is an algebra. Hence there is an isomorphism $A \cong e_A B e_A$ of algebras, where e_A is the idempotent of B corresponding to the idempotent $\begin{bmatrix} 1 & 0 \\ 0 & 0 \end{bmatrix}$ of the algebra $\begin{bmatrix} A & 0 \\ {}_C N_A & C \end{bmatrix}$. By (I.6.6) and (I.6.8), the idempotent e_A defines the restriction functor

$$\operatorname{res}_A : \operatorname{mod} B \longrightarrow \operatorname{mod} A \qquad (4.2)$$

that associates to each B-module X the A-module $\operatorname{res}_A X = X e_A$, called the restriction of X to A. Note that the ordinary quiver Q_A of A is a full convex subquiver of Q_B. When we view the module X as a K-linear representation of Q_B, then the A-module $\operatorname{res}_A X = X e_A$, viewed as a K-linear representation of Q_A, is just the restriction of the representation X of Q_B to the subquiver Q_A of Q_B.

The following theorem describes the structure of the module category $\operatorname{mod} B$ and homological properties of connected components of the Auslander–Reiten quiver $\Gamma(\operatorname{mod} B)$ of any algebra B that is a tubular extension of a concealed algebra A of Euclidean type. We recall from (2.10) that the rank $r_\lambda^B = \operatorname{rk} \mathcal{T}_\lambda^B$ of a ray tube $\mathcal{T}_\lambda^B$ is the number of rays the tube $\mathcal{T}_\lambda^B$ contains.

4.3. Theorem. *Assume that A is a concealed algebra of Euclidean type and let $\boldsymbol{T}^A = \{\mathcal{T}_\lambda^A\}_{\lambda \in \mathbb{P}_1(K)}$ be the complete $\mathbb{P}_1(K)$-family of all standard stable tubes of $\Gamma(\operatorname{mod} A)$. Let B be a tubular extension the algebra A. The Auslander–Reiten quiver $\Gamma(\operatorname{mod} B)$ of B has a disjoint union decomposition*

$$\Gamma(\operatorname{mod} B) = \boldsymbol{P}^B \cup \boldsymbol{T}^B \cup \boldsymbol{Q}^B$$

and the following conditions are satisfied.

(a) $\boldsymbol{T}^B = \{\mathcal{T}_\lambda^B\}_{\lambda \in \mathbb{P}_1(K)}$ *is a $\mathbb{P}_1(K)$-family of pairwise orthogonal standard ray tubes of $\Gamma(\operatorname{mod} B)$. It is obtained from the complete $\mathbb{P}_1(K)$-family $\boldsymbol{T}^A = \{\mathcal{T}_\lambda^A\}_{\lambda \in \mathbb{P}_1(K)}$ of pairwise orthogonal hereditary standard stable tubes $\mathcal{T}_\lambda^A$ of $\Gamma(\operatorname{mod} A)$ by iterated rectangle insertions. The family $\boldsymbol{T}^B$ consists of all indecomposable B-modules N such that the restriction $\operatorname{res}_A N$ (4.2) of N to A is a module in $\operatorname{add} \boldsymbol{T}^A$. Moreover, if B has the branch $\boldsymbol{T}^A$-extension form*

$$B = A[E_1, \mathcal{L}^{(1)}, \dots, E_s, \mathcal{L}^{(s)}],$$

where $E_1, \dots, E_s$ are A-modules lying on the mouth of some tubes of $\boldsymbol{T}^A$ and $\mathcal{L}^{(1)}, \dots, \mathcal{L}^{(s)}$ are branches, then

$$r_\lambda^B = r_\lambda^A + \sum_{j \in \mathcal{S}(\lambda)} |\mathcal{L}^{(j)}|,$$

for each $\lambda \in \mathbb{P}_1(K)$, where r_λ^B is the rank of the ray tube $\mathcal{T}_\lambda^B$, r_λ^A is the rank of the stable tube $\mathcal{T}_\lambda^A$, $|\mathcal{L}^{(j)}|$ is the number of vertices of the branch $\mathcal{L}^{(j)}$, and the sum is taken over the subset $\mathcal{S}(\lambda)$ of $\{1, 2, \dots, s\}$ consisting of all j such that the module E_j lies on $\mathcal{T}_\lambda^A$.

(b) $\boldsymbol{P}^B = \mathcal{P}(B) = \mathcal{P}(A)$ *is the unique postprojective component of $\Gamma(\operatorname{mod} B)$. An indecomposable B-module M belongs to $\mathcal{P}(B)$ if the restriction $\operatorname{res}_A M$ (4.2) of M to A admits an indecomposable direct summand from $\mathcal{P}(A)$.*

(c) $\boldsymbol{Q}^B$ *is a family of components of $\Gamma(\operatorname{mod} B)$ consisting of all B-modules X such that the restriction $\operatorname{res}_A X$ (4.2) of X to A is a module in $\operatorname{add} \mathcal{Q}(A)$. The family $\boldsymbol{Q}^B$ contains all modules from the component $\mathcal{Q}(A)$, and each of the indecomposable injective B-modules belongs to $\boldsymbol{Q}^B$.*

(d) *Every indecomposable projective B-module P lies in $\mathcal{P}(B) \cup \boldsymbol{T}^B$, and the radical $\operatorname{rad} P$ of P lies in $\operatorname{add}(\mathcal{P}(B) \cup \boldsymbol{T}^B)$.*

(e) *The family $\boldsymbol{T}^B$ separates the component $\mathcal{P}(B)$ from the family $\boldsymbol{Q}^B$ in the sense of (XII.3.3).*

(f) $\operatorname{pd} X \leq 1$, *for each module X in $\mathcal{P}(B) \cup \boldsymbol{T}^B$, and $\operatorname{id} Y \leq 1$, for each module Y in $\boldsymbol{Q}^B$.*

(g) $\operatorname{gl.dim} B \leq 2$.

Proof. Assume that A is a concealed algebra of Euclidean type. We recall from the structure theorem (XII.3.4) that the Auslander–Reiten quiver $\Gamma(\mathrm{mod}\,A)$ of A has a disjoint union decomposition

$$\Gamma(\mathrm{mod}\,A) = \mathcal{P}(A) \cup \boldsymbol{T}^A \cup \mathcal{Q}(A)$$

where $\mathcal{P}(A)$ is the unique postprojective component containing all the indecomposable projective A-modules, $\mathcal{Q}(A)$ is the unique preinjective component containing all the indecomposable injective A-modules, and

$$\boldsymbol{T}^A = \{\mathcal{T}_\lambda^A\}_{\lambda \in \mathbb{P}_1(K)}$$

is a $\mathbb{P}_1(K)$-family of pairwise orthogonal hereditary standard stable tubes $\mathcal{T}_\lambda^A$. The family $\boldsymbol{T}^A$ separates the component $\mathcal{P}(A)$ from $\mathcal{Q}(A)$.

Assume that B is a tubular extension of the algebra A. Then, by (2.9), there exists a sequence of algebras

$$B_0 = A, B_1, \ldots, B_m = B$$

such that, for each $i \in \{1, \ldots, m\}$, the algebra B_i is a line extension

$$B_i = B_{i-1}(X_i, t_i) = (B_{i-1} \times H_{t_i})[X_i \oplus P(t_i)]$$

of B_{i-1} with respect to an admissible ray module X_i lying in a standard stable tube $\mathcal{T}_\lambda^A$ or in a ray tube $\mathcal{T}_\lambda^{B_{i-1}}$ of $\Gamma(\mathrm{mod}\,B_{i-1})$, obtained from a standard stable tube $\mathcal{T}_\lambda^A$ of the family $\boldsymbol{T}^A$ by the rectangle insertions created by the line extensions done so far, see (2.5). It follows (2.12) that $\Gamma(\mathrm{mod}\,B)$ admits a $\mathbb{P}_1(K)$-family

$$\boldsymbol{T}^B = \{\mathcal{T}_\lambda^B\}_{\lambda \in \mathbb{P}_1(K)}$$

of pairwise orthogonal standard ray tubes $\mathcal{T}_\lambda^B$ of $\Gamma(\mathrm{mod}\,B)$ obtained from the complete $\mathbb{P}_1(K)$-family

$$\boldsymbol{T}^A = \{\mathcal{T}_\lambda^A\}_{\lambda \in \mathbb{P}_1(K)}$$

of tubes $\mathcal{T}_\lambda^A$ of $\Gamma(\mathrm{mod}\,A)$ by iterated rectangle insertions. Because the family $\boldsymbol{T}^A$ separates the component $\mathcal{P}(A)$ from $\mathcal{Q}(A)$ then $\mathrm{Hom}_A(\boldsymbol{T}^A, \mathcal{P}(A)) = 0$ and, by applying (1.7), we show inductively that, for each $i \in \{1, \ldots, m+1\}$,

- $\mathcal{P}(B_{i-1}) = \mathcal{P}(A)$ is a postprojective component of $\Gamma(\mathrm{mod}\,B_{i-1})$,
- $\mathrm{Hom}_{B_{i-1}}(X_i \oplus P(t_i), \mathcal{P}(A)) = \mathrm{Hom}_{B_{i-1}}(X_i \oplus P(t_i), \mathcal{P}(B_{i-1})) = 0$, and hence
- every indecomposable projective B_{i-1}-module P lies in $\mathcal{P}(B_{i-1})$ or in $\boldsymbol{T}^{B_{i-1}}$, and the radical $\mathrm{rad}\,P$ of P lies in $\mathrm{add}\,(\mathcal{P}(B_{i-1}) \cup \boldsymbol{T}^{B_{i-1}})$.

By applying this to $i = m + 1$, we conclude that

$$\mathcal{P}(B) = \mathcal{P}(B_m) = \mathcal{P}(A)$$

is a postprojective component of $\Gamma(\bmod B)$. Moreover, $\mathcal{P}(B)$ is a unique postprojective component of $\Gamma(\bmod B)$, because every indecomposable projective B-module P lies in $\mathcal{P}(B)$ or in $\boldsymbol{T}^B$.

To finish the proof, we show inductively, by applying (1.7), (1.8), and (2.7), that, for each $i \in \{1, \dots, m+1\}$,

- if M is an indecomposable B_{i-1}-module such that the restriction $\mathrm{res}_A M$ (4.2) of M to A has an indecomposable direct summand in $\mathcal{P}(B_{i-1}) = \mathcal{P}(A)$ then M belongs to $\mathcal{P}(B_{i-1}) = \mathcal{P}(A)$,
- if N is an indecomposable B_{i-1}-module such that the restriction $\mathrm{res}_A N$ (4.2) of N to A has an indecomposable direct summand in $\boldsymbol{T}^{B_{i-1}}$ then N belongs to $\boldsymbol{T}^{B_{i-1}}$.

Hence, we easily conclude that the Auslander–Reiten quiver $\Gamma(\bmod B)$ of B has a disjoint union decomposition

$$\Gamma(\bmod B) = \boldsymbol{P}^B \cup \boldsymbol{T}^B \cup \boldsymbol{Q}^B,$$

with $\boldsymbol{P}^B = \mathcal{P}(B)$, such that the conditions (a), (b), (c), and (d) are satisfied. Note also that the equality

$$r_\lambda^B = r_\lambda^A + \sum_{j \in \mathcal{S}(\lambda)} |\mathcal{L}^{(j)}|, \text{ with } \lambda \in \mathbb{P}_1(K),$$

stated in (a) follows from (2.11), (2.15), and (3.9).

Now we prove the statement (e). First we note that

$$\mathrm{Hom}_B(\boldsymbol{T}^B, \mathcal{P}(B)) = 0 \quad \text{and} \quad \mathrm{Hom}_B(\boldsymbol{Q}^B, \mathcal{P}(B)) = 0,$$

because the postprojective component $\mathcal{P}(B) = \mathcal{P}(A)$ is closed under predecessors in $\bmod B$, see (VIII.2.5).

To show that $\mathrm{Hom}_B(\boldsymbol{Q}^B, \boldsymbol{T}^B) = 0$, assume that M is an indecomposable B-module in $\boldsymbol{Q}^B$ and N is an indecomposable B-module in $\boldsymbol{T}^B$ such that $\mathrm{Hom}_B(M, N) \neq 0$. Let

$$M' = \mathrm{res}_A M \quad \text{and} \quad N' = \mathrm{res}_A N$$

in $\bmod A$ be the restriction of M and N to A, respectively. By our assumption, B is a tubular extension of A then B is a branch $\boldsymbol{T}^A$-extension of A and, hence, the ordinary quiver Q_A of A is a full convex subquiver of Q_B. Because the $\mathbb{P}_1(K)$-family $\boldsymbol{T}^B$ of pairwise orthogonal standard ray tubes is obtained from the family $\boldsymbol{\mathcal{T}}^A$ of tubes $\mathcal{T}^A_\lambda$ of $\Gamma(\mathrm{mod}\,A)$ by iterated rectangle insertions then, by (2.5), the class of all indecomposable B-modules Z in $\boldsymbol{\mathcal{T}}^B$ such that the restriction $\mathrm{res}_A Z$ to A is zero coincides with the class of all indecomposable directing modules of $\boldsymbol{\mathcal{T}}^B$ and is closed under predecessors in $\mathrm{mod}\,B$. It follows that $\mathrm{Hom}_A(M',N') \neq 0$, because $\mathrm{Hom}_B(M,N) \neq 0$, M belongs to $\boldsymbol{\mathcal{Q}}^B$, and N belongs to $\boldsymbol{\mathcal{T}}^B$.

On the other hand, the module $M' = \mathrm{res}_A M$ belongs to $\mathrm{add}\,\mathcal{Q}(A)$ and $N' = \mathrm{res}_A N$ belongs to $\mathrm{add}\,\boldsymbol{\mathcal{T}}^A$ and, hence, $\mathrm{Hom}_A(M',N') = 0$, because $\mathrm{Hom}_A(\mathcal{Q}(A),\boldsymbol{\mathcal{T}}^A)=0$. The contradiction implies that $\mathrm{Hom}_B(\mathcal{P}(B),\boldsymbol{\mathcal{T}}^B)=0$.

Let U be an indecomposable B-module in $\mathcal{P}(B) = \mathcal{P}(A)$, N an indecomposable B-module in $\boldsymbol{\mathcal{Q}}^B$, and let $f : U \longrightarrow V$ be a non-zero homomorphism of B-modules. Let

$$V' = \mathrm{res}_A V$$

be the restriction of V to A and let $u : V' \longrightarrow V$ be the canonical embedding of B-modules. Because U is an A-module then $\mathrm{Im}\,f \subseteq V'$ and, hence, there is a factorisation $f = ug$, where $g : U \longrightarrow V'$ is induced by f. Further, the family $\boldsymbol{\mathcal{T}}^A$ separates $\mathcal{P}(A)$ from $\mathcal{Q}(A)$, U belongs to $\mathcal{P}(B) = \mathcal{P}(A)$, and V' belongs to $\mathrm{add}\,\mathcal{Q}(A)$ then, for a fixed $\lambda \in \mathbb{P}_1(K)$, there exist a module R_λ in $\mathrm{add}\,\mathcal{T}^A_\lambda$ and two homomorphisms

$$u_\lambda : U \longrightarrow R_\lambda \quad \text{and} \quad v'_\lambda : R_\lambda \longrightarrow V'$$

such that $g = v'_\lambda u_\lambda$. Moreover, because every module of the tube $\mathcal{T}^A_\lambda$ lies on the tube $\mathcal{T}^B_\lambda$ then the module R_λ belongs to $\mathrm{add}\,\mathcal{T}^B_\lambda$. If we set $v_\lambda = u v'_\lambda : R_\lambda \longrightarrow V$, we get $f = ug = uv'_\lambda u_\lambda = v_\lambda u_\lambda$. This shows that the family $\boldsymbol{\mathcal{T}}^B$ separates $\boldsymbol{\mathcal{P}}^B = \mathcal{P}(B)$ from $\boldsymbol{\mathcal{Q}}^B$ and finishes the proof of (e).

To prove the statement (f), we recall from (c) that every indecomposable injective B-module belongs to $\boldsymbol{\mathcal{Q}}^B$ and, hence, the module $D(B)$ belongs to $\mathrm{add}\,\boldsymbol{\mathcal{Q}}^B$. Then the equality $\mathrm{Hom}_B(\boldsymbol{\mathcal{Q}}^B, \mathcal{P}(B) \cup \boldsymbol{\mathcal{T}}^B) = 0$ yields the equality $\mathrm{Hom}_B(D(B), \tau_B X) = 0$, for any indecomposable B-module X in $\mathcal{P}(B) \cup \boldsymbol{\mathcal{T}}^B$. In view of (IV.2.7)(a), it follows that $\mathrm{pd}\,X \leq 1$, for any B-module X in $\mathrm{add}\,(\mathcal{P}(B) \cup \boldsymbol{\mathcal{T}}^B)$.

Finally, we recall from (d) that every indecomposable projective B-module belongs to $\mathcal{P}(B) \cup \boldsymbol{\mathcal{T}}^B$ and, hence, the module B_B belongs to $\mathrm{add}\,(\mathcal{P}(B) \cup \boldsymbol{\mathcal{T}}^B)$. Then the equality $\mathrm{Hom}_B(\boldsymbol{\mathcal{Q}}^B, \mathcal{P}(B) \cup \boldsymbol{\mathcal{T}}^B) = 0$ yields the equality $\mathrm{Hom}_B(\tau_B^{-1} Y, B) = 0$, for any indecomposable B-module Y in $\boldsymbol{\mathcal{Q}}^B$. In view of (IV.2.7)(b), it follows that $\mathrm{id}\,Y \leq 1$, for any B-module Y in $\mathrm{add}\,\boldsymbol{\mathcal{Q}}^B$.

To prove the remaining statement (g), we recall from (d) that the radical $\mathrm{rad}\,P$ of any indecomposable projective B-module P belongs to the category add $(\mathcal{P}(B) \cup \boldsymbol{T}^B)$ and, according to (f), we have pd (rad P) ≤ 1. This shows that gl.dim $B \leq 2$, and finishes the proof of the theorem. $\square$

The following theorem is a coextension analogue of Theorem (4.3).

4.4. Theorem. *Assume that A is a concealed algebra of Euclidean type and let*

$$\boldsymbol{T}^A = \{\mathcal{T}_\lambda^A\}_{\lambda \in \mathbb{P}_1(K)}$$

be the complete $\mathbb{P}_1(K)$-family of all standard stable tubes of $\Gamma(\mathrm{mod}\,A)$. Let B be a tubular coextension of the algebra A. The Auslander–Reiten quiver $\Gamma(\mathrm{mod}\,B)$ of B has a disjoint union decomposition

$$\Gamma(\mathrm{mod}\,B) = \boldsymbol{P}^B \cup \boldsymbol{T}^B \cup \boldsymbol{Q}^B$$

and the following conditions are satisfied.

(a) *$\boldsymbol{T}^B = \{\mathcal{T}_\lambda^B\}_{\lambda \in \mathbb{P}_1(K)}$ is a $\mathbb{P}_1(K)$-family of pairwise orthogonal standard coray tubes of $\Gamma(\mathrm{mod}\,B)$. It is obtained from the complete $\mathbb{P}_1(K)$-family $\boldsymbol{T}^A = \{\mathcal{T}_\lambda^A\}_{\lambda \in \mathbb{P}_1(K)}$ of pairwise orthogonal hereditary standard stable tubes $\mathcal{T}_\lambda^A$ of $\Gamma(\mathrm{mod}\,A)$ by iterated rectangle coinsertions. The family $\boldsymbol{T}^B$ consists of all indecomposable B-modules N such that the restriction $\mathrm{res}_A N$ (4.2) of N to A is a module in add $\boldsymbol{T}^A$. Moreover, if B has the branch $\boldsymbol{T}^A$-coextension form*

$$B = [E_1, \mathcal{L}^{(1)}, \ldots, E_s, \mathcal{L}^{(s)}]A,$$

where $E_1, \ldots, E_s$ are A-modules lying on the mouth of some tubes of $\boldsymbol{T}^A$ and $\mathcal{L}^{(1)}, \ldots, \mathcal{L}^{(s)}$ are branches, then

$$r_\lambda^B = r_\lambda^A + \sum_{j \in \mathcal{S}(\lambda)} |\mathcal{L}^{(j)}|,$$

for each $\lambda \in \mathbb{P}_1(K)$, where r_λ^B is the rank of the coray tube $\mathcal{T}_\lambda^B$, r_λ^A is the rank of the stable tube $\mathcal{T}_\lambda^A$, $|\mathcal{L}^{(j)}|$ is the number of vertices of the branch $\mathcal{L}^{(j)}$, and the sum is taken over the subset $\mathcal{S}(\lambda)$ of $\{1, 2, \ldots, s\}$ consisting of all j such that the module E_j lies on $\mathcal{T}_\lambda^A$.

(b) *$\boldsymbol{Q}^B = \mathcal{Q}(B) = \mathcal{Q}(A)$ is the unique preinjective component of $\Gamma(\mathrm{mod}\,B)$. An indecomposable B-module M belongs to $\mathcal{Q}(B)$ if the restriction $\mathrm{res}_A M$ of M to A admits an indecomposable direct summand from $\mathcal{Q}(A)$.*

(c) *$\boldsymbol{P}^B$ is a family of components of $\Gamma(\mathrm{mod}\,B)$ consisting of all B-modules X such that the restriction $\mathrm{res}_A X$ of X to A is a module in add $\mathcal{P}(A)$. The family $\boldsymbol{P}^B$ contains all modules from the component $\mathcal{P}(A)$, and each of the indecomposable projective B-modules belongs to $\boldsymbol{P}^B$.*

(d) *Every indecomposable injective B-module I lies in $\mathcal{Q}(B) \cup \boldsymbol{T}^B$, and the top $\operatorname{top} I = I/\operatorname{rad} I$ of I lies in* $\operatorname{add}(\boldsymbol{T}^B \cup \mathcal{Q}(B))$.

(e) *The family $\boldsymbol{T}^B$ separates the family $\boldsymbol{P}^B$ from the component $\mathcal{Q}(B)$ in the sense of* (XII.3.3).

(f) $\operatorname{pd} X \leq 1$, *for each module X in $\boldsymbol{P}^B$, and* $\operatorname{id} Y \leq 1$, *for each module Y in $\boldsymbol{T}^B \cup \mathcal{Q}(B)$.*

(g) $\operatorname{gl.dim} B \leq 2$.

Proof. The arguments used in the proof of Theorem (4.3) dualise almost verbatim. The details are left to the reader. □

4.5. Definition. (a) Assume that B is a tubular extension of a concealed algebra A of Euclidean type and let

$$\boldsymbol{T}^B = \{\mathcal{T}_\lambda^B\}_{\lambda \in \mathbb{P}_1(K)}$$

be the $\mathbb{P}_1(K)$-family of all pairwise orthogonal standard ray tubes of the Auslander–Reiten quiver $\Gamma(\operatorname{mod} B)$ of B.

(i) The **tubular type** of the algebra B is defined to be the $\mathbb{P}_1(K)$-sequence

$$r^B = (r_\lambda^B)_{\lambda \in \mathbb{P}_1(K)},$$

where $r_\lambda^B = \operatorname{rk} \mathcal{T}_\lambda^B$ is the rank of the ray tube $\mathcal{T}_\lambda^B$, that is, the number of rays the tube $\mathcal{T}_\lambda^B$ contains.

(ii) We identify the sequence r^B with its reduced form defined to be the finite sequence

$$\widehat{r}^B = (r_{\lambda_1}^B, \dots, r_{\lambda_s}^B),$$

with $s \geq 2$, containing the ranks r_λ^B, with $\lambda \in \mathbb{P}_1(K)$, such that $r_\lambda^B \geq 2$, arranged in the non-decreasing order.

(b) Assume that B is a tubular coextension of a concealed algebra A of Euclidean type and let

$$\boldsymbol{T}^B = \{\mathcal{T}_\lambda^B\}_{\lambda \in \mathbb{P}_1(K)}$$

be the $\mathbb{P}_1(K)$-family of all pairwise orthogonal standard coray tubes of $\Gamma(\operatorname{mod} B)$.

(i) The **tubular type** of the algebra B is defined to be the $\mathbb{P}_1(K)$-sequence

$$r^B = (r_\lambda^B)_{\lambda \in \mathbb{P}_1(K)},$$

where $r_\lambda^B = \operatorname{rk} \mathcal{T}_\lambda^B$ is the rank of the coray tube $\mathcal{T}_\lambda^B$, that is, the number of corays the tube $\mathcal{T}_\lambda^B$ contains.

(ii) We identify the sequence r^B with its reduced form defined to be the finite sequence
$$\vec{r}^B = (r^B_{\lambda_1}, \ldots, r^B_{\lambda_s}),$$
with $s \geq 2$, containing the ranks r^B_λ, with $\lambda \in \mathbb{P}_1(K)$, such that $r^B_\lambda \geq 2$, arranged in the non-decreasing order.

4.6. Definition. Assume that A is a concealed algebra of Euclidean type.

(i) An algebra B is a **domestic tubular extension** of the algebra A if B is a tubular extension of A such that the tubular type $r^B = \hat{r}^B$ of B is of one of the following five forms

(p, q), with $1 \leq p \leq q$, $(2, 2, m{-}2)$, with $m \geq 4$, $(2, 3, 3)$, $(2, 3, 4)$, $(2, 3, 5)$

that correspond to the Dynkin diagrams $\mathbb{A}_{p+q-1}$, $\mathbb{D}_m$, $\mathbb{E}_6$, $\mathbb{E}_7$, $\mathbb{E}_8$.
   The tubular type $r^B = \hat{r}^B$ of such an algebra B is called the domestic tubular extension type of B.

(ii) An algebra B is a **domestic tubular coextension** of the algebra A if B is a tubular coextension of A such that the tubular type $r^B = \hat{r}^B$ of B is of one of the following five forms

(p, q), with $1 \leq p \leq q$, $(2, 2, m{-}2)$, with $m \geq 4$, $(2, 3, 3)$, $(2, 3, 4)$, $(2, 3, 5)$

The tubular type $r^B = \hat{r}^B$ of such an algebra B is called the domestic tubular coextension type of B.

The following theorem proved in (XII.3.4) shows that any concealed algebra of Euclidean type has (trivially) a domestic tubular extension type.

4.7. Theorem. *Let Q be an acyclic quiver whose underlying graph $\overline{Q}$ is Euclidean, and let B be a concealed algebra of type Q. Let*

$$\boldsymbol{\mathcal{T}}^B = \{\mathcal{T}^B_\lambda\}_{\lambda \in \mathbb{P}_1(K)}$$

be the $\mathbb{P}_1(K)$-family of all pairwise orthogonal standard stable tubes of the Auslander–Reiten quiver $\Gamma(\mathrm{mod}\, B)$ of B. The tubular type

$$\mathbf{m}_B = (m_1, \ldots, m_s)$$

of the $\mathbb{P}_1(K)$-family $\boldsymbol{\mathcal{T}}^B$ depends only on the Euclidean quiver Q and equals $\mathbf{m}_Q$, where

- $\mathbf{m}_Q = (p, q)$, *if $\overline{Q} = \widetilde{\mathbb{A}}_m$, $m \geq 1$, $p = \min\{p', p''\}$ and $q = \max\{p', p''\}$, where p' and p'' are the number of counterclockwise-oriented arrows in Q and clockwise-oriented arrows in Q, respectively,*

- $\mathbf{m}_Q = (2, 2, m - 2)$, *if* $\overline{Q} = \widetilde{\mathbb{D}}_m$ *and* $m \geq 4$,
- $\mathbf{m}_Q = (2, 3, 3)$, *if* $\overline{Q} = \widetilde{\mathbb{E}}_6$,
- $\mathbf{m}_Q = (2, 3, 4)$, *if* $\overline{Q} = \widetilde{\mathbb{E}}_7$, *and*
- $\mathbf{m}_Q = (2, 3, 5)$, *if* $\overline{Q} = \widetilde{\mathbb{E}}_8$.   □

Now we illustrate theorems (4.3) and (4.4) with the following two examples.

4.8. Example. Let B be the algebra given by the following quiver

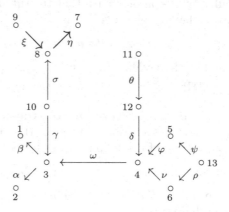

bound by the five relations

$$\gamma\beta = 0, \ \gamma\alpha = 0, \ \xi\eta = 0, \ \delta\omega = 0, \text{ and } \psi\varphi = \rho\nu.$$

Applying (XIII.2.6) and (XIII.2.9), we conclude that B is the tubular extension

$$B \cong C[F_1, \mathcal{L}^{(1)}, F_2, \mathcal{L}^{(2)}, F_3, \mathcal{L}^{(3)}]$$

of the path algebra $C = K\Delta(\widetilde{\mathbb{D}}_5)$ of the quiver

$$\Delta = \Delta(\widetilde{\mathbb{D}}_5) :$$

of the Euclidean type $\widetilde{\mathbb{D}}_5$, using the mouth modules

$$F_1 = F_1^{(1)} :$$

$$
F_2 = F_2^{(1)} : \qquad
\begin{array}{ccc}
0 & & 0 \\
\nwarrow & & \swarrow \\
0 \longleftarrow & & K \\
\swarrow & & \nwarrow \\
0 & & 0,
\end{array}
$$

$$
F_3 = F_3^{(1)} : \qquad
\begin{array}{ccc}
K & & K \\
\nwarrow{\scriptstyle 1} & & \swarrow{\scriptstyle 1} \\
K \xleftarrow{\ 1\ } & & K \\
\swarrow{\scriptstyle 1} & & \nwarrow{\scriptstyle 1} \\
K & & K
\end{array}
$$

from the unique stable tube $\mathcal{T}_1^{\Delta(\widetilde{\mathbb{D}}_5)}$ of rank 3 in $\Gamma(\operatorname{mod} C)$, and the branches $\mathcal{L}^{(1)}$, $\mathcal{L}^{(2)}$, and $\mathcal{L}^{(3)}$ of the capacities $|\mathcal{L}^{(1)}| = 4$, $|\mathcal{L}^{(2)}| = 2$, and $|\mathcal{L}^{(3)}| = 1$ described in (2.5). It follows that B is a tubular extension algebra of the tubular type

$$
r^B = \widehat{r}^B = (2, 2, 10).
$$

Moreover, by (4.3), the Auslander–Reiten quiver $\Gamma(\operatorname{mod} B)$ of B has a disjoint union decomposition

$$
\Gamma(\operatorname{mod} B) = \boldsymbol{P}^B \cup \boldsymbol{T}^B \cup \boldsymbol{Q}^B,
$$

where

- $\boldsymbol{P}^B = \mathcal{P}(B) = \mathcal{P}(C)$ is the postprojective component of $\Gamma(\operatorname{mod} C)$ and $C = K\Delta(\widetilde{\mathbb{D}}_5)$,
- $\boldsymbol{T}^B = \{\mathcal{T}_\lambda^B\}_{\lambda \in \mathbb{P}_1(K)}$ is a $\mathbb{P}_1(K)$-family of pairwise orthogonal standard ray tubes of $\Gamma(\operatorname{mod} B)$ of tubular type $r^B = \widehat{r}^B = (2, 2, 10)$,
- the tube $\mathcal{T}_1^B$ is described in (2.5) and the remaining tubes of $\boldsymbol{T}^B$ form the K-family

$$
\boldsymbol{T}^C \setminus \{\mathcal{T}_1^C\} = \{\mathcal{T}_\lambda^C\}_{\lambda \in \mathbb{P}_1(K) \setminus \{1\}}
$$

  of pairwise orthogonal standard stable tubes of $\Gamma(\operatorname{mod} C)$ described in (XIII.2.9),
- $\boldsymbol{Q}^B$ is a family of components of $\Gamma(\operatorname{mod} B)$ consisting of all indecomposable B-modules X such that the restriction $\operatorname{res}_C X$ of X to C is a module in $\mathcal{Q}(C)$.

The standard calculation technique shows that the family $\boldsymbol{Q}^B$ contains a connected component $\mathcal{Q}^*(B)$ such that the right hand part of $\mathcal{Q}^*(B)$ looks as follows

$$
\begin{array}{c}
0\ 0\\
0\ 0\\
-- \quad 10\ 01\\
0\ 1\ 1\ 1
\end{array} = I(1)
$$

$$\cdots \searrow \quad \nearrow I(2) \quad \searrow$$

$$
\cdots \to \circ \to
\begin{array}{c}0\ 0\\0\ 0\\00\ 01\\1\ 1\ 1\ 1\end{array} \to
\begin{array}{c}0\ 0\\0\ 0\\00\ 01\\0\ 1\ 1\ 1\end{array} --
\begin{array}{c}0\ 1\\1\ 0\\01\ 00\\0\ 0\ 0\end{array} = I(7)
$$

$$\cdots \nearrow \searrow \quad \nearrow \searrow \nearrow \searrow$$

$$
-- \quad \circ \qquad --
\begin{array}{c}0\ 1\\1\ 0\\01\ 01\\0\ 1\ 1\ 1\end{array} --
\begin{array}{c}0\ 0\\1\ 0\\01\ 00\\0\ 0\ 0\ 0\end{array} --
\begin{array}{c}1\ 0\\0\ 0\\00\ 00\\0\ 0\ 0\ 0\end{array} = I(9)
$$

$$\nearrow \searrow \quad \nearrow \searrow \nearrow \searrow \quad \nearrow$$

$$
-- \qquad \circ \quad --
\begin{array}{c}0\ 0\\1\ 0\\01\ 01\\0\ 1\ 1\ 1\end{array} --
\begin{array}{c}1\ 0\\1\ 0\\01\ 00\\0\ 0\ 0\end{array} = I(8)
$$

$$\nearrow \searrow \nearrow \searrow \nearrow$$

$$
-- \quad \circ \quad --
\begin{array}{c}1\ 0\\1\ 0\\01\ 01\\0\ 1\ 1\ 1\end{array} --
\begin{array}{c}0\ 0\\0\ 0\\01\ 00\\0\ 0\ 0\ 0\end{array} = I(10)
$$

$$\nearrow \searrow \nearrow \searrow \nearrow$$

$$
-- \quad \circ \quad --
\begin{array}{c}0\ 0\\0\ 0\\01\ 01\\0\ 1\ 1\ 1\end{array} = I(3)
$$

$$\nearrow \searrow \nearrow \searrow$$

$$
-- \quad \circ \qquad --
\begin{array}{c}0\ 0\\0\ 0\\00\ 01\\0\ 1\ 1\end{array} --
\begin{array}{c}0\ 0\\0\ 0\\00\ 10\\0\ 0\ 0\ 0\end{array} --
\begin{array}{c}0\ 0\\0\ 1\\00\ 00\\0\ 0\ 0\ 0\end{array} = I(11)
$$

$$\nearrow \searrow \quad \nearrow \searrow \nearrow$$

$$
-- \quad \circ \qquad --
\begin{array}{c}0\ 0\\0\ 0\\00\ 11\\0\ 1\ 1\end{array} --
\begin{array}{c}0\ 0\\0\ 1\\00\ 10\\0\ 0\ 0\ 0\end{array} = I(12)
$$

$$\nearrow \searrow \nearrow \searrow$$

$$
-- \quad \circ \quad --
\begin{array}{c}0\ 0\\0\ 1\\00\ 11\\0\ 1\ 1\ 1\end{array} = I(4)
$$

$$\nearrow \searrow \nearrow \searrow$$

$$
\cdots \quad \to \circ \to \circ \to \circ \to
\begin{array}{c}0\ 0\\0\ 0\\00\ 01\\0\ 0\ 1\ 1\end{array} \to I(5) \to I(13)
$$

$$\searrow \nearrow \searrow \nearrow \searrow \nearrow$$

$$
-- \quad \circ \quad -- \quad \circ \quad -- \quad I(6)
$$

where

$$
I(5) =
\begin{array}{c}0\ 0\\0\ 0\\00\ 01\\0\ 0\ 1\end{array},\quad
I(13) =
\begin{array}{c}0\ 0\\0\ 0\\00\ 00\\0\ 0\ 1\end{array},\quad
I(6) =
\begin{array}{c}0\ 0\\0\ 0\\00\ 00\\0\ 1\end{array},
$$

and the indecomposable modules are represented by their dimension vectors.

Observe that the component $\mathcal{Q}^*(B)$ contains all the indecomposable injective B-modules and a section $\Sigma_B \cong \Delta(\widetilde{\mathbb{D}}_{12}) \cong \Delta(\widetilde{\mathbb{D}}_{12})^{\mathrm{op}}$ of the form

$$
\begin{array}{l}
I(1) \\
\quad\searrow \\
I(2) \;\longrightarrow \tau_B I(7) \\
\qquad\qquad\searrow \\
\qquad\quad \tau_B^2 I(9) \\
\qquad\qquad\searrow \\
\qquad\qquad \tau_B I(8) \\
\qquad\qquad\quad\searrow \\
\qquad\qquad\quad \tau_B I(10) \\
\qquad\qquad\qquad\searrow \\
\qquad\qquad\qquad I(3) \\
\qquad\qquad\qquad\quad\searrow \\
\qquad\qquad\qquad\quad \tau_B^2 I(11) \\
\qquad\qquad\qquad\qquad\searrow \\
\qquad\qquad\qquad\qquad \tau_B I(12) \\
\qquad\qquad\qquad\qquad\quad\searrow \\
\qquad\qquad\qquad\qquad\quad I(4) \\
\qquad\qquad\qquad\qquad\qquad\searrow \\
\qquad\qquad\qquad\qquad\qquad \tau_B I(13) \;\longrightarrow\; I(5) \\
\qquad\qquad\qquad\qquad\qquad\qquad\searrow \\
\qquad\qquad\qquad\qquad\qquad\qquad I(6)
\end{array}
$$

Because all indecomposable projective B-modules lie in $\mathcal{P}(B) \cup \boldsymbol{T}^B$ then $\mathcal{Q}^*(B)$ is a preinjective component of $\Gamma(\mathrm{mod}\, B)$. It is shown in (XVII.2.5) that B is a tilted algebra of the Euclidean type

$$
\Delta(\widetilde{\mathbb{D}}_{12}) : \quad
\begin{array}{c}
\overset{1}{\circ} \qquad\qquad\qquad\qquad\qquad\qquad\qquad \overset{12}{\circ} \\
\nwarrow \qquad\qquad\qquad\qquad\qquad\qquad\nearrow \\
\circ\!\leftarrow\!\circ\!\leftarrow\!\circ\!\leftarrow\!\circ\!\leftarrow\!\circ\!\leftarrow\!\circ\leftarrow\!\circ\!\leftarrow\!\circ\leftarrow\!\circ \\
\swarrow\;\; 3 \quad 4 \quad 5 \quad 6 \quad 7 \quad 8 \quad 9 \quad 10 \quad 11 \searrow \\
\underset{2}{\circ} \qquad\qquad\qquad\qquad\qquad\qquad\qquad \underset{13.}{\circ}
\end{array}
$$

Moreover, it follows from (XVII.3.5) that $\boldsymbol{Q}^B = \mathcal{Q}^*(B)$.

We finish the example by observing that the algebra B^{op} opposite to B is the tubular cocxtension

$$
B^{\mathrm{op}} \cong [F_1, \mathcal{L}^{(1)}, F_2, \mathcal{L}^{(2)}, F_3, \mathcal{L}^{(3)}] C
$$

of the algebra $C \cong C^{\mathrm{op}}$ of the same tubular type $\widehat{r}^{B^{\mathrm{op}}} = \widehat{r}^B = (2, 2, 10)$.

4.9. Example. Let A be the path algebra of the Kronecker quiver

$$1 \circ \underset{\beta}{\overset{\alpha}{\rightleftarrows}} \circ 2.$$

It follows from (XI.4.6) that, for each $\lambda \in K$, the Kronecker module

$$E^{(\lambda)} = (\, K \underset{\lambda}{\overset{1}{\longleftarrow}} K \,)$$

is the mouth A-module of the tube $\mathcal{T}_\lambda^A$ of rank one.

Let λ_3, λ_4, λ_5, λ_6, and λ_7 be pairwise different elements of $K \setminus \{0\}$, and let $\mathcal{L}^{(3)}$, $\mathcal{L}^{(4)}$, $\mathcal{L}^{(5)}$, $\mathcal{L}^{(6)}$, and $\mathcal{L}^{(7)}$ be branches of capacity one, that is, the branch $\mathcal{L}^{(j)}$ consists of the vertex j, for $j \in \{3,4,5,6,7\}$. We form the tubular extension

$$B = A[E^{(\lambda_3)}, \mathcal{L}^{(3)}, E^{(\lambda_4)}, \mathcal{L}^{(4)}, E^{(\lambda_5)}, \mathcal{L}^{(5)}, E^{(\lambda_6)}, \mathcal{L}^{(6)}, E^{(\lambda_7)}, \mathcal{L}^{(7)}]$$

of A. It is easy to see that B is given by the quiver

and is bound by the following five relations $\lambda_3\gamma_3\alpha = \gamma_3\beta$, $\lambda_4\gamma_4\alpha = \gamma_4\beta$, $\lambda_5\gamma_5\alpha = \gamma_5\beta$, $\lambda_6\gamma_6\alpha = \gamma_6\beta$, and $\lambda_7\gamma_7\alpha = \gamma_7\beta$. It follows from (4.3) that the Auslander–Reiten quiver $\Gamma(\mathrm{mod}\, B)$ of B has a disjoint union decomposition

$$\Gamma(\mathrm{mod}\, B) = \boldsymbol{P}^B \cup \boldsymbol{T}^B \cup \boldsymbol{Q}^B,$$

where

- $\boldsymbol{P}^B = \mathcal{P}(B) = \mathcal{P}(A)$ is the postprojective component of $\Gamma(\mathrm{mod}\, A)$,
- $\boldsymbol{T}^B = \{\mathcal{T}_\lambda^B\}_{\lambda \in \mathbb{P}_1(K)}$ is a $\mathbb{P}_1(K)$-family of pairwise orthogonal standard ray tubes of $\Gamma(\mathrm{mod}\, B)$ of tubular type $r^B = \hat{r}^B = (2,2,2,2,2)$,
- $\mathcal{T}_\lambda^B = \mathcal{T}_\lambda^A$, for all $\lambda \in \mathbb{P}_1(K) \setminus \{\lambda_3, \lambda_4, \lambda_5, \lambda_6, \lambda_7\}$, and the tube $\mathcal{T}_{\lambda_j}^B$ is obtained from the tube $\mathcal{T}_{\lambda_j}^A$ by inserting the ray starting from the indecomposable projective B-module $P(j) = e_j B$ at the vertex j of the quiver Q, for each $j \in \{3,4,5,6,7\}$,
- $\boldsymbol{Q}^B$ is a family of components of $\Gamma(\mathrm{mod}\, B)$ consisting of all indecomposable B-modules X such that the restriction $\mathrm{res}_A X$ of X to A is a module in $\mathcal{Q}(A)$.

Here $e_j \in B$ is the primitive idempotent given by the vertex j of Q.
Let $H = K\Delta$ be the path algebra of the subquiver

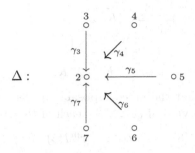

It is easy to see that there is an isomorphism of algebras $H \cong B/Be_1B$.
Then the epimorphism $B \longrightarrow H$ induces a full and faithful embedding
$\mathrm{mod}\, H \hookrightarrow \mathrm{mod}\, B$.

It follows from (VIII.2.3) that the Auslander–Reiten quiver $\Gamma(\mathrm{mod}\, H)$ of
H admits a unique preinjective component $\mathcal{Q}(H) \cong \mathbb{N}\Delta^{\mathrm{op}}$ containing all the
indecomposable injective H-modules. The standard calculation technique
shows that the family $\mathcal{Q}(B)$ contains a component $\mathcal{Q}^*(B)$ such that the
right hand part of $\mathcal{Q}^*(B)$ is of the form

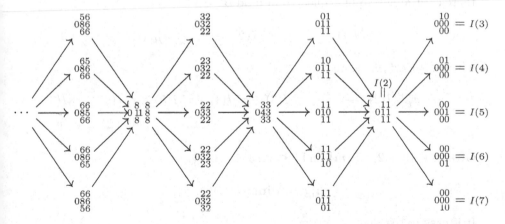

where the indecomposable modules are represented by their dimension
vectors.

Now we show that $\mathcal{Q}^*(B) = \mathcal{Q}(H)$, under the canonical embedding
$\mathrm{mod}\, H \hookrightarrow \mathrm{mod}\, B$. To see this we note that, in view of (III.2.6), the indecomposable injective B-module $I(1) = D(Be_1)$ at the vertex 1 of Q is the

module

$$
\begin{array}{ccc}
& K & K \\
& \left[\begin{smallmatrix}\lambda_3\\1\end{smallmatrix}\right]\downarrow \quad \swarrow\left[\begin{smallmatrix}\lambda_4\\1\\ \lambda_5\\1\end{smallmatrix}\right] & \\
K \; \underset{[0\,1]}{\overset{[1\,0]}{\rightleftarrows}} \; K^2 \longleftarrow K & & \\
& \left[\begin{smallmatrix}\lambda_7\\1\end{smallmatrix}\right]\uparrow \quad \nwarrow\left[\begin{smallmatrix}\lambda_6\\1\end{smallmatrix}\right] & \\
& K & K.
\end{array}
$$

It follows from (4.3) that the indecomposable projective B-modules lie in $\mathcal{P}(B)\cup \boldsymbol{T}^B$. Hence, in view of (IV.2.10), each of the B-modules

$$
\tau_B^m I(2), \quad \tau_B^m I(3), \quad \tau_B^m I(4), \quad \tau_B^m I(5), \quad \tau_B^m I(6), \text{ and } \quad \tau_B^m I(7),
$$

is indecomposable, for any $m \geq 0$.

Assume, to the contrary, that $\mathcal{Q}^*(B) \neq \mathcal{Q}(H)$. Then there exist an index $i_0 \in \{3, 4, 5, 6, 7\}$ and an integer $m_0 \geq 0$ such that $\tau_B^{m_0} I(i_0) \cong I(1)/\operatorname{soc} I(1)$ and, consequently, we get

$$
(\operatorname{\mathbf{dim}} \tau_B^{m_0} I(i_0))_2 = (\operatorname{\mathbf{dim}} I(1)/\operatorname{soc} I(1))_2 = 2.
$$

Hence, in view of our description of the right hand part of $\mathcal{Q}^*(B)$, we conclude that $m \geq 3$. On the other hand, for each $r \in \{0, 1, 2, \ldots, m-1\}$, there exist almost split sequences in mod B

$$
0 \longrightarrow \tau_B^{r+1}I(i) \longrightarrow \tau_B^r I(2) \longrightarrow \tau_B^r I(i) \longrightarrow 0
$$

with $i \in \{3, 4, 5, 6, 7\}$, and

$$
0 \longrightarrow \tau_B^{r+1}I(2) \longrightarrow \tau_B^r I(3) \oplus \tau_B^r I(4) \oplus \tau_B^r I(5) \oplus \tau_B^r I(6) \oplus \tau_B^r I(7)
$$
$$
\longrightarrow \tau_B^r I(2) \longrightarrow 0.
$$

Given $s \in \{0, 1, 2, \ldots, m\}$ and $j \in \{2, 3, 4, 5, 6, 7\}$, we set

$$
a_{j,s} = (\operatorname{\mathbf{dim}} \tau_B^s I(j))_2.
$$

It is easy to see that

- $a_{2,0} = 1$, $a_{2,1} = 4$, $a_{j,0} = 0$, and $a_{j,1} = 1$, for $j \in \{3, 4, 5, 6, 7\}$,
- $a_{j,r+1} = a_{2,r} - a_{j,r}$, for $j \in \{3, 4, 5, 6, 7\}$,
- $a_{2,r+1} = a_{3,r+1} + a_{4,r+1} + a_{5,r+1} + a_{6,r+1} + a_{7,r+1} - a_{2,r}$, for $r \in \{1, 2, \ldots, m-1\}$, and
- $a_{3,s} = a_{4,s} = a_{5,s} = a_{6,s} = a_{7,s}$, for $s \in \{0, 1, 2, \ldots, m\}$.

The final equality follows by induction on s. In particular, we get

$(*_r)$ $a_{2,r+1} - 5a_{3,r+1} - a_{2,r}$ and $a_{3,r+1} = 4a_{3,r} - a_{2,r-1}$,

for each $r \in \{1, 2, \ldots, m-1\}$. Now we show that the inequalities

$(**_r)$ $a_{2,r} > a_{2,r-1}$ and $a_{3,r} > a_{3,r-1}$

hold, for each $r \in \{1, \ldots, m\}$. Obviously, $(**_1)$ holds. Let $t \in \{1, \ldots, m-1\}$ be such that $(**_r)$ holds for all $r \in \{1, \ldots, t\}$. Then, by applying $(*_r)$, we get

$$
\begin{aligned}
a_{2,t+1} - a_{2,t} &= 5a_{3,t+1} - 2a_{2,t} \\
&= 5(4a_{3,t} - a_{2,t-1}) - 2(5a_{3,t} - a_{2,t-1}) \\
&= 10a_{3,t} - 3a_{2,t-1} \\
&> 3(2a_{3,t} - a_{2,t-1}) \\
&= 3(a_{3,t} + (a_{3,t} - a_{2,t-1})) = 3(a_{3,t} - a_{3,t-1}) > 0, \\
a_{3,t+1} - a_{3,t} &= 4(a_{3,t} - a_{2,t-1}) - a_{3,t} \\
&= 3a_{3,t} - a_{2,t-1} \\
&> 2a_{3,t} - a_{2,t-1} \\
&= a_{3,t} - a_{3,t-1} > 0.
\end{aligned}
$$

Because $m \geq 3$ then, for any $i \in \{2, 3, 4, 5, 6, 7\}$, we get the inequalities

$$(\mathbf{dim}\, \tau_B^m I(i))_2 = a_{i,m} > a_{i,m-1} > \ldots > a_{i,2} > 2.$$

This is a contradiction, because we have shown earlier that

$$(\mathbf{dim}\, \tau_B^{m_0} I(i_0))_2 = 2,$$

for some $i_0 \in \{3, 4, 5, 6, 7\}$ and $m_0 \geq 0$. Consequently, the required equality

$$\mathcal{Q}^*(B) = \mathcal{Q}(H)$$

holds. It follows that the indecomposable injective B-module $I(1)$ at the vertex 1 belongs to the family $\mathcal{Q}^B \supset \mathcal{Q}^*(B)$, but $I(1)$ does not belong to the preinjective component $\mathcal{Q}^*(B)$. We show later in (XVII.5.1) that B is not a tilted algebra of Euclidean type. This finishes the example.

XV.5. Exercises

1. Compute the Auslander–Reiten quiver of the algebra given by the quiver

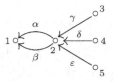

bound by $\gamma\alpha = 0$, $\varepsilon\beta = 0$ and $\delta\alpha = \delta\beta$. Show that this algebra is tilted.

2. Compute the Auslander–Reiten quiver of the algebra given by the quiver

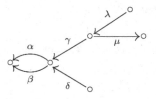

bound by $\gamma\alpha = 0$, $\delta\beta = 0$ and $\lambda\mu = 0$. Show that this algebra is tilted.

3. Let B be the path K-algebra of the following quiver

$$\Delta: \quad \circ \longrightarrow \circ \longrightarrow \circ \xrightarrow{\gamma} \circ \overset{1}{} \begin{array}{c} \overset{3}{\circ} \\ \swarrow \quad \searrow \overset{2}{} \end{array} \overset{\sigma}{} \circ \xleftarrow{} \circ \xleftarrow{\mu} \circ$$

bound by the zero relations $\gamma\alpha = 0$, $\sigma\beta = 0$, and $\mu\xi = 0$.

  (a) Show that the algebra B is a tubular extension of the path algebra $A = K\Delta'$ of the subquiver

$$\Delta': \quad 1 \circ \begin{array}{c} \overset{3}{\circ} \\ \swarrow \quad \searrow \\ \underset{\alpha}{} \quad \underset{\beta}{} \\ \underset{0}{\circ} \end{array} \circ 2$$

(of the Euclidean type $\tilde{\mathbb{A}}_3$) of Δ.

(b) Determine the connected components of the Auslander–Reiten quiver $\Gamma(\mathrm{mod}\,B)$ of B containing the simple modules $S(1)$ and $S(2)$.

4. Let B be the path K-algebra of the following quiver

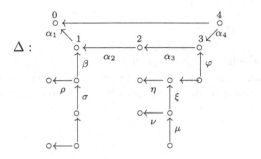

bound by the five zero relations $\beta\alpha_1 = 0$, $\sigma\rho = 0$, $\varphi\alpha_3 = 0$, $\xi\eta = 0$, and $\mu\nu = 0$.

(a) Show that B is a tubular extension of the path algebra $A = K\Delta'$ of the subquiver

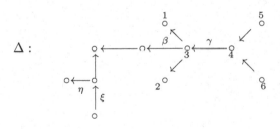

(of the Euclidean type $\widetilde{\mathbb{A}}_4$) of Δ.

(b) Determine the connected components of the Auslander–Reiten quiver $\Gamma(\mathrm{mod}\,B)$ of B containing the simple modules $S(1)$, $S(2)$ and $S(3)$.

5. Let B be the path K-algebra of the following quiver

Δ :

bound by two zero relations $\gamma\beta = 0$ and $\xi\eta = 0$.

(a) Show that B is a tubular coextension of the path algebra $A = K\Delta'$ of the subquiver

$$\Delta' : \qquad \begin{array}{c} \underset{1}{\circ} \qquad\qquad \underset{5}{\circ} \\[4pt] \nwarrow \qquad\qquad \swarrow \\[2pt] \underset{3}{\circ} \xleftarrow{\ \gamma\ } \underset{4}{\circ} \\[2pt] \swarrow \qquad\qquad \searrow \\[4pt] \underset{2}{\circ} \qquad\qquad \underset{6}{\circ} \end{array}$$

(of the Euclidean type $\widetilde{\mathbb{D}}_5$) of Δ.

(b) Determine the connected components of the Auslander–Reiten quiver $\Gamma(\mathrm{mod}\, B)$ of B containing the simple modules $S(3)$ and $S(4)$.

Chapter XVI

Branch algebras

A complete classification of all tilted algebras of the Dynkin type $\mathbb{A}_n$ is given in Chapter IX of Volume 1 by proving in (IX.6.12) that a basic algebra A is a tilted algebra of the Dynkin type $\mathbb{A}_n$ if and only if A is the bound quiver algebra KQ/I, where the bound quiver (Q, I) is a finite connected full bound subquiver of the infinite tree

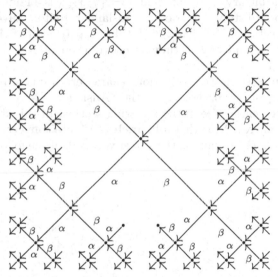

bound by all possible relations of the forms $\alpha\beta = 0$ and $\beta\alpha = 0$ and contains no full bound subquiver of the form

$$\underset{1}{\circ} \overset{\delta}{\longleftarrow} \underset{2}{\circ} \overset{\gamma}{\longleftarrow} \underset{3}{\circ} \quad \circ \quad \cdots \quad \circ \quad \underset{t-2}{\circ} \overset{\beta}{\longleftarrow} \underset{t-1}{\circ} \overset{\alpha}{\longleftarrow} \underset{t}{\circ}$$

with $t \geq 4$, $\alpha\beta = 0$, $\gamma\delta = 0$, all unoriented edges may be oriented arbitrarily; and there are no other zero relations between 2 and $t-1$. We also recall from (IX.6.8) that the Auslander–Reiten quiver $\Gamma(\text{mod } A)$ of a tilted algebra of the Dynkin type $\mathbb{A}_n$ is a finite connected quiver with a section Σ such that the underlying unoriented graph $\overline{\Sigma}$ of Σ is the Dynkin diagram $\mathbb{A}_n$. Each vertex of the quiver $\Gamma(\text{mod } A)$ is the source and the target of at most two arrows.

It turns out that in the representation theory of finite dimensional algebras an important rôle is played by a special class of tilted algebras of the Dynkin type $\mathbb{A}_n$, namely, by the tilted algebras A of the equioriented linear type $\Delta(\mathbb{A}_n) : \overset{1}{\circ}\!\leftarrow\!\overset{2}{\circ}\!\leftarrow\ldots\leftarrow\!\overset{n}{\circ}$.

One of the main aims of this chapter is to prove several characterisations of these algebras A, to provide a description of the module category $\mathrm{mod}\,A$, and to describe the structure of the Auslander–Reiten quiver $\Gamma(\mathrm{mod}\,A)$ of any such an algebra A.

XVI.1. Branches and finite line extensions

Let A be an arbitrary K-algebra and $\boldsymbol{\mathcal{T}} = \{\mathcal{T}_\lambda\}_{\lambda\in\Lambda}$ is a family of pairwise orthogonal standard stable tubes of the Auslander–Reiten quiver $\Gamma(\mathrm{mod}\,A)$. We proved in Section XV.3 that any $\boldsymbol{\mathcal{T}}$-tubular extension of A is a branch $\boldsymbol{\mathcal{T}}$-extension of A, and any $\boldsymbol{\mathcal{T}}$-tubular coextension of A is a branch $\boldsymbol{\mathcal{T}}$-coextension of A.

In the present section we give various characterisations of branches, the bound quiver algebras of branches, and their Auslander–Reiten quivers. The results play an essential rôle in the following sections of this chapter.

We recall from (XV.2.14) that a **branch** is a finite connected full bound subquiver $\mathcal{L} = (L, I)$, containing the lowest vertex 0, of the following infinite tree

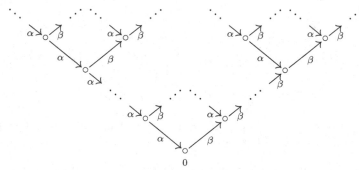

bound by all possible zero relations of the form $\alpha\beta = 0$.

Throughout we use the following definition.

1.1. Definition. Let $\mathcal{L} = (L, I)$ be a branch.
- The lowest vertex 0 of a branch $\mathcal{L}$ is called the **germ** of $\mathcal{L}$.
- The number of vertices of a branch $\mathcal{L}$ is called the **capacity** of $\mathcal{L}$.
- A **branch algebra** is the bound quiver algebra
$$K\mathcal{L} = KL/I$$
of a branch $\mathcal{L} = (L, I)$.

It follows from the classification of all tilted algebras of the Euclidean type $\widetilde{\mathbb{A}}_n$, given in (IX.6.12), that a branch algebra $K\mathcal{L}$ of a branch $\mathcal{L}$ of capacity $n \geq 1$ is a tilted algebra of the Dynkin type $\mathbb{A}_n$. We prove that a converse to the statement remains valid, by showing that the class of branch algebras $K\mathcal{L}$ of branches $\mathcal{L}$ of capacity $n \geq 1$ coincides with the class of tilted algebras of the Dynkin **equioriented linear quiver** type

$$\Delta(\mathbb{A}_n): \quad \overset{1}{\circ} \longleftarrow \overset{2}{\circ} \longleftarrow \overset{3}{\circ} \longleftarrow \ldots \longleftarrow \overset{n-1}{\circ} \longleftarrow \overset{n}{\circ}. \qquad (1.2)$$

Note that the path algebra $H_n = K\Delta(\mathbb{A}_n)$ of the quiver $\Delta(\mathbb{A}_n)$ is isomorphic with the subalgebra

$$H_n = \begin{bmatrix} K & 0 & \cdots & 0 \\ K & K & \cdots & 0 \\ \vdots & \vdots & \ddots & \vdots \\ K & K & \cdots & K \end{bmatrix} \qquad (1.3)$$

of $\mathbb{M}_n(K)$ consisting of all $n \times n$-lower triangular matrices with coefficients in the field K. We recall that, given $n \geq 1$, the Auslander–Reiten quiver $\Gamma(\operatorname{mod} H_n)$ of H_n is a cone in the following sense.

1.4. Definition. Let A be an algebra, let $\mathcal{C}$ be a component of the Auslander–Reiten quiver $\Gamma(\operatorname{mod} A)$ of A, and let M be an indecomposable module in $\mathcal{C}$.

A **cone of depth** $m \geq 1$ determined by the module M in $\mathcal{C}$ is the full translation subquiver $\mathcal{C}(M)$ of $\mathcal{C}$ of the form

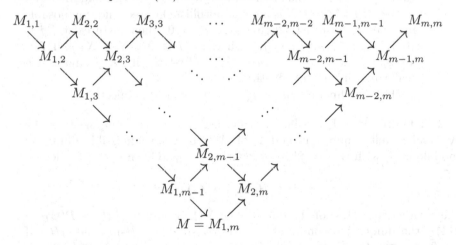

The module $M = M_{1,m}$ is called the **germ** of the cone $\mathcal{C}(M)$, and the depth m of $\mathcal{C}(M)$ is denoted by $\operatorname{depth}\mathcal{C}(M)$.

Note that the Auslander–Reiten quiver $\Gamma(\operatorname{mod} H_n)$ of the hereditary algebra H_n (1.2), with $n \geq 1$, is the cone $\mathcal{C}(P(n))$ of depth n determined by the unique indecomposable projective-injective H_n-module $M = P(n) \cong I(1)$ and, for each $j \in \{1, \dots, n\}$, we have

- the module $M_{1,j}$ is the indecomposable projective H_n-module $P(j) = e_j H_n$ at the vertex j of $\Delta(\mathbb{A}_n)$,
- the module $M_{j,n}$ is the indecomposable injective H_n-module $I(j) = D(H_n e_j)$ at the vertex j of $\Delta(\mathbb{A}_n)$, and
- $M_{j,j}$ is the simple H_n-module $S(j) = \operatorname{top} P(j)$ at the vertex j of the quiver $\Delta(\mathbb{A}_n)$.

1.5. Definition. Let A be an algebra and $\mathcal{C}$ a component of the Auslander–Reiten quiver $\Gamma(\operatorname{mod} A)$ of A.

(a) An indecomposable module X in $\mathcal{C}$ is defined to be a **finite ray module**, of $\mathcal{C}$, if there is a finite sectional path of **length** $s = s_X \geq 0$

$$X = X_0 \xrightarrow{u_1} X_1 \xrightarrow{u_2} X_2 \longrightarrow \dots \xrightarrow{u_1} X_{s-1} \xrightarrow{u_s} X_s, \qquad (1.6)$$

called the **ray of length** s_X **starting from** X, containing all sectional paths of the component $\mathcal{C}$ starting at X.

(b) A finite ray module X of $\mathcal{C}$, with the ray (1.6) of length $s = s_X$, is defined to be **admissible**, if the following three conditions are satisfied.

    (b1) If M is an indecomposable A-module in $\mathcal{C}$ such that $\operatorname{Hom}_A(X, M) \neq 0$, then there exists $j \in \{0, 1, \dots, s_X\}$ and $M \cong X_j$.

    (b2) If $f : M \longrightarrow N$ is a homomorphism between indecomposable modules in $\mathcal{C}$ such that $\operatorname{Hom}_A(X, f) \neq 0$, then there exist $i, j \in \{0, 1, \dots, s_X\}$, with $i \leq j$, such that $M \cong X_i$, $N \cong X_j$ and f is a scalar multiple of $u_j \dots u_{i+1} : X_i \longrightarrow X_j$, if $i < j$, and f is a scalar multiple of the identity, if $i = j$.

    (b3) The final module X_{s_X} of the ray (1.6) is injective.

1.7. Definition. Let A be an algebra, $\mathcal{C}$ a standard component of the Auslander–Reiten quiver $\Gamma(\operatorname{mod} A)$ of A, and X an admissible finite ray module in $\mathcal{C}$, with the ray (1.6) of length $s = s_X$. Given $t \geq 1$, we denote by

$$H = H_t = K\Delta(\mathbb{A}_t)$$

the path algebra (1.3) of the quiver $\Delta(\mathbb{A}_t)$ (1.2), and by $P(t) = P(t)_H \cong I(1)_H$ the unique indecomposable projective-injective H-module $e_t H$. If $t = 0$, we agree to denote by $H = H_0$ the zero algebra, and we set $P(t) = 0$.

(a) A **finite t-linear extension** of A at X is defined to be the one-point extension algebra

$$A(X, s_X, t) = (A \times H_t)[X \oplus P(t)] = \begin{bmatrix} A \times H_t & 0 \\ X \oplus P(t) & K \end{bmatrix}$$

of the product algebra $A \times H_t$ by $X \oplus P(t)$ endowed with the coordinate-wise $A \times H_t$-module structure.

(b) An algebra A' is defined to be a **finite line extension**, if there exist a standard component $\mathcal{C}$ of the Auslander–Reiten quiver $\Gamma(\operatorname{mod} A)$ of A, an admissible finite ray module X in $\mathcal{C}$, with a ray (1.6) of length s_X, and $t \geq 0$ such that $A' \cong A(X, s_X, t)$, that is, A' is a finite t-linear extension of A at X.

Thus, if $t = 0$ then the t-linear extension of A at X reduces to the one-point extension algebra $A[X]$ of A by X.

Note that the bound quiver of a t-linear extension $A(X, s_X, t)$ is of the form

where 0 denotes the extension point, the shaded part denotes the bound quiver of A, and there are possibly additional relations from 0 to Q_A so that X equals the summand of $\operatorname{rad} P(0)_{A(X,t)}$ which is an A-module.

Now we modify the procedure of a rectangle insertion, described in (XV.2.5), as follows.

1.8. A finite rectangle insertion. Assume that $t \geq 0$ is an integer, A is an algebra, $\mathcal{C}$ a standard component of the Auslander–Reiten quiver $\Gamma(\operatorname{mod} A)$ of A, and X is an indecomposable admissible finite ray module of $\mathcal{C}$, with the ray (1.6) of length $s = s_X$. We view $\mathcal{C}$ as a translation quiver. Let

$$A' = A(X, s_X, t)$$

be the t-linear extension of the algebra A by the admissible ray module X of $\mathcal{C}$.

Our purpose is to describe the component of $\Gamma(\operatorname{mod} A')$ that contains the indecomposable module X, viewed as an A'-module. To do it, we first construct a translation quiver

$$(\mathcal{C}', \tau') \subseteq \Gamma(\operatorname{mod} A'),$$

which we later show to equal this particular component in $\Gamma(\operatorname{mod} A')$.

We recall that $H = H_t = K\Delta(\mathbb{A}_t)$ is the path algebra of the **equiori-ented linear quiver**

$$\Delta(\mathbb{A}_t): \quad \overset{1}{\circ}\longleftarrow\overset{2}{\circ}\longleftarrow\overset{3}{\circ}\longleftarrow\ldots\longleftarrow\overset{t-1}{\circ}\longleftarrow\overset{t}{\circ}.$$

and the Auslander—Reiten quiver $\Gamma(\mathrm{mod}\,H)$ of $\mathrm{mod}\,H$ is the cone

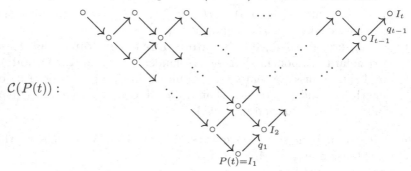

$\mathcal{C}(P(t)):$

where we denote by

$$I_1 = I(1)_H, I_2 = I(2)_H, \ldots, I_t = I(t)_H$$

the indecomposable injective H-modules, viewed as A'-modules; that is, we set

$$I_j = I(j) = D(e_j H),$$

for $j = 1, \ldots, t$. In particular,

$$I_1 = I(1)_H = D(e_1 H) \cong e_t H = P(t)$$

is the unique indecomposable projective-injective right H-module, and $I_t = I(t)_H = D(e_t H)$ is the unique simple injective right H-module.

On the other hand, the component $\mathcal{C}$ of $\Gamma(\mathrm{mod}\,A)$ being standard, the admissibility conditions on X imply that $\mathcal{C}$ looks as follows

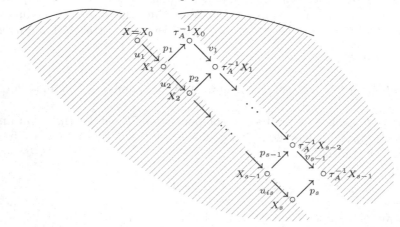

where $s = s_X$ is the length of the ray (1.6) and, to avoid ambiguity, we denote by τ_A the Auslander–Reiten translation in $\mathrm{mod}\,A$.

We associate to the component $\mathcal{C}$ of $\Gamma(\mathrm{mod}\,A)$ the following translation quiver $(\mathcal{C}', \tau')$ consisting of A'-modules, and homomorphisms of A'-modules

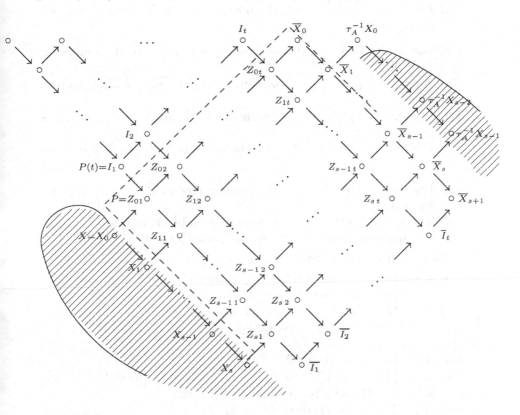

Here we set

- $\overline{X}_{s+1} = S(0) = \overline{I}_{t+1}$, where $S(0) = (K, 0, 0)$ is the simple injective A'-module at the extension vertex 0,
- $\overline{X}_i = (K, X_i, 1)$, for $i \in \{0, 1, \ldots, s\}$,
- $\overline{I}_j = (K, I_j, 1)$, for $j \in \{1, \ldots, t\}$, and
- $Z_{ij} = \left(K, X_i \oplus I_j, \begin{bmatrix} 1 \\ 1 \end{bmatrix}\right)$, for $i \in \{0, 1, \ldots, s\}$ and $j \in \{1, \ldots, t\}$,

It is easily seen that the modules Z_{ij}, $\overline{I}_j$, and $\overline{X}_i$ are indecomposable A'-modules. The homomorphisms are the obvious ones, defined as follows.

(i) For a fixed $j \in \{1, \ldots, t\}$, the homomorphism $Z_{ij} \longrightarrow Z_{i+1,j}$ is given by $\left(1, \begin{bmatrix} u_i & 0 \\ 0 & 1 \end{bmatrix}\right)$.

(ii) For a fixed $i \in \{0, 1, \ldots, s\}$, the homomorphism $Z_{ij} \longrightarrow Z_{i,j+1}$ is given by $\left(1, \begin{bmatrix} 1 & 0 \\ 0 & q_j \end{bmatrix}\right)$.

(iii) For any $i \in \{0, 1, \ldots, s\}$, the homomorphism $X_i \longrightarrow Z_{i1}$ is given by $\left(0, \begin{bmatrix} 1 \\ 0 \end{bmatrix}\right)$.

(iv) For any $i \in \{0, 1, \ldots, s\}$, the homomorphism $Z_{it} \longrightarrow \overline{X}_i$ is given by $(1, [1 \ 0])$.

(v) For any $i \in \{0, 1, \ldots, s\}$, the homomorphism $\overline{X}_i \longrightarrow \overline{X}_{i+1}$ is given by $(1, u_i)$.

(vi) For any $i \in \{0, 1, \ldots, s\}$, the homomorphism $\overline{X}_i \longrightarrow \tau_A^{-1} X_{i-1}$ is given by $(0, p_i)$.

(vii) For any $i \in \{0, 1, \ldots, t\}$, the homomorphism $I_j \longrightarrow Z_{0j}$ is given by $\left(0, \begin{bmatrix} 0 \\ 1 \end{bmatrix}\right)$.

(viii) For any $i \in \{0, 1, \ldots, t\}$, the homomorphism $Z_{sj} \longrightarrow \overline{I}_j$ is given by $(1, [0, 1])$.

(ix) The homomorphism $\overline{X}_s \longrightarrow S(0)$ is given by $(1, 0)$.

(x) The homomorphism $\overline{I}_t \longrightarrow S(0)$ is given by $(1, 0)$.

The translation τ' of $\mathcal{C}'$ is defined as follows.

(i) If $i \in \{1, \ldots, s\}$ and $j \in \{2, \ldots, t\}$, then $\tau' Z_{ij} = Z_{i-1,j-1}$.

(ii) If $i \in \{1, \ldots, s\}$, then $\tau' Z_{i1} = X_{i-1}$.

(iii) If $i \in \{2, \ldots, t\}$, then $\tau' Z_{0j} = I_{j-1}$.

(iv) The module $P = Z_{01}$ is projective.

(v) $\tau' \overline{X}_0 = I_t$.

(vi) If $i \in \{1, \ldots, s\}$, then $\tau' \overline{X}_i = Z_{i-1,t}$.

(vii) If $i \in \{0, \ldots, s-1\}$, then $\tau'(\tau_A^{-1} X_i) = \overline{X}_i$, if the module X_i is not injective; otherwise the module $\overline{X}_i$ is injective.

(viii) If $j \in \{2, \ldots, t+1\}$, then $\tau' \overline{I}_j = Z_{s,j-1}$.

(ix) $\tau' \overline{I}_1 = X_s$.

(x) The modules $\overline{I}_1, \overline{I}_2, \ldots, \overline{I}_t, \overline{I}_{t+1} = S(0)$ are injective.

For the remaining points of $\mathcal{C}$ (or of $\Gamma(\mathrm{mod}\, H)$), the translation τ' coincides with τ_A (or with the translation τ_H in $\Gamma(\mathrm{mod}\, H)$, respectively). This finishes the construction of the translation quiver $(\mathcal{C}', \tau')$.

The procedure presented above is called the **finite rectangle insertion**.

Intuitively, the construction of the translation quiver $(\mathcal{C}', \tau')$ may be understood as the following four step procedure:

(1°) take the standard component $\mathcal{C}$ of $\Gamma(\operatorname{mod} A)$ and an admissible finite ray module X in $\mathcal{C}$, with the ray (1.6) of length $s = s_X$.

(2°) 'cut it' along the arrows $p_1, p_2, \ldots, p_s$,

(3°) 'insert in it' the finite $(s+2) \times (t+1)$- rectangle consisting of the modules

- $\overline{X}_0, \overline{X}_1, \ldots, \overline{X}_s$,
- $\overline{X}_{s+1} = S(0) = \overline{I}_{t+1}$,
- $\overline{I}_1, \overline{I}_2, \ldots, \overline{I}_t$, and
- the modules Z_{ij}, with $i \in \{0, 1, \ldots, s\}$ and $j \in \{1, \ldots, t\}$,

and finally

(4°) 'glue' the quiver $\Gamma(\operatorname{mod} H)$ to the quiver constructed in (3°).

Now we establish one of the main properties of the translation quiver $(\mathcal{C}', \tau')$.

1.9. Proposition. *Assume that $t \geq 0$ is an integer, A is an algebra, $\mathcal{C}$ a standard component of the Auslander–Reiten quiver $\Gamma(\operatorname{mod} A)$ of A, and X is an indecomposable admissible finite ray module of $\mathcal{C}$, with the ray (1.6) of length $s = s_X$. Let*

$$A' = A(X, s_X, t)$$

be the t-linear extension of the algebra A at the admissible finite ray module X of $\mathcal{C}$, and let $(\mathcal{C}', \tau')$ be the translation quiver associated to $\mathcal{C}$ in the finite rectangle insertion (1.8).

(a) *The translation quiver $\mathcal{C}'$ is the Auslander–Reiten component of $\Gamma(\operatorname{mod} A')$ containing the ray module X, when viewed as an A'-module.*

(b) *The component $\mathcal{C}'$ is standard.*

Proof. We fix an integer $t \geq 0$, an algebra A, a component $\mathcal{C}$, and a module X, with the ray (1.6) of length $s = s_X$, as in the proposition. We use freely the notation introduced the finite rectangle insertion (1.8). We recall that, for $t \geq 1$,

$$H = H_t = K\Delta(\mathbb{A}_t)$$

is the path algebra of the equioriented linear quiver $\Delta(\mathbb{A}_t)$.

We denote by

$$H' = K\Delta'_{t+1}$$

the path algebra of the equioriented linear quiver

$$\Delta'_{t+1} = \Delta(\mathbb{A}_{t+1}): \quad \overset{1}{\circ} \longleftarrow \overset{2}{\circ} \longleftarrow \overset{3}{\circ} \longleftarrow \cdots \longleftarrow \overset{t-1}{\circ} \longleftarrow \overset{t}{\circ} \longleftarrow \overset{0}{\circ}$$

obtained from $\Delta(\mathbb{A}_t)$ by adding the arrow

$$t \; \circ \longleftarrow \circ \; 0,$$

where 0 is the extension vertex of the one-point extension algebra $A' = A(X, s_X, t)$, see (1.7).

First we note that the modules

$$\overline{I}_1, \overline{I}_2, \dots, \overline{I}_t, \text{ and } \overline{I}_{t+1} = S(0) = \overline{X}_{s+1}$$

are the pairwise non-isomorphic indecomposable injective H'-modules

$$I(1)_{H'}, I(2)_{H'}, \dots, I(t)_{H'}, I(0)_{H'},$$

respectively. Moreover, the A'-module

$$\overline{X}_s = (K, X_s, 1)$$

is injective, because the A-module X_s is injective, by our hypothesis.

Next, we note that the canonical exact sequence

$$0 \longrightarrow X_s \longrightarrow Z_{s1} \longrightarrow \overline{I}_1 \longrightarrow 0$$

is an almost split sequence in $\bmod A'$, because the A'-module $\overline{I}_1$ is the cokernel of the unique irreducible morphism

$$X_s \longrightarrow Z_{s1}$$

starting from X_s.

Hence, an easy induction on $j \in \{1, \dots, t\}$ implies that, for each $j \leq t$, there exists an almost split sequence of the form

$$0 \longrightarrow Z_{sj} \longrightarrow Z_{s\,j+1} \oplus \overline{I}_j \longrightarrow \overline{I}_{j+1} \longrightarrow 0,$$

in $\bmod A'$, where

$$Z_{s\,t+1} = \overline{X}_s.$$

Now, the proof of the proposition follows by a simple modification of the arguments given in the proof of (XV.2.7) and (XV.2.8). The details are left to the reader. $\qquad\square$

Now we illustrate the finite rectangle insertion with an example.

1.10. Example. Let $A = KQ$ be the path algebra of the oriented tree

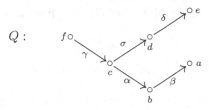

bound by two zero relations $\alpha\beta = 0$ and $\gamma\sigma = 0$.

It is clear that the algebra A is representation-finite, because A is a quotient of the path algebra KQ of the tree Q which is a quiver of the Dynkin type $\mathbb{E}_6$. Moreover, the algebra A is representation-directed. The standard calculation technique shows that the Auslander–Reiten quiver $\Gamma(\operatorname{mod} A)$ of A has the form

$$
P(f)= {}^{\ 0}_{1}{}^{0}_{1}{}^{\ 0}_{0} =I(b)
$$

$$
\begin{array}{l}
P(e)= {}^{\ \ 1}_{0}{}^{0}_{0}{}^{0}_{0}{}_{0} \quad {-}{-}{-} \quad {}^{\ \ 0}_{0}{}^{1}_{0}{}^{0}_{0}{}_{0} \quad {-}{-}{-} \quad {}^{\ \ 0}_{0}{}^{0}_{1}{}^{0}_{0}{}_{1} \quad {-}{-}{-} \quad {}^{\ \ 0}_{1}{}^{0}_{1}{}^{0}_{0}{}_{0} =I(c)
\end{array}
$$

where the indecomposable modules are represented by their dimension vectors.

Observe that the modules

$$
P(b),\ S(b) = \tau_A^{-1}P(a),\ P(c),\ \tau_A^{-1}P(d),\ \tau_A^{-2}P(e)\ \text{and}\ P(f)
$$

form a faithful section

$$
\Sigma: \quad \overset{b}{\circ} \longrightarrow \overset{a}{\circ} \longrightarrow \overset{c}{\circ} \longrightarrow \overset{d}{\circ} \longrightarrow \overset{e}{\circ} \longrightarrow \overset{f}{\circ}
$$

of the Dynkin type $\mathbb{A}_6$. Hence, by the criterion (VIII.5.6), A is a tilted algebra of type $\Sigma^{\operatorname{op}} = \Sigma$.

Note that the indecomposable A-module

$$X = S(d) = {}^0_{0\,0}{}^{0\,1\,0}_{0}$$

is an admissible finite ray module and

$$
\begin{array}{ccccc}
X = X_0 & & X_1 & & X_2 \\
\| & & \| & & \| \\
{}^{0\;1\;0}_{0\;0\;0}{}^{0}_{\;0} & \longrightarrow & {}^{0\;1\;0}_{\;1\;0}{}^{0}_{\;1} & \longrightarrow & {}^{0\;1\;0}_{\;1\;0}{}^{0}_{\;0}
\end{array}
$$

is the unique section in $\Gamma(\mathrm{mod}\,A)$ starting at X. It follows that the section is a ray of length $s_X = 3$. Take $t = 1$, $s_X = 3$, and form the t-line extension

$$A' = A(X, s_X, t) = A(X, 3, 1) = (A \times H_1)[X \oplus P(1)],$$

where

$$H_1 = K\Delta(\mathbb{A}_1) = K$$

is the path algebra of the one vertex quiver $\Delta(\mathbb{A}_1)$, and $P(1) = S(1) \cong K$ is the unique simple H_1-module.

It is easy to see that A' is given by the tree

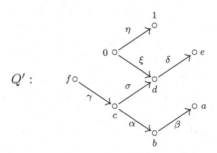

Q' :

bound by three zero relations $\alpha\beta = 0$, $\gamma\sigma = 0$, and $\xi\delta = 0$. Here, 0 is the extension vertex of the algebra

$$A' = A(X, 3, 1)$$

and the vertex 1 of Q' is the unique vertex of the quiver $\Delta(\mathbb{A}_1)$.

Note that the algebra A' is representation-finite, because, by (IX.6.12), A' is a tilted algebra of type $\mathbb{A}_8$. By applying (1.9) to the translation quiver

$\mathcal{C} = \Gamma(\mathrm{mod}\,A)$, we conclude that the translation quiver $\mathcal{C}' = \Gamma(\mathrm{mod}\,A')$ is of the form

where the indecomposable modules are represented by their dimension vectors,

$$P'(a),\; P'(b),\; P'(c),\; P'(d),\; P'(e),\; P'(f),\; P'(0),\; P'(1)$$

are the pairwise non-isomorphic indecomposable projective A'-modules, and

$$I'(a),\; I'(b),\; I'(c),\; I'(d),\; I'(e),\; I'(f),\; I'(0),\; I'(1)$$

are the pairwise non-isomorphic indecomposable injective A'-modules at the vertices $a, b, c, d, e, f, 0, 1$ of Q', respectively. Here we use the notation introduced in the finite rectangle insertion procedure (1.8).

It is easy to see that the Auslander–Reiten quiver $\Gamma(\operatorname{mod} A')$ of the algebra A' admits a faithful section

$$\Sigma' : \quad \overset{b}{\circ}\longrightarrow\overset{a}{\circ}\longrightarrow\overset{c}{\circ}\longrightarrow\overset{d}{\circ}\longrightarrow\overset{e}{\circ}\longrightarrow\overset{0}{\circ}\longrightarrow\overset{1}{\circ}\longrightarrow\overset{f}{\circ}$$

of the Dynkin type $\mathbb{A}_8$ given by the modules

$$P'(b) = I'(a),$$
$$\tau_{A'}^{-1}P'(a) = S'(b),$$
$$P'(c),$$
$$\tau_{A'}^{-1}P'(d) = X_1,$$
$$\tau_{A'}^{-2}P'(e) = Z_{11},$$
$$\tau_{A'}^{-1}P'(0) = \overline{X}_1,$$
$$\tau_{A'}^{-2}P'(1) = \tau_{A'}^{-1}\overline{X}_0, \text{ and}$$
$$I'(b) = P'(f).$$

Hence, by the criterion (VIII.5.6), A' is a tilted algebra of type $\Sigma'^{\mathrm{op}} = \Sigma'$. It is obvious that the section Σ' is obtained from the section Σ by an insertion of $t = 1$ arrows; precisely, by an insertion of the arrow $0\;\circ\!\longrightarrow\!\circ\;1$, and by replacing the arrow $e\;\circ\!\longrightarrow\!\circ\;f$ by two arrows $e\;\circ\!\longrightarrow\!\circ\;0$ and $1\;\circ\!\longrightarrow\!\circ\;f$.

XVI.2. Tilted algebras of an equioriented type $\mathbb{A}_m$

Our main objective of this section is to define a class of tilted algebras of an equioriented type $\mathbb{A}_m$ and characterise them, by applying the finite rectangle insertion procedure (1.8).

2.1. Definition. Let $m \geq 1$ be an integer. An algebra A is defined to be a **tilted algebra of an equioriented type** $\mathbb{A}_m$ if there exists a multiplicity-free tilting module T_{H_m} over the path algebra

$$H_m = K\Delta_m = K\Delta(\mathbb{A}_m) \cong \begin{bmatrix} K & 0 & \ldots & 0 \\ K & K & \ldots & 0 \\ \vdots & \vdots & \ddots & \vdots \\ K & K & \ldots & K \end{bmatrix} \subseteq \mathbb{M}_m(K)$$

of the equioriented linear quiver

$$\Delta_m = \Delta(\mathbb{A}_m) : \quad \overset{1}{\circ}\longleftarrow\overset{2}{\circ}\longleftarrow\overset{3}{\circ}\longleftarrow \ldots \longleftarrow\overset{m-1}{\circ}\longleftarrow\overset{m}{\circ}.$$

such that $A \cong \operatorname{End} T_{H_m}$.

Now we are able to formulate a characterisation of tilted algebras A of the equioriented type $\mathbb{A}_n$, where $n = \operatorname{rk} K_0(A) \geq 1$ is the rank of the Grothendieck group $K_0(A)$ of A.

2.2. Theorem. *Assume that A is an algebra and $n = \operatorname{rk} K_0(A) \geq 1$ is the rank of the Grothendieck group $K_0(A)$ of A. The following statements are equivalent.*

(a) *A is a tilted algebra of the equioriented type $\Delta(\mathbb{A}_n)$:* $\overset{1}{\circ}\leftarrow\overset{2}{\circ}\leftarrow \ldots \leftarrow\overset{n}{\circ}.$

(b) *The Auslander–Reiten translation quiver $\Gamma(\operatorname{mod} A)$ of A is finite and admits precisely one section Σ that is isomorphic to the equioriented linear quiver $\Delta_n = \Delta(\mathbb{A}_n)$.*

(c) *The category $\operatorname{mod} A$ admits precisely one multiplicity-free tilting module T_A such that the algebra $\operatorname{End} T_A$ is isomorphic to the path algebra $K\Delta(\mathbb{A}_n)$.*

(d) *The algebra A is isomorphic to a branch algebra*

$$KL = KL/I$$

of a branch $\mathcal{L} = (L, I)$ of capacity n.

(e) *There are an integer $r \geq 0$, a finite sequence*

$$m_0 < m_1 < \ldots < m_{r-1} < m_r = n$$

of integers, and a finite sequence of algebras

$$A_0, A_1, \ldots, A_r = A$$

such that

- *the algebra A_0 is the hereditary path algebra $K\Delta(\mathbb{A}_{m_0})$ of the equioriented linear quiver $\Delta(\mathbb{A}_{m_0})$,*
- *for each $j \in \{1, 2, \ldots, r\}$, the algebra A_j is a tilted algebra of the type $\Delta(\mathbb{A}_{m_j})$ and A_j is of the form*

$$A_j = A_{j-1}(X^{(j)}, s_{X^{(j)}}, t_j),$$

that is, A_j is a finite line extension of the algebra A_{j-1} by an admissible ray A_{j-1}-module $X^{(j)}$, with $t_j \geq 0$ and the ray of length $s_{X^{(j)}}$ in $\Gamma(\operatorname{mod} A_{j-1})$,
- *$m_j = 1 + m_{j-1} + t_j = m_0 + j + t_1 + \ldots + t_j$, and*
- *$n = m_r = m_0 + r + t_1 + \ldots + t_r$.*

Because the proof is rather long, we split it into several partial results, and then we present the proof in (1.9), at the end of the section, by applying Theorems (2.3) and (2.7).

In the following theorem we list the main properties of the tilted algebras of the equioriented type $\Delta(\mathbb{A}_m)$ and of their module categories. Later on, we prove in (2.7) that, in fact, some of the properties listed in (2.3) characterise this class of algebras.

2.3. Theorem. *Assume that $m \geq 1$ is an integer and A is a tilted algebra of the equioriented type $\Delta(\mathbb{A}_m)$: $\overset{1}{\circ}\longleftarrow\overset{2}{\circ}\longleftarrow\ldots\longleftarrow\overset{m}{\circ}$.*

(i) *There exist a branch $\mathcal{L} = (L, I)$ of capacity m and an isomorphism of algebras*
$$A \cong K\mathcal{L} = KL/I.$$

(ii) *The Auslander–Reiten translation quiver $\Gamma(\operatorname{mod} A)$ of A admits precisely one section Σ isomorphic to the equioriented linear quiver $\Delta(\mathbb{A}_m)$.*

(iii) *If A is a branch algebra $A = K\mathcal{L} = KL/I$ of a branch $\mathcal{L} = (L, I)$, 0 is the germ vertex of the branch $\mathcal{L}$, and Σ is a unique section of $\Gamma(\operatorname{mod} A)$ isomorphic to $\Delta(\mathbb{A}_m)$, then*

- *$m = \operatorname{rk} K_0(A)$ is the rank of the Grothendieck group $K_0(A)$ of the algebra A,*
- *the source module of Σ is the indecomposable projective A-module at the germ 0 of $\mathcal{L} = (L, I)$, and*
- *the sink module of Σ is the indecomposable injective A-module at the germ 0 of $\mathcal{L} = (L, I)$.*

(iv) *Assume that*
$$A = \operatorname{End} T_{H_m},$$

where T_{H_m} is a multiplicity-free tilting module over the algebra $H_m = K\Delta(\mathbb{A}_m)$. Let Σ_A be a unique section of $\Gamma(\operatorname{mod} A)$ isomorphic to $\Delta(\mathbb{A}_m)$, and let $(\mathcal{X}(T), \mathcal{Y}(T))$ be the torsion pair in $\operatorname{mod} A$ determined by the tilting module T_{H_m}, see (VI.3.6). Then

$$(\mathcal{X}(T), \mathcal{Y}(T)) = (\operatorname{add} \mathcal{T}(\Sigma_A), \operatorname{add} \mathcal{F}(\Sigma_A)),$$

where $\mathcal{T}_A = \mathcal{T}(\Sigma_A)$ is the set of all indecomposable proper successors of Σ_A in $\Gamma(\operatorname{mod} A)$ and $\mathcal{F}(\Sigma_A)$ is the set of all indecomposable predecessors of Σ_A in $\Gamma(\operatorname{mod} A)$.

Proof. We fix an integer $m \geq 1$ and assume that A is a tilted algebra of the equioriented type $\Delta(\mathbb{A}_m)$. Without loss of generality, we may assume that $m \geq 2$ and
$$A = \operatorname{End} T_H,$$

where T_H is a multiplicity-free tilting module over the path algebra $H = H_m = K\Delta(\mathbb{A}_m)$ of the equioriented linear quiver

$$\Delta_m = \Delta(\mathbb{A}_m) : \quad \overset{1}{\circ}\longleftarrow\overset{2}{\circ}\longleftarrow\overset{3}{\circ}\longleftarrow\ldots\longleftarrow\overset{m-1}{\circ}\longleftarrow\overset{m}{\circ}.$$

We recall that the Auslander–Reiten translation quiver $\Gamma(\mathrm{mod}\,H)$ of H has the form

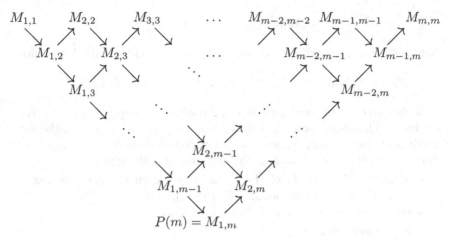

In other words, $\Gamma(\mathrm{mod}\,H) = \mathcal{C}(P(m))$ is the cone $\mathcal{C}(P(m))$ of depth $m \geq 2$ determined by the unique indecomposable projective-injective module

$$M_{1,m} = P(m) \cong I(1)$$

in $\mathcal{C}$. Moreover,

$$M_{1,j} = P(j),\ M_{j,m} = I(j),\ \text{and}\ M_{j,j} = S(j),$$

for all $j \in \{1, 2, \ldots, m\}$.

First we show that the indecomposable projective-injective module $P(m)$ is a direct summand of the tilting module T_H. For the proof, we note that, by the condition (T3) of the Definition (VI.2.1) of tilting H-module, there exists a short exact sequence

$$0 \longrightarrow H_H \longrightarrow T_H^{(1)} \longrightarrow T_H^{(2)} \longrightarrow 0$$

in $\mathrm{mod}\,H$, with $T_H^{(1)}$ and $T_H^{(2)}$ in $\mathrm{add}\,T$. Because the module $P(m)$ is projective, $P(m)$ is a direct summand of H_H. The fact that $P(m)$ is also injective yields that $P(m)$ is a direct summand of the module $T_H^{(1)}$. Consequently, $P(m)$ is a direct summand of the tilting module T_H. It follows from (VI.4.4) that there is a decomposition

$$T_H = T_0 \oplus T_1 \oplus \ldots \oplus T_{m-1},$$

where $T_0 = P(m)$, and $T_0, T_1, \ldots, T_{m-1}$ are pairwise non-isomorphic indecomposable H-modules. Because the algebra H is hereditary then, by

(IV.2.14), the tilting vanishing condition $\operatorname{Ext}_H^1(T,T) = 0$ is equivalent to the vanishing condition

$$\operatorname{Hom}_H(T_i, \tau_H T_j) = 0,$$

for all $i, j \in \{0, 1, 2, \dots, m-1\}$. It follows from (VI.3.1) that the A-modules

$$P(0)_A = \operatorname{Hom}_A(T, T_0), \ \dots, \ P(m-1)_A = \operatorname{Hom}_A(T, T_{m-1}),$$

form a complete set of pairwise non-isomorphic indecomposable projective A-modules. Therefore, we may assume that $\{0, 1, 2, \dots, m-1\}$ is the set of vertices of the ordinary quiver $L = Q_A$ of the algebra A.

Now we prove the statement (i), by showing that there exist

- a branch $\mathcal{L} = (L, I)$ of the capacity m, with the germ vertex 0 corresponding to the module $T_0 = P(m)$, and
- an algebra isomorphism $A \cong K\mathcal{L} = KL/I$.

We apply the induction on $m \geq 2$.

It follows from the preceding description of $\Gamma(\operatorname{mod} H)$ that it is a finite standard component. Hence, there is an equivalence of categories

$$\operatorname{ind} H \cong K\mathcal{C}(P(m)) = K\Gamma(\operatorname{mod} H),$$

where $\operatorname{ind} H$ is the full subcategory of the category $\operatorname{mod} H$ formed by the indecomposable modules, and $K\mathcal{C}(P(m))$ is the mesh category of the cone $\mathcal{C}(P(m)) = \Gamma(\operatorname{mod} H)$.

Because, by (VI.3.5), the tilted algebra A is connected and $P(m) = T_0$ is a direct summand of T, then T admits a direct summand T_i that is one of the modules

$$M_{1,1}, M_{1,2}, \dots, M_{1,m-1}, \ M_{2,m}, M_{3,m}, \dots, M_{m-1,m}, M_{m,m}.$$

We have three cases to consider.

Case 1°. Assume that the indecomposable summands $T_1, T_2, \dots, T_{m-1}$ lie in the cone $\mathcal{C}(M_{1,m-1}) = \mathcal{C}(P(m-1))$. Observe that the cone $\mathcal{C}(M_{1,m-1})$ is the Auslander–Reiten quiver $\Gamma(\operatorname{mod} H')$ of the path algebra

$$H' = H_{m-1} = K\Delta(\mathbb{A}_{m-1})$$

of the quiver $\Delta(\mathbb{A}_{m-1})$ obtained from $\Delta(\mathbb{A}_m)$ by deleting the vertex m. It follows that

$$T' = T_1 \oplus T_2 \oplus \dots \oplus T_{m-1},$$

is a tilting module in $\operatorname{mod} H'$. Note also that the H'-module $M_{1,m-1} = P(m-1)$ is a direct summand of T', because $P(m-1)$ is an indecomposable projective-injective H'-module.

By the inductive hypothesis, the tilted algebra $A' = \operatorname{End} T'_{H'}$ of the equioriented type $\mathbb{A}_{m-1}$ is isomorphic with the branch algebra $K\mathcal{L}' = KL'/I'$ of a branch $\mathcal{L}' = (L', I')$ of capacity $m-1$ with the germ vertex $0'$ corresponding to the direct summand $M_{1,m-1} = P(m-1)$ of T', where $L' = Q_{A'}$ is the ordinary quiver of the algebra A'.

We recall from (VI.3.1) that the functor $\operatorname{Hom}_H(T, -)$ restricts to the equivalence of categories

$$\operatorname{Hom}_H(T, -) : \operatorname{add} T \longrightarrow \operatorname{proj} A, \qquad (*)$$

where $\operatorname{proj} A$ is the category of finitely generated projective A-modules.

Assume that the indecomposable projective H-modules $M_{1,i} = P(i)$, $M_{1,j} = P(j)$, and $M_{1,k} = P(k)$ are direct summands of T, for some $i, j, k \in \{1, 2, \ldots, m\}$, and there exist non-zero homomorphisms

$$M_{1,i} \xrightarrow{\ f\ } M_{1,j} \xrightarrow{\ g\ } M_{1,k}$$

of H-modules. Because the algebra H is hereditary, then $i \le j \le k$, both f and g are monomorphisms and, hence, $gf \ne 0$. Because the functor $(*)$ is an equivalence of categories, then the induced homomorphisms

$$\operatorname{Hom}_H(T, M_{1,i}) \xrightarrow{\operatorname{Hom}_H(T,f)} \operatorname{Hom}_H(T, M_{1,j}) \xrightarrow{\operatorname{Hom}_H(T,g)} \operatorname{Hom}_H(T, M_{1,k})$$

between indecomposable projective modules over $A = \operatorname{End} T_H$ are also monomorphisms and their composition is non-zero.

Hence we conclude that the algebra A is isomorphic to the branch algebra $K\mathcal{L} = KL/I$ of the branch $\mathcal{L} = (L, I)$ of capacity m, with $L = Q_A$, obtained from the branch $\mathcal{L}' = (L', I')$ of capacity $m-1$ by adding one arrow $0 \longrightarrow 0'$ from the germ vertex 0 of $\mathcal{L}$, corresponding to the module $T_0 = M_{1,m}$, to the germ vertex $0'$ of $\mathcal{L}'$, corresponding to the direct summand $M_{1,m-1}$ of T'. Note that 0 is a source vertex of the quiver $L = Q_A$ of A.

Case 2°. Assume that the indecomposable summands $T_1, T_2, \ldots, T_{m-1}$ lie in the cone $\mathcal{C}(M_{2,m}) = \mathcal{C}(I(2))$.

Observe that the cone $\mathcal{C}(M_{2,m})$ is the Auslander–Reiten quiver $\Gamma(\operatorname{mod} H')$ of the path algebra $H'' = K\Delta'_{m-1}$ of the quiver Δ'_{m-1} obtained from $\Delta(\mathbb{A}_m)$ by deleting the vertex 1. It follows that

$$T'' = T_2 \oplus T_3 \oplus \ldots \oplus T_m,$$

is a tilting module in $\operatorname{mod} H''$. Because $M_{2,m} \cong I(2)$ is an indecomposable projective-injective H''-module, then $M_{2,m}$ is a direct summand of T''.

By the inductive hypothesis, the tilted algebra

$$A'' = \operatorname{End} T''_{H''}$$

of the equioriented type $\mathbb{A}_{m-1}$ is isomorphic with the branch algebra $K\mathcal{L}'' = KL''/I''$ of a branch $\mathcal{L}'' = (L'', I'')$ of capacity $m-1$ with the germ vertex $0''$ corresponding to the direct summand $M_{2,m} \cong I(2)$ of T'', where $L'' = Q_{A''}$ is the ordinary quiver of the algebra A''.

Further, assume that the indecomposable injective modules $M_{i,m} = I(i)$, $M_{j,m} = I(j)$, and $M_{k,m} = I(k)$ are direct summands of T, for some $i, j, k \in \{1, 2, \ldots, m\}$, and there exist non-zero homomorphisms

$$M_{i,m} \xrightarrow{\ f\ } M_{j,m} \xrightarrow{\ g\ } M_{k,m}$$

of H-modules. Because the algebra H is hereditary, then $i \leq j \leq k$, both f and g are epimorphisms and, hence, $gf \neq 0$. Because the functor $(*)$ is an equivalence of categories, then the composition of the induced homomorphisms

$$\operatorname{Hom}_H(T, M_{i,m}) \xrightarrow{\operatorname{Hom}_H(T,f)} \operatorname{Hom}_H(T, M_{j,m}) \xrightarrow{\operatorname{Hom}_H(T,g)} \operatorname{Hom}_H(T, M_{k,m})$$

between indecomposable projective A-modules is non-zero.

Hence we conclude that the algebra A is isomorphic to the branch algebra $K\mathcal{L} = KL/I$ of the branch $\mathcal{L} = (L, I)$ of capacity m, with $L = Q_A$, obtained from the branch $\mathcal{L}'' = (L'', I'')$ of capacity $m-1$ by adding one arrow $0'' \longrightarrow 0$ from the germ vertex $0''$ of $\mathcal{L}''$, corresponding to the direct summand $M_{2,m}$ of T'', to the germ vertex 0 of $\mathcal{L}$, corresponding to the module $T_0 = M_{1,m}$. Note that 0 is a sink vertex of the quiver $L = Q_A$ of A.

Case 3°. Assume that T admits a direct summand $M_{1,i} = P(i)$ and a direct summand $M_{j,m} = I(j)$, for some $i, j \in \{1, \ldots, m-1\}$.

Let $r \in \{1, \ldots, m-1\}$ be the maximal index such that the projective module $M_{1,r} = P(r)$ is a direct summand of T, and let $s \in \{1, \ldots, m-1\}$ be the minimal index such that the injective module $M_{s,m} = I(s)$ is a direct summand of T.

First, we show that the cones $\mathcal{C}(M_{1,r})$ and $\mathcal{C}(M_{s,m})$ in the Auslander–Reiten quiver $\Gamma(\operatorname{mod} H)$ of the path algebra $H = K\Delta(\mathbb{A}_m)$ are disjoint. To prove it, assume to the contrary that there exist two indices $u \in \{2, \ldots, r\}$ and $w \in \{s, \ldots, m-1\}$ such that the module

$$M_{u,r} = M_{s,w}$$

lies in $\mathcal{C}(M_{1,r})$ and in $\mathcal{C}(M_{s,m})$. It follows that there exist an H-module epimorphism $M_{1,r} \longrightarrow M_{u-1,r}$ and an H-module monomorphism $M_{u-1,r} \longrightarrow M_{s-1,m-1}$. Consequently, $\operatorname{Hom}_H(M_{1,r}, M_{s-1,m-1}) \neq 0$ and we get a contradiction with the tilting vanishing condition $\operatorname{Hom}_H(T_i, \tau_H T_j) = 0$, for all direct summands T_i and T_j of the module T, because the isomorphism $M_{s-1,m-1} \cong \tau_H M_{s,m}$ yields

$$0 \neq \operatorname{Hom}_H(M_{1,r}, M_{s-1,m-1}) \cong \operatorname{Hom}_H(M_{1,r}, \tau_H M_{s,m}) = 0.$$

Then, without loss of generality, we may assume that

$$T_1 = M_{1,r}, T_2, \ldots, T_p$$

are all indecomposable direct summands of T lying in the cone $\mathcal{C}(M_{1,r})$, and

$$T_{q+1} = M_{s,m}, T_{q+2}, \ldots, T_{m-1}$$

are all indecomposable direct summands of T lying in the cone $\mathcal{C}(M_{s,m})$, where $p \leq q$. Let

$$T^{(1)} = T_1 \oplus T_2 \oplus \ldots \oplus T_p \quad \text{and} \quad T^{(2)} = T_{q+1} \oplus T_{q+2} \oplus \ldots \oplus T_{m-1}.$$

Now we prove that $p = q$. Assume to the contrary, that $p < q$. Then the module

$$T^{(3)} = T_{p+1} \oplus T_{p+2} \oplus \ldots \oplus T_q$$

is non-zero and

$$T = T_0 \oplus T^{(1)} \oplus T^{(2)} \oplus T^{(3)}.$$

Now we show that the tilted algebra $A = \operatorname{End} T_H$ is not connected and then we get a contradiction with (VI.3.5). This is a consequence of the following observations:

- $\operatorname{Hom}_H(T_0, T^{(3)}) = 0$ and $\operatorname{Hom}_H(T^{(3)}, T_0) = 0$, because $T_0 = M_{1,m}$ and the direct summands of $T^{(3)}$ lie in the cone $\mathcal{C}(M_{2,m-1})$.
- $\operatorname{Hom}_H(T^{(3)}, T^{(1)}) = 0$ and $\operatorname{Hom}_H(T^{(2)}, T^{(3)}) = 0$, because the cone $\mathcal{C}(M_{1,r})$ is closed under predecessors in $\mathcal{C}(M_{1,m}) = \mathcal{C}(P(m)) = \Gamma(\operatorname{mod} H)$, and the cone $\mathcal{C}(M_{s,m})$ is closed under successors in $\mathcal{C}(M_{1,m}) = \mathcal{C}(I(1)) = \Gamma(\operatorname{mod} H)$.
- The equality $\operatorname{Ext}_H^1(T^{(3)}, T^{(1)}) = 0$ yields $\operatorname{Hom}_H(T^{(1)}, \tau_H T^{(3)}) = 0$, by (IV.2.14), because the algebra H is hereditary.
- The equality $\operatorname{Ext}_H^1(T^{(2)}, T^{(3)}) = 0$ yields $\operatorname{Hom}_H(T^{(3)}, \tau_H T^{(2)}) = 0$.

Hence, we get $\operatorname{Hom}_H(T^{(1)}, T^{(3)}) = 0$ and $\operatorname{Hom}_H(T^{(3)}, T^{(2)}) = 0$, and we conclude that A decomposes into a direct product of algebras

$$A = \operatorname{End} T_H \cong \operatorname{End}(T_0 \oplus T^{(1)} \oplus T^{(2)})_H \times \operatorname{End} T_H^{(3)},$$

that is, the tilted algebra A is not connected. This contradiction proves that $p = q$, that is, $T^{(3)} = 0$ and

$$T = T_0 \oplus T^{(1)} \oplus T^{(2)}.$$

Observe now that $T^{(1)}$ is a partial tilting module over the path algebra

$$H^{(1)} = K\Delta_r$$

of the quiver Δ_r obtained from Δ_m by deleting the vertices $r+1, \ldots, m$, and $T^{(2)}$ is a partial tilting module over the path algebra

$$H^{(2)} = K\Delta^*_{m-s+1}$$

of the quiver Δ^*_{m-s+1} obtained from Δ_m by deleting the vertices $1, \ldots, s-1$. Then (VI.4.4) yields $p \leq r$ and $m - 1 - p = m - 1 - q \leq m - s + 1$.

On the other hand, by the tilting vanishing condition, we have

$$\operatorname{Hom}_H(M_{1,r}, M_{s-1,m-1}) \cong \operatorname{Hom}_H(M_{1,r}, \tau_H M_{s,m}) = 0.$$

It follows that the cones $\mathcal{C}(M_{1,r})$ and $\tau_A \mathcal{C}(M_{s,m}) = \mathcal{C}(M_{s-1,m-1})$ are disjoint. Hence, the cones $\mathcal{C}(M_{1,r})$ and $\mathcal{C}(M_{s-1,m})$ are also disjoint, and we get the inequality $r + (m - s + 2) \leq m$, which yields $r + 2 \leq s$. Together with the inequalities $p \leq r$ and $m - 1 - p \leq m - s + 1$ proved earlier, this yields

- $p = r$, and
- $m - 1 - p = m - s + 1$.

Consequently, the partial tilting module $T^{(1)}$ is a tilting module in $\operatorname{mod} H^{(1)}$ and the partial tilting module $T^{(2)}$ is a tilting module in $\operatorname{mod} H^{(2)}$.

By the inductive hypothesis, the tilted algebra

$$A^{(1)} = \operatorname{End} T^{(1)}_{H^{(1)}}$$

is isomorphic with the branch algebra $K\mathcal{L}^{(1)} = KL^{(1)}/I^{(1)}$ of a branch $\mathcal{L}^{(1)} = (L^{(1)}, I^{(1)})$ of capacity r with the germ vertex $0^{(1)}$ corresponding to the direct summand $M_{1,r} = P(r)$ of $T^{(1)}$, and the tilted algebra

$$A^{(2)} = \operatorname{End} T^{(2)}_{H^{(2)}}$$

is isomorphic with the branch algebra $K\mathcal{L}^{(2)} = KL^{(2)}/I^{(2)}$ of a branch $\mathcal{L}^{(2)} = (L^{(2)}, I^{(2)})$ of capacity $m - 1 - r = m - s + 1$ with the germ vertex $0^{(2)}$ corresponding to the direct summand $M_{s,m} \cong M_{r+2,m}$ of $T^{(2)}$.

Then the standardness of the cone $\mathcal{C}(M_{s,m})$ yields

$$\operatorname{Hom}_H(T^{(1)}, T^{(2)}) = 0, \quad \operatorname{Hom}_H(T^{(2)}, T^{(1)}) = 0,$$

and there exist non-zero homomorphisms

$$M_{1,r} \xrightarrow{\ f\ } M_{1,m} \xrightarrow{\ g\ } M_{s,m}$$

of H-modules, where f is a monomorphism and g is an epimorphism. Hence, by applying again the equivalence $(*)$, we conclude that the algebra A is isomorphic to the branch algebra $K\mathcal{L} = KL/I$ of the branch $\mathcal{L} = (L, I)$ of capacity m obtained from the branches

$$\mathcal{L}^{(1)} = (L^{(1)}, I^{(1)}) \text{ and } \mathcal{L}^{(2)} = (L^{(2)}, I^{(2)})$$

by adding two arrows

$$0^{(2)} \xrightarrow{\alpha} 0 \xrightarrow{\beta} 0^{(1)}$$

and the zero relation $\alpha\beta = 0$. This finishes the proof of the statement (i).

For the proof of (ii), we note that the algebra $A = \operatorname{End} T_H$ is representation-finite, because T_H is a splitting tilting H-module and the algebra $H = K\Delta(\mathbb{A}_m)$ is representation-finite, see (VI.5.7) and (VII.5.10). Then, by (VIII.3.5), the Auslander–Reiten translation quiver $\Gamma(\operatorname{mod} A)$ of A coincides with the connecting component $\mathcal{C}_T$ determined by the tilting H-module T. It follows that $\Gamma(\operatorname{mod} A) = \mathcal{C}_T$ admits a section Σ of the form

$$N_1 \longrightarrow N_2 \longrightarrow N_3 \longrightarrow \ldots \longrightarrow N_{m-1} \longrightarrow N_m,$$

given by the images $\operatorname{Hom}_H(T, M_{j,m}) = \operatorname{Hom}_H(T, I(j)_H)$ of the indecomposable injective H-modules $M_{j,m} = I(j)_H$ via the functor $\operatorname{Hom}_H(T, -)$, for $j \in \{1, 2, \ldots, m\}$.

The A-module $N_1 = \operatorname{Hom}_H(T, M_{1,m})$ is isomorphic to the indecomposable projective A-module $P(0)_A$ at the germ vertex 0 of the branch $\mathcal{L}$, because the module $M_{1,m} = P(m)_H = T_0$ is the direct summand of T corresponding to 0.

By the connecting lemma (VI.4.9), the A-module

$$N_m = \operatorname{Hom}_H(T, M_{m,m}) = \operatorname{Hom}_H(T, I(m)_H)$$

is injective, because the projective H-module $M_{1,m} = P(m)_H$ is the direct summand T_0 of T.

Now we prove that the injective A-module $N_m = \operatorname{Hom}_H(T, I(m)_H)$ is isomorphic to the indecomposable injective A-module $I(0)_A$ of the germ vertex 0 of $\mathcal{L}$, by showing that there is a non-zero homomorphism $S(0)_A \longrightarrow N_m$ from the simple A-module $S(0)_A$ to N_m and, hence, $\operatorname{soc} N_m \cong S(0)_A$. To prove it, it is sufficient to show that $S(0)_A$ lies on the section Σ, because then the composition of irreducible morphisms corresponding to a sectional path starting from $S(0)_A$ is non-zero, see (IX.2.2).

To prove that the simple A-module $S(0)_A$ lies on the section Σ, we look at the shape of the branch $\mathcal{L}$ and we consider two cases.

Case (1). Assume that the germ vertex 0 of $\mathcal{L} = (L, I)$ is a sink of the quiver $L = Q_A$. Then $N_1 = P(0)_A = S(0)_A$ and, hence, $S(0)_A$ lies on Σ.

Case (2). Assume that the germ vertex 0 of $\mathcal{L} = (L, I)$ is not a sink of $L = Q_A$, that is, the quiver L admits an arrow $0 \xrightarrow{\gamma} a$. It is easy to see that
- γ is a unique arrow in L starting at 0,
- the radical rad $P(0)_A$ of $P(0)_A$ is isomorphic to the indecomposable projective A-module $P(a)_A$ at the vertex a, and
- the arrow $N_{m-1} \longrightarrow N_m$ is the unique arrow in $\Gamma(\mathrm{mod}\, A)$ with target N_m, because N_m is the target of the section Σ of type $\Delta(\mathbb{A}_m)$.

Hence we conclude that the Auslander–Reiten translation quiver $\Gamma(\mathrm{mod}\, A)$ of A admits a full translation subquiver of the form

$$
\begin{array}{ccccccc}
P(a)_A = \tau_A N_2 & \longrightarrow & \tau_A N_3 & \longrightarrow & \cdots & \longrightarrow & \tau_A N_t \\
\downarrow & & \downarrow & & & & \downarrow \\
P(0)_A = \ N_1 & \longrightarrow & N_2 & \longrightarrow & \cdots & \longrightarrow & N_{t-1} \xrightarrow{\gamma_t} N_t,
\end{array}
$$

where γ_t is the unique arrow in $\Gamma(\mathrm{mod}\, A)$ with target N_t and $t \geq 3$, because the A-module $P(a)_A$ is not injective. It follows that there are almost split sequences

$$
\begin{array}{ccccccccc}
0 & \longrightarrow & \tau_A N_2 & \longrightarrow & N_1 \oplus \tau_A N_3 & \longrightarrow & N_2 & \longrightarrow & 0 \\
0 & \longrightarrow & \tau_A N_3 & \longrightarrow & N_2 \oplus \tau_A N_4 & \longrightarrow & N_3 & \longrightarrow & 0 \\
& & \vdots & & \vdots & & \vdots & & \\
0 & \longrightarrow & \tau_A N_{t-1} & \longrightarrow & N_{t-2} \oplus \tau_A N_t & \longrightarrow & N_{t-1} & \longrightarrow & 0 \\
0 & \longrightarrow & \tau_A N_t & \longrightarrow & N_{t-1} & \longrightarrow & N_t & \longrightarrow & 0.
\end{array}
$$

Because there exists an isomorphism rad $P(0)_A \cong P(a)_A$ of A-modules, we have $\dim_K P(0)_A = 1 + \dim_K P(a)_A$, that is,

$$
\dim_K N_1 = 1 + \dim_K \tau_A N_2,
$$

and an obvious induction shows that the exact sequences yield the equalities

$$
\begin{aligned}
\dim_K N_2 &= 1 + \dim_K \tau_A N_3, \\
\dim_K N_3 &= 1 + \dim_K \tau_A N_4,
\end{aligned}
$$

$$
\vdots \qquad\qquad \vdots
$$

$$
\begin{aligned}
\dim_K N_{t-1} &= 1 + \dim_K \tau_A N_t, \\
\dim_K N_t &= \dim_K N_{t-1} - \dim_K \tau_A N_t = 1.
\end{aligned}
$$

Consequently, the A-module N_t is simple. The existence of a sectional path from $P(0)_A = N_1$ to N_t yields the existence of a surjection $P(0)_A \longrightarrow N_t$.

It follows that $N_t \cong \operatorname{top} P(0)_A = S(0)_A$. This shows that the module $S(0)_A$ lies on Σ and $N_t \cong I(0)_A$. Note also that there are quiver isomorphisms $\Sigma \cong \Delta(\mathbb{A}_m) \cong \Delta(\mathbb{A}_m)^{\mathrm{op}}$.

To finish the proof of (ii), it remains to show that Σ is the unique section of $\Gamma(\operatorname{mod} A)$ that is isomorphic to the equioriented quiver $\Delta(\mathbb{A}_m)$. To prove it, we assume that Σ' is a section of $\Gamma(\operatorname{mod} A)$ isomorphic to the quiver $\Delta(\mathbb{A}_m)$. Because Σ' intersects every τ_A-orbit of $\Gamma(\operatorname{mod} A)$ then there exist integers $p \geq 0$ and $q \geq 0$ such that the modules $\tau_A^{-p} P(0)$ and $\tau_A^q I(0)$ lie on Σ'. Because Σ' is a sectional path from $\tau_A^{-p} P(0)$ to $\tau_A^q I(0)$ then $p = q = 0$, that is, $\Sigma' = \Sigma$. This finishes the proof of the statement (ii).

(iii) The statement immediately follows from the proof of (ii).

(iv) By (VIII.3.5), the set of the indecomposable A-modules in the torsion-free part $\mathcal{Y}(T)$ of the torsion pair $(\mathcal{X}(T), \mathcal{Y}(T))$ of $\operatorname{mod} A$ determined by T coincides with the set $\mathcal{F}(\Sigma_A)$, and the set of the indecomposable A-modules in the torsion part $\mathcal{X}(T)$ coincides with the set $\mathcal{T}(\Sigma_A)$. Hence (v) follows. $\qquad\square$

2.4. Corollary. *Assume that $m \geq 1$ is an integer, A is a tilted algebra of the equioriented type $\Delta(\mathbb{A}_m)$, and $H_m = K\Delta(\mathbb{A}_m)$ is the path algebra of $\Delta(\mathbb{A}_m)$.*

(i) *The category $\operatorname{mod} H_m$ admits precisely one tilting module T_{H_m} such that $\operatorname{End} T_{H_m} \cong A$.*

(ii) *The category $\operatorname{mod} A$ admits precisely one multiplicity-free tilting module T_A such that $\operatorname{End} T_A \cong K\Delta(\mathbb{A}_m)$.*

Proof. Fix an integer $m \geq 1$ and assume that A is a tilted algebra of the equioriented type $\Delta(\mathbb{A}_m)$.

(i) Let $H = H_m = K\Delta(\mathbb{A}_m)$ and let T_H be a tilting H-module such that $\operatorname{End} T_H \cong A$. It follows from (2.3) that $\Gamma(\operatorname{mod} H)$ admits a unique section Σ isomorphic to the equioriented quiver $\Delta(\mathbb{A}_m)$ (2.2). The section Σ is given by the images

$$\operatorname{Hom}_H(T_H, I(1)_H), \ldots, \operatorname{Hom}_H(T_H, I(m)_H)$$

of the indecomposable injective H-modules $I(1), \ldots, I(m)$ via the functor $\operatorname{Hom}_H(T_H, -) : \operatorname{mod} H \longrightarrow \operatorname{mod} A$.

Let R be the direct sum of all indecomposable H-modules lying on the section Σ. It follows from the criterion (VIII.5.6) that R_A is a tilting right A-module, $H \cong \operatorname{End} R_A$, $R^* = D({}_H R)$ is a tilting H-module, and $\Gamma(\operatorname{mod} A)$ is the connecting component $\mathcal{C}_{R^*}$ determined by R^*.

On the other hand, there is an isomorphism

$$D(H)_H \cong I(1)_H \oplus \ldots \oplus I(m)_H$$

of H-modules, and hence, there is an isomorphism

$$R \cong \mathrm{Hom}_H(T_H, D(H))$$

of A-modules. It follows from (VI.3.3) that $_AT$ is a tilting left A-module and there is an isomorphism $D(_AT) \cong \mathrm{Hom}_H(T_H, D(H))$ of right A-modules. Hence, the A-modules R and $D(_AT)$ are isomorphic and, consequently, there are isomorphisms

$$R^* \cong D(_HR) \cong DD(T_H) \cong T_H$$

of H-modules. Hence, the statement (i) follows, because we know from (2.3) that Σ is the unique section of $\Gamma(\mathrm{mod}\,A)$ isomorphic to the equioriented quiver $\Delta(\mathbb{A}_m) : \overset{1}{\circ} \longleftarrow \overset{2}{\circ} \longleftarrow \ldots \longleftarrow \overset{m}{\circ}$.

(ii) Note that the tilting right A-module R_A constructed above is the unique multiplicity-free tilting A-module such that $\mathrm{End}\,R_A \cong K\Delta(\mathbb{A}_m)$. Hence the statement (ii) follows. $\square$

The preceding corollary allows us to divide naturally the indecomposable modules over a tilted algebra of the equioriented type $\Delta(\mathbb{A}_m)$ into two parts as follows.

2.5. Definition. Assume that $m \geq 1$ is an integer, A is a tilted algebra of the equioriented type $\Delta(\mathbb{A}_m) : \overset{1}{\circ} \longleftarrow \overset{2}{\circ} \longleftarrow \ldots \longleftarrow \overset{m}{\circ}$. Let Σ_A be a unique section of $\Gamma(\mathrm{mod}\,A)$ isomorphic to the quiver $\Delta(\mathbb{A}_m)$.

  (i) The **torsion-free part** of $\Gamma(\mathrm{mod}\,A)$ is defined to be the set $\mathcal{F}_A = \mathcal{F}(\Sigma_A)$ of all indecomposable proper successors of Σ_A in $\Gamma(\mathrm{mod}\,A)$.
 (ii) The **torsion part** of $\Gamma(\mathrm{mod}\,A)$ is defined to be the set $\mathcal{T}_A = \mathcal{T}(\Sigma_A)$ of all indecomposable predecessors of Σ_A in $\Gamma(\mathrm{mod}\,A)$.
(iii) The modules in $\mathrm{add}\,\mathcal{T}(\Sigma_A)$ are called **torsion** A-modules, and the modules in $\mathrm{add}\,\mathcal{F}(\Sigma_A)$ are called **torsion-free** A-modules

We recall from (2.3) that the equality

$$(\mathcal{X}(T_A), \mathcal{Y}(T_A)) = (\mathrm{add}\,\mathcal{T}(\Sigma_A), \mathrm{add}\,\mathcal{F}(\Sigma_A)),$$

holds, where $(\mathcal{X}(T_A), \mathcal{Y}(T_A))$ is the torsion pair in $\mathrm{mod}\,A$ determined by the unique tilting module T_A such that $\mathrm{End}\,T_A \cong K\Delta(\mathbb{A}_m)$, see also (VI.3.6).

Now we illustrate the statements of the preceding theorem with $m = 8$ with an example.

2.6. Example. Let $H = K\Delta(\mathbb{A}_8)$ be the path algebra of the quiver

$$\Delta(\mathbb{A}_8): \quad \overset{1}{\circ} \longleftarrow \overset{2}{\circ} \longleftarrow \overset{3}{\circ} \longleftarrow \overset{4}{\circ} \longleftarrow \overset{5}{\circ} \longleftarrow \overset{6}{\circ} \longleftarrow \overset{7}{\circ} \longleftarrow \overset{8}{\circ}.$$

Then the Auslander–Reiten quiver $\Gamma(\mathrm{mod}\,H)$ of H is the cone $\mathcal{C}(P(8))$ of depth 8

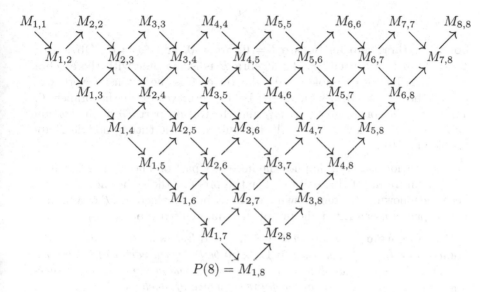

determined by the unique projective-injective H-module

$$P(8) \cong I(1).$$

Recall that $M_{1,j} = P(j)$, $M_{j,8} = I(j)$, and $M_{j,j} = S(j)$, for all $j \in \{1, 2, 3, 4, 5, 6, 7, 8\}$, and consider the following eight indecomposable H-modules

$T_0 = M_{1,8} = P(8)$, $T_1 = M_{1,1} = S(1)$, $T_2 = M_{3,8} = I(3)$, $T_3 = M_{8,8} = S(8)$, $T_4 = M_{3,6}$, $T_5 = M_{3,3} = S(3)$, $T_6 = M_{5,6}$, and $T_7 = M_{5,5} = S(5)$.

One easily checks that the tilting vanishing condition $\mathrm{Hom}_H(T_i, \tau_H T_j) = 0$ holds, for all $i, j \in \{1, 2, 3, 4, 5, 6, 7, 8\}$. It then follows that

$$T_H = T_0 \oplus T_1 \oplus T_2 \oplus T_3 \oplus T_4 \oplus T_5 \oplus T_6 \oplus T_7 \oplus T_8$$

is a tilting H-module, and the associated tilted algebra $H = \mathrm{End}\,T_H$ is given by the bound quiver $\mathcal{L} = (L, I)$, where L is the quiver

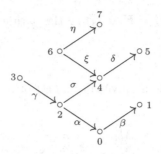

bound by three zero relations $\alpha\beta = 0$, $\gamma\sigma = 0$, and $\xi\delta = 0$. Obviously, $\mathcal{L}$ is a branch of capacity 8, the algebra B is isomorphic with the algebra $A' = A(X, 3, 1)$ of Example (1.10), the vertex 6 is the extension vertex of the algebra $A' = A(X, 3, 1)$, and 0 is the germ vertex of the branch $\mathcal{L}$. Therefore, the torsion-free part $\mathcal{F}_B$ and the torsion part $\mathcal{T}_B$ of the torsion pair $(\mathcal{T}_B, \mathcal{F}_B)$ in $\operatorname{mod} B \cong \operatorname{mod} A'$ determined by the tilting module T are visible in (1.10).

The previous theorem implies the necessity part of Theorem (2.2) which is the main result of this section. For the proof of the sufficiency part, we need the following theorem showing that any branch algebra $K\mathcal{L}$ of a branch $\mathcal{L}$ of capacity n is a tilted algebra of the equioriented type $\Delta(\mathbb{A}_n)$.

2.7. Theorem. *Assume that $A = K\mathcal{L} = KL/I$ is a branch algebra of a branch $\mathcal{L} = (L, I)$ of capacity $n \geq 1$. Let 0 be the germ vertex of the branch $\mathcal{L}$. Then there exist an integer $r \geq 0$, an increasing sequence of integers $m_0 < m_1 < \ldots < m_r = n$, and a finite sequence of algebras*

$$A_0, A_1, \ldots, A_r = A$$

with the following properties.

(i) *$A_0 = K\Delta(\mathbb{A}_{m_0})$ is the path algebra of the equioriented quiver $\Delta(\mathbb{A}_{m_0})$, for some $m_0 \in \{1, 2, \ldots, n\}$.*

(ii) *For each $j \in \{1, 2, \ldots, r\}$, the algebra A_j is a tilted algebra of the equioriented type $\Delta(\mathbb{A}_{m_j})$ and*

$$A_j = K\mathcal{L}^{(j)}$$

is the branch algebra of a branch $\mathcal{L}^{(j)} = (L^{(j)}, I^{(j)})$ of capacity m_j.

(iii) *For each $j \in \{1, 2, \ldots, r\}$, the algebra A_j is a finite line extension*

$$A_j = A_{j-1}(X^{(j)}, s_{X^{(j)}}, t_j)$$

of the algebra A_{j-1} by a torsion-free admissible finite ray A_{j-1}-module $X^{(j)}$, with $t_j \geq 0$ and the ray of length $s_{X^{(j)}}$ in $\Gamma(\operatorname{mod} A_{j-1})$. Moreover,

- $m_j = 1 + m_{j-1} + t_j = m_0 + j + t_1 + \ldots + t_j,$
- $n = m_r = m_0 + r + t_1 + \ldots + t_r,$

and the germ vertex $0^{(j)}$ of the branch $\mathcal{L}^{(j)}$ of A_j is the minimal vertex m_j of the quiver $\Delta(\mathbb{A}_{m_j})$, that is, the germ vertex of the quiver $\Delta(\mathbb{A}_{m_j})$, viewed as a branch.

(iv) For each $j \in \{1, 2, \ldots, r\}$, the Auslander–Reiten quiver $\Gamma(\operatorname{mod} A_j)$ of A_j is a standard translation quiver and $\Gamma(\operatorname{mod} A_j)$ is a finite rectangle insertion of $\Gamma(\operatorname{mod} A_{j-1})$.

Proof. We fix an integer $n \geq 1$. Assume that $A = K\mathcal{L} = KL/I$ is a branch algebra of a branch $\mathcal{L} = (L, I)$ of capacity $n \geq 1$, and let 0 be the germ vertex of the branch $\mathcal{L}$. It follows from the Definition (2.1) that there is a maximal linear full convex subquiver $L^{(0)}$ of L of the form

$$L^{(0)} : \quad \underset{\beta}{\overset{0 \,=\, a_{m_0}}{\circ}} \xrightarrow{} \underset{\beta}{\overset{a_{m_0-1}}{\circ}} \xrightarrow{} \underset{\beta}{\overset{a_{m_0-2}}{\circ}} \xrightarrow{} \ldots \ldots \xrightarrow{} \underset{\beta}{\overset{a_2}{\circ}} \xrightarrow{} \underset{\beta}{\overset{a_1}{\circ}}$$

starting from the germ vertex 0 of L and consisting of the β-arrows of L, where $m_0 \in \{1, \ldots, n\}$. Obviously, the quiver $L^{(0)}$ is isomorphic to the equioriented quiver $\Delta(\mathbb{A}_{m_0})$ (2.2).

Consider the branch $\mathcal{L}^{(0)} = (L^{(0)}, I^{(0)})$ of capacity $m_0 \geq 1$ with the germ vertex $a_{m_0} = 0$, where we set $I^{(0)} = (0)$. Let $A_0 = K\mathcal{L}^{(0)}$ be the branch algebra of $\mathcal{L}^{(0)}$. Note that $A_0 = K\mathcal{L}^{(0)} = KL^{(0)}$ is the path algebra of the quiver $L^{(0)}$, because $I^{(0)}$ is the zero ideal of $KL^{(0)}$.

It is easy to see that

- the Auslander–Reiten quiver $\Gamma(\operatorname{mod} A_0)$ of A_0 is the cone $\mathcal{C}(P(a_{m_0}))$ of depth m_0 determined by the unique indecomposable projective-injective A_0-module $P(a_{m_0}) \cong I(a_1)$, see (2.4),
- the indecomposable injective A_0-modules form the unique section

$$\Sigma_0 : \quad I(a_1) \longrightarrow I(a_2) \longrightarrow \ldots \longrightarrow I(a_{m_0-1}) \longrightarrow I(a_{m_0}) = S(a_{m_0})$$

of $\Gamma(\operatorname{mod} A_0)$ of the equioriented type $\Delta(\mathbb{A}_{m_0})$,
- all indecomposable A_0-modules are A_0-torsion-free, in the sense of (2.5),
- the translation quiver $\Gamma(\operatorname{mod} A_0)$ is standard, that is, the category ind A_0 of indecomposable A_0-modules is equivalent to the mesh category $K\Gamma(\operatorname{mod} A_0) = K\mathcal{C}(P(a_{m_0}))$ of $\Gamma(\operatorname{mod} A_0) = \mathcal{C}(P(a_{m_0}))$, by (2.9),
- the simple A_0-modules $S(a_1), S(a_2), \ldots, S(a_{m_0}) = I(a_{m_0})$ coincide with the finite ray modules in $\Gamma(\operatorname{mod} A_0)$, and
- each of the simple A_0-modules $S(a_1), S(a_2), \ldots, S(a_{m_0}) = I(a_{m_0})$ is an admissible finite ray module, in the sense of (2.5).

Now we proceed to the proof of the theorem by induction on $n - m_0$. If $n = m_0$, we are done, because

$$\mathcal{L} = \mathcal{L}_0 \text{ and } A = K\mathcal{L} = K\mathcal{L}^{(0)} = KL^{(0)} = A_0.$$

Assume that $n - m_0 \geq 1$. It follows that $L \neq L^{(0)}$ and the quiver L contains a maximal full convex subquiver Δ of the form

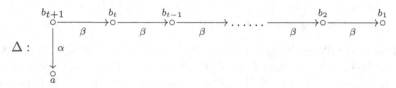

such that

- $t \geq 0$,
- the vertex b_1 is a sink of the quiver L,
- the vertex b_{t+1} is a source of L, and
- each of the vertices $b_t, b_{t-1}, \dots, b_3, b_2$ has precisely two neighbours in L, that is, the arrows of Δ are the only arrows of L starting or ending at the vertices $b_{t+1}, b_t, \dots, b_2, b_1$.

Let $\mathcal{L}^{\vee} = (L^{\vee}, I^{\vee})$ be the branch obtained from $\mathcal{L}$ by removing the subquiver

$$
\underset{b_{t+1}}{\circ} \xrightarrow{\ \ \beta\ \ } \underset{b_t}{\circ} \xrightarrow{\ \ \beta\ \ } \underset{b_{t-1}}{\circ} \xrightarrow{\ \ \beta\ \ } \cdots \cdots \xrightarrow{\ \ \beta\ \ } \underset{b_2}{\circ} \xrightarrow{\ \ \beta\ \ } \underset{b_1}{\circ}.
$$

Note that germ vertex 0 of the branch $\mathcal{L}$ remains the germ vertex of the branch $\mathcal{L}^{\vee}$, the branch $\mathcal{L}$ contains the branch $\mathcal{L}^{(0)}$, and the capacity of $\mathcal{L}^{\vee}$ equals $n - t - 1$. It follows that the inductive hypothesis applies to the associated branch algebra

$$A^{\vee} = K\mathcal{L}^{\vee} = KL^{\vee}/I^{\vee},$$

because the assumption $t + 1 > 0$ yields $n - t - 1 - m_0 < n - m_0$.

By the inductive hypothesis, there exist an integer $r_1 \geq 0$, an increasing finite sequence of integers $m_0 < m_1 < \dots < m_{r_1} = n - t - 1$, and a finite sequence of algebras

$$A_0 = K\Delta(\mathbb{A}_{m_0}), \ A_1, \dots, A_{r_1} = A^{\vee}$$

such that the conditions (ii)–(iv) of the theorem are satisfied, with r and r_1 interchanged. In particular, for each $j \in \{1, 2, \dots, r_1\}$, the algebra A_j is a finite line extension

$$A_j = A_{j-1}(X^{(j)}, s_{X^{(j)}}, t_j)$$

of the algebra A_{j-1} by a torsion-free admissible finite ray A_{j-1}-module $X^{(j)}$, with $t_j \geq 0$ and the ray of length $s_{X^{(j)}}$ in $\Gamma(\mathrm{mod}\, A_{j-1})$. Moreover, $m_j = 1 + m_{j-1} + t_j = m_0 + j + t_1 + \ldots + t_j$.

Hence, the ordinary quiver $Q_{A_j} = L^{(j)}$ of A_j is obtained from the quiver $Q_{A_{j-1}} = L^{(j-1)}$ of A_{j-1} by adding the extension vertex $0^{(j)}$ of the one-point extension algebra

$$A_{j-1}(X^{(j)}, s_{X^{(j)}}, t_j) = (A_{j-1} \times H^{(j)})[X^{(j)} \oplus P(t_j^{(j)})_{H^{(j)}}]$$

as follows

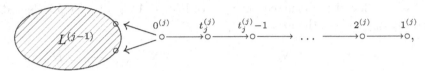

where $t_j^{(j)} = t_j$, $H^{(j)}$ is the path algebra of the linear subquiver

$$\underset{t_j^{(j)}}{\circ} \longrightarrow \underset{t_j^{(j)}-1}{\circ} \longrightarrow \cdots \longrightarrow \underset{2^{(j)}}{\circ} \longrightarrow \underset{1^{(j)}}{\circ},$$

of $L^{(j)}$, $P(t_j^{(j)})_{H^{(j)}}$ is the unique projective-injective $H^{(j)}$-module, and we set $H^{(j)} = 0$, and $P(t_j^{(j)})_{H^{(j)}} = 0$, if $t_j = 0$.

Further, for each $j \in \{1, 2, \ldots, r_1\}$, we have

- the Auslander–Reiten quiver $\Gamma(\mathrm{mod}\, A_j)$ of A_j is a standard translation quiver and $\Gamma(\mathrm{mod}\, A_j)$ is a finite rectangle insertion that creates $1 + t_j$ additional finite rays in $\Gamma(\mathrm{mod}\, A_{j-1})$, with injective targets, as described in (2.8),
- every torsion-free finite ray module in $\Gamma(\mathrm{mod}\, A_j)$, except the simple $H^{(j)}$-module $X^{(j)}$, is either a finite ray module in $\Gamma(\mathrm{mod}\, A_{j-1})$ or is the module $\overline{X}^{(j)} = (K, X^{(j)}, 1)$,
- the unique section Σ_j of $\Gamma(\mathrm{mod}\, A_j)$ of type $\Delta(\mathbb{A}_{m_j})$ is obtained from the unique section Σ_{j-1} of $\Gamma(\mathrm{mod}\, A_{j-1})$ of type $\Delta(\mathbb{A}_{m_{j-1}})$ by inserting the quiver $\Delta(\mathbb{A}_{t_j+1})$, and
- the source of the section Σ_j is the indecomposable projective A_j-module $P(0^{(j)})_{A_j}$ at the germ vertex $0^{(j)}$ of $\mathcal{L}^{(j)}$, and the target of Σ_j is the indecomposable injective A_j-module $I(0^{(j)})_{A_j}$ at the germ vertex $0^{(j)}$, by (2.3).

We set $r = 1 + r_1$. Then $r_1 = r - 1$, $A_{r-1} = A_{r_1}$ and, by the inductive hypothesis, we may assume that every torsion-free finite ray module in

$\Gamma(\text{mod } A_{r-1}) = \Gamma(\text{mod } A_{r_1})$ is admissible and is one of the following two forms:

1° The simple module over one of the hereditary algebras

$$H^{(0)} = A_0, H^{(1)}, \dots, H^{(r-1)},$$

except the simple modules in the family $X^{(1)}, X^{(2)}, \dots, X^{(r-1)}$.

2° For each $j \in \{1, \dots, r-1\}$, the module $\overline{X}^{(j)} = (K, X^{(j)}, 1)$ over the finite line extension algebra $A_j = A_{j-1}(X^{(j)}, s_{X^{(j)}}, t_j)$, provided $\overline{X}^{(j)}$ is not taken for the extension ray module $X^{(s)}$, for some $s > j$.

Assume that the unique section Σ_{r-1} of $\Gamma(\text{mod } A_{r-1})$ of the equioriented type $\Delta(\mathbb{A}_{m_{r-1}})$ is of the form

$$N_1 \longrightarrow N_2 \longrightarrow N_3 \longrightarrow \dots \longrightarrow N_{m_{r-1}-1} \longrightarrow N_{m_{r-1}}.$$

Then the class of torsion-free modules of $\Gamma(\text{mod } A_{r-1})$ coincides with the class of modules of the form $\tau_{A_{r-1}}^s N_j$, with $s \geq 0$ and $j \in \{1, \dots, m_{r-1}\}$.

Let $H^{(r)}$ be the path algebra of the subquiver

$$\underset{\beta}{\overset{b_t}{\circ}} \longrightarrow \underset{\beta}{\overset{b_{t-1}}{\circ}} \longrightarrow \dots \dots \underset{\beta}{\overset{b_2}{\circ}} \longrightarrow \underset{\beta}{\overset{b_1}{\circ}}.$$

of the quiver $L^\vee$ defining the branch $\mathcal{L}$.

We set $t_r = t$ and we denote by $P(b_{t_r})$ the unique indecomposable projective-injective $H^{(r)}$-module, if $t_r > 0$; otherwise, we set $H^{(r)} = 0$ and $P(b_{t_r}) = 0$.

Then the radical rad $P(b_{t+1})$ of the indecomposable projective A-module $P(b_{t+1})$ at the vertex b_{t+1} of the quiver L has a decomposition

$$\text{rad } P(b_{t+1}) = X^{(r)} \oplus P(b_{t_r}),$$

where $X^{(r)}$ is a uniserial A_{r-1}-module such that $X^{(r)}/\text{rad } X^{(r)}$ is isomorphic to the simple A_{r-1}-module at the vertex a of $L^\vee$. Observe that a is a source of the branch $\mathcal{L}^{(r-1)} = \mathcal{L}^\vee = (L^\vee, I^\vee)$ defining the algebra $A_{r-1} = A^\vee$.

By the preceding observations, we have

- the module $X^{(r)}$ is a simple module over some of the hereditary algebras $H^{(0)} = A_0, H^{(1)}, \dots, H^{(r-1)}$, or
- $X^{(r)}$ is a module of the form $\overline{X}^{(j)} = (K, X^{(j)}, 1)$ over the finite line extension algebra $A_j = A_{j-1}(X^{(j)}, s_{X^{(j)}}, t_j)$ of the sequence $A_0, A_1, \dots, A_{r-1} = A^\vee$ of line extensions leading from the hereditary algebra $A_0 = K\Delta(\mathbb{A}_{m_0})$ to the algebra $A_{r-1} = A^\vee$.

Because the vertex a is a source of the quiver $L^{(r-1)} = L^{\vee}$ and $X^{(r)}$ is not a module $\overline{X}^{(i)}$ taken for the extension ray module of a finite line extension

$$A_i = A_{i-1}(X^{(i)}, s_{X^{(i)}}, t_i),$$

for some $i \in \{1, \ldots, r-1\}$ then, by the inductive hypothesis, $X^{(r)}$ is a torsion-free admissible finite ray module in $\Gamma(\operatorname{mod} A_{r-1})$, with a ray of length $s_{X^{(r)}}$. It follows that $A = A_r$ is a finite line extension of A_{r-1} of the form

$$A = A_r = A_{r-1}(X^{(r)}, s_{X^{(r)}}, t_r) = (A_{r-1} \times H^{(r)})[X^{(r)} \oplus P(b_{t_r})].$$

Let

$$X^{(r)} = X_0 \longrightarrow X_1 \longrightarrow X_2 \longrightarrow \cdots \longrightarrow X_{s-1} \longrightarrow X_s, \quad (*)$$

with $s = s_{X^{(r)}}$, be the unique finite ray in $\Gamma(\operatorname{mod} A_{r-1})$ starting at $X^{(r)}$. Because $X^{(r)}$ is a torsion-free module in $\Gamma(\operatorname{mod} A_{r-1})$, in the sense of (2.5), then $X^{(r)}$ is a predecessor of the section Σ_{r-1} of $\Gamma(\operatorname{mod} A_{r-1})$. It follows that the section Σ_r of $\Gamma(\operatorname{mod} A_r)$ contains exactly one module of the ray $(*)$ starting at $X^{(r)}$; say a module

$$N_p = X_q.$$

By (2.9), the Auslander–Reiten quiver $\Gamma(\operatorname{mod} A) = \Gamma(\operatorname{mod} A_r)$ of $A = A_r$ is obtained from $\Gamma(\operatorname{mod} A_{r-1})$ by a finite rectangle insertion procedure described in (2.8). Hence we conclude that $\Gamma(\operatorname{mod} A)$ admits a section Σ_r of the form

$$N_1 \to \cdots \to N_p \to Z_{q_1} \to \cdots \to Z_{q_t} \to \overline{X}_q \to N_{p+1} \to \cdots \to N_{m_{r-1}}.$$

obtained from the section Σ_{r-1} by inserting $t + 1 = t_r + 1$ modules.

Observe that $n = m_r = 1 + m_{r_1} + t_r$ and

$$\Sigma_r \cong \Delta(\mathbb{A}_{m_r}) = \Delta(\mathbb{A}_n),$$

because $\Sigma_{r-1} \cong \Delta(\mathbb{A}_{m_{r-1}})$, the section Σ_{r-1} is of length m_{r-1} and n is the capacity of the branch

$$\mathcal{L} = \mathcal{L}^{(r)} = (L^{(r)}, I^{(r)}).$$

Moreover, by (2.9), the category $\operatorname{ind} A = \operatorname{ind} A_r$ of indecomposable A-modules is equivalent to the mesh category $K\Gamma(\operatorname{mod} A)$ of $\Gamma(\operatorname{mod} A)$. We also note that

   (a) the simple $H^{(r)}$-modules, together with the module

$$\overline{X}^{(r)} = (K, X^{(r)}, 1),$$

form a complete set of the isomorphism classes of indecomposable finite ray A-modules, that are not A_{r-1}-modules,

   (b) each of the modules listed in (a) is an admissible finite ray module,

   (c) $\Gamma(\operatorname{mod} A)$ admits exactly $n = m_r$ finite rays starting from torsion-free finite ray modules and containing all modules of $\Gamma(\operatorname{mod} A)$.

The statement (c) follows from the iterated finite line extension procedure leading from the hereditary algebra A_0 to the algebra $A_r = A$.

Now we show that the section Σ_r of type $\Delta(\mathbb{A}_n)$ is a faithful section of the Auslander–Reiten quiver $\Gamma(\operatorname{mod} A)$ of $A = A_r$. We recall that

- $n = \operatorname{rk} K_0(A)$ is the rank of the Grothendieck group $K_0(A)$ of A,
- the number of pairwise non-isomorphic indecomposable modules lying on Σ_r equals n,
- every indecomposable injective A-module is a successor of Σ_r in $\Gamma(\operatorname{mod} A)$, and
- every indecomposable projective A-module is a predecessor of Σ_r in the quiver $\Gamma(\operatorname{mod} A)$.

Let T be the direct sum of all the indecomposable modules lying on the section Σ_r. We show that T is a faithful A-module, by showing that the module A_A is cogenerated by T_A. Let P be an indecomposable projective A-module and let $u : P \longrightarrow E(P)$ be an injective envelope of P in $\operatorname{mod} A$. By the preceding observations, u admits a factorisation

$$P \xrightarrow{\ \ u\ \ } E(P)$$
$$g \searrow \qquad \nearrow f$$
$$M$$

in $\operatorname{mod} A$, where M is a module in $\operatorname{add} T_A$. It follows that $g : P \longrightarrow M$ is a monomorphism, because the map u is injective. This shows that the module A_A is cogenerated by T_A. Hence, by (VI.2.2), T_A is a faithful A-module and, equivalently, Σ_r is a faithful section of the quiver $\Gamma(\operatorname{mod} A)$.

   Because T is a direct sum of indecomposable modules lying on the section Σ_r of $\Gamma(\operatorname{mod} A)$, then $\Gamma(\operatorname{mod} A)$ is directed and, hence, $\operatorname{Hom}_A(T_A, \tau_A T_A) = 0$. Hence, by the criterion (VIII.5.6) and the observations made above, we conclude that

- T_A is a tilting A-module,
- the algebra $B = \operatorname{End} T_A$ is isomorphic to the path algebra $K\Delta(\mathbb{A}_n)$ of the equioriented quiver $\Delta(\mathbb{A}_n)$,

- $T^* = D(T_A)$ is a tilting B-module such that $A \cong \operatorname{End} T_B^*$, and
- the quiver $\Gamma(\operatorname{mod} A)$ is the connecting component $\mathcal{C}_{T^*}$ determined by the module T^*.

In particular, A is a tilted algebra of the equioriented type $\Delta(\mathbb{A}_n)$. This finishes the proof of the theorem. $\qquad \square$

As a consequence of the preceding proof, we get an inductive description of the Auslander–Reiten quiver $\Gamma(\operatorname{mod} A)$ of any tilted algebra A of the equioriented type $\Delta(\mathbb{A}_n)$: $\overset{1}{\circ}\longleftarrow\overset{2}{\circ}\longleftarrow \ldots \longleftarrow\overset{n}{\circ}$.

2.8. Corollary. *Assume that A is a tilted algebra of the equioriented type $\Delta(\mathbb{A}_n)$, with $n \geq 1$. There exists a hereditary algebra*

$$H = K\Delta(\mathbb{A}_m), \quad \text{with} \quad m \leq n,$$

such that the Auslander–Reiten translation quiver $\Gamma(\operatorname{mod} A)$ of A is obtained from the quiver $\Gamma(\operatorname{mod} H)$ of H by a sequence of finite rectangle insertions.

Proof. By (2.3), any tilted algebra A of the equioriented type $\Delta(\mathbb{A}_n)$ is isomorphic to a branch algebra $K\mathcal{L}$. Then Theorem (2.7) and its proof apply. $\qquad \square$

2.9. Proof of Theorem 2.2. The implication (a)$\Rightarrow$(d) is a consequence of Theorem (2.3)(i), and the converse implication (d)$\Rightarrow$(a) follows from Theorem (2.7). The implications (a)$\Rightarrow$(b) and (a)$\Rightarrow$(c) follow from Theorem (2.3) and Corollary (2.4).

To prove the implication (b)$\Rightarrow$(a), assume that the Auslander–Reiten translation quiver $\Gamma(\operatorname{mod} A)$ of A is finite and admits precisely one section Σ that is isomorphic to the equioriented linear quiver $\Delta(\mathbb{A}_n)$. Let T_A be the direct sum of all modules in the section Σ. By applying the same type of arguments as in the proof of Theorem (2.7), we show that right A-module T is faithful. Then, by (VIII.5.6), T_A is a tilting A-module, the tilted algebra $H = \operatorname{End} T_A$ is isomorphic to the path algebra $K\Delta(\mathbb{A}_n)$ of $\Delta(\mathbb{A}_n)$, and there is a tilting H-module T'_H such that

$$A \cong \operatorname{End} T'_H.$$

Then A is a tilted algebra of the equioriented type $\Delta(\mathbb{A}_n)$ and (a) follows.

To prove the implication (c)$\Rightarrow$(a), assume that $\operatorname{mod} A$ admits a multiplicity-free tilting module T_A such that the algebra

$$H = \operatorname{End} T_A$$

is isomorphic to the path algebra $K\Delta(\mathbb{A}_n)$ of $\Delta(\mathbb{A}_n)$. Then, by (VI.3.3), $_H T$ is a left tilting H-module in $\operatorname{mod} H$ and

$$A \cong (\operatorname{End}_H T)^{\mathrm{op}} \cong \operatorname{End}(T_H^*).$$

This shows that A is a tilted algebra of type $\Delta(\mathbb{A}_n)$ and (a) follows.

Because the implication (d)$\Rightarrow$(e) is a consequence of Theorem (2.7)(e), and the converse implication (e)$\Rightarrow$(d) is obvious, then the proof of Theorem (2.2) is complete. $\square$

XVI.3. Exercises

1. Describe all branches of capacity 15.

2. Let A be an algebra given by the quiver

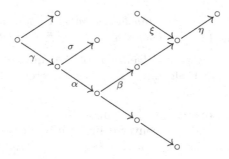

bound by three zero relations $\alpha\beta = 0$, $\gamma\sigma = 0$, and $\xi\eta = 0$. Find a tilting module over the path algebra $H = K\Delta(\mathbb{A}_{11})$ such that $A \cong \operatorname{End} T_H$, where $\Delta(\mathbb{A}_{11})$ is the equioriented quiver (2.2) with $n = 11$.

3. Assume that A is an algebra given by the quiver

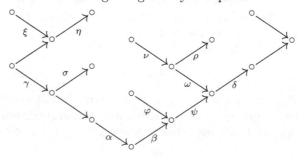

bound by six zero relations $\alpha\beta = 0$, $\gamma\sigma = 0$, $\xi\eta = 0$, $\varphi\psi = 0$, $\nu\rho = 0$, and $\omega\delta = 0$. Find a tilting module over the path algebra $H = K\Delta(\mathbb{A}_{17})$ such that

$$A \cong \operatorname{End} T_H,$$

where $\Delta(\mathbb{A}_{17})$ is the equioriented quiver (1.2) with $n = 17$.

4. Let $n \geq 2$ be an integer and let $H_n = K\Delta(\mathbb{A}_n)$ be the path algebra of the equioriented quiver $\Delta(\mathbb{A}_n)$ (2.2). Show that the Auslander–Reiten quiver $\Gamma(\mathrm{mod}\,H_n)$ of H_n admits at least n different sections.

5. Let A be a tilted algebra of the equioriented type $\Delta(\mathbb{A}_n)$ (2.2). Prove that the Auslander–Reiten quiver $\Gamma(\mathrm{mod}\,A)$ of A has precisely one section if and only if there is a pair of indecomposable projective-injective A-modules $P(i_1)$ and $I(i_2)$ such that

$$P(i_1)/\mathrm{rad}\,P(i_1) \cong \mathrm{soc}\,I(i_2).$$

6. Let $A = KQ$ be the path algebra of the quiver

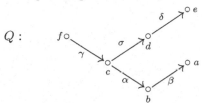

$$Q:$$

bound by two zero relations $\alpha\beta = 0$ and $\gamma\sigma = 0$. We recall from (1.10) that A is a tilted algebra of the equioriented type $\Delta(\mathbb{A}_6)$, see (2.2).

Show that the Auslander–Reiten quiver $\Gamma(\mathrm{mod}\,A)$ of A has precisely one section. **Hint:** Apply (1.10) or Exercise 5.

7. Let $A' = KQ'$ be the path algebra of the quiver

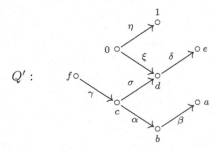

$$Q':$$

bound by three zero relations $\alpha\beta = 0$, $\gamma\sigma = 0$, and $\xi\delta = 0$. We recall from (1.10) that A' is a tilted algebra of the equioriented type $\Delta(\mathbb{A}_8)$, see (2.2).

Show that the Auslander–Reiten quiver $\Gamma(\mathrm{mod}\,A')$ of A' has precisely one section. **Hint:** Apply Exercise 5.

8. Let $B = KQ$ be the path algebra of the quiver

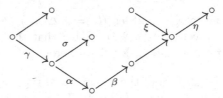

bound by three zero relations $\alpha\beta = 0$, $\gamma\sigma = 0$, and $\xi\eta = 0$. Prove that the Auslander–Reiten quiver $\Gamma(\bmod A')$ of A' has precisely two sections.

Chapter XVII

Tilted algebras of Euclidean type

We have already described in Chapter XII the structure and the combinatorial invariants of the module categories mod A of concealed algebras A of Euclidean type, while in Chapter XIV we give a complete classification of these algebras A by means of quivers and relations.

The main objective of this chapter is to describe the structure and the combinatorial invariants of the module category mod B of an arbitrary representation-infinite tilted algebra B of Euclidean type. Moreover, we show that these algebras B are domestic branch extensions or domestic branch coextensions of concealed algebras A of Euclidean type. In Section 1, we study the distribution of indecomposable direct summands of a splitting tilting module among the hereditary standard stable tubes of the Auslander–Reiten quiver $\Gamma(\text{mod } B)$ of an arbitrary algebra B, while in Section 2 we show how the structure of the hereditary standard stable tubes in $\Gamma(\text{mod } B)$ is changed under the related tilting process of B.

The main result of Section 3 asserts that every representation-infinite tilted algebra B of Euclidean type is a domestic tubular (branch) extension or a domestic tubular (branch) coextension of a concealed algebra A of Euclidean type. The inverse implication is proved in Section 4 by showing that every domestic tubular (branch) extension and every domestic tubular (branch) coextension of a concealed algebra of Euclidean type is a representation-infinite tilted algebra of Euclidean type.

In Section 5, we present a characterisation of representation-infinite tilted algebras B of Euclidean type, and we exhibit their module categories mod B. We also show that the number of the isomorphism classes of basic tilted algebras B of any fixed Euclidean type is finite.

In the final Section 6, we show that the module category mod B of an arbitrary tilted algebra B of Euclidean type is link controlled by the Euler quadratic form

$$q_B : K_0(B) \longrightarrow \mathbb{Z}$$

of the algebra B, where $K_0(B) \cong \mathbb{Z}^n$ is the Grothendieck group of B and $n \geq 1$ is the number of pairwise non-isomorphic simple B-modules.

XVII.1. Stone cones in hereditary standard stable tubes

Throughout this section, we assume that A is an arbitrary algebra, $n = \operatorname{rk} K_0(A)$ is the rank of the Grothendieck group of A, and

$$\mathcal{T} = \{\mathcal{T}_\lambda\}_{\lambda \in \Lambda} \tag{1.1}$$

is a Λ-family of standard stable tubes of the Auslander–Reiten quiver $\Gamma(\operatorname{mod} A)$ of A, where Λ is a non-empty index set. Given a tube $\mathcal{T}_\lambda$, we denote by $r_\lambda \geq 1$ the rank of $\mathcal{T}_\lambda$.

Throughout we need the following definition.

1.2. Definition. Let A be an algebra. An A-module N is defined to be a **stone**, if N is indecomposable and $\operatorname{Ext}_A^1(N, N) = 0$.

In the literature, the stone modules (introduced by Kerner in [346]) are also called **rigid modules**, or **exceptional modules**.

It is clear that every indecomposable direct summand N of a partial tilting A-module T is a stone. We recall from (XI.3.3) that, for any concealed algebra B of Euclidean type, all tubes in $\Gamma(\operatorname{mod} B)$ are hereditary.

The main objective of this section is to study stone cones in a hereditary standard stable tube $\mathcal{T}_\lambda$ and their connection with partial tilting A-modules in the abelian subcategory $\operatorname{add} \mathcal{T}_\lambda$ of $\operatorname{mod} A$.

Throughout, we assume that $\mathcal{T}_\lambda$ is a fixed hereditary standard stable tube of the Λ-family $\mathcal{T}$ (1.1), $r = r_\lambda$ is the rank of $\mathcal{T}_\lambda$, and the modules $E_1, E_2, \ldots, E_r$ lying on the mouth of $\mathcal{T}_\lambda$ form a τ_A-cycle $(E_1, E_2, \ldots, E_r)$, that is, they are ordered in such a way that

$$\tau_A E_2 \cong E_1, \tau_A E_3 \cong E_2, \ldots, \tau_A E_r \cong E_{r-1}, \tau_A E_1 \cong E_r.$$

Because $\mathcal{T}_\lambda$ is a hereditary standard stable tube of rank $r = r_\lambda$ in $\Gamma(\operatorname{mod} A)$, then

- the modules $E_1, E_2, \ldots, E_r$ are pairwise orthogonal bricks in $\operatorname{mod} A$,
- the additive subcategory $\operatorname{add} \mathcal{T}_\lambda$ of $\operatorname{mod} A$ is hereditary; in particular $\operatorname{Ext}_A^2(E_i, E_j) = 0$, for each pair $i, j \in \{1, 2, \ldots, r\}$,
- the category $\operatorname{add} \mathcal{T}_\lambda$ has the extension category form

  $$\operatorname{add} \mathcal{T}_\lambda = \mathcal{EXT}_A(E_1, E_2, \ldots, E_r). \tag{1.3}$$

  Therefore $\operatorname{add} \mathcal{T}_\lambda$ is an abelian category and the modules $E_1, \ldots, E_r$ form a complete set of pairwise non-isomorphic simple objects of the category $\operatorname{add} \mathcal{T}_\lambda$, see (X.2.1) and (X.2.6),
- any indecomposable object X of $\mathcal{T}_\lambda$ is a uniserial object of the category $\operatorname{add} \mathcal{T}_\lambda$ of the form $X \cong E_i[j]$, with $j \geq 1$ and $i \in \{1, 2, \ldots, r\}$,
- the module $E_i = E_i[1]$ is the unique simple subobject of $X \cong E_i[j]$, and j is the length $\ell_\lambda(X)$ of $X \cong E_i[j]$ in $\operatorname{add} \mathcal{T}_\lambda$.

The length $\ell_\lambda(X)$ of an object X in the uniserial abelian category add $\mathcal{T}_\lambda$ is called the $\mathcal{T}_\lambda$-**length** of the A-module X.

Here we freely use the terminology and notation introduced in (X.2.2). In particular, we assume that the indecomposable objects of the category add $\mathcal{T}_\lambda = \mathcal{E}\mathcal{X}\mathcal{T}_A(E_1, E_2, \ldots, E_r)$ are the objects $E_i[j]$, with $j \geq 1$ and $i \in \{1, 2, \ldots, r\}$, constructed in (X.2.2). Here we set

$$E_1[1] = E_1, E_2[1] = E_2, \ldots, E_r[1] = E_r, \quad \text{and} \quad E_{i+kr}[1] = E_i[1],$$

for all $i \in \{1, 2, \ldots, r\}$ and $k \in \mathbb{Z}$.

For any indecomposable A-module $M \cong E_i[j]$ we denote by $\mathcal{C}(M)$ the cone in the tube $\mathcal{T}_\lambda$ determined by M in the sense of (XV.1.4). The cone $\mathcal{C}(M)$ is called a **stone cone** if all A-modules in $\mathcal{C}(M)$ are stones.

In the notation of (XV.1.3), given $j \geq 1$ and $i \in \{1, 2, \ldots, r\}$, the cone $\mathcal{C}(E_i[j])$ determined by the module $E_i[j]$ of the tube $\mathcal{T}_\lambda$ consists of the indecomposable A-modules $E_s[u]$, with $s \in \{i, i+1, \ldots, i+j-1\}$ and $u \leq i + j - s$.

Given a module $M \cong E_i[j]$, we visualise the cone $\mathcal{C}(M) = \mathcal{C}(E_i[j])$ as the diagram

$$
\begin{array}{ccccccccc}
E_i[1] & & E_{i+1}[1] & & E_{i+2}[1] & \cdots & & E_{i+j-2}[1] & E_{i+j-1}[1] \\
& \searrow^{u_{i2}} & \nearrow^{p_{i2}} & \searrow^{u_{i+1,2}} & \nearrow^{p_{i+1,2}} \searrow & & \nearrow & \searrow^{u_{i+j-2,2}} \nearrow^{p_{i+j-2,2}} \\
& E_i[2] & & E_{i+1}[2] & & \ddots & \reflectbox{$\ddots$} & E_{i+j-2}[2] \\
& & \searrow^{u_{i3}} & \nearrow^{p_{i3}} & \searrow & & & \nearrow \\
& & E_i[3] & & \ddots & & \reflectbox{$\ddots$} & \\
& & & \searrow & & \searrow \nearrow & & \\
& & & & E_{i+1}[j{-}2] & & & \\
& & & \searrow & \nearrow^{p_{i,j-1}} \searrow^{u_{i+1,j-1}} \nearrow & & & \\
& & & E_i[j{-}1] & & E_{i+1}[j{-}1] & & \\
& & & & \searrow^{u_{ij}} \nearrow^{p_{ij}} & & & \\
& & & & M = E_i[j] & & &
\end{array}
\tag{1.4}
$$

where

- we set $E_i[1] = E_{i+kr}[1]$, for all $k \in \mathbb{Z}$,
- the module $M = E_i[j]$ is the germ of the cone $\mathcal{C}(M)$,
- all the arrows pointing down represent monomorphisms,
- all the arrows pointing up represent epimorphisms, and
- each of the squares represents an almost split sequence (as well as each of the triangles on the top).

We start with three technical lemmata.

1.5. Lemma. *Let A be an algebra and let $\mathcal{T}_\lambda$ be a hereditary tube in $\Gamma(\operatorname{mod} A)$. For any pair of modules M and N in $\operatorname{add}\mathcal{T}_\lambda$, there exist K-linear isomorphisms*

$$D\operatorname{Hom}_A(N, \tau_A M) \cong \operatorname{Ext}_A^1(M, N) \cong D\operatorname{Hom}_A(\tau_A^{-1} N, M)$$

that are functorial with respect to homomorphisms $M \to M'$ and $N \to N'$ of A-modules.

Proof. By our assumption, $\operatorname{pd} X \leq 1$ and $\operatorname{id} X \leq 1$, for any module X in $\operatorname{add}\mathcal{T}_\lambda$. Then the lemma follows from (IV.2.14).   □

1.6. Lemma. *Let A be an algebra and M an indecomposable A-module in a hereditary standard stable tube $\mathcal{T}_\lambda$ of $\Gamma(\operatorname{mod} A)$ of rank $r_\lambda \geq 1$. The following conditions are equivalent.*

(a) *M is a stone.*
(b) *The cone $\mathcal{C}(M)$ (1.4) determined by M is a stone cone.*
(c) *There exists a module $E_i[j]$ of (X.2.2), with the indices $i \in \{1, \dots, r_\lambda\}$ and $j \in \{1, \dots, r_\lambda - 1\}$, such that $M \cong E_i[j]$.*
(d) *$r_\lambda \geq 2$ and M is a brick such that $\ell_\lambda(M) \leq r_\lambda - 1$, where $\ell_\lambda(M)$ is the $\mathcal{T}_\lambda$-length of M in $\operatorname{add}\mathcal{T}_\lambda$.*
(e) *There is an equivalence of categories $\operatorname{add}\mathcal{C}(M) \cong \operatorname{mod} K\Delta(\mathbb{A}_m)$, where $m = \ell_\lambda(M)$.*

Proof. Assume that M is an indecomposable A-module in a hereditary standard stable tube $\mathcal{T}_\lambda$ of $\Gamma(\operatorname{mod} A)$. By (X.2.2), there exist indices $i \in \{1, \dots, r_\lambda\}$ and $j \geq 1$, and an isomorphism $M \cong E_i[j]$ of A-modules, where $r_\lambda \geq 1$ is the rank of the tube $\mathcal{T}_\lambda$.

Because the tube $\mathcal{T}_\lambda$ is standard, the full subcategory of $\operatorname{mod} A$ formed by the modules in $\mathcal{T}_\lambda$ is equivalent to mesh category $K(\mathcal{T}_\lambda)$ of $\mathcal{T}_\lambda$. By (1.5), for any indecomposable module X in $\mathcal{T}_\lambda$, the equality $\operatorname{Ext}_A^1(X, X) = 0$ is equivalent to the equality $\operatorname{Hom}_A(X, \tau_A X) = 0$. Hence we conclude that the A-module $M \cong E_i[j]$, with $i \in \{1, \dots, r_\lambda\}$, is a stone if and only if $j \in \{1, \dots, r_\lambda - 1\}$. Moreover, if an A-module $E_s[u]$ lies in the cone $\mathcal{C}(E_i[j])$ then $i \leq s \leq i + j - 1$ and $u \leq i + j - s$. It follows that the conditions (a), (b), (c), and (d) are equivalent.

To prove that (c) implies (e), it is enough to note that $\ell_\lambda(M) \leq r_\lambda - 1$ and the tube $\mathcal{T}_\lambda$ is standard.

To finish the proof, we note that the implication (e)$\Rightarrow$(a) follows from the fact that the existence of an equivalence of categories $\operatorname{add}\mathcal{C}(M) \cong \operatorname{mod} K\Delta(\mathbb{A}_m)$ implies the equality $\operatorname{Ext}_A^1(M, M) = 0$.   □

The upper bound $\ell_\lambda(M) \leq r_\lambda - 1$ given in (1.6)(d) for the $\mathcal{T}_\lambda$-length of stones M in $\mathcal{T}_\lambda$ is the best possible, because any indecomposable module X

in $\mathcal{T}_\lambda$ such that $\ell_\lambda(X) = r_\lambda$ is a brick and

$$\mathrm{Ext}^1_A(X, X) \cong \mathrm{Hom}_A(X, \tau_A X) \cong K.$$

The full translation subquiver $\mathcal{Cr}(\mathcal{T}^A_\lambda)$ of a hereditary tube $\mathcal{T}_\lambda$ formed by the stone modules is defined to be a **stone crown** of $\mathcal{T}_\lambda$. It follows from (1.6) that the stone crown of $\mathcal{T}_\lambda$ consists of the indecomposable A-modules M in $\mathcal{T}_\lambda$ such that $\ell_\lambda(M) \leq r_\lambda - 1$. Note that any tube $\mathcal{T}_\lambda$ of rank one has no stone modules.

1.7. Lemma. *Let A be an algebra and $\mathcal{T}_\lambda$ a hereditary standard stable tube in $\Gamma(\mathrm{mod}\,A)$. For any pair of indecomposable A-modules M and N in add $\mathcal{T}_\lambda$ such that*

$$\mathrm{Ext}^1_A(M \oplus N, M \oplus N) = 0,$$

one of the following three conditions is satisfied:
  (a) $\mathcal{C}(M) \subseteq \mathcal{C}(N)$,
  (b) $\mathcal{C}(M) \supseteq \mathcal{C}(N)$, *or*
  (c) $\mathcal{C}(M) \cap \mathcal{C}(N) = \emptyset$, $\mathcal{C}(M) \cap \tau_A \mathcal{C}(N) = \emptyset$, *and* $\mathcal{C}(N) \cap \tau_A \mathcal{C}(M) = \emptyset$.

Proof. Because the modules M and N are in $\mathcal{T}_\lambda$, there exist indices $i, s \in \{1, \dots, r\}$ and $j, u \geq 1$, and isomorphisms of A-modules

$$M \cong E_i[j] \quad \text{and} \quad N \cong E_s[u].$$

Moreover, we have $\tau_A \mathcal{C}(M) = \mathcal{C}(\tau_A M)$ and $\tau_A \mathcal{C}(N) = \mathcal{C}(\tau_A N)$. In view of (1.5), our assumption yields

$$D\mathrm{Hom}_A(M \oplus N, \tau_A M \oplus \tau_A N) \cong \mathrm{Ext}^1_A(M \oplus N, M \oplus N) = 0.$$

Hence, the modules M and N are stones such that

$$\mathrm{Hom}_A(M, \tau_A N) = 0 \text{ and } \mathrm{Hom}_A(N, \tau_A M) = 0.$$

It follows from (1.6) that $M \cong E_i[j]$, $N \cong E_s[u]$, and $i, s \in \{1, \dots, r\}$, $j, u \in \{1, \dots, r - 1\}$.

We assume that the inclusions $\mathcal{C}(M) \subseteq \mathcal{C}(N)$ and $\mathcal{C}(M) \supseteq \mathcal{C}(N)$ do not hold; and we prove that the three conditions listed in (c) are satisfied.

First we observe that the inequality $\mathcal{C}(M) \cap \mathcal{C}(N) \neq \emptyset$, implies that $\mathcal{C}(M) \cap \tau_A \mathcal{C}(N) \neq \emptyset$, or $\mathcal{C}(N) \cap \tau_A \mathcal{C}(M) \neq \emptyset$. Indeed, the assumption yields $j \geq 2$, $u \geq 2$ and, hence, $\mathcal{C}(M) \cap \mathcal{C}(N)$ contains a module $E_k[1]$ such that the module $\tau_A E_k[1] \cong E_{k-1}[1]$ belongs to $\mathcal{C}(M)$, or the module $\tau_A E_k[1] \cong E_{k-1}[1]$ belongs to $\mathcal{C}(N)$; and our claim follows.

Suppose that $\mathcal{C}(M) \cap \tau_A \mathcal{C}(N) \neq \emptyset$. Then there exist $p \in \{0, 1, \dots, j-1\}$ and $q \in \{1, \dots, u\}$ such that $E_{i+p}[j-p] = E_{s-1}[q] \cong \tau_A E_s[q]$. Because the tube $\mathcal{T}_\lambda$ is standard, there is an epimorphism $f : E_i[j] \longrightarrow E_{i+p}[j-p]$ and a monomorphism $g : E_{s-1}[q] \longrightarrow E_{s-1}[u] \cong \tau_A E_s[u]$. Consequently, the composite homomorphism $gf \in \mathrm{Hom}_A(E_i[j], \tau_A E_s[u]) \cong \mathrm{Hom}_A(M, \tau_A N)$ is non-zero, and we get a contradiction. This shows that $\mathcal{C}(M) \cap \tau_A \mathcal{C}(N) = \emptyset$. Analogously, the equality $\mathrm{Hom}_A(N, \tau_A M) = 0$ yields $\mathcal{C}(N) \cap \tau_A \mathcal{C}(M) = \emptyset$. This completes the proof of the lemma. $\square$

1.8. Lemma. *Assume that A is an algebra, $\mathcal{T}_\lambda$ is a hereditary standard stable tube of $\Gamma(\mathrm{mod}\,A)$, and M is a stone in $\mathcal{T}_\lambda$. Let $\{X_1, \dots, X_p\}$ be an arbitrary non-empty family of pairwise non-isomorphic indecomposable modules in the cone $\mathcal{C}(M)$ determined by M such that the module $X = X_1 \oplus \dots \oplus X_p$ has no self-extensions, that is, $\mathrm{Ext}^1_A(X, X) = 0$. Then*

$$p \leq \mathrm{depth}\,\mathcal{C}(M),$$

where $\mathrm{depth}\,\mathcal{C}(M)$ is the depth of the cone $\mathcal{C}(M)$.

Proof. Assume that M is a stone in $\mathcal{T}_\lambda$ and $\{X_1, \dots, X_p\}$ is a fixed non-empty family of pairwise non-isomorphic indecomposable modules in the cone $\mathcal{C}(M)$ such that $\mathrm{Ext}^1_A(X, X) = 0$, where $X = X_1 \oplus \dots \oplus X_p$. By (1.5), the equality $\mathrm{Ext}^1_A(X, X) = 0$ is equivalent to the equalities $\mathrm{Ext}^1_A(X_i, \tau_A X_j) = 0$, for all $i, j \in \{1, \dots, p\}$.

Let $H = K\Delta(\mathbb{A}_m)$ be the path algebra $K\Delta(\mathbb{A}_m)$ of the equioriented quiver $\Delta(\mathbb{A}_m) : \overset{1}{\circ} \longleftarrow \overset{2}{\circ} \longleftarrow \dots \longleftarrow \overset{m}{\circ}$, where $m = \mathrm{depth}\,\mathcal{C}(M)$. Because the tube $\mathcal{T}_\lambda$ is standard and the module M is a stone in $\mathcal{T}_\lambda$ then, by (1.6), there is an equivalence of categories

$$F : \mathrm{add}\,\mathcal{C}(M) \longrightarrow \mathrm{mod}\,H.$$

For each $j \in \{1, \dots, p\}$, we set $Y_j = F(X_j)$. Then the H-modules $Y_1, \dots, Y_p$ are indecomposable, pairwise non-isomorphic, and $Y = Y_1 \oplus \dots \oplus Y_p$ is a partial tilting H-module. Hence, by applying (VI.4.4), we get the required inequality $p \leq m$, because $m = \mathrm{rk}\,K_0(H)$ is the rank of the Grothendieck group $K_0(H)$ of the algebra $H = K\Delta(\mathbb{A}_m)$. $\square$

XVII.2. Tilting with hereditary standard stable tubes

We show in this section how the shape of a hereditary standard stable tube $\mathcal{T}_\lambda$ of the Auslander–Reiten quiver $\Gamma(\mathrm{mod}\,A)$ of an algebra A changes when we pass from A to a tilted algebra $B = \mathrm{End}\,T_A$ defined by a splitting tilting A-module T_A with non-zero direct summand in $\mathrm{add}\,\mathcal{T}_\lambda$.

2.1. Proposition. *Assume that A is an algebra and M is an indecomposable module of $\mathcal{T}_\lambda$-length $m = \ell_\lambda(M)$ in a hereditary standard stable tube $\mathcal{T}_\lambda$ of $\Gamma(\mathrm{mod}\,A)$. If T_A is a tilting A-module such that M is a direct* *summand of T_A, then T_A admits exactly $m = \ell_\lambda(M)$ indecomposable direct summands lying in the cone $\mathcal{C}(M)$ (1.4) of $\mathcal{T}_\lambda$ determined by M.*

Proof. Assume that T_A is a tilting A-module such that M is a direct summand of T_A. It follows that the module M is a stone, because $\mathrm{Ext}_A^1(T_A, T_A) = 0$ and M is an indecomposable direct summand of T_A. By (1.6), $M \cong M_i[m]$, where $i \in \{1, \ldots, r_\lambda\}$, $m = \ell_\lambda(M) \in \{1, \ldots, r_\lambda - 1\}$, and $r_\lambda \geq 1$ is the rank of the tube $\mathcal{T}_\lambda$.

Assume that $X_1, \ldots, X_{p-1}, X_p = M$ are all indecomposable direct summands of T_A lying in the cone $\mathcal{C}(M)$, and set

$$X = X_1 \oplus \ldots \oplus X_p.$$

The equality $\mathrm{Ext}_A^1(T_A, T_A) = 0$ yields $\mathrm{Hom}_A(X, X) = 0$. Hence, by applying (1.8), we get $p \leq m = \ell_\lambda(M)$.

Now we prove that $p = m$. Assume, to the contrary, that $p \leq m - 1$. Let $H = K\Delta(\mathbb{A}_m)$ be the path algebra $K\Delta(\mathbb{A}_m)$ of the equioriented quiver

$$\Delta(\mathbb{A}_m) : \overset{1}{\circ} \longleftarrow \overset{2}{\circ} \longleftarrow \ldots \longleftarrow \overset{m}{\circ}.$$

Because the tube $\mathcal{T}_\lambda$ is standard, the module M is a stone in $\mathcal{T}_\lambda$, and $m = \ell_\lambda(M) = \mathrm{depth}\,\mathcal{C}(M)$ then, according to (1.6), there is an equivalence of categories

$$F : \mathrm{add}\,\mathcal{C}(M) \longrightarrow \mathrm{mod}\,H.$$

For each $j \in \{1, \ldots, p\}$, we set $Y_j = F(X_j)$. Then the H-modules $Y_1, \ldots, Y_p$ are indecomposable, pairwise non-isomorphic, and $Y_p = F(X_p) = F(M)$ is isomorphic to the unique indecomposable projective-injective H-module. By (1.5), the equality $\mathrm{Ext}_A^1(X, X) = 0$ is equivalent to the equalities $\mathrm{Hom}_A(X_i, \tau_A X_j) = 0$, for all $i, j \in \{1, \ldots, p\}$. Hence we get $\mathrm{Hom}_A(Y_i, \tau_H Y_j) = 0$, for all $i, j \in \{1, \ldots, p\}$ and, consequently,

$$\mathrm{Ext}_H^1(Y, Y) \cong D\mathrm{Hom}_H(Y, \tau_H Y) = 0.$$

This shows that $Y = Y_1 \oplus \ldots \oplus Y_p$ is a partial tilting H-module.

Because we assume that $p \leq m - 1$, where $m = \mathrm{rk}\,K_0(H)$ is the rank of the Grothendieck group $K_0(H)$ of the algebra $H = K\Delta(\mathbb{A}_m)$, then, by Bongartz's lemma (VI.2.4) and (VI.4.4), there exists an H-module V such that $Z = V \oplus Y$ is a tilting H-module and V is a direct sum of pairwise non-isomorphic indecomposable H-modules $V_1, \ldots, V_q$, where $q = m - p$.

Let $U_1, \ldots, U_q$ be indecomposable A-modules in $\mathcal{C}(M)$ such that

$$V_1 \cong F(U_1), \ \ldots, \ V_q \cong F(U_q).$$

We set $U = U_1 \oplus \ldots \oplus U_q$ and $N = U \oplus X$. The equalities $D\mathrm{Hom}_H(Z, \tau_H Z) = \mathrm{Ext}_H^1(Z, Z) = 0$ yield $\mathrm{Hom}_H(Z, \tau_H Z) = 0$, and consequently we get $\mathrm{Ext}_A^1(N, N) \cong D\mathrm{Hom}_A(N, \tau_A N) = 0$.

By the choice of X, the tilting A-module has a decomposition $T = X \oplus X'$, where X' is an A-module having no indecomposable direct summands lying in the cone $\mathcal{C}(M)$.

We show that

$$T' = U \oplus T$$

is a partial tilting A-module. To prove it, we note that U belongs to $\mathrm{add}\,\mathcal{C}(M)$, and hence $\mathrm{pd}_A U \leq 1$, because the tube $\mathcal{T}_\lambda$ is assumed to be hereditary. Hence we conclude that $\mathrm{pd}_A T' \leq 1$, because T is a tilting A-module. Then we have

- $T' = U \oplus T = N \oplus X'$,
- $\mathrm{Ext}^1_A(N, N) = 0$, and
- $\mathrm{Ext}^1_A(X', X') = 0$,

because X' is a direct summand of T. Then, it remains to show that

$$\mathrm{Ext}^1_A(U, X') = 0 \quad \text{and} \quad \mathrm{Ext}^1_A(X', U) = 0.$$

Assume, to the contrary, that $\mathrm{Ext}^1_A(U, X') \neq 0$. Hence, by the Auslander–Reiten formula (IV.2.13), we get $\mathrm{Hom}_A(X', \tau_A U) \neq 0$. It follows that there is an $s \in \{1, \dots, q\}$ and an indecomposable direct summand L of X' such that $\mathrm{Hom}_A(L, \tau_A U_s) \neq 0$. Because the module L is not in the cone $\mathcal{C}(M)$ then, by applying (IV.5.1), we infer that there exists a module $E_i[j]$, with $j \in \{1, \dots, m\}$, lying on the left border of the cone $\mathcal{C}(M)$, see (1.4), such that $\mathrm{Hom}_A(L, \tau_A E_i[j]) \neq 0$. On the other hand, the module $\tau_A E_i[j] \cong E_{i-1}[j]$ is isomorphic to a submodule of $E_{i-1}[m] \cong \tau_A E_i[m] \cong \tau_A M$. As a consequence, we get $\mathrm{Hom}_A(L, \tau_A M) \neq 0$. Because of the assumption $\mathrm{pd}_A M \leq 1$, (IV.2.14) yields

$$\mathrm{Ext}^1_A(M, L) \cong \mathrm{Hom}_A(L, \tau_A M) \neq 0.$$

On the other hand, because each of the modules M and L is a direct summand of the tilting module T, then $\mathrm{Ext}^1_A(M, L) = 0$, and we get a contradiction.

As a consequence, we get $\mathrm{Ext}^1_A(U, X') = 0$. Similarly, by applying the inequality $\mathrm{id}_A M \leq 1$ and (IV.2.14), we prove that $\mathrm{Ext}^1_A(X', U) = 0$.

Then, we have proved that $T' = U \oplus T$ is a partial tilting A-module. Hence, by (VI.4.4), the number of pairwise non-isomorphic indecomposable direct summands of T' is less than or equal to the rank $n = \mathrm{rk}\,K_0(A)$ of the Grothendieck group $K_0(A)$. On the other hand, by applying (VI.4.4) and the fact that T is a tilting module, we conclude that T has exactly $n = \mathrm{rk}\,K_0(A)$ pairwise non-isomorphic indecomposable direct summands. This forces the module U to be a direct summand of T and we get a contradiction, because the modules $X_1, \dots, X_{p-1}, X_p = M$ are all indecomposable direct summands of T_A lying in the cone $\mathcal{C}(M)$ and $X_i \not\cong U_j$, for all $i \in \{1, \dots, p\}$ and $j \in \{1, \dots, q\}$. This finishes the proof. $\square$

Given a partial tilting A-module V in mod A, we consider the following two classes of A-modules

$$\begin{aligned} \mathcal{F}(V) &= \{M_A; \ \mathrm{Hom}_A(V, M) = 0\}, \\ \mathcal{T}(V) &= \{M_A; \ \mathrm{Ext}_A^1(V, M) = 0\}. \end{aligned} \qquad \textbf{(2.2)}$$

We recall from (VI.2.3) that $\mathcal{F}(V)$ is a torsion-free class in mod A with a torsion class $\mathrm{Gen}\, V$, and $\mathcal{T}(V)$ is a torsion class in mod A with a torsion-free class $\mathrm{Cogen}\, \tau_A V$.

The following theorem is the main result of this section.

2.3. Theorem. *Let A be an algebra and let $\mathcal{T}_\lambda$ be a hereditary standard stable tube of rank $r_\lambda \geq 1$ in $\Gamma(\mathrm{mod}\, A)$. Assume that T_A is a splitting tilting module in* mod A *with a decomposition*

$$T_A = U \oplus V$$

such that $V \in \mathrm{add}\, \mathcal{T}_\lambda$, $V \neq 0$, and $\mathrm{Hom}_A(\mathcal{T}_\lambda, U) = 0$. Let $\mathcal{T}(T)$ and $\mathcal{F}(V)$ be the classes defined in (2.2), and we set

$$B = \mathrm{End}\, T_A \quad \text{and} \quad C = \mathrm{End}\, U_A.$$

(a) *The direct summand U of T_A is non-zero and $\dim_K C \geq 1$.*

(b) *$\mathcal{T}_\lambda' = \mathrm{Hom}_A(T_A, \mathcal{T}_\lambda \cap \mathcal{T}(T))$ is a standard ray tube of rank r_λ in $\Gamma(\mathrm{mod}\, B)$.*

(c) *$\mathcal{T}_\lambda'' = \mathrm{Hom}_A(U_A, \mathcal{T}_\lambda \cap \mathcal{T}(T) \cap \mathcal{F}(V))$ is a standard stable tube of rank $r_\lambda - s_V$ in $\Gamma(\mathrm{mod}\, C)$, where s_V is the number of all pairwise non-isomorphic indecomposable direct summands of V.*

(d) *The tube $\mathcal{T}_\lambda'$ is an iterated rectangle insertion of the stable tube $\mathcal{T}_\lambda''$.*

(e) *The algebra B is a $\mathcal{T}_\lambda''$-tubular extension of the algebra C.*

Proof. Assume that T_A is a tilting module in mod A with a decomposition $T_A = U \oplus V$ such that $V \in \mathrm{add}\, \mathcal{T}_\lambda$, $V \neq 0$, and $\mathrm{Hom}_A(U, \mathcal{T}_\lambda) = 0$. Fix a decomposition

$$V = V_1 \oplus \ldots \oplus V_q$$

of the A-module V into pairwise non-isomorphic indecomposable A-modules $V_1, \ldots, V_q$, where $q \geq 1$. For each $j \in \{1, \ldots, q\}$, we consider the cone $\mathcal{C}(V_j)$ of $\mathcal{T}_\lambda$ determined by V_j, and we denote by

$$m_j = \mathrm{depth}\, \mathcal{C}(V_j)$$

the depth of the cone $\mathcal{C}(V_j)$.

We view the family $\mathfrak{A} = \{\mathcal{C}(V_1), \ldots, \mathcal{C}(V_q)\}$ of cones of $\mathcal{T}_\lambda$ as a partially ordered set, with respect to the inclusion.

Choose an integer $p \in \{1, \ldots, q\}$ such that the subfamily

$$\{\mathcal{C}(V_{i_1}), \ldots, \mathcal{C}(V_{i_p})\}$$

of $\mathfrak{A}$ consists of all maximal elements of $\mathfrak{A}$. Without loss of generality, we may suppose that $i_1 = 1, \ldots, i_p = p$, that is,

$$\{\mathcal{C}(V_{i_1}), \ldots, \mathcal{C}(V_{i_p})\} = \{\mathcal{C}(V_1), \ldots, \mathcal{C}(V_p)\}.$$

Then, (1.7) and the choice of the subfamily $\{\mathcal{C}(V_1), \ldots, \mathcal{C}(V_p)\}$ of $\mathfrak{A}$ yield

- each of the indecomposable direct summands $V_1, \ldots, V_p, V_{p+1}, \ldots, V_q$ of V belongs to one of the cones $\mathcal{C}(V_1), \ldots, \mathcal{C}(V_p)$,
- $\mathcal{C}(V_j) \cap \mathcal{C}(V_k) = \emptyset$, and $\mathcal{C}(V_j) \cap \tau_A \mathcal{C}(V_k) = \emptyset$, for all $j, k \in \{1, \ldots, p\}$ such that $j \neq k$.

It follows from (2.1) that, for each $j \in \{1, \ldots, p\}$, the cone $\mathcal{C}(V_j)$ consists of $m_j = \operatorname{depth} \mathcal{C}(V_j)$ indecomposable direct summands of V. We assume that they are the modules $W_1^{(j)}, \ldots, W_{m_j}^{(j)} = V_j$, and we set

$$W_j = W_1^{(j)} \oplus \ldots \oplus W_{m_j}^{(j)}.$$

It follows that

- the A-module V has the decomposition

$$V = W_1 \oplus W_2 \oplus \ldots \oplus W_p,$$

- the module W_j belongs to $\operatorname{add} \mathcal{C}(V_j)$, for each $j \in \{1, \ldots, p\}$,
- m_j is the number of the mouth modules of $\mathcal{T}_\lambda$ lying in the cone $\mathcal{C}(V_j)$, because $m_j = \operatorname{depth} \mathcal{C}(V_j)$,
- the number s_V of all pairwise non-isomorphic indecomposable direct summands of V has the form

$$s_V = m_1 + m_2 + \ldots + m_p,$$

- $\operatorname{Hom}_A(W_j, W_k) = 0$, for all $j, k \in \{1, \ldots, p\}$ such that $j \neq k$, by the choice of $V_1, \ldots, V_p$ and the standardness of the tube $\mathcal{T}_\lambda$.

It follows that, for each $k \in \{1, \ldots, p\}$, there exists an integer $i_k \in \{1, \ldots, r_\lambda\}$ such that the A-modules $E_{i_k}, E_{i_k+1}, \ldots, E_{i_k+m_k-1}$ are all modules of the cone $\mathcal{C}(V_k)$ lying on the mouth of the tube $\mathcal{T}_\lambda$.

Without loss of generality, we may suppose that $i_1 < i_2 \ldots < i_p$. It follows from (1.6) that $m_k \leq r_\lambda - 1$. Hence, we get

$$s_V = m_1 + m_2 + \ldots + m_p \leq r_\lambda - p,$$

because $\mathcal{C}(V_j) \cap \mathcal{C}(V_k) = \emptyset$, and $\mathcal{C}(V_j) \cap \tau_A \mathcal{C}(V_k) = \emptyset$, for all $j, k \in \{1, \ldots, p\}$ such that $j \neq k$. In particular, this yields $U \neq 0$, $\dim_K \operatorname{End} U_A \geq 1$, and (a) follows.

To prove (b) and (c), we describe first the torsion part $\mathcal{T}_\lambda \cap \mathcal{T}(T_A)$ of the standard stable tube $\mathcal{T}_\lambda$. We recall that

$$\mathcal{T}(T_A) = \{M_A; \ \operatorname{Ext}_A^1(T_A, M_A) = 0\} = \{M_A; \ \operatorname{Hom}_A(M, \tau_A T) = 0\}.$$

Because $T = U \oplus V$, $V \in \text{add} \, \mathcal{T}_\lambda$, $V \neq 0$, and $\text{Hom}_A(U, \mathcal{T}_\lambda) = 0$ then we get
$$\mathcal{T}_\lambda \cap \mathcal{T}(T_A) = \mathcal{T}_\lambda \cap \mathcal{T}(V).$$
By our assumption and notation, there are isomorphisms $V_k \cong E_{i_k}[m_k]$ and $\tau_A V_k \cong E_{i_k-1}[m_k]$ of A-modules, for each $k \in \{1, \ldots, p\}$.

Given $k \in \{1, \ldots, p\}$, we consider the set
$$\Sigma_k = \{i_k - 1, i_k, \ldots, i_k + m_k - 2\}.$$
It is easy to see that the modules $E_i = E_i[1]$, with $i \in \Sigma_k$, are all modules of the mouth of $\mathcal{T}_\lambda$ lying in the cone $\mathcal{C}(\tau_A V_k) = \tau_A \mathcal{C}(V_k)$. We set

$$\Sigma(V) = \Sigma_1 \cup \Sigma_2 \cup \ldots \cup \Sigma_k \quad \text{and} \quad \Omega(V) = \{1, \ldots, r_\lambda\} \setminus \Sigma(V),$$

and note that $i_k - 2 \in \Omega(V)$ and $i_k + m_k - 1 \in \Omega(V)$, for any $k \in \{1, \ldots, p\}$.

Because $\mathcal{T}_\lambda$ is a standard stable tube of rank $r = r_\lambda \geq 1$ with the mouth modules $E_1, E_2, \ldots, E_r$ then, for each $k \in \{1, \ldots, r\}$, there is a coray

$(\mathfrak{c}_i)$ $\qquad \cdots \longrightarrow [j{+}1]E_i \longrightarrow [j]E_i \longrightarrow \cdots \longrightarrow [2]E_i \longrightarrow [1]E_i = E_i$

ending at the coray module E_i, where the irreducible morphisms corresponding to the arrows $[j{+}1]E_i \longrightarrow [j]E_i$ of the coray $(\mathfrak{c}_i)$ are surjective, see (X.2.2) and (X.2.6).

Now we prove that, for each $i \in \Omega(V)$, the coray $(\mathfrak{c}_i)$ is entirely contained in the torsion part $\mathcal{T}_\lambda \cap \mathcal{T}(T_A) = \mathcal{T}_\lambda \cap \mathcal{T}(V)$ of the tube $\mathcal{T}_\lambda$.

Fix $i \in \Omega(V)$, $j \geq 1$, and consider the A-module $X = [j]E_i$ of the coray $(\mathfrak{c}_i)$. We recall from (X.2.2) and (X.2.6) that X is a uniserial object of the abelian category

$$\text{add} \, \mathcal{T}_\lambda = \mathcal{EXT}_A(E_1, E_2, \ldots, E_r)$$

and $E_i = [1]E_i$ is the unique simple quotient object of X in $\text{add} \, \mathcal{T}_\lambda$. On the other hand, for each $k \in \{1, \ldots, p\}$, the modules $E_{i_k-1}, E_{i_k}, \ldots, E_{i_k+m_k-2}$ are the simple composition factors of the objects in the cone

$$\mathcal{C}(E_{i_k-1}[m_k]) = \tau_A \mathcal{C}(V_k).$$

Hence we conclude that $\text{Hom}_A(X, \tau_A Y) = 0$, for any indecomposable module Y in $\mathcal{C}(V_k)$. In particular, we get $\text{Hom}_A(X, W_k) = 0$. It follows that $\text{Hom}_A(X, \tau_A V) = 0$, that is, the module X belongs to $\mathcal{T}_\lambda \cap \mathcal{T}(V)$, as we required.

Next we prove that, for each $k \in \{1, \ldots, p\}$ and each $i \in \Sigma_k$, if the module $X = [j]E_i$ of the coray $(\mathfrak{c}_i)$ lies in the torsion part $\mathcal{T}(V)$, then X lies in the cone $\mathcal{C}(V_k)$. Assume, to the contrary, that X does not belong to the cone $\mathcal{C}(V_k)$. There is a unique pair of integers $u \in \{1, \ldots, j\}$ and $t \in \{1, \ldots, m_k\}$ such that $[u]E_i = E_{i_k-1}[t]$. If we set

$$Y = [u]E_i = E_{i_k-1}[t],$$

then there exist an epimorphism $f : X \longrightarrow Y$ and a monomorphism $g : Y \longrightarrow E_{i_k-1}[m_k]$. Then the homomorphism $gf : X \longrightarrow E_{i_k-1}[m_k]$ is non-zero. Because of the isomorphisms

$$E_{i_k-1}[m_k] \cong \tau_A E_{i_k}[m_k] \cong \tau_A V_k,$$

we get $\mathrm{Hom}_A(X, \tau_A V_k) \neq 0$. This implies that $\mathrm{Ext}_A^1(V, X) \neq 0$, because V_k is a direct summand of V. Consequently, the module X does not belong to the torsion part $\mathcal{T}(V)$, contrary to our assumption. Hence we conclude that $i \in \Sigma_k \setminus \{i_k - 1\}$ and

$$X \in \mathcal{C}(E_{i_k}[m_k-1]) \subseteq \mathcal{C}(E_{i_k}[m_k]) = \mathcal{C}(V_k),$$

as we claimed.

Fix an index $k \in \{1, \ldots, p\}$ and observe that the sectional path

$$V_k = E_{i_k}[m_k] \longrightarrow E_{i_k+1}[m_k-1] \longrightarrow \ldots \longrightarrow E_{i_k+m_k-2}[2] \longrightarrow E_{i_k+m_k-1}[1]$$

in $\mathcal{T}_\lambda$ is the common part of the cone $\mathcal{C}(V_k)$ and the coray $(\mathfrak{c}_{i_k+m_k-1})$, and it is contained in the torsion part $\mathcal{T}_\lambda \cap \mathcal{T}(T_A) = \mathcal{T}_\lambda \cap \mathcal{T}(V)$ of the tube $\mathcal{T}_\lambda$, because $i_k + m_k - 1 \in \Omega(V)$.

Now we show that

$$\mathcal{C}(V_k) \cap \mathcal{T}(V) = \mathcal{C}(V_k) \cap \mathcal{T}(W_k).$$

We recall that $V = W_1 \oplus W_2 \oplus \ldots \oplus W_p$, and the module W_j belongs to $\mathrm{add}\,\mathcal{C}(V_j)$, for each $j \in \{1, \ldots, p\}$. Further, by the choice of the modules $V_1, \ldots, V_p$, we have $\mathcal{C}(V_j) \cap \mathcal{C}(V_u) = \emptyset$, and $\mathcal{C}(V_j) \cap \tau_A \mathcal{C}(V_u) = \emptyset$, for all $j, u \in \{1, \ldots, p\}$ such that $j \neq u$. Then, for each $j \in \{1, \ldots, p\} \setminus \{k\}$ and any X in $\mathcal{C}(V_k)$, we have

$$\mathrm{Ext}_A^1(W_j, X) \cong D\mathrm{Hom}_A(X, \tau_A W_j) = 0.$$

It follows that, for any module X in the cone $\mathcal{C}(V_k)$, the equality $\mathrm{Ext}_A^1(V, X) = 0$ is equivalent to the equality $\mathrm{Ext}_A^1(W_k, X) = 0$. Hence the required equality $\mathcal{C}(V_k) \cap \mathcal{T}(V) = \mathcal{C}(V_k) \cap \mathcal{T}(W_k)$ follows.

Summarising, we have proved that an indecomposable A-module Z belongs to the torsion part $\mathcal{T}_\lambda \cap \mathcal{T}(T_A) = \mathcal{T}_\lambda \cap \mathcal{T}(V)$ of the tube $\mathcal{T}_\lambda$ if and only if Z lies on a coray $(\mathfrak{c}_i)$, with $i \in \Omega(V)$, or Z belongs to a torsion part $\mathcal{C}(V_k) \cap \mathcal{T}(W_k)$ of the cone $\mathcal{C}(V_k)$ with respect to a partial tilting module $W_k \in \mathrm{add}\,\mathcal{C}(V_k)$, for some $k \in \{1, \ldots, p\}$.

By (1.6), in view of our choice of the modules $V_1, \ldots, V_p$ and the standardness of the tube $\mathcal{T}_\lambda$, for each $k \in \{1, \ldots, p\}$, there exists an equivalence of categories

$$G_k : \mathrm{mod}\, H_{m_k} \longrightarrow \mathrm{add}\,\mathcal{C}(V_k),$$

where $H_{m_k} = K\Delta(\mathbb{A}_{m_k})$ is the path algebra of the equioriented quiver $\Delta(\mathbb{A}_{m_k})$ of length m_k, and m_k is the number of the mouth modules of $\mathcal{T}_\lambda$

lying in the cone $\mathcal{C}(V_k)$. Let $R_1^{(k)}, R_2^{(k)}, \ldots, R_{m_k}^{(k)}$ be indecomposable modules in $\operatorname{mod} H_{m_k}$ such that

$$G_k(R_1^{(k)}) = W_1^{(k)}, G_k(R_2^{(k)}) = W_2^{(k)}, \ldots, G_k(R_{m_k}^{(k)}) = W_{m_k}^{(k)}.$$

Then the H_{m_k}-module

$$R_k = R_1^{(k)} \oplus R_2^{(k)} \oplus \ldots \oplus R_{m_k}^{(k)}$$

is a tilting H_{m_k}-module such that $G_k(R_k) = W_k$ and $\mathcal{C}(V_k) \cap \mathcal{T}(W_k)$ is the image of the torsion part

$$\mathcal{T}(R_k) = \left\{ X_{H_{m_k}}; \ \operatorname{Ext}^1_{H_{m_k}}(R_k, X) = 0 \right\}$$

of $\operatorname{mod} H_{m_k}$ under the equivalence G_k. Moreover, the sectional path

$$V_k = E_{i_k}[m_k] \longrightarrow E_{i_k+1}[m_k-1] \longrightarrow \ldots \longrightarrow E_{i_k+m_k-2}[2] \longrightarrow E_{i_k+m_k-1}[1]$$

in $\mathcal{T}_\lambda$ is isomorphic to the image, under the equivalence G_k, of the sectional path

$$I(1)_{H_{m_k}} \longrightarrow I(2)_{H_{m_k}} \longrightarrow \ldots \longrightarrow I(m_k-1)_{H_{m_k}} \longrightarrow I(m_k)_{H_{m_k}}$$

in $\Gamma(\operatorname{mod} H_{m_k})$ formed by the indecomposable injective H_{m_k}-modules. It is clear that each of the modules $I(1)_{H_{m_k}}, \ldots, I(m_k)_{H_{m_k}}$ belongs to $\mathcal{T}(R_k)$ and, hence, is generated by the tilting H_{m_k}-module R_k, by (VI.2.5). It follows that, for each $s \in \{0, 1, \ldots, m_k - 1\}$, the A-module $E_{i_k+s}[m_k-s]$ is generated by the partial tilting module W_k of $\operatorname{add} \mathcal{C}(V_k)$.

Now we describe the part

$$\mathcal{T}_\lambda \cap \mathcal{T}(T_A) \cap \mathcal{F}(V) = \mathcal{T}_\lambda \cap \mathcal{T}(V) \cap \mathcal{F}(V)$$

of the stable tube $\mathcal{T}_\lambda$. Recall that, for each $i \in \{1, \ldots, r\}$, there is a ray

$$(\mathfrak{r}_i) \qquad E_i = E_i[1] \longrightarrow E_i[2] \longrightarrow \ldots \longrightarrow E_i[j] \longrightarrow E_i[j+1] \longrightarrow \ldots$$

starting at the ray module E_i, where the irreducible morphisms corresponding to the arrows $E_i[j] \longrightarrow E_i[j+1]$ of the ray $(\mathfrak{r}_i)$ are injective.

Fix an index $k \in \{1, \ldots, p\}$. Then, for each $s \in \{0, 1, \ldots, m_k - 1\}$ and $j \geq m_k - s$, there is a monomorphism $E_{i_k+s}[m_k-s] \longrightarrow E_{i_k+s}[j]$. It follows that $\operatorname{Hom}_A(W_k, E_{i_k+s}[j]) \neq 0$, because the A-module $E_{i_k+s}[m_k-s]$ is generated by W_k. Because every module X in $\mathcal{C}(V_k) \cap \mathcal{T}(W_k)$ is generated by W_k then $\operatorname{Hom}_A(W_k, X) \neq 0$. Moreover, in view of the decomposition $V = W_1 \oplus W_2 \oplus \ldots \oplus W_p$, we get

$$\mathcal{F}(V) = \mathcal{F}(W_1) \cap \mathcal{F}(W_2) \cap \ldots \cap \mathcal{F}(W_p).$$

For each $k \in \{1, \ldots, p\}$, we consider the set

$$\Theta_k = \{i_k, i_k + 1, \ldots, i_k + m_k - 1\},$$

and we set

$$\Theta(V) = \Theta_1 \cup \Theta_2 \cup \ldots \cup \Theta_p.$$

Then, by applying the preceding description of the torsion part $\mathcal{T}_\lambda \cap \mathcal{T}(T_A) = \mathcal{T}_\lambda \cap \mathcal{T}(V)$ of the tube $\mathcal{T}_\lambda$, we infer that $\mathcal{T}_\lambda \cap \mathcal{T}(T_A) \cap \mathcal{F}(V)$ consists of all modules lying on the corays $(\mathfrak{c}_i)$, with $i \in \Omega(V)$, except the modules lying on the rays $(\mathfrak{r}_j)$, with $j \in \Theta(V)$.

Recall from Chapter VI that the tilting A-module T determines the torsion pair $(\mathcal{T}(T), \mathcal{F}(T))$ in $\operatorname{mod} A$ and the torsion pair $(\mathcal{X}(T), \mathcal{Y}(T))$ in $\operatorname{mod} B$, where

$$B = \operatorname{End} T_A.$$

By the Brenner-Butler theorem (VI.3.8), we know that

- the functor $\operatorname{Hom}_A(T, -)$ induces an equivalence $\mathcal{T}(T) \xrightarrow{\simeq} \mathcal{Y}(T)$, while
- the functor $\operatorname{Ext}_A^1(T, -)$ induces an equivalence $\mathcal{F}(T) \xrightarrow{\simeq} \mathcal{X}(T)$.

Because the tilting module T is assumed to be splitting then, by (VI.5.2), any almost split sequence in $\operatorname{mod} B$ lies entirely in either $\mathcal{X}(T)$ or $\mathcal{Y}(T)$, or else it is of the form

$$0 \to \operatorname{Hom}_A(T, I) \longrightarrow \operatorname{Hom}_A(T, I/\operatorname{soc} I) \oplus \operatorname{Ext}_A^1(T, \operatorname{rad} P) \longrightarrow \operatorname{Ext}_A^1(T, P) \to 0,$$

where P is an indecomposable projective module not lying in $\operatorname{add} T$ and I is the indecomposable injective module such that $P/\operatorname{rad} P \cong \operatorname{soc} I$.

Because the tube $\mathcal{T}_\lambda$ is stable then, for any indecomposable A-module X lying in the torsion part $\mathcal{T}_\lambda \cap \mathcal{T}(T)$ of $\mathcal{T}_\lambda$, the indecomposable B-module $\operatorname{Hom}_A(T, X)$ is the left hand term of an almost split sequence in $\operatorname{mod} B$ contained entirely in the torsion-free part $\mathcal{Y}(T)$. In particular, by (VI.5.3), if

$$0 \longrightarrow L \xrightarrow{f} M \xrightarrow{g} N \longrightarrow 0$$

is an almost split sequence in $\operatorname{mod} A$, with L, M, and N in $\operatorname{add} \mathcal{T}_\lambda \cap \mathcal{T}(T)$, then the induced exact sequence

$$0 \longrightarrow \operatorname{Hom}_A(T, L) \xrightarrow{\operatorname{Hom}_A(T,f)} \operatorname{Hom}_A(T, M) \xrightarrow{\operatorname{Hom}_A(T,g)} \operatorname{Hom}_A(T, N) \longrightarrow 0$$

is an almost split sequence in $\operatorname{mod} B$. Moreover, if $h : X \longrightarrow Y$ is an irreducible morphism in $\operatorname{mod} A$, with X and Y in $\operatorname{add} \mathcal{T}_\lambda \cap \mathcal{T}(T)$, then

$$\operatorname{Hom}_A(T, f) : \operatorname{Hom}_A(T, X) \longrightarrow \operatorname{Hom}_A(T, Y)$$

is an irreducible morphism in $\operatorname{mod} B$.

Now we show that

$$\mathcal{T}_\lambda' = \mathrm{Hom}_A(T, \mathcal{T}_\lambda \cap \mathcal{T}(T))$$

is a standard ray tube of rank $r = r_\lambda$ in $\Gamma(\mathrm{mod}\, B)$. Because the tube $\mathcal{T}_\lambda$ is hereditary then (X.2.2) applies and, for each $i \in \{1, \ldots, , r\}$ and $j \geq 1$, we have

- a canonical irreducible monomorphism $u_{ij} : E_i[j-1] \longrightarrow E_i[j]$,
- a canonical irreducible epimorphism $p_{ij} : E_i[j] \longrightarrow E_{i+1}[j-1]$, and
- two almost split sequences in mod A

$$0 \longrightarrow E_i[1] \xrightarrow{\ u_{i1}\ } E_i[2] \xrightarrow{\ p_{i1}\ } E_{i+1}[1] \longrightarrow 0$$

$$0 \longrightarrow E_i[j] \xrightarrow{\begin{bmatrix} u_{i,j+1} \\ p_{ij} \end{bmatrix}} E_i[j+1] \oplus E_{i+1}[j-1] \xrightarrow{[\, p_{i,j+1}\ u_{i+1j}\,]} E_{i+1}[j] \longrightarrow 0,$$

where we set $E_{i+kr}[1] = E_i[1]$, for all $k \in \mathbb{Z}$.

Fix an index $k \in \{1, \ldots, p\}$. Because $i_k - 2 \in \Omega(V)$ and $i_k + m_k - 1 \in \Omega(V)$, the corays $(\mathfrak{c}_{i_k-2})$ and $(\mathfrak{c}_{i_k+m_k-1})$ are entirely contained in $\mathcal{T}_\lambda \cap \mathcal{T}(T)$. Further, because the arrow $E_{i_k-1}[m_k+1] \longrightarrow E_{i_k}[m_k]$ lies on the coray $(\mathfrak{c}_{i_k+m_k-1})$ and the module $E_{i_k}[m_k]$ is a direct summand of T, then the indecomposable B-module $\mathrm{Hom}_A(T, E_{i_k}[m_k])$ is projective, the homomorphism

$$\mathrm{Hom}_A(T, p_{i_k-1,m_k+1}) : \mathrm{Hom}_A(T, E_{i_k-1}[m_k+1]) \longrightarrow \mathrm{Hom}_A(T, E_{i_k}[m_k])$$

is an irreducible morphism in mod B, and $\mathrm{Hom}_A(T, p_{i_k-1,m_k+1})$ is a monomorphism.

Note also that the indecomposable B-module $\mathrm{Hom}_A(T, E_{i_k-1}[m_k+1])$ is not projective, because the A-module $E_{i_k-1}[m_k+1]$ is not isomorphic to a direct summand of T.

We show that the composite monomorphism

$$\varphi_k = u_{i_k-2,m_k+2} \ldots u_{i_k-2,2} : E_{i_k-2}[1] \longrightarrow E_{i_k-2}[m_k+2]$$

is a minimal left almost split morphism in $\mathcal{T}(T)$. Let $f : E_{i_k-2}[1] \longrightarrow Z$ be a non-zero homomorphism, with Z in $\mathcal{T}(T)$, and assume that f is not a section.

In case Z lies on the tube $\mathcal{T}_\lambda$, there is an isomorphism $Z \cong E_{i_k-2}[j]$, where $j \geq m_k + 2$, and therefore f admits a factorisation

(*)
$$
\begin{array}{ccc}
E_{i_k-2}[1] & \xrightarrow{\ \ f\ \ } & Z \\
{\scriptstyle \varphi_k} \searrow & & \nearrow {\scriptstyle f'} \\
& E_{i_k-2}[m_k+2] &
\end{array}
$$

in mod A. In case Z does not lie on the tube $\mathcal{T}_\lambda$, (X.2.8) applies and the homomorphism f admits a factorisation $(*)$.

Next we observe that there exist an irreducible epimorphism

$$p_{i_k-2,m_k+2} : E_{i_k-2}[m_k+2] \longrightarrow E_{i_k-1}[m_k+1],$$

and an exact sequence

$$0 \longrightarrow E_{i_k-2}[1] \xrightarrow{\varphi_k} E_{i_k-2}[m_k+2] \xrightarrow{p_{i_k-2,m_k+2}} E_{i_k-1}[m_k+1] \longrightarrow 0$$

in mod A. The induced exact sequence of B-modules

$$0 \longrightarrow \operatorname{Hom}_A(T, E_{i_k-2}[1]) \xrightarrow{\operatorname{Hom}_A(T,\varphi_k)} \operatorname{Hom}_A(T, E_{i_k-2}[m_k+2])$$

$$\xrightarrow{\operatorname{Hom}_A(T,p_{i_k-2,m_k+2})} \operatorname{Hom}_A(T, E_{i_k-1}[m_k+1]) \longrightarrow 0$$

is then an almost split sequence in mod B.

For each $j \geq 1$, there is an exact sequence

$$(\eta_j): \quad 0 \longrightarrow [j+1]E_{i_k-2} \xrightarrow{\left[\begin{smallmatrix} \varphi_{k,j+1} \\ w_{k,j+1} \end{smallmatrix}\right]} [m_k+2+j]E_{i_k+m_k-1} \oplus [j]E_{i_k-2}$$

$$\xrightarrow{[\pi_{k,j+1} \ \varphi_{k,j}]} [m_k+1+j]E_{i_k+m_k-1} \longrightarrow 0,$$

where

$$w_{k,j+1} : [j+1]E_{i_k-2} \longrightarrow [j]E_{i_k-2}, \text{ and}$$

$$\pi_{k,j+1} : [m_k+2+j]E_{i_k+m_k-1} \longrightarrow [m_k+1+j]E_{i_k+m_k-1}$$

are irreducible epimorphisms, and

$$\varphi_{k,t} : [t]E_{i_k-2} \longrightarrow [m_k+1+t]E_{i_k+m_k-1}, \text{ with } t \geq 1,$$

is the composition of the corresponding irreducible monomorphisms. Note that $\varphi_{k,1} = \varphi_k$.

An obvious induction on $j \geq 1$ shows that the exact sequence (η_j) is almost split in $\operatorname{add} \mathcal{T}_\lambda \cap \mathcal{T}(T)$. In the proof, we apply the fact that the homomorphisms $w_{k,j+1}$, $\pi_{k,j+1}$, and φ_k are irreducible morphisms in the category $\operatorname{add} \mathcal{T}_\lambda \cap \mathcal{T}(T)$. The details are left to the reader.

It is easy to see that

- if $i_k + m_k \neq i_{k+1} - 1$ then the almost split sequence

$$0 \longrightarrow [1]E_{i_k+m_k-1} \longrightarrow [2]E_{i_k+m_k} \longrightarrow [1]E_{i_k+m_k} \longrightarrow 0$$

in mod A is also an almost split sequence in $\mathcal{T}(T)$,

- if $i_k + m_k \neq i_{k+1} - 1$ then, for each $j \geq 1$, the almost split sequence

$$0 \longrightarrow [j+1]E_{i_k+m_k-1} \longrightarrow [j+2]E_{i_k+m_k} \oplus [j]E_{i_k+m_k-1} \longrightarrow [j+1]E_{i_k+m_k} \longrightarrow 0$$

in mod A is also an almost split sequence in $\mathcal{T}(T)$,
- if $i_{k-1} + m_{k-1} \neq i_k - 1$ then the almost split sequence

$$0 \longrightarrow [1]E_{i_k-3} \longrightarrow [2]E_{i_k-2} \longrightarrow [1]E_{i_k-2} \longrightarrow 0$$

in mod A is also an almost split sequence in $\mathcal{T}(T)$,
- if $i_{k-1}+m_{k-1} \neq i_k-1$ then, for each $j \geq 1$, the almost split sequence

$$0 \longrightarrow [j+1]E_{i_k-3} \longrightarrow [j+2]E_{i_k-2}\oplus[j]E_{i_k-3} \longrightarrow [j+1]E_{i_k-2} \longrightarrow 0$$

in mod A is also an almost split sequence in $\mathcal{T}(T)$,
- the functor $\mathrm{Hom}_A(T, -)$ carries such an almost split sequence (η_j) in mod A to the almost split sequence $\mathrm{Hom}_A(T, (\eta_j))$ in mod B, and
- $\mathrm{Hom}_A(T, M) = \mathrm{Hom}_A(W_k, M)$, for any M in $\mathrm{add}\,\mathcal{C}(V_k)$.

We recall that, for each $k \in \{1, \dots, p\}$, the torsion part

$$\mathcal{C}(V_k) \cap \mathcal{T}(T) = \mathcal{C}(V_k) \cap \mathcal{T}(W_k)$$

is the image of the torsion part $\mathcal{T}(R_k)$ of the tilting module R_k over the path algebra $H_{m_k} = K\Delta(\mathbb{A}_{m_k})$ via the functor $G_k : \mathrm{mod}\,H_{m_k} \longrightarrow \mathrm{add}\,\mathcal{C}(V_k)$, and the sectional path

$$V_k = E_{i_k}[m_k] \longrightarrow E_{i_k+1}[m_k-1] \longrightarrow \dots \longrightarrow E_{i_k+m_k-2}[2] \longrightarrow E_{i_k+m_k-1}[1]$$

in $\mathcal{T}_\lambda$ is the image of the sectional path in $\Gamma(\mathrm{mod}\,H_{m_k})$ formed by the indecomposable injective H_{m_k}-modules.

For each $k \in \{1, \dots, p\}$, consider the algebra

$$D_k = \mathrm{End}_A(W_k) = \mathrm{End}_A(G_k(R_k)) \cong \mathrm{End}_{H_{m_k}}(R_k).$$

It follows that D_k is a tilted algebra of type $\Delta(\mathbb{A}_{m_k})$ given by the tilting H_{m_k}-module R_k. By (XVI.2.2), D_k is a branch algebra $K\mathcal{L}^{(k)}$ of a branch $\mathcal{L}^{(k)} = (L^{(k)}, I^{(k)})$ of capacity m_k, with the germ vertex $0^{(k)}$ corresponding to the direct summand $R_{m_k}^{(k)}$ such that $G_k(R_{m_k}^{(k)}) = W_{m_k}^{(k)} = V_k$. Moreover,

$$\mathrm{Hom}_A(T, \mathcal{C}(V_k) \cap \mathcal{T}(T)) = \mathrm{Hom}_A(W_k, \mathcal{C}(V_k) \cap \mathcal{T}(W_k))$$

is equivalent to the torsion-free part $\mathcal{Y}(R_k)$ of mod D_k.

Finally, we recall from (XVI.2.8) that there exists a hereditary algebra

$$H^{(k)} = K\Delta(\mathbb{A}_{n_k}),$$

with $n_k \leq m_k$, such that the Auslander–Reiten translation quiver $\Gamma(\mathrm{mod}\,H_{m_k})$ of H_{m_k} is obtained from the translation quiver $\Gamma(\mathrm{mod}\,H^{(k)})$ of $H^{(k)}$ by a sequence of finite rectangle insertions. It follows that for any inde-composable module X in $\mathcal{C}(V_k) \cap \mathcal{T}(T)$ such that $X \ncong E_{i_k+s}[m_k-s]$, for $s \in \{0, 1, \ldots, m_k - 1\}$, there exists a unique sectional path in $\Gamma(\mathrm{mod}\,B)$ starting from the indecomposable B-module $\mathrm{Hom}_A(T, X)$ to a module $\mathrm{Hom}_A(T, E_{i_k+t}[m_k-t])$, for some $t \in \{0, 1, \ldots, m_k - 1\}$.

Summing up, we have proved that

$$\mathcal{T}'_\lambda = \mathrm{Hom}_A(T, \mathcal{T}_\lambda \cap \mathcal{T}(T))$$

is a ray tube of $\Gamma(\mathrm{mod}\,B)$, with $r = r_\lambda$ rays; hence of rank r_λ. More precisely, the tube $\mathcal{T}'_\lambda$, viewed as a translation quiver, is obtained from the stable tube $\mathcal{T}_\lambda$ by the following three operations:

(i) deleting all the corays $(\mathfrak{c}_i)$, with $i \in \Sigma(V)$,
(ii) shrinking the sectional path

$$[j]E_{i_k-2} \longrightarrow [1+j]E_{i_k-1} \longrightarrow \cdots \longrightarrow [m_k+1+j]E_{i_k+m_k-1},$$

to one arrow $[j]E_{i_k-2} \longrightarrow [m_k+1+j]E_{i_k+m_k-1}$, for each $k \in \{1, \ldots, p\}$ and $j \geq 1$,
(iii) glueing the torsion-free part $\mathcal{Y}(R_k)$ of $\Gamma(\mathrm{mod}\,D_k)$ along the sectional path

$$E_{i_k}[m_k] \longrightarrow E_{i_k+1}[m_k-1] \longrightarrow \cdots \longrightarrow E_{i_k+m_k-2}[2] \longrightarrow E_{i_k+m_k-1}[1]$$

to the translation quiver constructed in (i) and (ii), for each $k \in \{1, \ldots, p\}$.

The ray tube $\mathcal{T}'_\lambda$ is standard, because $\mathcal{T}_\lambda$ is a standard stable tube of $\Gamma(\mathrm{mod}\,A)$ and $\mathcal{T}'_\lambda$ is the image of the torsion part $\mathcal{T}_\lambda \cap \mathcal{T}(T)$ of the hereditary tube $\mathcal{T}_\lambda$ via the equivalence $\mathrm{Hom}_A(T, -) : \mathcal{T}(T) \xrightarrow{\cong} \mathcal{Y}(T)$.

Because the right A-module T_A has a decomposition $T = U \oplus V$ in $\mathrm{mod}\,A$ such that $\mathrm{Hom}_A(V, U) = 0$, the tilted algebra $B = \mathrm{End}\,T_A$ is of the lower triangular matrix form

$$B \cong \begin{bmatrix} \mathrm{End}\,U_A & \mathrm{Hom}_A(V, U) \\ \mathrm{Hom}_A(U, V) & \mathrm{End}\,V_A \end{bmatrix} = \begin{bmatrix} C & 0 \\ {}_D M_C & D \end{bmatrix},$$

where $C = \mathrm{End}\,U_A$, $D = \mathrm{End}\,V_A$, and ${}_D M_C = \mathrm{Hom}_A(U, V)$ is viewed as a D-C-bimodule in an obvious way. Note also that

- the algebra D is the product $D_1 \times D_2 \times \ldots \times D_p$ of the branch algebras $D_1 = K\mathcal{L}^{(1)}, D_2 = K\mathcal{L}^{(2)}, \ldots, D_p = K\mathcal{L}^{(p)}$,

- C is a quotient algebra of B, and
- the canonical surjection $B \longrightarrow C$ of algebras induces the fully faithful exact embedding $\mod C \hookrightarrow \mod B$.

Now we prove that $\operatorname{Hom}_A(T, \mathcal{T}_\lambda \cap \mathcal{T}(T) \cap \mathcal{F}(V))$ is the family of all indecomposable C-modules lying in the ray tube

$$\mathcal{T}_\lambda' = \operatorname{Hom}_A(T, \mathcal{T}_\lambda \cap \mathcal{T}(T)).$$

First, we observe that

$$\operatorname{Hom}_A(T, \mathcal{T}_\lambda \cap \mathcal{T}(T) \cap \mathcal{F}(V)) = \operatorname{Hom}_A(U, \mathcal{T}_\lambda \cap \mathcal{T}(T) \cap \mathcal{F}(V)),$$

because $T = U \oplus V$ and $\operatorname{Hom}_A(V, \mathcal{F}(V)) = 0$. Next, we recall from (VI.3.1) that the functor $\operatorname{Hom}_A(T, -)$ restricts to the equivalence of categories

$$\operatorname{Hom}_A(T, -) : \operatorname{add} T \xrightarrow{\;\simeq\;} \operatorname{proj} B,$$

where $\operatorname{proj} B$ is the category of finitely generated projective B-modules. Then every indecomposable projective B-module that is not a C-module is of the form $P = \operatorname{Hom}_A(T, V')$, where V' is an indecomposable direct summand of V. Hence, by (VI.3.8), given an indecomposable A-module X in $\mathcal{T}_\lambda \cap \mathcal{T}(T)$, the B-module $N = \operatorname{Hom}_A(T, X)$ is indecomposable and we get isomorphisms

$$\operatorname{Hom}_B(P, N) = \operatorname{Hom}_B(\operatorname{Hom}_A(T, V'), \operatorname{Hom}_A(T, X)) \cong \operatorname{Hom}_A(V', X).$$

It follows that the B-module $N = \operatorname{Hom}_A(T, X)$, with X in $\mathcal{T}_\lambda \cap \mathcal{T}(T)$, lies in the category

$$\mod C \hookrightarrow \mod B$$

if and only if $\operatorname{Hom}_A(V, X) = 0$, that is, if and only if X lies in $\mathcal{T}_\lambda \cap \mathcal{T}(T) \cap \mathcal{F}(V)$. Observe also that $\operatorname{Hom}_A(T, \mathcal{T}_\lambda \cap \mathcal{T}(T) \cap \mathcal{F}(V))$ is obtained from the tube $\mathcal{T}_\lambda'$ by deleting all rays in $\mathcal{T}_\lambda'$ passing through the B-modules

$$\operatorname{Hom}_A(T, E_{i_k}[m_k]), \ \operatorname{Hom}_A(T, E_{i_k+1}[m_k-1]), \ \ldots \ \operatorname{Hom}_A(T, E_{i_k+m_k-1}[1]),$$

for all $k \in \{1, \ldots, p\}$. This means that the number of rays of the tube $\mathcal{T}_\lambda'$, that are not removed, equals $r_\lambda - s_V = r_\lambda - (m_1 + \ldots + m_p)$.

It follows that

$$\mathcal{T}_\lambda'' = \operatorname{Hom}_A(T, \mathcal{T}_\lambda \cap \mathcal{T}(T) \cap \mathcal{F}(V)) = \operatorname{Hom}_A(U, \mathcal{T}_\lambda \cap \mathcal{T}(T) \cap \mathcal{F}(V)),$$

is a stable tube of $\Gamma(\text{mod}\,C)$ of rank $r_\lambda - s_V$, obtained from the tube $\mathcal{T}_\lambda'$ by deleting all the B-modules that are not in the category $\text{mod}\,C \hookrightarrow \text{mod}\,B$, and by shrinking each of the sectional paths in $\mathcal{T}_\lambda'$ of the form

$$\text{Hom}_A(T, E_{i_k-1}[m_k+2+j]) \longrightarrow \cdots \longrightarrow \text{Hom}_A(T, E_{i_k+m_k}[j]),$$

with $k \in \{1, \ldots, p\}$, to one arrow

$$\text{Hom}_A(U, E_{i_k-1}[m_k+2+j]) \longrightarrow \text{Hom}_A(U, E_{i_k+m_k}[j]).$$

To finish the proof of the theorem, we note that the following properties of the tube $\mathcal{T}_\lambda''$ follow from the preceding considerations.

- $\mathcal{T}_\lambda''$ is a standard stable tube of $\Gamma(\text{mod}\,C)$ of rank

$$r_\lambda - s_V = r_\lambda - (m_1 + \ldots + m_p).$$

- Each of the indecomposable C-modules

$$F_k = \text{Hom}_A(T, E_{i_k-1}[m_k+1]) = \text{Hom}_A(U, E_{i_k-1}[m_k+1]),$$

with $k \in \{1, \ldots, p\}$, lies on the mouth of the stable tube $\mathcal{T}_\lambda''$.
- The C-modules $F_1, F_2, \ldots, F_p$ are pairwise non-isomorphic.
- The tilted algebra $B = \text{End}\,T_A$ is a $\mathcal{T}_\lambda''$-branch extension of the algebra C, and B has the form, see (XV.3.4),

$$B = C[F_1, \mathcal{L}^{(1)}, F_2, \mathcal{L}^{(2)}, \ldots, F_p, \mathcal{L}^{(p)}].$$

- B is a $\mathcal{T}_\lambda''$-tubular extension of the algebra C, by (XV.3.9).
- The ray tube $\mathcal{T}_\lambda'$ of $\Gamma(\text{mod}\,B)$ is obtained from the stable tube $\mathcal{T}_\lambda''$ of $\Gamma(\text{mod}\,A)$ by an iterated rectangle insertion, that creates $s_V = m_1 + \ldots + m_p$ new rays.

This finishes the proof of the theorem.   □

The following theorem is an analogue of Theorem (2.3).

2.4. Theorem. *Let A be an algebra and let $\mathcal{T}_\lambda$ be a hereditary standard stable tube of rank $r_\lambda \geq 1$ in $\Gamma(\text{mod}\,A)$. Assume that T_A is a splitting tilting module in $\text{mod}\,A$ with a decomposition*

$$T_A = U \oplus V$$

such that $V \in \text{add}\,\mathcal{T}_\lambda$, $V \neq 0$, and $\text{Hom}_A(U, \mathcal{T}_\lambda) = 0$. Let $\mathcal{F}(T)$ and $\mathcal{T}(V)$ be the classes defined in (2.2), and we set

$$B = \text{End}\,T_A \quad \text{and} \quad C = \text{End}\,U_A.$$

(a) *The direct summand U of T_A is non-zero and $\dim_K C \geq 1$.*

(b) $\mathcal{T}_\lambda' = \mathrm{Hom}_A(T_A, \mathcal{T}_\lambda \cap \mathcal{F}(T))$ *is a standard coray tube of corank r_λ in* $\Gamma(\mathrm{mod}B)$.

(c) $\mathcal{T}_\lambda'' = \mathrm{Hom}_A(U_A, \mathcal{T}_\lambda \cap \mathcal{F}(T) \cap T(V))$ *is a standard stable tube of rank $r_\lambda - s_V$ in* $\Gamma(\mathrm{mod}\,C)$, *where s_V is the number of pairwise non-isomorphic indecomposable direct summands of V.*

(d) *The tube $\mathcal{T}_\lambda'$ is an iterated rectangle coinsertion of the stable tube $\mathcal{T}_\lambda''$.*

(e) *The algebra B is a $\mathcal{T}_\lambda''$-tubular coextension of the algebra C.*

Proof. The arguments used in the proof of Theorem (2.3) modify almost verbatim. The details are left to the reader. □

Now we illustrate the preceding considerations with an example.

2.5. Example. Let $A = K\Delta(\widetilde{\mathbb{D}}_{12})$ be the path algebra of the quiver

of the Euclidean type $\widetilde{\mathbb{D}}_{12}$. It follows from (XII.3.4) that the Auslander–Reiten quiver $\Gamma(\mathrm{mod}\,A)$ of A consists of three parts:

- a postprojective component $\mathcal{P}(A)$ containing all the indecomposable projective A-modules,
- a preinjective component $\mathcal{Q}(A)$ containing all the indecomposable injective A-modules, and
- a $\mathbb{P}_1(K)$-family

$$\boldsymbol{T}^A = \{\mathcal{T}_\lambda^A\}_{\lambda \in \mathbb{P}_1(K)}$$

of stable tubes $\mathcal{T}_\lambda^A$ of tubular type $(2, 2, 10)$ separating $\mathcal{P}(A)$ from $\mathcal{Q}(A)$.

Moreover, by (XIII.2.9), the mouth of the unique stable tube

$$\mathcal{T}_1^A = \mathcal{T}_1^{\Delta(\widetilde{\mathbb{D}}_{12})}$$

of rank 10 of $\Gamma(\mathrm{mod}\,A)$ consists of the nine simple A-modules

$E_1 = S(3), E_2 = S(4), E_3 = S(5), E_4 = S(6),$
$E_5 = S(7), E_6 = S(8), E_7 = S(9), E_8 = S(10), E_9 = S(11)$

at the vertices $3, 4, 5, 6, 7, 8, 9, 10,$ and $11,$ and the following module

$$E_{10}: \qquad K \xleftarrow{1} K \xleftarrow{1} K \xleftarrow{1} K \xleftarrow{1} K \xleftarrow{1} K \xleftarrow{1} K \xleftarrow{1} K \xleftarrow{1} K$$

We also recall that there are isomorphisms of A-modules

$$E_1 \cong \tau_A E_2, \ E_2 \cong \tau_A E_3, \ \ldots\ldots, \ E_9 \cong \tau_A E_{10}, \text{ and } E_{10} \cong \tau_A E_1.$$

Then, in the notation of (X.2.2), the upper part of the tube $\mathcal{T}_1^A$ is of the form

and we identify the modules along the vertical dotted lines.

Consider the indecomposable projective A-modules

$$T_1 = P(1),\ T_2 = P(2),\ T_3 = P(7),\ T_4 = P(10),\ T_5 = P(12),\ \text{and}\ T_6 = P(13)$$

at the vertices $1, 2, 7, 10, 12,$ and $13,$ and the following seven indecomposable A-modules

$$T_7 = E_2[1], T_8 = E_2[3], T_9 = E_4[1], T_{10} = E_2[4],$$
$$T_{11} = E_8[1], T_{12} = E_7[2], T_{13} = E_{10}[1].$$

We set

$$T_A = U \oplus V,$$

where

$$U = T_1 \oplus T_2 \oplus T_3 \oplus T_4 \oplus T_5 \oplus T_6 \in \operatorname{add} \mathcal{P}(A), \text{ and}$$

$$V = T_7 \oplus T_8 \oplus T_9 \oplus T_{10} \oplus T_{11} \oplus T_{12} \oplus T_{13} \in \operatorname{add} \mathcal{T}_1^A.$$

The following properties of the modules T, U, and V are easily verified.

   (i) $\operatorname{Hom}_A(\mathcal{T}_1^A, U) = 0$.

   (ii) $\operatorname{Ext}_A^1(U, T) = 0$, because the module U is projective.

   (iii) The modules T_7, T_8, T_9, and T_{10} belong to the cone $\mathcal{C}(E_2[4])$ of depth 4.

   (iv) The modules T_{11} and T_{12} belong to the cone $\mathcal{C}(E_7[2])$ of depth 2.

   (v) The module T_{13} forms the cone $\mathcal{C}(E_{10}[1])$ of depth 1.

   (vi) The modules $T_7, T_8, T_9, T_{11}, T_{12}$, and T_{13} satisfy the tilting vanishing condition $\operatorname{Hom}_A(T_i, \tau_A T_j) = 0$, for all $i, j \in \{7, 8, 9, 10, 11, 12, 13\}$.

  (vii) $\operatorname{Ext}_A^1(V, V) \cong D\operatorname{Hom}_A(V, \tau_A V) = 0$.

 (viii) $\operatorname{Ext}_A^1(V, U) \cong D\operatorname{Hom}_A(U, \tau_A V) = 0$.

The statement (vii) is an immediate consequence of (vi). To prove (viii) we observe that there is an isomorphism

$$\tau_A V \cong E_1[1] \oplus E_1[3] \oplus E_3[1] \oplus E_1[4] \oplus E_7[1] \oplus E_6[2] \oplus E_9[1],$$

and each of the following seven A-modules

$$E_1[1] = S(3), \quad E_3[1] = S(5), \quad E_7[1] = S(9), \quad E_9[1] = S(11),$$

$E_1[3]$:

$$0 \qquad\qquad\qquad\qquad\qquad\qquad\qquad\qquad 0$$
$$\nwarrow \qquad\qquad\qquad\qquad\qquad\qquad\qquad\qquad \swarrow$$
$$K \xleftarrow{1} K \xleftarrow{1} K \xleftarrow{} 0 \xleftarrow{} 0 \xleftarrow{} 0 \xleftarrow{} 0 \xleftarrow{} 0 \xleftarrow{} 0$$
$$\swarrow \qquad\qquad\qquad\qquad\qquad\qquad\qquad\qquad \nwarrow$$
$$0 \qquad\qquad\qquad\qquad\qquad\qquad\qquad\qquad 0,$$

$E_1[4]$:

$$0 \qquad\qquad\qquad\qquad\qquad\qquad\qquad\qquad 0$$
$$\nwarrow \qquad\qquad\qquad\qquad\qquad\qquad\qquad\qquad \swarrow$$
$$K \xleftarrow{1} K \xleftarrow{1} K \xleftarrow{1} K \xleftarrow{} 0 \xleftarrow{} 0 \xleftarrow{} 0 \xleftarrow{} 0 \xleftarrow{} 0$$
$$\swarrow \qquad\qquad\qquad\qquad\qquad\qquad\qquad\qquad \nwarrow$$
$$0 \qquad\qquad\qquad\qquad\qquad\qquad\qquad\qquad 0,$$

$E_6[2]$:

$$0 \qquad\qquad\qquad\qquad\qquad\qquad\qquad\qquad 0$$
$$\nwarrow \qquad\qquad\qquad\qquad\qquad\qquad\qquad\qquad \swarrow$$
$$0 \xleftarrow{} 0 \xleftarrow{} 0 \xleftarrow{} 0 \xleftarrow{} 0 \xleftarrow{} K \xleftarrow{1} K \xleftarrow{} 0 \xleftarrow{} 0$$
$$\swarrow \qquad\qquad\qquad\qquad\qquad\qquad\qquad\qquad \nwarrow$$
$$0 \qquad\qquad\qquad\qquad\qquad\qquad\qquad\qquad 0$$

has no simple composition factor isomorphic to any of the six simple modules

$$S(1) \cong P(1)/\operatorname{rad} P(1), \qquad S(2) \cong P(2)/\operatorname{rad} P(2),$$
$$S(7) \cong P(7)/\operatorname{rad} P(7), \qquad S(10) \cong P(10)/\operatorname{rad} P(10),$$

$$S(12) \cong P(12)/\operatorname{rad} P(12), \quad S(13) \cong P(13)/\operatorname{rad} P(13).$$

Because the algebra A is hereditary then, by (VI.4.4), T_A is a multiplicity-free tilting A-module. One easily observes that the algebra

$$C = \operatorname{End} U_A$$

is isomorphic to the path algebra $K\Delta(\widetilde{\mathbb{D}}_5)$ of the quiver

$$\Delta = \Delta(\widetilde{\mathbb{D}}_5) :$$

of the Euclidean type $\widetilde{\mathbb{D}}_5$. The ordering 1, 2, 3, 4, 5, 6 of the vertices of the quiver $\Delta(\widetilde{\mathbb{D}}_5)$ corresponds to the ordering T_1, T_2, T_3, T_4, T_5, T_6 of the indecomposable direct summands of the module U_A.

To determine the tilted algebra $B = \operatorname{End} T_A$, we note that $T_7 = S(4)$, $T_9 = S(6)$, $T_{11} = S(10)$, and the modules T_8, T_{10}, T_{12}, and $T_{13} = E_{10}$ are of the forms

Then a direct calculation shows that the tilted algebra $B = \operatorname{End} T_A$ is given by the quiver

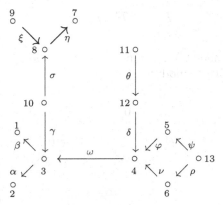

bound by the following five relations $\gamma\beta = 0$, $\gamma\alpha = 0$, $\xi\eta = 0$, $\delta\omega = 0$, and $\psi\varphi = \rho\nu$ considered in (XV.4.8). The ordering 7, 8, 9, 10, 11, 12, 13 of the vertices of the quiver corresponds to the ordering T_7, T_8, T_9, T_{10}, T_{11}, T_{12}, and T_{13} of the indecomposable direct summands of the module V_A.

Observe that the algebra $D = \operatorname{End} V_A$ is isomorphic to the product $D_1 \times D_2 \times D_3$, where

- $D_1 = K\mathcal{L}^{(1)}$ is the branch algebra of the branch $\mathcal{L}^{(1)}$ given by the quiver

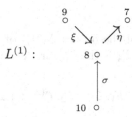

   bound by the zero relation $\xi\eta = 0$,
- $D_2 = K\mathcal{L}^{(2)}$ is the branch algebra of the branch $\mathcal{L}^{(2)}$ given by the quiver

$$L^{(2)}: \qquad \begin{array}{c} 11 \circ \\ \theta \downarrow \\ 12 \circ \end{array}$$

- $D_3 = K\mathcal{L}^{(3)}$ is the branch algebra of the branch $\mathcal{L}^{(3)}$ given by the one vertex quiver consisting of the vertex 13.

It follows from (XIII.2.9) that the tubular $\mathbb{P}_1(K)$-family

$$\mathcal{T}^C = \{\mathcal{T}_\lambda^C\}_{\lambda \in \mathbb{P}_1(K)}$$

of stable tubes in $\Gamma(\operatorname{mod} C)$ has a unique stable tube $\mathcal{T}_1^C$ of rank 3 with the following three mouth C-modules

$$F_1 = F_1^{(1)}: \quad
\begin{array}{ccc}
0 & & 0 \\
\nwarrow & & \swarrow \\
& K \longleftarrow 0 & \\
\swarrow & & \nwarrow \\
0 & & 0,
\end{array}$$

$$F_2 = F_2^{(1)}: \quad
\begin{array}{ccc}
0 & & 0 \\
\nwarrow & & \swarrow \\
& 0 \longleftarrow K & \\
\swarrow & & \nwarrow \\
0 & & 0,
\end{array}$$

$$F_3 = F_3^{(1)}: \quad
\begin{array}{ccc}
K & & K \\
\nwarrow {\scriptstyle 1} & & {\scriptstyle 1}\swarrow \\
& K \xleftarrow{\;1\;} K & \\
\swarrow {\scriptstyle 1} & & \nwarrow {\scriptstyle 1} \\
K & & K
\end{array}$$

viewed as representations of the quiver $\Delta(\widetilde{\mathbb{D}}_5)$ of the algebra C. Therefore the tilted algebra $B = \operatorname{End} T_A$ is the $\mathcal{T}_1^C$-branch extension (and $\mathcal{T}_1^C$-tubular extension)

$$B \cong C[F_1, \mathcal{L}^{(1)}, F_2, \mathcal{L}^{(2)}, F_3, \mathcal{L}^{(3)}]$$

of the algebra C. According to (XV.4.3), the Auslander–Reiten quiver $\Gamma(\operatorname{mod} B)$ of B admits a $\mathbb{P}_1(K)$-family

$$\mathcal{T}^B = \{\mathcal{T}_\lambda^B\}_{\lambda \in \mathbb{P}_1(K)}$$

of pairwise orthogonal standard ray tubes $\mathcal{T}_\lambda^B$ such that

- $\mathcal{T}_\lambda^B = \mathcal{T}_\lambda^C$, for each $\lambda \in \mathbb{P}_1(K) \setminus \{1\}$, and
- $\mathcal{T}_1^B$ is obtained from the stable tube $\mathcal{T}_1^C$ by the rectangle insertions corresponding to the branches $\mathcal{L}^{(1)}$, $\mathcal{L}^{(2)}$, and $\mathcal{L}^{(3)}$.

The tube $\mathcal{T}_1^B$ looks as follows:

where $F_3 = F_3[1]$, $F_2 = F_2[1]$, and we identify the modules along the vertical dotted lines.

We easily observe that, in the notation introduced in the proof of (2.3), we have

- $p = 3$, $m_1 = 4$, $m_2 = 2$, $m_3 = 1$, $s_V = m_1 + m_2 + m_3 = 7$,
- $V_1 = E_2[4] = T_{10}$, $V_2 = E_7[2] = T_{12}$, $V_3 = E_{10}[1] = T_{13}$,
- the sets Σ_1, Σ_2, Σ_3 are of the forms
  $\Sigma_1 = \{1, 2, 3, 4\}$, $\Sigma_2 = \{6, 7\}$, $\Sigma_3 = \{9\}$,
- $\Sigma(V) = \Sigma_1 \cup \Sigma_2 \cup \Sigma_3 = \{1, 2, 3, 4, 6, 7, 9\}$,

- $\Omega(V) = \{1, 2, 3, 4, 5, 6, 7, 8, 9, 10\} \setminus \Sigma(V) = \{5, 8, 10\}$,
- the sets Θ_1, Θ_2, Θ_3 are of the forms
  $$\Theta_1 = \{2, 3, 4, 5\}, \quad \Theta_2 = \{7, 8\}, \quad \Theta_3 = \{10\},$$
  and hence
- $\Theta(V) = \Theta_1 \cup \Theta_2 \cup \Theta_3 = \{2, 3, 4, 5, 7, 8, 10\}$.

It follows that the family of indecomposable A-modules in

$$\mathcal{T}_1^A \cap \mathcal{T}(T) \cap \mathcal{F}(V)$$

consists of all the modules lying on the corays $(\mathfrak{c}_5)$, $(\mathfrak{c}_8)$, and $(\mathfrak{c}_{10})$ of the tube $\mathcal{T}_1^A$ ending at the modules E_5, E_8, and E_{10}, except those modules that lie on the rays $(\mathfrak{r}_2)$, $(\mathfrak{r}_3)$, $(\mathfrak{r}_4)$, $(\mathfrak{r}_5)$, $(\mathfrak{r}_7)$, $(\mathfrak{r}_8)$, and $(\mathfrak{r}_{10})$ of the tube $\mathcal{T}_1^A$ starting at the modules E_2, E_3, E_4, E_5, E_7, E_8, and E_{10}. Hence

$$\mathcal{T}_1'' = \mathrm{Hom}_A(U_A, \mathcal{T}_1^A \cap \mathcal{T}(T) \cap \mathcal{F}(V))$$

is the unique stable tube $\mathcal{T}_1^C$ of $\Gamma(\mathrm{mod}\, C)$ of rank 3.

The partial tilting A-modules W_1, W_2, and W_3 are of the forms

$$W_1 = T_7 \oplus T_8 \oplus T_9 \oplus T_{10}, \quad W_2 = T_{11} \oplus T_{12}, \quad \text{and} \quad W_3 = T_{13}.$$

Then the torsion part $\mathcal{T}_1^A \cap \mathcal{T}(T) = \mathcal{T}_1^A \cap \mathcal{T}(V)$ of the tube $\mathcal{T}_1^A$ consists of all indecomposable modules lying on the corays $(\mathfrak{c}_5)$, $(\mathfrak{c}_8)$, and $(\mathfrak{c}_{10})$, and the indecomposable modules of the torsion parts $\mathcal{C}(V_1) \cap \mathcal{T}(W_1)$, $\mathcal{C}(V_2) \cap \mathcal{T}(W_2)$, and $\mathcal{C}(V_3) \cap \mathcal{T}(W_3)$ of the maximal cones $\mathcal{C}(V_1)$, $\mathcal{C}(V_2)$, and $\mathcal{C}(V_3)$ determined by the indecomposable direct summands of the partial tilting A-module V. Finally, we also observe that

- the indecomposable modules of $\mathcal{C}(V_1) \cap \mathcal{T}(W_1)$ are the modules
  $$T_7 = E_2[1], \quad T_8 = E_2[3], \quad E_3[2], \quad T_9 = E_4[1],$$
  $$T_{10} = E_2[4] = V_1, \quad E_3[3], \quad E_4[2], \text{ and } E_5[1],$$
- the indecomposable modules of $\mathcal{C}(V_2) \cap \mathcal{T}(W_2)$ are the modules
  $$T_{12} = E_7[2] = V_2, \quad \text{and} \quad T_{11} = E_8[1],$$
- $\mathcal{C}(V_3) \cap \mathcal{T}(W_3)$ contains only one indecomposable module
  $$T_{13} = E_{10}[1] = V_3.$$

It follows that

$$\mathcal{T}_1' = \mathrm{Hom}_A(T_A, \mathcal{T}_1^A \cap \mathcal{T}(T))$$

is the ray tube $\mathcal{T}_1^B$ of the tubular $\mathbb{P}_1(K)$-family $\boldsymbol{T}^B = \{\mathcal{T}_\lambda^B\}_{\lambda \in \mathbb{P}_1(K)}$ described above. This finishes the example.

XVII.3. Representation-infinite tilted algebras of Euclidean type

The aim of this section is to show that an arbitrary representation-infinite tilted algebra of Euclidean type is a domestic tubular extension or a domestic tubular coextension of a concealed algebra of Euclidean type in the sense of Definition (VI.4.6).

Throughout we use the following notation introduced in Chapter XII. We assume that $Q = (Q_0, Q_1)$ is an acyclic quiver whose underlying graph $\overline{Q}$ is Euclidean, $A = KQ$ is the path algebra of Q, and

$$n = \operatorname{rk} K_0(A) = |Q_0|$$

is the rank of the Grothendieck group $K_0(A) \cong \mathbb{Z}^n$ of A.

We recall from the structure theorem (XII.3.4) that the Auslander–Reiten quiver $\Gamma(\operatorname{mod} A)$ of A has a disjoint union decomposition

$$\Gamma(\operatorname{mod} A) = \mathcal{P}(A) \cup \mathcal{R}(A) \cup \mathcal{Q}(A),$$

where $\mathcal{P}(A)$ is a unique postprojective component containing all the indecomposable projective A-modules, $\mathcal{Q}(A)$ is a unique preinjective component containing all the indecomposable injective A-modules, and the regular part $\mathcal{R}(A)$ is a $\mathbb{P}_1(K)$-family

$$\boldsymbol{\mathcal{T}}^A = \{\mathcal{T}_\lambda^A\}_{\lambda \in \mathbb{P}_1(K)}$$

of pairwise orthogonal standard stable tubes. The family $\boldsymbol{\mathcal{T}}^A$ is of the tubular type $\mathbf{m}_Q$ (XV.4.7), separates $\mathcal{P}(A)$ from $\mathcal{Q}(A)$, and the following equality holds

$$\sum_{\lambda \in \mathbb{P}_1(K)} (r_\lambda^A - 1) = n - 2, \tag{3.1}$$

where r_λ^A is the rank of the tube $\mathcal{T}_\lambda^A$, for each $\lambda \in \mathbb{P}_1(K)$.

Assume that T_A is a multiplicity-free tilting A-module and

$$B = \operatorname{End} T_A$$

is the associated tilted algebra. We decompose the module T_A as follows

$$T_A = T_A^{pp} \oplus T_A^{rg} \oplus T_A^{pi}, \tag{3.2}$$

where T_A^{pp} is a postprojective A-module, T_A^{rg} is a regular A-module, and T_A^{pi} is a preinjective A-module.

We begin with two useful lemmata.

3.3. Lemma. *Let Q be an acyclic quiver whose underlying graph $\overline{Q}$ is Euclidean, $A = KQ$ the path algebra of Q, and T_A a multiplicity-free tilting A-module with a fixed decomposition (3.2).*

(i) *The tilted algebra $B = \operatorname{End} T_A$ is representation-finite if and only if $T_A^{pp} \neq 0$ and $T_A^{pi} \neq 0$.*

(ii) *If $B = \operatorname{End} T_A$ is not a concealed algebra and B is representation-infinite then $T_A^{rg} \neq 0$.*

Proof. Apply (VIII.4.3), (VIII.4.4), (VIII.4.6) and (XI.5.2). $\qquad\square$

3.4. Lemma. *Let Q be an acyclic quiver whose underlying graph $\overline{Q}$ is Euclidean, $A = KQ$ the path algebra of Q, $n = \operatorname{rk} K_0(A)$, and T_A a multiplicity-free tilting A-module with a fixed decomposition (3.2). Then the regular summand T_A^{rg} of T_A has at most $n - 2$ pairwise non-isomorphic indecomposable direct summands.*

Proof. Given $\lambda \in \mathbb{P}_1(K)$, we denote by r_λ^A the rank of the standard stable tube $\mathcal{T}_\lambda^A$ of the $\mathbb{P}_1(K)$-family $\boldsymbol{\mathcal{T}}^A = \{\mathcal{T}_\lambda^A\}_{\lambda \in \mathbb{P}_1(K)}$. Let T_A be a multiplicity-free tilting A-module with a decomposition $T_A = T_A^{pp} \oplus T_A^{rg} \oplus T_A^{pi}$ (3.2).

Assume that X is an indecomposable direct summand of the regular summand T_A^{rg} of T_A, and assume that X belongs to a standard stable tube $\mathcal{T}_\lambda^A$. Because T_A is a tilting A-module, the module X is a stone and, hence, $r_\lambda \geq 2$. It follows from (1.6) that $\ell_\lambda(X) \leq r_\lambda - 1$, or equivalently, the regular length $r\ell(X)$ of X is at most $r_\lambda - 1$. We also recall from (1.6) that the indecomposable direct summands of T_A lie in the stone cones of tube $\mathcal{T}_\lambda^A$ and satisfy the condition (1.7)(c). It follows that the number of pairwise non-isomorphic indecomposable direct summands of T_A^{rg} lying in the tube $\mathcal{T}_\lambda^A$ is less than or equal to $r_\lambda - 1$. Hence, in view of the formula (3.1), the number of pairwise non-isomorphic indecomposable direct summands of T_A^{rg} is at most $\sum_{\lambda \in \mathbb{P}_1(K)} (r_\lambda^A - 1) = n - 2$, and the lemma follows. $\qquad\square$

3.5. Theorem. *Let Q be an acyclic quiver whose underlying graph $\overline{Q}$ is Euclidean, $A = KQ$ the path algebra of Q, $n = \operatorname{rk} K_0(A)$, and T_A a multiplicity-free tilting A-module with a decomposition $T_A = T_A^{pp} \oplus T_A^{rg} \oplus T_A^{pi}$ (3.2) such that $T_A^{pi} = 0$. Let $B = \operatorname{End} T_A$, $C = \operatorname{End} T_A^{pp}$, and let $(\mathcal{T}(T), \mathcal{F}(T))$ be the torsion pair in $\operatorname{mod} A$ induced by the tilting module T.*

(a) *C is a concealed algebra of Euclidean type, and $2 \leq \operatorname{rk} K_0(C) \leq n$.*

(b) *The tilted algebra B is a domestic tubular extension of C.*

(c) *$\operatorname{gl.dim} B \leq 2$.*

(d) *The Auslander–Reiten quiver $\Gamma(\operatorname{mod} B)$ of B has a disjoint union decomposition*

$$\Gamma(\text{mod } B) = \mathcal{P}(B) \cup \boldsymbol{\mathcal{T}}^B \cup \mathcal{Q}(B),$$

with the following properties.

• $\mathcal{P}(B)$ *is a unique postprojective component of* $\Gamma(\text{mod } B)$ *and has the form*

$$\mathcal{P}(B) = \text{Hom}_A(T, \mathcal{T}(T) \cap \mathcal{P}(A)) = \mathcal{P}(C).$$

Every indecomposable projective C-*module lies in* $\mathcal{P}(B)$, *every indecomposable projective* B-*module* P *lies in* $\mathcal{P}(B) \cup \boldsymbol{\mathcal{T}}^B$, *and the radical* $\text{rad } P$ *of* P *lies in* $\text{add}\,(\mathcal{P}(B) \cup \boldsymbol{\mathcal{T}}^B)$.

• $\mathcal{Q}(B)$ *is a unique preinjective component of* $\Gamma(\text{mod } B)$. $\mathcal{Q}(B)$ *admits a section* $\Sigma_B \cong Q^{op}$ *of the Euclidean type, contains all the indecomposable injective* B-*modules, and contains the translation quiver* $\text{Hom}_A(T, \mathcal{Q}(A))$. *The section* Σ_B *is formed by the images* $\text{Hom}_A(T, I(a))$ *of the indecomposable injective* A-*modules* $I(a)$, *with* $a \in Q_0$. *The set of all proper successors of the section* Σ_B *is finite.*

• *The preinjective component* $\mathcal{Q}(B)$ *is a glueing along the section* Σ_B *of the translation quiver* $\text{Hom}_A(T, \mathcal{Q}(A))$ *and the full translation subquiver* $\Gamma(\text{ind } \mathcal{X}(T))$ *of* $\Gamma(\text{mod } B)$ *whose vertices are the indecomposable* B-*modules of* $\text{ind } \mathcal{X}(T) = \text{Ext}_A^1(T, \text{ind } \mathcal{F}(T))$.

• *The regular part* $\boldsymbol{\mathcal{T}}^B$ *of* $\Gamma(\text{mod } B)$ *has the form*

$$\boldsymbol{\mathcal{T}}^B = \text{Hom}_A(T, \mathcal{T}(T) \cap \boldsymbol{\mathcal{T}}^A)$$

and is a $\mathbb{P}_1(K)$-*family* $\boldsymbol{\mathcal{T}}^B = \{\mathcal{T}_\lambda^B\}_{\lambda \in \mathbb{P}_1(K)}$ *of pairwise orthogonal standard ray tubes* $\mathcal{T}_\lambda^B = \text{Hom}_A(T, \mathcal{T}(T) \cap \mathcal{T}_\lambda^A)$, *with* $r_\lambda^B = r_\lambda^A$. *The ray tube* $\mathcal{T}_\lambda^B$ *is stable if the tube* $\mathcal{T}_\lambda^A$ *contains no indecomposable direct summand of* T. *In this case* $\mathcal{T}_\lambda^B = \mathcal{T}_\lambda^C$ *is a standard stable tube of the* $\mathbb{P}_1(K)$-*family* $\boldsymbol{\mathcal{T}}^C = \{\mathcal{T}_\lambda^C\}_{\lambda \in \mathbb{P}_1(K)}$ *in* $\Gamma(\text{mod } C)$.

• *The family* $\boldsymbol{\mathcal{T}}^B$ *is of the tubular type* $\mathbf{m}_Q$ (XV.4.7), *and separates* $\mathcal{P}(B)$ *from* $\mathcal{Q}(B)$ *in the sense of* (XII.3.3).

(e) $\text{pd}\, X \leq 1$, *for all indecomposable* B-*modules* X, *except a finite number of modules lying in* $\mathcal{Q}(B) \cap \mathcal{X}(T)$.

(f) $\text{id}\, Y \leq 1$, *for each indecomposable module* Y *in* $\mathcal{Q}(B)$.

Proof. Assume that $A = KQ$ is the path algebra of a Euclidean quiver $Q = (Q_0, Q_1)$, $n = \text{rk}\, K_0(A) = |Q_0|$, T_A a multiplicity-free tilting A-module and $T_A = T_A^{pp} \oplus T_A^{rg} \oplus T_A^{pi}$ is a decomposition (3.2) such that $T_A^{pi} = 0$. We set

$$B = \text{End}\, T_A \quad \text{and} \quad C = \text{End}\, T_A^{pp}.$$

If $T_A^{rg} = 0$, then $B = C$ is a concealed algebra of the Euclidean type Q, and the statements (a)–(f) follow from the structure theorems (XII.3.4) and (VIII.4.5).

Assume that $T_A^{rg} \neq 0$. Then, by (3.3) and (VI.4.4), the postprojective A-module T_A^{pp} is non-zero and has at least two non-isomorphic indecomposable direct summands. It follows that the vector space $C = \operatorname{End} T_A^{pp}$ is non-zero, C is an algebra of finite dimension, and $2 \leq q = \operatorname{rk} K_0(C) \leq n$.

Because we assume that $T_A^{pi} = 0$ and $T_A^{rg} \neq 0$, the module T_A admits a direct sum decomposition

$$T_A = T_1 \oplus \ldots \oplus T_q \oplus T_{q+1} \oplus \ldots \oplus T_n,$$

where $n = \operatorname{rk} K_0(A) = |Q_0|$, $2 \leq q \leq n - 1$, and $T_1, \ldots, T_q, T_{q+1}, \ldots, T_n$ are pairwise non-isomorphic indecomposable A-modules such that

$$T_A^{pp} = T_1 \oplus \ldots \oplus T_q \quad \text{and} \quad T_A^{rg} = T_{q+1} \oplus \ldots \oplus T_n.$$

We recall from Chapter VI that the tilting A-module determines the torsion pair $(\mathcal{T}(T), \mathcal{F}(T))$ in $\operatorname{mod} A$ and the torsion pair $(\mathcal{X}(T), \mathcal{Y}(T))$ in $\operatorname{mod} B$. The Brenner-Butler theorem (VI.3.8) yields

- the functor $\operatorname{Hom}_A(T, -) : \operatorname{mod} A \longrightarrow \operatorname{mod} B$ restricts to an equivalence of categories $\operatorname{Hom}_A(T, -) : \mathcal{T}(T) \xrightarrow{\simeq} \mathcal{Y}(T)$, and
- the functor $\operatorname{Ext}_A^1(T, -) : \operatorname{mod} A \longrightarrow \operatorname{mod} B$ restricts to an equivalence of categories $\operatorname{Ext}_A^1(T, -) : \mathcal{F}(T) \xrightarrow{\simeq} \mathcal{X}(T)$.

Because the algebra A is hereditary then, by (VI.5.7), the tilting A-module is splitting, that is, every indecomposable module of $\operatorname{mod} B$ lies either in $\mathcal{X}(T)$ or in $\mathcal{Y}(T)$. We recall that $\mathcal{F}(T) = \{M_A; \operatorname{Hom}_A(T, M) = 0\}$, and (IV.2.4) yields

$$\mathcal{T}(T) = \left\{ M_A; \operatorname{Ext}_A^1(T, M) = 0 \right\} = \left\{ M_A; \operatorname{Hom}_A(M, \tau_A T) = 0 \right\}.$$

Because $\operatorname{Hom}_A(\mathcal{R}(A), \mathcal{P}(A)) = 0$ then $\operatorname{Hom}_A(T_A^{rg}, T_A^{pp}) = 0$ and, in view of the decomposition $T_A = T_A^{pp} \oplus T_A^{rg}$ of T_A the tilted algebra $B = \operatorname{End} T_A$ is of the lower triangular matrix form

$$B \cong \begin{bmatrix} \operatorname{End} T_A^{pp} & \operatorname{Hom}_A(T_A^{rg}, T_A^{pp}) \\ \operatorname{Hom}_A(T_A^{pp}, T_A^{rg}) & \operatorname{End} T_A^{rg} \end{bmatrix} = \begin{bmatrix} C & 0 \\ {}_D M_C & D \end{bmatrix},$$

where $C = \operatorname{End} T_A^{pp}$, $D = \operatorname{End} T_A^{rg}$, and ${}_D M_C = \operatorname{Hom}_A(T_A^{pp}, T_A^{rg})$ is viewed as a D-C-bimodule in an obvious way. The canonical surjection of algebras $B \to C$ induces a full and faithful embedding $\operatorname{mod} C \hookrightarrow \operatorname{mod} B$. It follows from (VI.3.1) that the functor $\mathrm{Hom}_A(T, -) : \mathrm{mod}\,A \longrightarrow \mathrm{mod}\,B$ restricts to an equivalence of categories $\mathrm{Hom}_A(T, -) : \mathrm{add}\,T_A \xrightarrow{\simeq} \mathrm{proj}\,B$, where $\mathrm{proj}\,B$ is the category of finite dimensional projective B-modules. Hence, we derive that

- the indecomposable B-modules

$$\mathrm{Hom}_A(T, T_1), \ldots, \mathrm{Hom}_A(T, T_q), \mathrm{Hom}_A(T, T_{q+1}), \ldots, \mathrm{Hom}_A(T, T_n)$$

form a complete set of pairwise non-isomorphic indecomposable projective B-modules, and
- the indecomposable B-modules

$$\mathrm{Hom}_A(T, T_1) = \mathrm{Hom}_A(T^{pp}, T_1), \ldots, \mathrm{Hom}_A(T, T_q) = \mathrm{Hom}_A(T^{pp}, T_q)$$

lie in $\mathrm{mod}\,C \hookrightarrow \mathrm{mod}\,B$ and form a complete set of pairwise non-isomorphic indecomposable projective C-modules.

To see the preceding statement, we note that, by applying the equality $\mathrm{Hom}_A(T_A^{rg}, T_A^{pp}) = 0$, we get

$$\mathrm{Hom}_A(T, T_j) = \mathrm{Hom}_A(T_A^{pp} \oplus T_A^{rg}, T_j) = \mathrm{Hom}_A(T^{pp}, T_j),$$

for each $j \in \{1, \ldots, q\}$, and therefore the B-module $\mathrm{Hom}_A(T, T_j)$ is a module over the algebra $C = \mathrm{End}\,T_A^{pp}$, in an obvious way.

We recall that the regular part $\mathcal{R}(A)$ of $\Gamma(\mathrm{mod}\,A)$ is a $\mathbb{P}_1(K)$-family

$$\boldsymbol{T}^A = \{\mathcal{T}_\lambda^A\}_{\lambda \in \mathbb{P}_1(K)}$$

of pairwise orthogonal standard stable tubes $\mathcal{T}_\lambda^A$ of $\Gamma(\mathrm{mod}\,A)$. Because the algebra A is hereditary, each of the tubes $\mathcal{T}_\lambda^A$ is hereditary. We denote by $r_\lambda^A = \mathrm{rk}\,\mathcal{T}_\lambda^A$ the rank of the stable tube $\mathcal{T}_\lambda^A$.

Because the proof of the theorem is rather long, we split it into several steps.

Step 1°. We construct a $\mathbb{P}_1(K)$-family

$$\boldsymbol{T}^B = \{\mathcal{T}_\lambda^B\}_{\lambda \in \mathbb{P}_1(K)}$$

of pairwise orthogonal standard ray tubes $\mathcal{T}_\lambda^B$ of $\Gamma(\mathrm{mod}\,B)$ such that

- (t1) $r_\lambda^B = r_\lambda^A$, for each $\lambda \in \mathbb{P}_1(K)$, and
- (t2) the tube $\mathcal{T}_\lambda^B$ lies in $\mathrm{mod}\,C \hookrightarrow \mathrm{mod}\,B$ if and only if the tube $\mathcal{T}_\lambda^A$ does not contain an indecomposable direct summand of the regular A-module T_A^{rg}.

We denote by $\Lambda(T)$ the subset of $\mathbb{P}_1(K)$ containing all $\lambda \in \mathbb{P}_1(K)$ such that the tube $\mathcal{T}_\lambda^A$ contains an indecomposable direct summand of the regular A-module T_A^{rg}.

Note that if $\mathcal{T}_\mu^A$ is a stable tube of $\boldsymbol{\mathcal{T}}^A$ of rank 1 then $\mu \notin \Lambda(T)$, because $\mathrm{Ext}_A^1(X, X) \neq 0$, for any indecomposable A-module of the tube $\mathcal{T}_\mu^A$.

Hence, we easily conclude that

- $r_\lambda^A \geq 2$, for each $\lambda \in \Lambda(T)$,
- $|\Lambda(T)| \leq 3$, by (XIII.3.4), and
- for each $\lambda \in \Lambda(T)$, the image

$$(+) \qquad \mathcal{T}_\lambda^B = \mathrm{Hom}_A(T, \mathcal{T}_\lambda^A \cap \mathcal{T}(T))$$

of $\mathcal{T}_\lambda^A \cap \mathcal{T}(T)$ under the functor $\mathrm{Hom}_A(T, -) : \mathrm{mod}\, A \longrightarrow \mathrm{mod}\, B$ is a standard ray tube of rank $r_\lambda^B = r_\lambda^A$, by (2.3).

Assume now that $\mu \in \mathbb{P}_1(K) \backslash \Lambda(T)$. Because the tubes in $\boldsymbol{\mathcal{T}}^A$ are pairwise orthogonal then

$$\mathrm{Hom}_A(\mathcal{T}_\mu^A, \mathcal{T}_\lambda^A) = 0 \quad \text{and} \quad \mathrm{Hom}_A(\mathcal{T}_\lambda^A, \mathcal{T}_\mu^A) = 0,$$

for each $\lambda \in \Lambda(T)$. It follows that

$$\mathrm{Hom}_A(T_A^{rg}, \mathcal{T}_\mu^A) = 0,$$

because $T_A^{rg} = T_{q+1} \oplus \ldots \oplus T_n$ and each of the indecomposable modules $T_{q+1}, \ldots, T_n$ belongs to a tube $\mathcal{T}_\lambda^A$, with $\lambda \in \Lambda(T)$.

Moreover, because the family $\boldsymbol{\mathcal{T}}^A$ separates $\mathcal{P}(A)$ from $\mathcal{Q}(A)$ then

$$\mathrm{Hom}_A(\mathcal{T}_\mu^A, \mathcal{P}(A)) = 0.$$

It follows that, for any indecomposable module X of the tube $\mathcal{T}_\mu^A$, with $\mu \notin \Lambda(T)$, there are isomorphisms

$$\mathrm{Ext}_A^1(T, X) \cong D\mathrm{Hom}_A(X, \tau_A T) \cong D\mathrm{Hom}_A(X, \tau_A T_A^{pp} \oplus \tau_A T_A^{rg}) = 0.$$

This shows that any module X of the tube $\mathcal{T}_\mu^A$, with $\mu \notin \Lambda(T)$, belongs to the torsion class $\mathcal{T}(T)$ and, consequently, the tube $\mathcal{T}_\mu^A$ is entirely contained in $\mathcal{T}(T)$. Because the tilting A-module T_A is splitting then, by (VI.5.2), the image

$$(++) \qquad \mathcal{T}_\mu^B = \mathrm{Hom}_A(T, \mathcal{T}_\mu^A)$$

of the tube $\mathcal{T}_\mu^A$ under the functor $\mathrm{Hom}_A(T, -) : \mathrm{mod}\, A \longrightarrow \mathrm{mod}\, B$ is a stable ray tube of $\Gamma(\mathrm{mod}\, B)$ of rank $r_\mu^B = r_\mu^A$, for each $\mu \notin \Lambda(T)$. The tube

$\mathcal{T}_\mu^B$ of $\Gamma(\mathrm{mod}\,B)$ is standard, because $\mathcal{T}_\mu^B$ is contained in the full subcategory $\mathcal{Y}(T)$ of $\mathrm{mod}\,B$ and is the image of the standard stable tube $\mathcal{T}_\mu^A$ of $\Gamma(\mathrm{mod}\,A)$ via the equivalence of categories $\mathrm{Hom}_A(T,-) : \mathcal{T}(T) \xrightarrow{\;\cong\;} \mathcal{Y}(T)$.

Now we show that the tube $\mathcal{T}_\mu^B$, with $\mu \notin \Lambda(T)$, consists entirely of C-modules. To see this, we take an indecomposable A-module in $\mathcal{T}_\mu^A$ and note that

$$\mathrm{Hom}_A(T,X) = \mathrm{Hom}_A(T_A^{pp} \oplus T_A^{rg}, X)$$
$$\cong \mathrm{Hom}_A(T_A^{pp}, X) \oplus \mathrm{Hom}_A(T_A^{rg}, X)$$
$$= \mathrm{Hom}_A(T^{pp}, X),$$

because we have observed earlier that $\mathrm{Hom}_A(T_A^{rg}, \mathcal{T}_\mu^A) = 0$. It follows that the B-module $\mathrm{Hom}_A(T,X)$ is a module over the algebra $C = \mathrm{End}\,T_A^{pp}$, in a natural way.

Consequently, the tube $\mathcal{T}_\mu^B$ lies in $\mathrm{mod}\,C \hookrightarrow \mathrm{mod}\,B$ and, therefore, $\mathcal{T}_\mu^C = \mathcal{T}_\mu^B$ is a standard stable tube in $\Gamma(\mathrm{mod}\,C)$ of rank $r_\mu^C = r_\mu^B$.

Summarising, the formulae (+) and (++), for $\lambda \in \Lambda(T)$ and $\mu \notin \Lambda(T)$, define a $\mathbb{P}_1(K)$-family

$$\boldsymbol{T}^B = \{\mathcal{T}_\lambda^B\}_{\lambda \in \mathbb{P}_1(K)}$$

of standard ray tubes $\mathcal{T}_\lambda^B$ of $\Gamma(\mathrm{mod}\,B)$ satisfying the properties (t1) and (t2). The tubes of $\boldsymbol{T}^B$ are pairwise orthogonal, because

$$\boldsymbol{T}^B = \mathrm{Hom}_A(T, \boldsymbol{T}^A \cap \mathcal{T}(T))$$

and the tubes of the family $\boldsymbol{T}^A$ are pairwise orthogonal.

Step 2°. Now we show that the algebra $C = \mathrm{End}\,T_A^{pp}$ is connected. Fix a homogeneous tube $\mathcal{T}_\mu^A$ of $\Gamma(\mathrm{mod}\,A)$, with $\mu \in \mathbb{P}_1(K)$, and an indecomposable A-module E lying on the mouth of $\mathcal{T}_\mu^A$. Then $\mu \notin \Lambda(T)$,

$$\mathcal{T}_\mu^C = \mathcal{T}_\mu^B = \mathrm{Hom}_A(T, \mathcal{T}_\mu^A \cap \mathcal{T}(T))$$

is a standard stable tube of $\Gamma(\mathrm{mod}\,C)$ of rank $r_\mu^C = r_\mu^A = 1$, and the indecomposable C-module $\mathrm{Hom}_A(T,E)$ lies on the mouth of $\mathcal{T}_\mu^C$.

Because $T_A^{pp} = T_1 \oplus \ldots \oplus T_q$ then, by applying (XII.3.6) and the tilting theorem (VI.3.8), we get

$$\mathrm{Hom}_B(\mathrm{Hom}_A(T,T_j), \mathrm{Hom}_A(T,E)) \cong \mathrm{Hom}_A(T_j, E) \neq 0$$

for each $j \in \{1, \ldots, q\}$. This shows that the C-module $\mathrm{Hom}_A(T^{pp}, E) = \mathrm{Hom}_A(T,E)$ is sincere, because the C-modules

$$\mathrm{Hom}_A(T,T_1) = \mathrm{Hom}_A(T^{pp}, T_1), \quad \ldots \quad , \mathrm{Hom}_A(T,T_q) = \mathrm{Hom}_A(T^{pp}, T_q)$$

form a complete set of pairwise non-isomorphic indecomposable projective C-modules. Hence easily follows that the algebra C is connected.

Step 3°. Next we show that

- the preinjective component $\mathcal{Q}(A)$ of $\Gamma(\text{mod } A)$ is entirely contained in the torsion part $\mathcal{T}(T)$, and
- the torsion-free part $\mathcal{F}(T) = \{M_A; \ \text{Hom}_A(T, M) = 0\}$ of mod A has only a finite number of indecomposable modules, up to isomorphism.

The inclusion $\mathcal{Q}(A) \subseteq \mathcal{T}(T)$ follows immediately from the description

$$\mathcal{T}(T) = \{M_A; \ \text{Ext}^1_A(T, M) = 0\} = \{M_A; \ \text{Hom}_A(M, \tau_A T) = 0\}$$

of the torsion part $\mathcal{T}(T)$, because $\text{Hom}_A(\mathcal{Q}(A), \mathcal{P}(A) \cup \mathcal{R}(A)) = 0$ and $T^{pi}_A = 0$, by our assumption.

To prove the second statement of Step 3°, we note that the torsion-free part $\mathcal{F}(T)$ does not contain non-zero preinjective modules, because they are contained in $\mathcal{T}(T)$ and $\mathcal{T}(T) \cap \mathcal{F}(T)$ is zero.

We recall from the beginning of the proof that $T^{pp}_A \neq 0$,

$$T^{pp}_A = T_1 \oplus \ldots \oplus T_q, \quad q \geq 2,$$

and $T_1, \ldots, T_q$ are indecomposable modules. Then each of the modules T_j, with $1 \leq j \leq q$, lies in the postprojective component $\mathcal{P}(A)$ and there is an isomorphism $T_j \cong \tau_A^{-m_j} P(a_j)$, for some vertex $a_j \in Q_0$ and some integer $m_j \geq 0$. Hence, in view of (IV.2.15), there is an isomorphism $\text{Hom}_A(T_j, M) \cong \text{Hom}_A(P(a_j), \tau_A^{m_j} M)$, for any indecomposable postprojective A-module M. If, in addition, the module M lies in $\mathcal{F}(T)$ then $\text{Hom}_A(T, M) = 0$ and, hence,

$$\text{Hom}_A(P(a_j), \tau_A^{m_j} M) \cong \text{Hom}_A(T_j, M) = 0,$$

for $j = 1, \ldots, q$. This means that any such an indecomposable postprojective A-module $\tau_A^{m_j} M$ is not sincere.

On the other hand, it follows from (IX.5.6) that all but a finite number of indecomposable postprojective A-modules in $\mathcal{P}(A)$ are sincere. Hence we easily conclude that the intersection $\mathcal{F}(T) \cap \mathcal{P}(A)$ has a finite number of indecomposable modules.

To finish the proof of Step 3°, we show that the intersection $\mathcal{F}(T) \cap \mathcal{T}^A$ has also a finite number of indecomposable modules. To show this, we fix $\lambda \in \mathbb{P}_1(K)$. It follows from (XII.3.6) that $\text{Hom}_A(T_j, X) \neq 0$, for any indecomposable module of the tube $\mathcal{T}^A_\lambda$ such that $r\ell(X) \geq r^A_\lambda$, where $r\ell(X) = \ell_\lambda(X)$ is the regular length of X. This shows that the tube $\mathcal{T}^A_\lambda$, with $r^A_\lambda \geq 2$, contains a finite number of indecomposable modules from

$\mathcal{F}(T)$, and any tube $\mathcal{T}_\mu^A$, with $r_\mu^A = 1$, does not contain indecomposable modules from $\mathcal{F}(T)$. It follows that the intersection $\mathcal{F}(T) \cap \boldsymbol{\mathcal{T}}^A$ contains only a finite number of indecomposable modules, because the $\mathbb{P}_1(K)$-family $\boldsymbol{\mathcal{T}}^A$ admits at most 3 tubes $\mathcal{T}_\lambda^A$ with $r_\lambda^A \geq 2$.

Step 4°. Our next aim is to show that the Auslander–Reiten quiver $\Gamma(\operatorname{mod} B)$ of $B = \operatorname{End} T_A$ admits a unique preinjective component $\mathcal{Q}(B)$ such that

(i) $\mathcal{Q}(B)$ contains a section Σ_B of the Euclidean type Q^{op} that is formed by the images $\operatorname{Hom}_A(T, I(a))$ of the indecomposable injective A-modules $I(a)$, with $a \in Q_0$,

(ii) $\mathcal{Q}(B)$ contains all indecomposable injective B-modules,

(iii) $\mathcal{Q}(B)$ does not contain an indecomposable projective B-module, and

(iv) the set $\mathcal{Q}(B) \cap \mathcal{X}(T)$ is finite and consists of all proper successors of Σ_B,

(v) $\mathcal{Q}(B)$ contains the image $\operatorname{Hom}_A(T, \mathcal{Q}(A))$ of the preinjective component $\mathcal{Q}(A)$ of $\Gamma(\operatorname{mod} A)$ under the functor

$$\operatorname{Hom}_A(T, -) : \operatorname{mod} A \longrightarrow \operatorname{mod} B.$$

More precisely, we show that the connecting component $\mathcal{C}_T$ of $\Gamma(\operatorname{mod} B)$ determined by T has the properties (i)–(v), and we set $\mathcal{Q}(B) = \mathcal{C}_T$.

Because the component $\mathcal{Q}(A)$ is contained in the torsion part $\mathcal{F}(T)$ and the tilting module T_A is splitting then, according to (VI.5.2), the image $\operatorname{Hom}_A(T, \mathcal{Q}(A))$ of the preinjective component $\mathcal{Q}(A)$ under the functor $\operatorname{Hom}_A(T, -) : \operatorname{mod} A \longrightarrow \operatorname{mod} B$ is a full translation subquiver of $\Gamma(\operatorname{mod} B)$ that is closed under predecessors in $\Gamma(\operatorname{mod} B)$. On the other hand, it follows from (VIII.3.5) that the class of the B-modules of the form $\operatorname{Hom}_A(T, I)$, which runs through pairwise non-isomorphic indecomposable injective A-modules, forms a section Σ_B in the connecting component $\mathcal{C}_T$ of $\Gamma(\operatorname{mod} B)$ determined by T (in the sense of Chapter VIII, page 322) such that

- all predecessors of Σ_B in $\mathcal{C}_T$ lie in $\mathcal{Y}(T)$,
- all proper successors of Σ_B in $\mathcal{C}_T$ lie in $\mathcal{X}(T)$, and
- there is a quiver isomorphism $\Sigma_B \cong Q^{\mathrm{op}}$.

We show that $\mathcal{C}_T$ is a unique preinjective component of $\Gamma(\operatorname{mod} B)$ such that the conditions (i)–(v) are satisfied. Because $\operatorname{Hom}_A(T, \mathcal{Q}(A))$ is closed under predecessors in $\Gamma(\operatorname{mod} B)$ and contains the section Σ_B then

$$\operatorname{Hom}_A(T, \mathcal{Q}(A)) = \mathcal{Y}(T) \cap \mathcal{C}_T.$$

Hence, by applying (VIII.4.1) and the assumption $T_A^{pi} = 0$, we conclude that $\mathcal{C}_T$ contains no indecomposable projective B-modules.

Note that $\mathcal{X}(T)$ has only a finite number of pairwise non-isomorphic indecomposable B-modules, because the functor

$$\mathrm{Ext}_A^1(T,-) : \mathcal{F}(T) \xrightarrow{\;\simeq\;} \mathcal{X}(T)$$

is an equivalence of categories and the torsion-free class $\mathcal{F}(T)$ has only a finite number of pairwise non-isomorphic indecomposable A-modules, by Step 3°. It follows that the set $\mathcal{C}_T \cap \mathcal{X}(T)$ of all proper successors of the section Σ_B in $\mathcal{C}_T$ is finite. Hence, for any indecomposable injective A-module I, there exists an integer $m \geq 0$ such that $\tau_B^m \mathrm{Hom}_A(T,I)$ is an indecomposable injective B-module. Because $n = |Q_0|$ is the rank of the Grothendieck group $K_0(A)$ and, by (VIII.4.3), there is a group isomorphism $K_0(A) \xrightarrow{\;\simeq\;} K_0(B)$ then, by the preceding consideration, the component $\mathcal{C}_T$ contains all indecomposable injective B-modules.

This shows that $\mathcal{Q}(B) = \mathcal{C}_T$ is a preinjective component of $\Gamma(\mathrm{mod}\,B)$ such that the conditions (i), (ii), and (iii) are satisfied. Because obviously $\mathcal{Q}(B) = \mathcal{C}_T$ is a unique preinjective component of $\Gamma(\mathrm{mod}\,B)$ then the statement of Step 4° is proved.

Step 5°. In this step we show that the image

$$\mathcal{P}(B) = \mathrm{Hom}_A(T, \mathcal{T}(T) \cap \mathcal{P}(A))$$

of $\mathcal{T}(T) \cap \mathcal{P}(A)$ under the functor $\mathrm{Hom}_A(T,-) : \mathrm{mod}\,A \longrightarrow \mathrm{mod}\,B$ is a postprojective component of $\Gamma(\mathrm{mod}\,B)$ satisfying the conditions stated in (d) of the theorem.

First we observe that, because $\mathrm{Hom}_A(\mathcal{R}(A), \mathcal{P}(A)) = 0$, then, for any indecomposable A-module Z in $\mathcal{T}(T) \cap \mathcal{P}(A)$, we have $\mathrm{Hom}_A(T_A^{rg}, Z) = 0$. It follows that

$$\mathrm{Hom}_A(T_A, Z) = \mathrm{Hom}_A(T_A^{pp}, Z) \oplus \mathrm{Hom}_A(T_A^{rg}, Z) = \mathrm{Hom}_A(T_A^{pp}, Z).$$

Thus the B-module $\mathrm{Hom}_A(T_A, Z)$ lies in $\mathrm{mod}\,C \hookrightarrow \mathrm{mod}\,B$. This shows that $\mathcal{P}(B) = \mathrm{Hom}_A(T, \mathcal{T}(T) \cap \mathcal{P}(A))$ lies entirely in $\mathrm{mod}\,C \hookrightarrow \mathrm{mod}\,B$.

Next, we show that the Auslander–Reiten quiver $\Gamma(\mathrm{mod}\,C)$ of C admits a family of postprojective components such that their union contains all indecomposable projective C-modules. By (IX.5.1), it is sufficient to show that there is a common bound of the length of the paths in $\mathrm{mod}\,C$ (in the sense of Section IX.1) ending at the indecomposable projective C-modules.

Let $P_i = \mathrm{Hom}_A(T, T_i)$ be an indecomposable projective C-module, with $i \in \{1, \ldots, q\}$, and let

$$(*) \qquad U_0 \xrightarrow{\;g_1\;} U_1 \xrightarrow{\;g_2\;} U_2 \longrightarrow \cdots \longrightarrow U_{s-1} \xrightarrow{\;g_s\;} U_s = P_i$$

be a path of indecomposable modules in $\operatorname{mod} C \hookrightarrow \operatorname{mod} B$. The indecomposable C-modules $U_0, U_1, U_2, \ldots, U_{s-1}, U_s$, viewed as B-modules, are obviously indecomposable. The path $(*)$ lies in the torsion-free part $\mathcal{Y}(T)$ of $\operatorname{mod} B$, because the torsion pair $(\mathcal{X}(T), \mathcal{Y}(T))$ is splitting, the module P belongs to $\mathcal{Y}(T)$, and $\mathcal{Y}(T)$ is closed under predecessors in $\operatorname{mod} B$. Because the functor $\operatorname{Hom}_A(T, -) : \mathcal{T}(T) \xrightarrow{\simeq} \mathcal{Y}(T)$ is an equivalence of categories then there exists a path

$$(**) \qquad Z_0 \xrightarrow{f_1} Z_1 \xrightarrow{f_2} Z_2 \longrightarrow \cdots \longrightarrow Z_{s-1} \xrightarrow{f_s} Z_s = T_i$$

of indecomposable A-modules in $\mathcal{T}(T)$ such that the path $(*)$ is the image of the path $(**)$ under the functor $\operatorname{Hom}_A(T, -)$.

In other words, $U_j = \operatorname{Hom}_A(T, Z_j)$ and $g_{j+1} = \operatorname{Hom}_A(T, f_{j+1})$, for all $j \in \{0, 1, \ldots, q-1\}$. Because the A-modules $T_1, \ldots, T_q$ lie in $\mathcal{P}(A)$, the algebra A is hereditary and Q is a Euclidean quiver then, according to (VIII.2.3), there is a common bound of the length of the paths in $\operatorname{mod} A$ (in the sense of Section IX.1) ending at the postprojective modules $T_1, \ldots, T_q$. It follows that there is a common bound of the length of the paths in $\operatorname{mod} C$ ending at the indecomposable projective C-modules $P_1, \ldots, P_q$. This finishes the proof of our claim.

Because the algebra C is connected and, according to (VIII.2.5), different postprojective components of $\Gamma(\operatorname{mod} C)$ are orthogonal then any family of postprojective components of $\Gamma(\operatorname{mod} C)$ such that their union contains all indecomposable projective C-modules has only one component. It then follows that $\Gamma(\operatorname{mod} C)$ admits a unique postprojective component $\mathcal{P}(C)$ containing all the indecomposable projective C-modules.

Now we show that

$$\mathcal{P}(C) = \mathcal{P}(B) = \operatorname{Hom}_A(T, \mathcal{T}(T) \cap \mathcal{P}(A)).$$

To prove the inclusion $\mathcal{P}(C) \supseteq \mathcal{P}(B) = \operatorname{Hom}_A(T, \mathcal{T}(T) \cap \mathcal{P}(A))$, take an indecomposable module $N = \operatorname{Hom}_A(T, M)$ in $\mathcal{P}(B)$, where M is an indecomposable A-module in $\mathcal{T}(T) \cap \mathcal{P}(A)$. Note that $\operatorname{Hom}_A(T^{rg}, M) = 0$ and, hence, $N = \operatorname{Hom}_A(T, M) = \operatorname{Hom}_A(T_A^{pp}, M)$, that is, the B-module N lies in $\operatorname{mod} C \hookrightarrow \operatorname{mod} B$.

To see that N belongs to $\mathcal{P}(C)$, we show first that there exists an indecomposable projective C-module P such that $\operatorname{Hom}_C(P, N) \neq 0$.

By (VI.2.5), the torsion class $\mathcal{T}(T)$ of $\operatorname{mod} A$ coincides with the class $\operatorname{Gen} T_A$ of all A-modules generated by T_A. Then there is an epimorphism $h : T^t \longrightarrow Z$ in $\operatorname{mod} A$, for some $t \geq 1$. Because $T_A = T_A^{pp} \oplus T_A^{rg}$, $\operatorname{Hom}_A(T_A^{rg}, \mathcal{P}(A)) = 0$, and Z lies in $\mathcal{P}(A)$ then h restricts to the epimorphism $(T_A^{pp})^t \longrightarrow M$ in $\mathcal{T}(T)$ and, consequently, there are a $j \in \{1, \ldots, q\}$

and a non-zero homomorphism $f : T_j \longrightarrow M$ in $\mathcal{T}(T)$. We recall that the modules T_A^{pp}, $T_1, \ldots, T_q$, and M lie in $\mathcal{T}(T) \cap \mathcal{P}(A)$. Because the functor

$$\operatorname{Hom}_A(T, -) : \mathcal{T}(T) \overset{\simeq}{\longrightarrow} \mathcal{Y}(T)$$

is an equivalence of categories then the induced homomorphism

$$f' = \operatorname{Hom}_A(T, f) : \operatorname{Hom}_A(T, T_j) \longrightarrow \operatorname{Hom}_A(T, M) = N$$

of B-modules is non-zero. Because $P_j = \operatorname{Hom}_A(T, T_j) = \operatorname{Hom}_A(T_A^{pp}, T_j)$ is an indecomposable projective C-module, our claim follows.

Assume, to the contrary, that the indecomposable C-module N does not belong to the postprojective component $\mathcal{P}(C)$ of $\Gamma(\operatorname{mod} C)$. Because $\operatorname{Hom}_C(P_j, N) \neq 0$, for an indecomposable projective C-module

$$P_j = \operatorname{Hom}_A(T, T_j),$$

and P_j belongs to $\mathcal{P}(C)$ then (IV.5.1) applies. It follows that there exist an infinite sequence

$$P_j = N_0 \overset{g_1}{\longrightarrow} N_1 \overset{g_2}{\longrightarrow} N_2 \longrightarrow \cdots \longrightarrow N_{t-1} \overset{g_t}{\longrightarrow} N_t \longrightarrow \cdots$$

of irreducible morphisms between indecomposable C-modules, and an infinite sequence $h_1 : N_1 \longrightarrow N, h_2 : N_2 \longrightarrow N, \ldots, h_t : N_t \longrightarrow N, \ldots$ of non-zero homomorphisms in $\operatorname{mod} C$ such that $h_t g_t \ldots g_1 \neq 0$, for each $t \geq 1$. Note that each h_t is not an isomorphism, because N does not belong to $\mathcal{P}(C)$ and the projective C-module P_j belongs to $\mathcal{P}(C)$. Because the algebra A is hereditary then torsion pair $(\mathcal{X}(T), \mathcal{Y}(T))$ of $\operatorname{mod} C$ is splitting and it follows from (VIII.3.2)(b) and its proof that the torsion part $\mathcal{Y}(T)$ of $\operatorname{mod} C$ is closed under predecessors. Hence the infinite sequence lies in $\mathcal{Y}(T)$, because the module N belongs to $\mathcal{Y}(T)$ and each $h_t : N_t \longrightarrow N$ is a non-zero non-isomorphism. Because the functor $\operatorname{Hom}_A(T, -) : \mathcal{T}(T) \overset{\simeq}{\longrightarrow} \mathcal{Y}(T)$ is an equivalence of categories then there exist an infinite sequence

$$T_j = M_0 \overset{f_1}{\longrightarrow} M_1 \overset{f_2}{\longrightarrow} M_2 \longrightarrow \cdots \longrightarrow M_{t-1} \overset{f_t}{\longrightarrow} M_t \longrightarrow \cdots$$

of irreducible morphisms between indecomposable A-modules in $\mathcal{T}(A) \hookrightarrow \operatorname{mod} A$, and a sequence

$$h_1' : M_1 \longrightarrow M, h_2' : M_2 \longrightarrow M, \ldots, h_t' : M_t \longrightarrow M, \ldots$$

of non-zero non-isomorphisms in $\mathcal{T}(A) \hookrightarrow \operatorname{mod} A$ such that

$$N_t = \operatorname{Hom}_A(T, M_t), \ f_t = \operatorname{Hom}_A(T, g_t), \ h_t' = \operatorname{Hom}_A(T, h_t), \ \text{for all } t \geq 1,$$

and $h_t g_t \ldots g_1 \neq 0$, for each $t \geq 1$.

Because the module M belongs to $\mathcal{P}(A)$, the component $\mathcal{P}(A)$ is closed under predecessors, and each $h'_t : M_t \longrightarrow M$ is a non-zero non-isomorphism then the modules $M_1, M_2, \dots, M_t, \dots$ belong to $\mathcal{P}(A)$. Because the homomorphisms $f_1, f_2, \dots, f_t, \dots$ are irreducible morphisms, they are non-zero non-isomorphisms. It follows that the postprojective A-module M has infinitely many predecessors, and we get a contradiction. This finishes the proof of the inclusion $\mathcal{P}(C) \supseteq \mathcal{P}(B) = \operatorname{Hom}_A(T, \mathcal{T}(T) \cap \mathcal{P}(A))$.

To prove the inverse inclusion $\mathcal{P}(C) \subseteq \mathcal{P}(B) = \operatorname{Hom}_A(T, \mathcal{T}(T) \cap \mathcal{P}(A))$, we take an indecomposable module N in $\mathcal{P}(C)$. Because the embedding $\operatorname{mod} C \hookrightarrow \operatorname{mod} B$ is full and faithful and the torsion pair $(\mathcal{X}(T), \mathcal{Y}(T))$ is splitting then N is an indecomposable B-module and N belongs either to $\mathcal{X}(T)$ or to $\mathcal{Y}(T)$.

Note also that

- $\mathcal{P}(C)$ is the postprojective component of $\Gamma(\operatorname{mod} B)$,
- $\mathcal{P}(C) \neq \mathcal{Q}(B)$, because $\mathcal{Q}(B)$ does not contain any indecomposable projective B-module,
- $\mathcal{P}(C)$ does not contain any indecomposable injective B-module, because $\mathcal{Q}(B)$ contains all indecomposable injective B-modules, and
- the number of pairwise non-isomorphic modules of $\mathcal{X}(T)$ is finite.

It follows that the module N of $\mathcal{P}(C)$ is a predecessor of an indecomposable module N' from $\mathcal{Y}(T)$ and, hence, N belongs to $\mathcal{Y}(T)$, because $\mathcal{Y}(T)$ is closed under predecessors in $\operatorname{mod} B$, by (VIII.3.2).

Because the functor $\operatorname{Hom}_A(T, -) : \mathcal{T}(T) \xrightarrow{\simeq} \mathcal{Y}(T)$ is an equivalence of categories then there exists an indecomposable A-module M in $\mathcal{T}(T)$ such that $N = \operatorname{Hom}_A(T, M)$. It follows that the module M lies in $\mathcal{P}(A)$, because we have proved in Step 1° that

$$\boldsymbol{T}^B = \operatorname{Hom}_A(T, \boldsymbol{T}^A \cap \mathcal{T}(T))$$

is a $\mathbb{P}_1(K)$-family of standard ray tubes, and in Step 4° that the image $\operatorname{Hom}_A(T, \mathcal{Q}(A))$ of $\mathcal{Q}(A)$ under $\operatorname{Hom}_A(T, -)$ is contained in the preinjective component $\mathcal{Q}(B)$ of $\Gamma(\operatorname{mod} B)$. This finishes the proof of the equality $\mathcal{P}(C) = \mathcal{P}(B) = \operatorname{Hom}_A(T, \mathcal{T}(T) \cap \mathcal{P}(A))$.

Step 6°. The Auslander–Reiten translation quiver $\Gamma(\operatorname{mod} B)$ of B has the disjoint union decomposition

$$\Gamma(\operatorname{mod} B) = \mathcal{P}(B) \cup \boldsymbol{T}^B \cup \mathcal{Q}(B),$$

every indecomposable projective B-module P lies in $\mathcal{P}(B) \cup \boldsymbol{T}^B$, and its radical $\operatorname{rad} P$ belongs to $\operatorname{add}(\mathcal{P}(B) \cup \boldsymbol{T}^B)$.

First we note that the considerations in Steps 1°–5° show that the three parts $\mathcal{P}(B)$, $\boldsymbol{T}^B$, and $\mathcal{Q}(B)$ are pairwise disjoint. To show that the quiver

$\Gamma(\bmod B)$ admits the decomposition $\Gamma(\bmod B) = \mathcal{P}(B) \cup \boldsymbol{T}^B \cup \mathcal{Q}(B)$, we take an indecomposable module N in $\bmod B$. We recall from (VIII.3.2) that the torsion pair $(\mathcal{X}(T), \mathcal{Y}(T))$ is splitting. Moreover, the functors

$$\operatorname{Hom}_A(T, -) : \mathcal{T}(T) \xrightarrow{\simeq} \mathcal{Y}(T) \text{ and } \operatorname{Ext}_A^1(T, -) : \mathcal{F}(T) \xrightarrow{\simeq} \mathcal{X}(T)$$

are equivalences of categories. Then the indecomposable module N lies either in $\mathcal{X}(T)$ or in $\mathcal{Y}(T)$ and therefore N is of one of the forms

- $N = \operatorname{Hom}_A(T, M)$, where M is an indecomposable A-module in $\mathcal{T}(A)$, or
- $N = \operatorname{Ext}_A^1(T, M)$, where M is an indecomposable A-module in $\mathcal{F}(A)$.

Because the Auslander–Reiten translation quiver $\Gamma(\bmod A)$ of A admits the disjoint union decomposition $\Gamma(\bmod A) = \mathcal{P}(A) \cup \boldsymbol{T}^A \cup \mathcal{Q}(A)$, then the module M belongs to one of the parts $\mathcal{P}(A)$, $\boldsymbol{T}^A$, $\mathcal{Q}(A)$ of $\Gamma(\bmod A)$. Hence, the description of the parts $\mathcal{P}(B)$, $\boldsymbol{T}^B$, and $\mathcal{Q}(B)$ of $\Gamma(\bmod B)$ presented in Steps 1°–5° implies that the indecomposable B-module N belongs to the disjoint union $\mathcal{P}(B) \cup \boldsymbol{T}^B \cup \mathcal{Q}(B)$ and, consequently, we get the decomposition $\Gamma(\bmod B) = \mathcal{P}(B) \cup \boldsymbol{T}^B \cup \mathcal{Q}(B)$.

Finally, if P is an indecomposable projective B-module then P has the form $P \cong \operatorname{Hom}_A(T, T_j)$, for some $j \in \{1, \ldots, n\}$, and it follows from the construction of $\mathcal{P}(B)$ and $\boldsymbol{T}^B$ that the module P belongs to $\mathcal{P}(B) \cup \boldsymbol{T}^B$, because T_j lies in $\mathcal{P}(A) \cup \boldsymbol{T}^A$. Hence we conclude that the radical $\operatorname{rad} P$ of an indecomposable projective B-module belongs to $\operatorname{add}(\mathcal{P}(B) \cup \boldsymbol{T}^B)$, because the embedding $\operatorname{rad} P \hookrightarrow P$ is an irreducible morphism, any indecomposable direct summand of $\operatorname{rad} P$ is a predecessor of P, and $\mathcal{P}(B) \cup \boldsymbol{T}^B$ is closed under predecessors.

Step 7°. We complete the proof of the statement (d) of the theorem by showing that

- the torsion part $\mathcal{X}(T)$ of $\bmod B$ has only finitely many indecomposable modules, up to isomorphism, and they lie in $Q(B)$, and
- the preinjective component $\mathcal{Q}(B)$ is a glueing along the section Σ_B of the translation subquiver $\operatorname{Hom}_A(T, \mathcal{Q}(A))$ of $\mathcal{Q}(B)$ and the full translation subquiver $\Gamma(\operatorname{ind}\mathcal{X}(T))$ of $\Gamma(\bmod B)$ whose vertices are the indecomposable modules of

$$\operatorname{ind} \mathcal{X}(T) = \operatorname{Ext}_A^1(T, \operatorname{ind}\mathcal{F}(T)).$$

We recall from Step 4° that $\mathcal{Q}(B)$ is just the connecting component $\mathcal{C}_T$ of $\Gamma(\bmod B)$ determined by T, and the indecomposable B-modules of the torsion part $\mathcal{X}(T)$ of $\bmod B$ that lie in $\mathcal{C}_T = \mathcal{Q}(B)$ are just the proper successors of the section Σ_B of $\mathcal{C}_T$.

Because the quiver Q is connected and $A = KQ$ then, by (VI.3.5), the tilted algebra $B = \operatorname{End} T_A$ is also connected. Hence, by applying (IV.5.4), we infer that every component of $\Gamma(\operatorname{mod} B)$ is infinite, because Q is a Euclidean quiver and $A = KQ$ is representation-infinite, by (VII.2.7). We recall from Step 3° that the torsion-free part $\mathcal{F}(T)$ of mod A has only finitely many indecomposable modules, up to isomorphism. In view of the equivalence of categories $\operatorname{Ext}_A^1(T, -) : \mathcal{F}(T) \overset{\simeq}{\longrightarrow} \mathcal{X}(T)$, the torsion part $\mathcal{X}(T)$ of mod B has only finitely many indecomposable modules, up to isomorphism. It follows from the description of $\mathcal{P}(B)$ and $\boldsymbol{T}^B$ given in Steps 1° and 5° that $\mathcal{P}(B) \cup \boldsymbol{T}^B$ does not contain indecomposable module from the torsion part $\mathcal{X}(T)$ of mod B. Then, by the disjoint union decomposition $\Gamma(\operatorname{mod} B) = \mathcal{P}(B) \cup \boldsymbol{T}^B \cup \mathcal{Q}(B)$, all indecomposable modules of $\mathcal{X}(T)$ lie in $\mathcal{Q}(B) = \mathcal{C}_T$.

It is easy to see that the full translation subquiver $\Gamma(\operatorname{ind}(\mathcal{Q}(B) \cap \mathcal{Y}(T)))$ of $\Gamma(\operatorname{mod} B)$, whose vertices are the indecomposable preinjective B-modules that belong to $\mathcal{Y}(T)$, is just the translation subquiver $\operatorname{Hom}_A(T, \mathcal{Q}(A))$ of $\mathcal{Q}(B)$.

Then, by the construction of the section Σ_B, the translation quiver $\mathcal{Q}(B)$ is a glueing of the translation subquivers $\Gamma(\operatorname{ind} \mathcal{X}(T))$ and $\operatorname{Hom}_A(T, \mathcal{Q}(A))$ of $\mathcal{Q}(B)$ along the section Σ_B formed by the images $\operatorname{Hom}_A(T, I(a))$ of indecomposable injective A-modules $I(a)$, with $a \in Q_0$.

Step 8°. We prove that $C = \operatorname{End} T_A^{pp}$ is a concealed algebra of Euclidean type, by showing that C is minimal representation-infinite, in the sense of (XIV.2.1). We recall from the earlier steps that C is connected, representation-infinite, and $\Gamma(\operatorname{mod} C)$ admits a unique postprojective component containing all the indecomposable projective C-modules. The direct sum decomposition

$$T_A^{pp} = T_1 \oplus \ldots \oplus T_q$$

of T_A^{pp} into the direct sum of pairwise non-isomorphic indecomposable A-modules $T_1, \ldots, T_q$ yields the right ideal decomposition

$$C = \operatorname{End} T_A^{pp} \cong \operatorname{Hom}_A(T_A^{pp}, T_1) \oplus \ldots \oplus \operatorname{Hom}_A(T_A^{pp}, T_q)$$

of the algebra C into a direct sum of pairwise non-isomorphic indecomposable projective right C-modules. Let $c_1, \ldots, c_q$ be the pairwise orthogonal primitive idempotents of C defined by the decomposition $T_A^{pp} = T_1 \oplus \ldots \oplus T_q$, see (VI.3.10). Then $1_C = e_1 + \ldots + e_q$ and

$$e_j C = \operatorname{Hom}_A(T_A^{pp}, T_j) = \operatorname{Hom}_A(T_A, T_j),$$

for any $j \in \{1, \ldots, q\}$. We recall that there is a full and faithful embedding $\operatorname{mod} C \hookrightarrow \operatorname{mod} B$ induced by the canonical epimorphism $B \overset{\simeq}{\longrightarrow} C$ of algebras. The tilting A-module T_A determines the torsion pair $(\mathcal{X}(T), \mathcal{Y}(T))$ in mod B such that

- $(\mathcal{X}(T), \mathcal{Y}(T))$ is a splitting torsion pair,
- the number of indecomposable modules in $\mathcal{X}(T)$ is finite, up to isomorphism,
- all but finitely many indecomposable B-modules belong to $\mathcal{Y}(T)$,
- every indecomposable B-module N in $\mathcal{Y}(T)$ is of the form $N = \mathrm{Hom}_A(T, M)$, where M is an indecomposable A-module in $\mathcal{T}(T)$, and
- for each $j \in \{1, \dots, q\}$, there are isomorphisms

$$
\begin{aligned}
N e_j &\cong \mathrm{Hom}_C(e_j C, M) \\
&\cong \mathrm{Hom}_B(\mathrm{Hom}_A(T, T_j), \mathrm{Hom}_A(T, M)) \\
&\cong \mathrm{Hom}_A(T_j, M).
\end{aligned}
$$

To prove that C is a minimal representation-infinite algebra, it is sufficient to show that each of the quotient algebras $C/Ce_1C, \dots, C/Ce_qC$ is representation-finite, because we have shown in Step 1° that C is representation-infinite.

Fix $j \in \{1, \dots, q\}$. Note that indecomposable C/Ce_jC-modules are just the indecomposable C-modules N such that $Ne_j = 0$. In view of the isomorphism $Ne_j \cong \mathrm{Hom}_A(T_j, M)$, for a C-module $N = \mathrm{Hom}_A(T, M)$ with M an indecomposable A-module in $\mathcal{T}(T)$, the C-module N is a C/Ce_jC-module if and only if $\mathrm{Hom}_A(T_j, M) = 0$.

It follows from the preceding observations that the algebra C/Ce_jC is representation-finite, if $\mathrm{Hom}_A(T_j, M) \neq 0$, for all but a finite number of indecomposable A-modules M, up to isomorphism.

It is already shown in Steps 1° and 5° that $\mathrm{Hom}_A(T_j, M) \neq 0$, for all but a finite number of indecomposable A-modules M in $\mathcal{P}(A) \cup \boldsymbol{T}^A$. Then, it remains to show this for indecomposable A-modules in $\mathcal{Q}(A)$.

We recall that each of the indecomposable A-modules $T_1, \dots, T_q$ lies in the postprojective component $\mathcal{P}(A)$ and, for any $j \in \{1, \dots, q\}$, there is an isomorphism $T_j \cong \tau_A^{-m_j} P(a_j)$, for some vertex $a_j \in Q_0$ and some integer $m_j \geq 0$.

Let M be an indecomposable A-module in $\mathcal{Q}(A)$. Then, in view of (IV.2.15), there is an isomorphism

$$
\mathrm{Hom}_A(T_j, M) \cong \mathrm{Hom}_A(P(a_j), \tau_A^{m_j} M),
$$

the module $\tau_A^{m_j} M$ is indecomposable, and lies in $\mathcal{Q}(A)$. On the other hand, it follows from (IX.5.6) that all but a finite number of indecomposable A-modules in $\mathcal{Q}(A)$ are sincere. This means that $\mathrm{Hom}_A(P(a_j), Y) \neq 0$, for all but a finite number of indecomposable A-modules Y in $\mathcal{Q}(A)$. Hence, we easily conclude that $\mathrm{Hom}_A(T_j, M) \neq 0$, for all but a finite number of indecomposable A-modules M in $\mathcal{Q}(A)$. Consequently, the algebra C is minimal representation-infinite.

Because the algebra C has a postprojective component, it follows from the criterion (XIV.2.4) that C is either a concealed algebra of Euclidean type, or else C is the path algebra of the enlarged Kronecker quiver

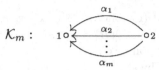

with $m \geq 3$ arrows $\alpha_1, \ldots, \alpha_m$.

To exclude the later case, we recall from (VI.4.7) that $K_0(B) \cong K_0(A) \cong \mathbb{Z}^n$, where $n = |Q_0|$, and the Euler quadratic forms $q_B, q_A : \mathbb{Z}^n \longrightarrow \mathbb{Z}$ of the algebras B and A are $\mathbb{Z}$-congruent, because $B = \mathrm{End}\,T_A$ is a tilted algebra of the Euclidean type Q. By (VII.4.2), the quadratic form q_A is positive semidefinite and, hence, the quadratic form q_B is also positive semidefinite.

On the other hand, because B has the lower triangular matrix form

$$B \cong \begin{bmatrix} C & 0 \\ {}_D M_C & D \end{bmatrix}$$

then the Cartan matrix $\mathbf{C}_B \in \mathbb{M}_n(\mathbb{Z})$ of B has the upper triangular form

$$\mathbf{C}_B = \begin{bmatrix} \mathbf{C}_C & * \\ 0 & \mathbf{C}_D \end{bmatrix},$$

where $\mathbf{C}_C \in \mathbb{M}_q(\mathbb{Z})$ is the Cartan matrix of C, $q = \mathrm{rk}\,K_0(C)$, and $\mathbf{C}_D \in \mathbb{M}_{n-q}(\mathbb{Z})$ is the Cartan matrix of D. It follows from the definition of the Euler quadratic form (III.3.11) that $q_B(\mathbf{x}) = \mathbf{x}^t(\mathbf{C}_B^{-1})^t\mathbf{x}$ and $q_C(\mathbf{x}) = \overline{\mathbf{x}}^t(\mathbf{C}_C^{-1})^t\overline{\mathbf{x}}$, where

$\mathbf{x}^t = [x_1 \ \ldots \ x_q \ \ldots \ x_n] \in \mathbb{Z}^n = K_0(B)$,
$\overline{\mathbf{x}}^t = [x_1 \ \ldots \ x_q] \in \mathbb{Z}^q = K_0(C) \hookrightarrow \mathbb{Z}^n = K_0(B)$, and
$\mathbb{Z}^q \hookrightarrow \mathbb{Z}^n$ is the embedding of abelian groups defined by the formula $[x_1 \ \ldots \ x_q] \mapsto [x_1 \ \ldots \ x_q \ 0 \ \ldots \ 0]$.

Hence easily follows that $q_C : \mathbb{Z}^q \longrightarrow \mathbb{Z}$ is the restriction of $q_B : \mathbb{Z}^n \longrightarrow \mathbb{Z}$ to $\mathbb{Z}^q \hookrightarrow \mathbb{Z}^n$. Because q_B is positive semidefinite then q_C is also positive semidefinite.

It follows that C is not isomorphic to the path algebra $H = K\mathcal{K}_m$ of the enlarged Kronecker quiver $\mathcal{K}_m$, with $m \geq 3$, because the Euler quadratic form $q_H : \mathbb{Z}^2 \longrightarrow \mathbb{Z}$ of H is given by the formula

$$q_H([x_1 \ x_2]^t) = x_1^2 + x_2^2 - mx_1x_2$$

and we have $q_H([1 \ 1]) = 1 + 1 - m = -m + 2 < 0$, when $m \geq 3$.

Consequently, the minimal representation-infinite algebra C is a concealed algebra of Euclidean type. This finishes the proof of the statement (a) of the theorem.

It remains to prove the statement (b), (e), and (f), because B is a tilted algebra and, according to (VI.4.2), we get

$$\text{gl.dim } B - \text{gl.dim } A = \text{gl.dim } B - 1 \leq 1.$$

Hence, gl.dim $B \leq 2$ and (c) follows. Moreover, it follows from (XV.4.3) that $\boldsymbol{T}^B$ separates $\mathcal{P}(B)$ from $\mathcal{Q}(B)$, because we prove in the following step that B is a tubular extension of C.

Step 9°. We prove (b) by showing that the tilted algebra B is a domestic tubular extension of C of tubular type $\mathbf{m}_Q$.

We recall that T_A has the decomposition $T_A = T_A^{pp} \oplus T_A^{rg}$ (3.2), and $\Lambda(T)$ is the subset of $\mathbb{P}_1(K)$ consisting of all $\lambda \in \mathbb{P}_1(K)$ such that the tube $\mathcal{T}_\lambda^A$ contains an indecomposable direct summand of the regular A-module T_A^{rg}.

Fix $\lambda \in \Lambda(T)$ and consider the decomposition

$$T_A^{rg} = T_A^\lambda \oplus \widehat{T}_A^\lambda$$

of the regular A-module T_A^{rg}, where T_A^λ is the direct sum of all indecomposable direct summands of T_A^{rg} lying in the stable tube $\mathcal{T}_\lambda^A$ of the $\mathbb{P}_1(K)$-family $\boldsymbol{T}^A$. Denote by s_λ the number of pairwise non-isomorphic indecomposable direct summands of T_A^λ. Further, we set

$$U_A^\lambda = T_A^{pp} \oplus \widehat{T}_A^\lambda \quad \text{and} \quad B^\lambda = \text{End } U_A^\lambda.$$

It is clear that $T_A = U_A^\lambda \oplus T_A^\lambda$ and $\text{Hom}_A(T_A^\lambda, U_A^\lambda) = 0$, because different tubes of the family $\boldsymbol{T}^A$ are orthogonal and $\text{Hom}_A(\boldsymbol{T}^A, \mathcal{P}(A)) = 0$. By applying (2.3), we infer that

- the translation subquiver $\mathcal{T}_\lambda^\bullet = \text{Hom}_A(T, \mathcal{T}_\lambda^A \cap \mathcal{T}(T) \cap \mathcal{F}(T_A^\lambda))$ is a standard stable tube in $\Gamma(\text{mod } B^\lambda)$ of rank $s_\lambda^\bullet = r_\lambda^A - s_\lambda$,
- the standard ray tube $\mathcal{T}_\lambda^B = \text{Hom}_A(T, \mathcal{T}_\lambda^A \cap \mathcal{T}(T))$ in $\Gamma(\text{mod } B)$ of rank $s_\lambda^B = r_\lambda^A$ is obtained from the stable tube $\mathcal{T}_\lambda^\bullet$ by an iterated rectangle insertion, and
- the tilted algebra B is a $\mathcal{T}_\lambda^\bullet$-tubular extension of the algebra B^λ.

Because $\text{Hom}_A(\widehat{T}_A^\lambda, T_A^{pp}) = 0$ then, in view of the decomposition $U_A^\lambda = T_A^{pp} \oplus \widehat{T}_A^\lambda$ of the A-module U_A^λ, the algebra $B^\lambda = \text{End } U_A^\lambda$ is of the lower triangular matrix form

$$B^\lambda \cong \begin{bmatrix} C & 0 \\ {}_{\widehat{D}^\lambda}\widehat{M}_C & \widehat{D}^\lambda \end{bmatrix},$$

where $C = \operatorname{End} T_A^{pp}$, $\widehat{D}^\lambda = \operatorname{End} \widehat{T}_A^\lambda$, and $_{\widehat{D}^\lambda}\widehat{M}_C = \operatorname{Hom}_A(T_A^{pp}, \widehat{T}_A^\lambda)$ is viewed as a $\widehat{D}^\lambda$-C bimodule in an obvious way. The canonical surjection of algebras $B^\lambda \to C$ induces a full and faithful embedding $\operatorname{mod} C \hookrightarrow \operatorname{mod} B^\lambda$.

Analogously, in view of the decomposition $T_A = U_A^\lambda \oplus T_A^\lambda$ of the A-module T_A, the algebra $B = \operatorname{End} T_A = \operatorname{End}(U_A^\lambda \oplus T_A^\lambda)$ is of the lower triangular matrix form

$$B \cong \begin{bmatrix} B^\lambda & 0 \\ _{D^\lambda}M_{B^\lambda} & D^\lambda \end{bmatrix},$$

where $B^\lambda = \operatorname{End} U_A^\lambda$, $D^\lambda = \operatorname{End} T_A^\lambda$, and $_{D^\lambda}M_{B^\lambda} = \operatorname{Hom}_A(U_A^\lambda, T_A^\lambda)$ is viewed as a D^λ-B^λ-bimodule in an obvious way. The canonical surjection of algebras $B \to B^\lambda$ induces a full and faithful embedding $\operatorname{mod} B^\lambda \hookrightarrow \operatorname{mod} B$.

Now we show that the stable tube $\mathcal{T}_\lambda^\bullet$ consists entirely of C-modules in the subcategory $\operatorname{mod} C \hookrightarrow \operatorname{mod} B^\lambda \hookrightarrow \operatorname{mod} B$ of $\operatorname{mod} B$, that is, $\mathcal{T}_\lambda^\bullet$ coincides with the stable standard tube $\mathcal{T}_\lambda^C$ of $\Gamma(\operatorname{mod} C)$.

Let L be an indecomposable B^λ-module lying in the tube $\mathcal{T}_\lambda^\bullet$. Then L is of the form $L = \operatorname{Hom}_A(T_A, Z)$, where Z is an indecomposable A-module from $\mathcal{T}_\lambda^A \cap \mathcal{T}(T) \cap \mathcal{F}(T_A^\lambda)$. Then $\operatorname{Hom}_A(T_A^\lambda, Z) = 0$, because Z lies in $\mathcal{F}(T_A^\lambda)$, and $\operatorname{Hom}_A(\widehat{T}_A^\lambda, Z) = 0$, because Z lies in the tube $\mathcal{T}_\lambda^A$ and $\widehat{T}_A^\lambda$ belongs to $\operatorname{add}(\mathcal{T}^A \setminus \mathcal{T}_\lambda^A)$, and different tubes are orthogonal. The decomposition $T_A = T_A^{pp} \oplus T_A^\lambda \oplus \widehat{T}_A^\lambda$ yields

$$L = \operatorname{Hom}_A(T_A, Z)$$
$$\cong \operatorname{Hom}_A(T_A^{pp}, Z) \oplus \operatorname{Hom}_A(T_A^\lambda, Z) \oplus \operatorname{Hom}_A(\widehat{T}_A^\lambda, Z)$$
$$= \operatorname{Hom}_A(T_A^{pp}, Z).$$

It follows that the B-module $L \cong \operatorname{Hom}_A(T_A^{pp}, Z)$ has a natural structure of C-module, that is, L belongs to $\operatorname{mod} C \hookrightarrow \operatorname{mod} B^\lambda \hookrightarrow \operatorname{mod} B$. Hence, $\mathcal{T}_\lambda^\bullet$ coincides with the standard stable tube $\mathcal{T}_\lambda^C$ of $\Gamma(\operatorname{mod} C)$.

It follows that the Auslander–Reiten quiver $\Gamma(\operatorname{mod} C)$ of C admits a unique $\mathbb{P}_1(K)$-family $\boldsymbol{\mathcal{T}}^C = \{\mathcal{T}_\lambda^C\}_{\lambda \in \mathbb{P}_1(K)}$ of pairwise orthogonal standard stable tubes of the form

$$\mathcal{T}_\lambda^C = \begin{cases} \operatorname{Hom}_A(T, \mathcal{T}_\lambda^A), & \text{for } \lambda \in \mathbb{P}_1(K) \setminus \Lambda(T), \\ \operatorname{Hom}_A(T, \mathcal{T}_\lambda^A \cap \mathcal{T}(T) \cap \mathcal{F}(T_A^\lambda)), & \text{for } \lambda \in \Lambda(T), \end{cases}$$

the algebra B is a $\boldsymbol{\mathcal{T}}^C$-tubular extension of the algebra C, and the $\mathbb{P}_1(K)$-family $\boldsymbol{\mathcal{T}}^B = \{\mathcal{T}_\lambda^B\}_{\lambda \in \mathbb{P}_1(K)}$ of pairwise orthogonal ray tubes is obtained from the family $\boldsymbol{\mathcal{T}}^C$ by rectangle insertions. The tubular type $r^C = (r_\lambda^C)_{\lambda \in \mathbb{P}_1(K)}$ of the $\mathbb{P}_1(K)$-family $\boldsymbol{\mathcal{T}}^C = \{\mathcal{T}_\lambda^C\}_{\lambda \in \mathbb{P}_1(K)}$ is given by the formula

$$r^C_\lambda = \begin{cases} r^B_\lambda = r^A_\lambda, & \text{for } \lambda \in \mathbb{P}_1(K) \setminus \Lambda(T), \\ r^\bullet_\lambda = r^A_\lambda - s_\lambda, & \text{for } \lambda \in \Lambda(T), \end{cases}$$

where s_λ is the number of pairwise non-isomorphic indecomposable direct summands of T^λ_A. Because it is shown in Steps 1° and 2° that the tubular type $r^B = (r^B_\lambda)_{\lambda \in \mathbb{P}_1(K)}$ of B coincides with the tubular type $r^A = \mathbf{m}_Q$ (XV.4.7) of the family $\boldsymbol{\mathcal{T}}^A$ then the algebra B is a domestic tubular extension of the algebra C.

<u>Step 10°.</u> We complete the proof of the theorem by proving the statements (e) and (f). We recall from Step 8° that gl.dim $B \leq 2$, that is, pd $X \leq 2$ and id $X \leq 2$, for any indecomposable B-module X.

Now we show that the indecomposable B-modules X such that pd $X = 2$ are contained in $\mathcal{Q}(B) \cap \mathcal{X}(T)$ and the number of pairwise non-isomorphic such modules X is finite. To prove the claim, assume that X is an indecomposable B-module such that pd $X = 2$. It follows from (IV.2.7) that then $\operatorname{Hom}_B(D(B), \tau_B X) \neq 0$. This implies that X is a proper successor of Σ_B, because Σ_B is the section of $\mathcal{Q}(B) = \mathcal{C}_T$, and $\mathcal{Q}(B)$ contains all the indecomposable injective B-modules, that is, the indecomposable direct summands of $D(B)$, which are then successors of Σ_B. By Step 4°, the set $\mathcal{Q}(B) \cap \mathcal{X}(T)$ consisting of all proper successors of Σ_B is finite, and our claim follows. Hence we derive the statement (e) asserting that pd $X \leq 1$, for all indecomposable modules in mod B, except finitely many indecomposable modules lying in $\mathcal{Q}(B) \cap \mathcal{X}(T)$.

To prove the statement (f), assume that Y is an indecomposable B-module in $\mathcal{Q}(B)$. It follows that $\operatorname{Hom}_B(\tau^{-1}_B Y, B) = 0$, because $\tau^{-1}_B Y$ is zero or is an indecomposable B-module from $\mathcal{Q}(B)$, the indecomposable direct summands of the projective B-module B lie in $\mathcal{P}(B) \cup \boldsymbol{\mathcal{T}}^B$ (see Step 6°) and the separation property established in Step 8° yields the equality $\operatorname{Hom}_B(\mathcal{Q}(B), \mathcal{P}(B) \cup \boldsymbol{\mathcal{T}}^B) = 0$. Then (IV.2.7) applies, and we get id $Y \leq 1$. This completes the proof of the theorem. $\square$

We may visualise the structure of the Auslander–Reiten quiver $\Gamma(\operatorname{mod} B)$ of the tilted algebra $B = \operatorname{End} T_A$ of Euclidean type Q studied in Theorem (3.5) in the following picture

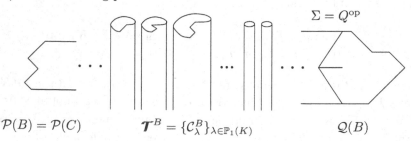

$$\mathcal{P}(B) = \mathcal{P}(C) \qquad \boldsymbol{\mathcal{T}}^B = \{\mathcal{C}^B_\lambda\}_{\lambda \in \mathbb{P}_1(K)} \qquad \mathcal{Q}(B)$$

The following tubular coextension analogue of Theorem (3.5) is of importance

3.6. Theorem. *Let Q be an acyclic quiver whose underlying graph $\overline{Q}$ is Euclidean, $A = KQ$ the path algebra of Q, $n = \operatorname{rk} K_0(A)$, and T_A a multiplicity-free tilting A-module with a decomposition* (3.2)

$$T_A = T_A^{pp} \oplus T_A^{rg} \oplus T_A^{pi}$$

such that $T_A^{pp} = 0$. Let $B = \operatorname{End} T_A$, $C = \operatorname{End} T_A^{pi}$, and let $(\mathcal{T}(T), \mathcal{F}(T))$ be the torsion pair in $\operatorname{mod} A$ induced by the tilting module T.

(a) *C is a concealed algebra of Euclidean type, $T^{pi} \neq 0$, and $2 \leq \operatorname{rk} K_0(C) \leq n$.*

(b) *The tilted algebra B is a domestic tubular coextension of C.*

(c) *gl.dim $B \leq 2$.*

(d) *The Auslander–Reiten quiver $\Gamma(\operatorname{mod} B)$ of B has a disjoint union decomposition*

$$\Gamma(\operatorname{mod} B) = \mathcal{P}(B) \cup \boldsymbol{T}^B \cup \mathcal{Q}(B),$$

with the following properties.

• *$\mathcal{Q}(B)$ is a unique preinjective component of $\Gamma(\operatorname{mod} B)$ and has the form*

$$\mathcal{Q}(B) = \operatorname{Ext}_A^1(T, \mathcal{F}(T) \cap \mathcal{Q}(A)) = \mathcal{Q}(C).$$

Every indecomposable injective C-module lies in $\mathcal{Q}(B)$, every indecomposable injective B-module I lies in $\boldsymbol{T}^B \cup \mathcal{Q}(B)$, and the top $\operatorname{top} I = I/\operatorname{rad} I$ of I lies in $\operatorname{add}(\boldsymbol{T}^B \cup \mathcal{Q}(B))$.

• *$\mathcal{P}(B)$ is a unique postprojective component of $\Gamma(\operatorname{mod} B)$. $\mathcal{P}(B)$ admits a section $\Sigma_B \cong Q^{op}$ of the Euclidean type, contains all the indecomposable projective B-modules, and contains the translation quiver $\operatorname{Ext}_A^1(T, \mathcal{P}(A))$. The section Σ_B is formed by the images $\operatorname{Ext}_A^1(T, P(a))$ of the indecomposable projective A-modules $P(a)$, with $a \in Q_0$. The set of all predecessors of the section Σ_B is finite.*

• *The postprojective component $\mathcal{P}(B)$ is a glueing along the section Σ_B of the translation quiver $\operatorname{Ext}_A^1(T, \mathcal{P}(A))$ and the full translation subquiver $\Gamma(\operatorname{ind} \mathcal{Y}(T))$ of $\Gamma(\operatorname{mod} B)$ whose vertices are the indecomposable modules of $\operatorname{ind} \mathcal{Y}(T) = \operatorname{Hom}_A(T, \operatorname{ind} \mathcal{T}(T))$.*

• *The regular part $\boldsymbol{T}^B$ of $\Gamma(\operatorname{mod} B)$ has the form*

$$\boldsymbol{T}^B = \operatorname{Ext}_A^1(T, \mathcal{F}(T) \cap \boldsymbol{T}^A)$$

and is a $\mathbb{P}_1(K)$-family $\boldsymbol{T}^B = \{\mathcal{T}_\lambda^B\}_{\lambda \in \mathbb{P}_1(K)}$ of pairwise orthogonal standard coray tubes $\mathcal{T}_\lambda^B = \operatorname{Hom}_A(T, \mathcal{T}_\lambda^A)$, with $r_\lambda^B = r_\lambda^A$. The coray tube $\mathcal{T}_\lambda^B$ is stable if the tube $\mathcal{T}_\lambda^A$ contains no indecomposable direct *summand of T. In this case $\mathcal{T}_\lambda^B = \mathcal{T}_\lambda^C$ is a standard stable tube of the $\mathbb{P}_1(K)$-family $\boldsymbol{\mathcal{T}}^C = \{\mathcal{T}_\lambda^C\}_{\lambda \in \mathbb{P}_1(K)}$ in $\Gamma(\operatorname{mod} C)$.*

- *The family $\boldsymbol{\mathcal{T}}^B$ is of the tubular type $\mathbf{m}_Q$ (XV.4.7), and separates $\mathcal{P}(B)$ from $\mathcal{Q}(B)$ in the sense of (XII.3.3).*

(e) $\operatorname{id} X \leq 1$, *for all indecomposable B-modules X, except a finite number of modules lying in $\mathcal{P}(B) \cap \mathcal{Y}(T)$.*

(f) $\operatorname{pd} Y \leq 1$, *for each indecomposable module Y in $\mathcal{P}(B)$.*

Proof. The arguments applied in the proof of Theorem (3.5) modify almost verbatim. The details of the proof are left to the reader. $\square$

We may visualise the structure of the Auslander–Reiten quiver $\Gamma(\operatorname{mod} B)$ of the tilted algebra $B = \operatorname{End} T_A$ of Euclidean type Q studied in Theorem (3.6) in the following picture

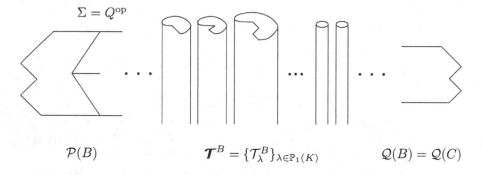

$$\mathcal{P}(B) \qquad\qquad \boldsymbol{\mathcal{T}}^B = \{\mathcal{T}_\lambda^B\}_{\lambda \in \mathbb{P}_1(K)} \qquad\qquad \mathcal{Q}(B) = \mathcal{Q}(C)$$

We illustrate the main idea of Theorem (3.5) with an example.

3.7. Example. Let $A = KQ$ be the path algebra of the four subspace quiver

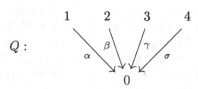

The indecomposable A-modules and the Auslander–Reiten quiver $\Gamma(\operatorname{mod} A)$ are completely described in Section XIII.3. We recall that the unique postprojective component $\mathcal{P}(A)$ of $\Gamma(\operatorname{mod} A)$ is of the form

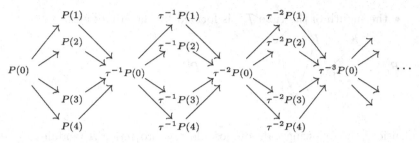

and the unique preinjective component $\mathcal{Q}(A)$ of $\Gamma(\operatorname{mod} A)$ is of the form

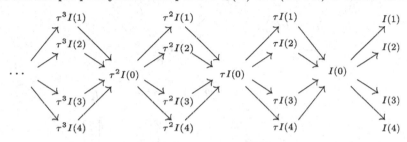

The regular part $\mathcal{R}(A)$ of $\Gamma(\operatorname{mod} A)$ is a $\mathbb{P}_1(K)$-family

$$\boldsymbol{T}^A = \{\mathcal{T}_\lambda^A\}_{\lambda \in \mathbb{P}_1(K)}$$

of standard stable tubes $\mathcal{T}_\lambda^A$ of tubular type $r^A = \hat{r}^A = (2,2,2)$, that is, $\boldsymbol{T}^A$ has three rank two tubes $\mathcal{T}_\infty^A = \mathcal{T}_\infty^Q$, $\mathcal{T}_0^A = \mathcal{T}_0^Q$, and $\mathcal{T}_1^A = \mathcal{T}_1^Q$ presented in (XIII.3.17), and the remaining tubes of $\boldsymbol{T}^A$ are of rank one. We recall from (XIII.3.11) that

- the mouth of the tube $\mathcal{T}_\infty^A$ is formed by the A-modules

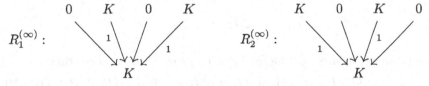

- the mouth of the tube $\mathcal{T}_0^A$ is formed by the A-modules

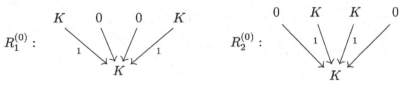

- the mouth of the tube $\mathcal{T}_1^A$ is formed by the A-modules

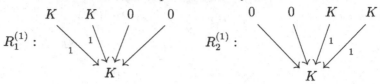

Consider the following indecomposable postprojective A-modules

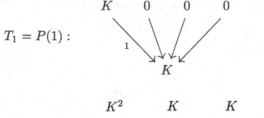

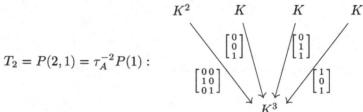

see (XIII.3.6)(c), and the three indecomposable regular A-modules

$$T_3 = R_2^{(\infty)}, \quad T_4 = R_1^{(0)}, \quad \text{and} \quad T_5 = R_1^{(1)}$$

presented above. Now we show that the A-module

$$T_A = T_1 \oplus T_2 \oplus T_3 \oplus T_4 \oplus T_5$$

is a (multiplicity-free) tilting module. To prove it, we observe that

- $\operatorname{Ext}_A^1(T_2, T_1) \cong D\operatorname{Hom}_A(T_1, \tau_A T_2) \cong D\operatorname{Hom}_A(P(1), P(1,1)) = 0$, because the module $\tau_A T_2 = \tau_A^{-1} P(1) = P(1,1)$ has the form

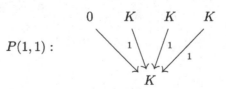

- $\operatorname{Ext}_A^1(T_1, T) = 0$, because the module T_1 is projective,

- $\mathrm{Ext}_A^1(T_2, T_2) \cong D\mathrm{Hom}_A(T_2, \tau_A T_2) = 0$, because T_2 and $\tau_A T_2$ belong to the acyclic component $\mathcal{P}(A)$.
- $\mathrm{Ext}_A^1(T_i, T_j) \cong D\mathrm{Hom}_A(T_j, \tau_A T_i) = 0$, for all $i, j \in \{3, 4, 5\}$, because T_3, T_4, and T_5 lie on the mouth of pairwise orthogonal standard stable tubes $\mathcal{T}_\infty^A$, $\mathcal{T}_0^A$, and $\mathcal{T}_1^A$, respectively,
- $\mathrm{Ext}_A^1(T_2, T_3 \oplus T_4 \oplus T_5) \cong D\mathrm{Hom}_A(T_3 \oplus T_4 \oplus T_5, \tau_A T_2) = 0$, because T_3, T_4, T_5 belong to $\boldsymbol{T}^A$, $\tau_A T_2$ lies on $\mathcal{P}(A)$, and $\mathrm{Hom}_A(\boldsymbol{T}^A, \mathcal{P}(A)) = 0$; and hence
- $\mathrm{Ext}_A^1(T_2, T) = 0$.

Further, there are isomorphisms

$$
\begin{aligned}
\mathrm{Ext}_A^1(T_3 \oplus T_4 \oplus T_5, T) &\cong \mathrm{Ext}_A^1(T_3 \oplus T_4 \oplus T_5, T_1 \oplus T_2) \\
&\cong D\mathrm{Hom}_A(T_1 \oplus T_2, \tau_A T_3 \oplus \tau_A T_4 \oplus \tau_A T_5) \\
&\cong D\mathrm{Hom}_A(T_1 \oplus T_2, R_1^{(\infty)} \oplus R_2^{(0)} \oplus R_2^{(1)}) \\
&\cong D\mathrm{Hom}_A(T_2, R_1^{(\infty)} \oplus R_2^{(0)} \oplus R_2^{(1)}) \\
&\cong D\mathrm{Hom}_A(\tau_A^{-2} P(1), R_1^{(\infty)} \oplus R_2^{(0)} \oplus R_2^{(1)}) \\
&\cong D\mathrm{Hom}_A(P(1), \tau_A^2 R_1^{(\infty)} \oplus \tau_A^2 R_2^{(0)} \oplus \tau_A^2 R_2^{(1)}) \\
&\cong D\mathrm{Hom}_A(P(1), R_1^{(\infty)} \oplus R_2^{(0)} \oplus R_2^{(1)}) = 0,
\end{aligned}
$$

by the shape of the modules $R_1^{(\infty)}$, $R_2^{(0)}$, and $R_2^{(1)}$. Hence $\mathrm{Ext}_A^1(T, T) = 0$ and, according to (VI.4.4), T_A is a tilting A-module, because the algebra A is hereditary and the number 5 of pairwise non-isomorphic direct summands of T equals the rank of the Grothendieck group $K_0(A)$ of A.

In the notation of (3.2), we have $T_A^{pi} = 0$ and

$$
T_A = T_A^{pp} \oplus T_A^{rg}, \text{ where } T_A^{pp} = T_1 \oplus T_2 \text{ and } T_A^{rg} = T_3 \oplus T_4 \oplus T_5.
$$

Let $B = \mathrm{End}\, T_A$ be the associated tilted algebra of the Euclidean type $\widetilde{\mathbb{D}}_4$, and we set $C = \mathrm{End}\, T_A^{pp}$. We show that C is isomorphic to the Kronecker algebra, and we describe B by means of a bound quiver.

First we recall from (VI.3.10) that the following B-modules

$$
\begin{aligned}
P_1 &= \mathrm{Hom}_A(T, T_1), \quad P_2 = \mathrm{Hom}_A(T, T_2), \quad P_3 = \mathrm{Hom}_A(T, T_3), \\
P_4 &= \mathrm{Hom}_A(T, T_4), \quad \text{and} \quad P_5 = \mathrm{Hom}_A(T, T_5),
\end{aligned}
$$

form a complete set of pairwise non-isomorphic indecomposable projective B-modules, the following two C-modules

$$
\begin{aligned}
P_1 &= \mathrm{Hom}_A(T, T_1) = \mathrm{Hom}_A(T^{pp}, T_1), \\
P_2 &= \mathrm{Hom}_A(T, T_2) = \mathrm{Hom}_A(T^{pp}, T_2),
\end{aligned}
$$

form a complete set of pairwise non-isomorphic indecomposable projective C-modules, and there are isomorphisms of K-vector spaces

$$\text{Hom}_B(P_i, P_j) \cong \text{Hom}_A(T_i, T_j) \quad \text{and} \quad \text{Hom}_C(P_a, P_b) \cong \text{Hom}_A(T_a, T_b)$$

for all $i, j \in \{1, 2, 3, 4, 5\}$ and $a, b \in \{1, 2\}$. Moreover, it is easy to see that there exist isomorphisms of K-vector spaces

- $\text{End}_A(T_j) \cong K$, for any $j \in \{1, 2, 3, 4, 5\}$,
- $\text{Hom}_A(T_i, T_j) = 0$, for any $i \in \{3, 4, 5\}$ and $j \in \{1, 2, 3, 4, 5\} \setminus \{i\}$,
- $\text{Hom}_A(T_2, T_1) = 0$,
- $\text{Hom}_A(T_1, T_2) \cong \text{Hom}_A(P(1), P(2, 1)) \cong K$,
- $\text{Hom}_A(T_2, T_3) \cong \text{Hom}_A(T_1, T_3) \cong \text{Hom}_A(P(1), R_2^{(\infty)}) \cong K$,
- $\text{Hom}_A(T_2, T_4) \cong \text{Hom}_A(T_1, T_4) \cong \text{Hom}_A(P(1), R_1^{(0)}) \cong K$, and
- $\text{Hom}_A(T_2, T_5) \cong \text{Hom}_A(T_1, T_5) \cong \text{Hom}_A(P(1), R_1^{(1)}) \cong K$, because $T_3 \cong \tau^2 T_3$, $T_4 \cong \tau^2 T_4$, and $T_5 \cong \tau^2 T_5$.

It follows that the algebra $C = \text{End}\, T_A^{pp} = \text{End}_A(T_1 \oplus T_2)$ is isomorphic to the path algebra of the Kronecker quiver $1 \circ \underset{\beta}{\overset{\alpha}{\rightleftarrows}} \circ 2$, where the vertex 1 corresponds to the module T_1 and the vertex 2 corresponds to the module T_2. Take for the basis $\{\alpha, \beta\}$ of the K-vector space $\text{Hom}_C(P_1, P_2) \cong \text{Hom}_A(T_1, T_2)$ the following two linearly independent compositions of irreducible morphisms

$$f_\alpha = (T_1 = P(1) \xrightarrow{u} \tau_A^{-1} P(0) \xrightarrow{g} \tau_A^{-1} P(3) \xrightarrow{h} \tau_A^{-2} P(0) \xrightarrow{v} \tau_A^{-2} P(1) = T_2),$$
$$f_\beta = (T_1 = P(1) \xrightarrow{u} \tau_A^{-1} P(0) \xrightarrow{e} \tau_A^{-1} P(4) \xrightarrow{w} \tau_A^{-2} P(0) \xrightarrow{v} \tau_A^{-2} P(1) = T_2).$$

Because $\text{Hom}_A(\mathcal{R}(A), \mathcal{P}(A)) = 0$ then $\text{Hom}_A(T_A^{rg}, T_A^{pp}) = 0$ and, in view of the decomposition $T_A = T_A^{pp} \oplus T_A^{rg}$ of T_A, the tilted algebra $B = \text{End}\, T_A$ is of the lower triangular matrix form

$$B \cong \begin{bmatrix} \text{End}\, T_A^{pp} & \text{Hom}_A(T_A^{rg}, T_A^{pp}) \\ \\ \text{Hom}_A(T_A^{pp}, T_A^{rg}) & \text{End}\, T_A^{rg} \end{bmatrix} = \begin{bmatrix} C & 0 \\ {}_D M_C & D \end{bmatrix},$$

where $C = \text{End}\, T_A^{pp}$, $D = \text{End}\, T_A^{rg}$, and ${}_D M_C = \text{Hom}_A(T_A^{pp}, T_A^{rg})$ is viewed as a D-C-bimodule in an obvious way. It follows from (3.5) that the algebra B is a domestic tubular extension of the Kronecker algebra

$$C \cong \begin{bmatrix} K & 0 \\ K^2 & K \end{bmatrix}$$

of tubular type $r^B = \hat{r}^B = (2, 2, 2)$. It is clear that there are isomorphisms of algebras

$$D \cong \text{End}_A(T_3) \times \text{End}_A(T_4) \times \text{End}_A(T_5) \cong K \times K \times K.$$

Hence, we easily conclude that the tilted algebra B is a tubular extension

$$B \cong C[E_3, \mathcal{L}^{(3)}, E_4, \mathcal{L}^{(4)}, E_5, \mathcal{L}^{(5)}],$$

where E_3, E_4, E_5 are the mouth C-modules of three rank one different tubes $\mathcal{T}_{\lambda_3}^A$, $\mathcal{T}_{\lambda_4}^A$, $\mathcal{T}_{\lambda_5}^C$ of the $\mathbb{P}_1(K)$-family $\boldsymbol{\mathcal{T}}^C = \{\mathcal{T}_\lambda^C\}_{\lambda \in \mathbb{P}_1(K)}$ of standard stable tubes in $\Gamma(\mathrm{mod}\, C)$, and each of the three branches $\mathcal{L}^{(3)}, \mathcal{L}^{(4)}, \mathcal{L}^{(5)}$ consists of one vertex. Then, in the notation of (XI.4.3), we have $E_3 = \mathcal{T}_{\lambda_3}^A$, $E_4 = \mathcal{T}_{\lambda_4}^A$, and $E_5 = \mathcal{T}_{\lambda_5}^A$. Hence, we conclude that the ordinary quiver Q_B of B is of the form

If the points λ_3, λ_4, and λ_5 of the projective space $\mathbb{P}_1(K)$ arc of the form $\lambda_3 = (a_3 : b_3)$, $\lambda_4 = (a_4 : b_4)$, and $\lambda_5 = (a_5 : b_5)$ then we have the following relations: $b_3\gamma\alpha = a_3\gamma\beta$, $b_4\gamma\alpha = a_4\gamma\beta$, and $b_5\gamma\alpha = a_5\gamma\beta$. Because $\mathrm{Hom}_A(T_2, T_3)$, $\mathrm{Hom}_A(T_2, T_4)$, and $\mathrm{Hom}_A(T_2, T_5)$ are one-dimensional vector spaces then

$$\mathrm{Hom}_A(T_2, T_3) = Kf_\gamma, \ \mathrm{Hom}_A(T_2, T_4) = Kf_\sigma, \ \text{and} \ \mathrm{Hom}_A(T_2, T_5) = Kf_\delta,$$

for some non-zero homomorphisms $f_\gamma : T_2 \longrightarrow T_3$, $f_\sigma : T_2 \longrightarrow T_4$, and $f_\delta : T_2 \longrightarrow T_5$. Because

$$\mathrm{Hom}_A(\tau_A^{-1}P(3), T_3) \cong \mathrm{Hom}_A(P(3), \tau_A T_3) \cong \mathrm{Hom}_A(P(3), R_1^{(\infty)}) = 0,$$
$$\mathrm{Hom}_A(\tau_A^{-1}P(4), T_4) \cong \mathrm{Hom}_A(P(4), \tau_A T_4) \cong \mathrm{Hom}_A(P(4), R_2^{(0)}) = 0$$

then $f_\gamma f_\alpha = 0$ and $f_\sigma f_\beta = 0$, and the relations $\gamma\alpha = 0$ and $\sigma\beta = 0$ hold in the algebra B. It follows that $\lambda_3 = (0 : 1) = \infty$, $\lambda_4 = (1 : 0) = 1$, under the identification $\mathbb{P}_1(K) = K \cup \{\infty\}$, and we get

$$E_3 \cong E_\infty : \ K \ \underset{1}{\overset{0}{\rightleftarrows}} \ K, \ \text{and} \ \ E_4 \cong E_0 : \ K \ \underset{0}{\overset{1}{\rightleftarrows}} \ K.$$

Then the module E_5 lies in the tube $\mathcal{T}_{\lambda_5}^C$, where $\lambda_5 \in \mathbb{P}_1(K) \setminus \{0, \infty\}$ is of the form $\lambda_5 = (1 : \lambda)$, for some $\lambda \in K \setminus \{0\}$. Hence, the module E_5 has the form

$$E_5 \cong E_\lambda : \ K \ \underset{\lambda}{\overset{1}{\rightleftarrows}} \ K.$$

It follows that the relation $\lambda f_\delta f_\alpha = f_\delta f_\beta$ holds in $\mathrm{Hom}_A(T_1, T_5)$ and, consequently, the relation $\lambda\delta\alpha = \delta\beta$ holds in the algebra B. This shows that the algebra B is isomorphic to an algebra $B(\lambda)$ given by the quiver

$$
\begin{array}{c}
 \\
1 \circ \underset{\beta}{\overset{\alpha}{\rightrightarrows}} \circ 2 \overset{\sigma}{\leftarrow} \circ 4 \\
\end{array}
$$

and bound by the relations $\gamma\alpha = 0$, $\sigma\beta = 0$, and $\lambda\delta\alpha = \delta\beta$, for some $\lambda \in K \setminus \{0\}$. It is easy to see that, for any $\lambda \in K \setminus \{0\}$, there is an isomorphism of algebras $\varphi : B(1) \xrightarrow{\ \simeq\ } B(\lambda)$ given on the basis vectors of $B(1)$ by $\varphi(\alpha) = \alpha$, $\varphi(\beta) = \lambda^{-1}\beta$, $\varphi(\gamma) = \gamma$, $\varphi(\sigma) = \sigma$, $\varphi(\delta) = \delta$, $\varphi(\gamma\beta) = \lambda^{-1}\gamma\beta$, $\varphi(\sigma\alpha) = \sigma\alpha$, and $\varphi(\delta\beta) = \lambda^{-1}\delta\beta$.

It follows that the algebras B and $B(1)$ are isomorphic and, hence, B is isomorphic to the tubular extension

$$C[E_\infty, E_0, E_1] = C[E_\infty, \mathcal{L}^{(\infty)}, E_0, \mathcal{L}^{(0)}, E_1, \mathcal{L}^{(1)}],$$

where $\mathcal{L}^{(\infty)}$, $\mathcal{L}^{(0)}$, and $\mathcal{L}^{(1)}$ are branches of capacity one. Throughout, we make the identification $B = B(1)$. The remaining part of the example is split into four steps.

Step 1°. We determine the torsion-free part $\mathcal{F}(T)$ of $\operatorname{mod} A$. It follows from the description of the postprojective component $\mathcal{P}(A)$ of $\Gamma(\operatorname{mod} A)$ given in (XIII.3.7) that $\operatorname{Hom}_A(T_1, X) = \operatorname{Hom}_A(P(1), X) \neq 0$, for each indecomposable module X that is not isomorphic to one of the five modules

$$P(0), \quad P(2), \quad P(3), \quad P(4), \quad \text{and} \quad \tau_A^{-1}P(1) = P(1,1).$$

In fact, these are just all indecomposable modules lying in $\mathcal{P}(A) \cap \mathcal{F}(T)$, because the modules $T_2 = \tau_A^{-2}P(1)$, T_3, T_4, and T_5 lie in $\boldsymbol{T}^A$,

$$\operatorname{Hom}_A(\boldsymbol{T}^A, \mathcal{P}(A)) = 0,$$

and the quiver $\mathcal{P}(A)$ is acyclic. Further, it follows from (XII.3.6) that $\operatorname{Hom}_A(T_1, X) = \operatorname{Hom}_A(P(1), X) \neq 0$, for each indecomposable module X lying in a tube $\mathcal{T}_\lambda^A$ of regular length $r\ell(X) \geq r_\lambda$, where r_λ is the rank of $\mathcal{T}_\lambda^A$. Because $\mathcal{T}_\infty^A$, $\mathcal{T}_0^A$, and $\mathcal{T}_1^A$ are the only tubes of $\Gamma(\operatorname{mod} A)$ of rank greater than one, then $\boldsymbol{T}^A \cap \mathcal{F}(T)$ consists of the indecomposable modules

$$R_1^{(\infty)} = \tau_A T_3, \quad R_2^{(0)} = \tau_A T_4, \quad \text{and} \quad R_2^{(1)} = \tau_A T_5.$$

Moreover, because $T^{pp} = 0$ then the preinjective component $\mathcal{Q}(A)$ is entirely contained in the torsion class $\mathcal{T}(T)$. Therefore the modules

$$P(0), \quad P(2), \quad P(3), \quad P(4), \quad P(1,1), \quad R_1^{(\infty)}, \quad R_2^{(0)} \text{ and } R_2^{(1)}$$

form a complete list of pairwise non-isomorphic indecomposable A-modules lying in $\mathcal{F}(T)$.

Step 2°. We determine the torsion part $\mathcal{P}(A) \cap \mathcal{T}(T)$ of $\mathcal{P}(A)$. First we note that $\mathcal{P}(A) \cap \mathcal{T}(T)$ consists of all indecomposable A-modules X satisfying the condition

$$\operatorname{Hom}_A(X, P(1,1) \oplus R_1^{(\infty)} \oplus R_2^{(0)} \oplus R_2^{(1)}) = 0,$$

because $\operatorname{Ext}_A^1(T, X) \cong D\operatorname{Hom}_A(X, \tau_A T)$ and $\tau_A T \cong P(1,1) \oplus R_1^{(\infty)} \oplus R_2^{(0)} \oplus R_2^{(1)}$.

Next, we recall that any indecomposable A-module in $\mathcal{P}(A)$ is of the form $\tau_A^{-j} P(i)$, for some $i \in \{0, 1, 2, 3, 4\}$ and $j \geq 0$, and there is an isomorphism

$$\operatorname{Hom}_A(\tau_A^{-j} P(i), R_1^{(\infty)} \oplus R_2^{(0)} \oplus R_2^{(1)}) \cong \operatorname{Hom}_A(\tau_A^{-j-2} P(i), R_1^{(\infty)} \oplus R_2^{(0)} \oplus R_2^{(1)}),$$

because $R_1^{(\infty)} \cong \tau_A^2 R_1^{(\infty)}$, $R_2^{(0)} \cong \tau_A^2 R_2^{(0)}$, and $R_2^{(1)} \cong \tau_A^2 R_2^{(1)}$.

Finally, it is easy to see that:

- $\operatorname{Hom}_A(P(0), R_1^{(\infty)}) \neq 0$, $\operatorname{Hom}_A(P(2), R_1^{(\infty)}) \neq 0$, $\operatorname{Hom}_A(P(3), R_2^{(0)}) \neq 0$, and $\operatorname{Hom}_A(P(4), R_1^{(\infty)}) \neq 0$,
- $\operatorname{Hom}_A(\tau_A^{-1} P(0), R_1^{(\infty)}) \cong \operatorname{Hom}_A(P(0), \tau_A R_1^{(\infty)}) \cong \operatorname{Hom}_A(P(0), R_2^{(\infty)}) \neq 0$,
- $\operatorname{Hom}_A(\tau_A^{-1} P(1), R_1^{(\infty)}) \cong \operatorname{Hom}_A(P(1), \tau_A R_1^{(\infty)}) \cong \operatorname{Hom}_A(P(1), R_2^{(\infty)}) \neq 0$,
- $\operatorname{Hom}_A(\tau_A^{-1} P(3), R_1^{(\infty)}) \cong \operatorname{Hom}_A(P(3), \tau_A R_1^{(\infty)}) \cong \operatorname{Hom}_A(P(3), R_2^{(\infty)}) \neq 0$,
- $\operatorname{Hom}_A(\tau_A^{-1} P(2), R_2^{(1)}) \cong \operatorname{Hom}_A(P(2), \tau_A R_2^{(1)}) \cong \operatorname{Hom}_A(P(2), R_1^{(1)}) \neq 0$,
- $\operatorname{Hom}_A(\tau_A^{-1} P(4), R_2^{(0)}) \cong \operatorname{Hom}_A(P(4), \tau_A R_2^{(0)}) \cong \operatorname{Hom}_A(P(4), R_1^{(0)}) \neq 0$,
- $\operatorname{Hom}_A(P(1), R_1^{(\infty)} \oplus R_2^{(0)} \oplus R_2^{(1)}) \cong \operatorname{Hom}_A(T_1, \tau_A T_3 \oplus \tau_A T_4 \oplus \tau_A T_5) = 0$.

It follows that the torsion part $\mathcal{P}(A) \cap \mathcal{T}(T)$ of $\mathcal{P}(A)$ consists of the modules $P(2m, 1) \cong \tau_A^{-2m} P(1)$, with $m \geq 0$, constructed in (XIII.3.6)(c). Moreover, for each $m \geq 0$, there are isomorphisms

$$\operatorname{Hom}_A(P(2m, 1), P(2m+2, 1)) \cong \operatorname{Hom}_A(P(1), \tau_A^{-2} P(1))$$
$$\cong \operatorname{Hom}_A(T_1, T_2) \cong K^2.$$

It follows that the image $\mathcal{P}(B) = \operatorname{Hom}_A(T, \mathcal{P}(A) \cap \mathcal{T}(T))$ of the torsion part $\mathcal{P}(A) \cap \mathcal{T}(T)$ of $\mathcal{P}(A)$ under the functor $\operatorname{Hom}_A(T, -) : \operatorname{mod} A \longrightarrow \operatorname{mod} B$ is indeed the postprojective component

of the Auslander–Reiten quiver $\Gamma(\operatorname{mod} C)$ of the Kronecker algebra $C = \operatorname{End} T_A^{pp}$.

Step 3°. We determine the torsion part

$$\boldsymbol{\mathcal{T}}^A \cap \mathcal{T}(T) = \{\mathcal{T}_\lambda^A \cap \mathcal{T}(T)\}_{\lambda \in \mathbb{P}_1(K)}$$

of the $\mathbb{P}_1(K)$-family $\boldsymbol{\mathcal{T}}^A = \{\mathcal{T}_\lambda^A\}_{\lambda \in \mathbb{P}_1(K)}$. In the notation of the proof of (3.5), the subset $\Lambda(T) = \{0, 1, \infty\}$ of the projective line $\mathbb{P}_1(K)$ consists of all $\lambda \in \mathbb{P}_1(K)$ such that the tube $\mathcal{T}_\lambda^A$ contains an indecomposable direct summand of T. It follows that the set $\mathbb{P}_1(K) \backslash \Lambda(T)$ consists of all $\lambda \in \mathbb{P}_1(K)$ such that $r_\lambda^A = 1$. Hence, each stable tube $\mathcal{T}_\lambda^A$, with $\lambda \in \mathbb{P}_1(K) \setminus \Lambda(T)$, is of rank one, is contained entirely in $\mathcal{T}(T)$, and its image

$$\mathcal{T}_\lambda^B = \operatorname{Hom}_A(T, \mathcal{T}_\lambda^A)$$

under the functor $\operatorname{Hom}_A(T, -) : \operatorname{mod} A \longrightarrow \operatorname{mod} B$ is a standard stable tube in $\Gamma(\operatorname{mod} B)$ of rank $r_\lambda^B = r_\lambda^A = 1$. Moreover, it follows from the proof of (3.5) that the tubes $\mathcal{T}_\lambda^B$, with $\lambda \in \mathbb{P}_1(K) \setminus \Lambda(T)$, consist entirely of C-modules, that is, they form the family

$$\{\mathcal{T}_\lambda^C\}_{\lambda \in \mathbb{P}_1(K) \backslash \Lambda(T)} = \{\mathcal{T}_\lambda^B\}_{\lambda \in \mathbb{P}_1(K) \backslash \Lambda(T)}$$

of standard stable tubes in $\Gamma(\operatorname{mod} C)$. On the other hand, it follows from the proof of (2.3) that

- $\mathcal{T}_\infty^A \cap \mathcal{T}(T)$ is formed by all modules of the coray

$$\cdots \longrightarrow [j+1]E_2^{(\infty)} \longrightarrow [j]E_2^{(\infty)} \longrightarrow \cdots \longrightarrow [2]E_2^{(\infty)} \longrightarrow [1]E_2^{(\infty)} = E_2^{(\infty)}$$

of the tube $\mathcal{T}_\infty^A$ ending at the coray module $E_2^{(\infty)} = T_3$,
- $\mathcal{T}_0^A \cap \mathcal{T}(T)$ is formed by all modules of the coray

$$\cdots \longrightarrow [j+1]E_2^{(0)} \longrightarrow [j]E_2^{(0)} \longrightarrow \cdots \longrightarrow [2]E_2^{(0)} \longrightarrow [1]E_2^{(0)} = E_2^{(0)}$$

of the tube $\mathcal{T}_0^A$ ending at the coray module $E_2^{(0)} = T_4$, and
- $\mathcal{T}_1^A \cap \mathcal{T}(T)$ is formed by all modules of the coray

$$\cdots \longrightarrow [j+1]E_2^{(1)} \longrightarrow [j]E_2^{(1)} \longrightarrow \cdots \longrightarrow [2]E_2^{(1)} \longrightarrow [1]E_2^{(1)} = E_2^{(1)}$$

of the tube $\mathcal{T}_1^A$ ending at the coray module $E_2^{(1)} = T_5$.

Therefore
$$\mathcal{T}_\infty^B = \mathrm{Hom}_A(T, \mathcal{T}_\infty^A), \; \mathcal{T}_0^B = \mathrm{Hom}_A(T, \mathcal{T}_0^A), \text{ and } \mathcal{T}_1^B = \mathrm{Hom}_A(T, \mathcal{T}_1^A)$$
are the ray tubes in $\Gamma(\mathrm{mod}\, B)$ of rank 2, and they look as follows

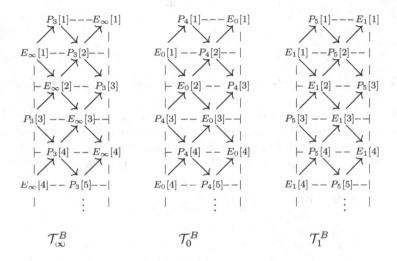

$$\mathcal{T}_\infty^B \qquad\qquad \mathcal{T}_0^B \qquad\qquad \mathcal{T}_1^B$$

where $E_\infty[1] = E_\infty$, $E_0[1] = E_0$, and $E_1[1] = E_1$ are the mouth modules from the homogeneous tubes $\mathcal{T}_\infty^C$, $\mathcal{T}_0^C$, and $\mathcal{T}_1^C$ of $\Gamma(\mathrm{mod}\, C)$, and
$$P_3[1] = P(3) = \mathrm{Hom}_A(T, T_3),$$
$$P_4[1] = P(4) = \mathrm{Hom}_A(T, T_4),$$
$$P_5[1] = P(5) = \mathrm{Hom}_A(T, T_5),$$
are the indecomposable projective B-modules at the vertices 3, 4, and 5 of Q_B.

Note that the tubes $\mathcal{T}_\infty^B$, $\mathcal{T}_0^B$, and $\mathcal{T}_1^B$ are obtained from the tubes $\mathcal{T}_\infty^C$, $\mathcal{T}_0^C$, and $\mathcal{T}_1^C$ by one ray insertions creating the rays starting from the projective modules $P(3)$, $P(4)$, and $P(5)$.

Step 4°. Finally, we describe the preinjective component $\mathcal{Q}(B)$ of the Auslander–Reiten quiver $\Gamma(\mathrm{mod}\, B)$ of the algebra $B = B(1)$. We recall from (3.5) that $\mathcal{Q}(B)$ is a glueing of

$$\mathcal{Q}(B) \cap \mathcal{Y}(T) = \mathrm{Hom}_A(T, \mathcal{Q}(A))$$

and the quiver $\Gamma(\mathcal{X}(T))$ of $\mathcal{X}(T)$ along the section Σ_B formed by the images $I_j = \mathrm{Hom}_A(T, I(j))$ of the indecomposable injective A-modules $I(j)$, with $j \in \{0, 1, 2, 3, 4\}$. A standard calculation technique shows that the component $\mathcal{Q}(B)$ looks as follows

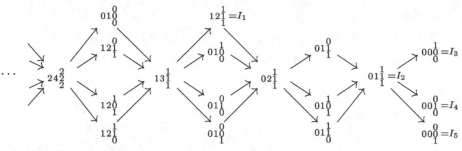

where the indecomposable modules are represented by their dimension vectors.

We note that the module $T_1 = P(1)$ is the unique indecomposable projective direct summand of T_A and $I(1)$ is the injective envelope of the simple module $S(1) \cong P(1)/\operatorname{rad} P(1)$ in $\operatorname{mod} A$. Because T_A is a splitting tilting A-module then, by applying (VI.5.8), we conclude that the B-modules

- $I_1 = \operatorname{Hom}_A(T, I(1))$,
- $I_2 = \operatorname{Ext}_A^1(T, \tau_A T_2)$,
- $I_3 = \operatorname{Ext}_A^1(T, \tau_A T_3)$,
- $I_4 = \operatorname{Ext}_A^1(T, \tau_A T_4)$,
- $I_5 = \operatorname{Ext}_A^1(T, \tau_A T_5)$,

form a complete set of indecomposable injective B-modules. Then the section Σ_B of the connecting component $\mathcal{Q}(B) = \mathcal{C}_T$ is of the form

$$12\genfrac{}{}{0pt}{}{1}{1} = \operatorname{Hom}_A(T, I(1)) = I_1$$

$$01\genfrac{}{}{0pt}{}{1}{0} = \operatorname{Hom}_A(T, I(2))$$

$$\operatorname{Hom}_A(T, I(0)) = 13\genfrac{}{}{0pt}{}{1}{1}$$

$$01\genfrac{}{}{0pt}{}{0}{0} = \operatorname{Hom}_A(T, I(3))$$

$$01\genfrac{}{}{0pt}{}{0}{1} = \operatorname{Hom}_A(T, I(4))$$

Further, by (VI.5.2), we have the following connecting almost split sequences in $\operatorname{mod} B$

$$0 \to \operatorname{Hom}_A(T, I(0)) \to \operatorname{Hom}_A(T, I(0)/S(0)) \oplus \operatorname{Ext}_A^1(T, \operatorname{rad} P(0))$$
$$\to \operatorname{Ext}_A^1(T, P(0)) \to 0,$$

$$0 \to \operatorname{Hom}_A(T, I(2)) \to \operatorname{Hom}_A(T, I(2)/S(2)) \oplus \operatorname{Ext}_A^1(T, \operatorname{rad} P(2))$$
$$\to \operatorname{Ext}_A^1(T, P(2)) \to 0,$$

$$0 \to \operatorname{Hom}_A(T, I(3)) \to \operatorname{Hom}_A(T, I(3)/S(3)) \oplus \operatorname{Ext}_A^1(T, \operatorname{rad} P(3))$$
$$\to \operatorname{Ext}_A^1(T, P(3)) \to 0,$$

$$0 \to \operatorname{Hom}_A(T, I(4)) \to \operatorname{Hom}_A(T, I(4)/S(4)) \oplus \operatorname{Ext}_A^1(T, \operatorname{rad} P(4))$$
$$\to \operatorname{Ext}_A^1(T, P(4)) \to 0.$$

It is easy to see that:

- $I(0)/S(0) \cong I(1) \oplus I(2) \oplus I(3) \oplus I(4)$,
- $\operatorname{rad} P(0) = 0$,
- $I(2)/S(2) = 0$, $I(3)/S(3) = 0$, and
- $I(4)/S(4) = 0$.

It follows that the translation quiver $\Gamma(\mathcal{X}(T)) = \mathcal{Q}(B) \cap \mathcal{X}(T)$ is of the form

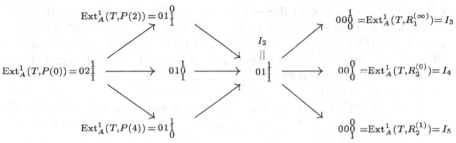

where $I_2 = \operatorname{Ext}_A^1(T, \tau_A^{-1} P(1))$ and $10101 = \operatorname{Ext}_A^1(T, P(3))$.

Finally, we remark that the simple injective B-modules I_3, I_4, and I_5 exhaust the image $\operatorname{Ext}_A^1(T, \boldsymbol{T}^A \cap \mathcal{F}(T))$ of the torsion-free part

$$\boldsymbol{T}^A \cap \mathcal{F}(T) = (\mathcal{T}_\infty^A \cap \mathcal{F}(T)) \cup (\mathcal{T}_0^A \cap \mathcal{F}(T)) \cup (\mathcal{T}_1^A \cap \mathcal{F}(T)) = \left\{ R_1^{(\infty)}, R_2^{(0)}, R_2^{(1)} \right\}$$

of $\boldsymbol{T}^A$ under the functor $\operatorname{Ext}_A^1(T, -) : \operatorname{mod} A \longrightarrow \operatorname{mod} B$. This finishes the example.

XVII.4. Domestic tubular extensions and domestic tubular coextensions of concealed algebras of Euclidean type

The main objective of this section is to prove a converse to Theorems (3.5) and (3.6). We show that every domestic tubular (branch) extension and every domestic tubular (branch) coextension of a concealed algebra of Euclidean type is a representation-infinite tilted algebra of Euclidean type. We begin with two preparatory propositions.

4.1. Proposition. *Let C be a concealed algebra of Euclidean type and let $B = C[E_1, \mathcal{L}^{(1)}, E_2, \mathcal{L}^{(2)}, \ldots, E_s, \mathcal{L}^{(s)}]$ be a domestic branch extension of C.*

(a) *The algebra B is isomorphic to a branch extension*

$$B = C[E_1', \mathcal{L}^{(1)}, E_2', \mathcal{L}^{(2)}, \ldots, E_s', \mathcal{L}^{(s)}]$$

of C such that $r^B = r^{B'}$ and the modules $E_1', E_2', \ldots, E_s'$ lie in $\mathcal{T}_\infty^C \cup \mathcal{T}_0^C \cup \mathcal{T}_1^C$.

(b) *If the algebra B is of tubular type (p, q), with $1 \leq p \leq q$, then the modules $E_1', E_2', \ldots, E_s'$ lie in $\mathcal{T}_\infty^C \cup \mathcal{T}_0^C$.*

Proof. Because we assume that $B = C[E_1, \mathcal{L}^{(1)}, E_2, \mathcal{L}^{(2)}, \ldots, E_s, \mathcal{L}^{(s)}]$ is a domestic branch extension of C then the tubular type $r^B = \hat{r}^B$ of B is of one of the following five forms (p, q), with $1 \leq p \leq q$, $(2, 2, m - 2)$, with $m \geq 4$, $(2, 3, 3)$, $(2, 3, 4)$, and $(2, 3, 5)$. It follows that the $\mathbb{P}_1(K)$-family $\mathcal{T}^B = \{\mathcal{T}_\lambda^B\}_{\lambda \in \mathbb{P}_1(K)}$ of ray tubes of $\Gamma(\mathrm{mod}\, B)$ admits at most 3 tubes of rank bigger than 1. Hence, if C is any of the concealed algebras $C(2, 2, m - 2)$, with $m \geq 4$, $C(2, 3, 3)$, $C(2, 3, 4)$, $C(2, 3, 5)$ then, by (XII.2.12), the three tubes $\mathcal{T}_\infty^C$, $\mathcal{T}_0^C$, and $\mathcal{T}_1^C$ are all stable tubes of $\Gamma(\mathrm{mod}\, C)$ of rank bigger than 1, and therefore the modules $E_1, E_2, \ldots, E_s$ lie in $\mathcal{T}_\infty^C \cup \mathcal{T}_0^C \cup \mathcal{T}_1^C$. In this case we set $E_1' = E_1, E_2' = E_2, \ldots, E_s' = E_s$ and we are done.

Assume that $C = C(p, q)$, where (p, q) is a pair of integers such that $1 \leq p \leq q$. Then C is the path algebra $C = K\Delta(p, q)$ of the following acyclic Euclidean quiver

$$\Delta(p, q) = \Delta(\widetilde{\mathbb{A}}_{p,q}) :$$

By (XII.2.8), the $\mathbb{P}_1(K)$-family $\mathcal{T}^C = \{\mathcal{T}_\lambda^C\}_{\lambda \in \mathbb{P}_1(K)}$ of standard stable tubes of $\Gamma(\mathrm{mod}\, C)$ has the following properties:

(a) $\mathcal{T}_\infty^C$ is a stable tube of rank $r_\infty^C = p$ with the mouth modules $E_1^{(\infty)}, \ldots, E_p^{(\infty)}$,

(b) $\mathcal{T}_0^C$ is a stable tube of rank $r_0^C = q$ with the mouth modules $E_1^{(0)}, \ldots, E_q^{(0)}$,

(c) for each $\lambda \in \mathbb{P}_1(K)$, $\mathcal{T}_\lambda^C$ is a stable tube of rank $r_\lambda^C = 1$ and with the unique mouth module $E^{(\lambda)}$ described in the tables (XII.2.5).

Let E be a module lying on the mouth of a tube of the family $\mathcal{T}^C$. Let $D = C[E]$ be the one-point extension of C by E, and let d be the extension vertex of D. Then (XII.2.5) yields

(a) If $E = E_i^{(\infty)}$ is one of the modules $E_1^{(\infty)}, \ldots, E_p^{(\infty)}$ then D is the bound quiver algebra $D \cong KQ_D/I_D$, where Q_D is the quiver obtained from the quiver $Q_C = \Delta(p, q)$ by adding one arrow $d \xrightarrow{\sigma} a_i$ and I_D is the ideal of KQ_D generated by the path $\sigma\alpha_i$. Here we set $a_p = \omega$.

(b) If $E = E_j^{(0)}$ is one of the modules $E_1^{(0)}, \ldots, E_q^{(0)}$ then D is the bound quiver algebra $D \cong KQ_D/I_D$, where Q_D is the quiver obtained from

the quiver $Q_C = \Delta(p,q)$ by adding one arrow $d \xrightarrow{\sigma} b_j$ and I_D is
the ideal of KQ_D generated by the path $\sigma\beta_j$. Here we set $b_q = \omega$.

(c) If $E = E^{(\lambda)}$ and $\lambda \in K \setminus \{0\}$ then D is the bound quiver algebra $D \cong KQ_D/I_D$, where Q_D is the quiver obtained from the quiver $Q_C = \Delta(p,q)$ by adding one arrow $d \xrightarrow{\sigma} \omega$ and I_D is the ideal of KQ_D generated by the commutativity relation $\lambda\sigma\alpha_p \ldots \alpha_1 - \sigma\beta_q \ldots \beta_1$.

Given $k \in \{1,\ldots,s\}$, we denote by $\mathcal{L}^{(k)} = (L^{(k)}, I^{(k)})$ the branch, with the germ d_k. Note that d_k is the extension vertex of the algebra $C[E_k]$. Then the quiver Q_B is obtained from the quiver $Q_C = \Delta(p,q)$ and the quivers $L^{(1)},\ldots,L^{(s)}$ of the branches $\mathcal{L}^{(1)},\ldots,\mathcal{L}^{(s)}$ by adding the arrows

$$d_1 \xrightarrow{\sigma_1} u_1, \quad d_2 \xrightarrow{\sigma_2} u_2, \quad \ldots, \quad d_s \xrightarrow{\sigma_s} u_s,$$

connecting each of the quivers $L^{(1)},\ldots,L^{(s)}$ with the quiver Q_C, where

$$u_1,\ldots,u_s \in \{a_1,\ldots,a_{p-1}, b_1,\ldots,b_{q-1}, \omega = a_p = b_q\}.$$

Given $k \in \{1,\ldots,s\}$, we choose an index $i_k \in \{1,\ldots,p\}$ such that $u_k = a_{i_k}$, if E_k lies in the tube $\mathcal{T}_\infty^C$, and an index $j_k \in \{1,\ldots,q\}$ such that $u_k = b_{j_k}$, if E_k lies in the tube $\mathcal{T}_0^C$.

Now we split the proof into several cases.

Case 1°. Assume that the tubular type $r^B = \hat{r}^B$ of B is of one of the four domestic types $(2,2,m-2)$, with $m \geq 4$, $(2,3,3)$, $(2,3,4)$, and $(2,3,5)$. We have three subcases to consider.

Case 1.1°. Assume that $q \geq p \geq 2$. Then each of the tubes $\mathcal{T}_\infty^C$ and $\mathcal{T}_0^C$ of $\Gamma(\mathrm{mod}\,C)$ is of rank at least 2, and there is precisely one module E in the set $\{E_1,\ldots,E_s\}$ such that E does not belong to $\mathcal{T}_\infty^C \cup \mathcal{T}_0^C$. It follows that the module E is the mouth module $E_1^{(\lambda)}$ of a stable rank one tube $\mathcal{T}_\lambda^C$, with $\lambda \in K \setminus \{0\}$. Without loss of generality, we may assume that $E = E_s$. Then the algebra

$$B = C[E_1, \mathcal{L}^{(1)}, E_2, \mathcal{L}^{(2)}, \ldots, E_s, \mathcal{L}^{(s)}]$$

is isomorphic to the bound quiver algebra KQ_B/I_B, where I_B is an admissible ideal of the path algebra KQ_B generated by the following elements:

- all the paths of length 2 that generate the ideals $I^{(1)},\ldots,I^{(s)}$ of the branches $\mathcal{L}^{(1)},\ldots,\mathcal{L}^{(s)}$,
- the paths $\sigma_k\alpha_{i_k}$, for all $k \in \{1,\ldots,s-1\}$ such that $E_k \in \mathcal{T}_\infty^C$,
- the paths $\sigma_k\beta_{j_k}$, for all $k \in \{1,\ldots,s-1\}$ such that $E_k \in \mathcal{T}_0^C$, and
- the commutativity relation $\lambda\sigma_s\alpha_s \ldots \alpha_1 - \sigma_s\beta_q \ldots \beta_1$.

Now we define the surjective homomorphism of algebras

$$f : KQ_B \longrightarrow KQ_B/I_B$$

by setting $f(\alpha_p) = \lambda\alpha_p + I_B$, and $f(\gamma) = \gamma + I_B$, for all arrows γ of Q_B that are different from the arrow α_p. Note that the ideal $I_B' = \mathrm{Ker}\, f$ is obtained from the ideal I_B by interchanging the commutativity relation $\lambda\sigma\alpha_s\ldots\alpha_1 - \sigma\beta_q\ldots\beta_1$ with the commutativity relation $\sigma_s\alpha_p\ldots\alpha_1 - \sigma_s\beta_q\ldots\beta_1$. Then f induces the isomorphism $KQ_B/I'_B \xrightarrow{\simeq} KQ_B/I_B$ and, consequently, B is isomorphic to the bound quiver algebra $B' = KQ_B/I'_B$ and B' is the branch extension

$$B' = C[E'_1, \mathcal{L}^{(1)}, E'_2, \mathcal{L}^{(2)}, \ldots, E'_s, \mathcal{L}^{(s)}],$$

where $E'_1 = E_1, E'_2 = E_2, \ldots, E'_{s-1} = E_{s-1}$, and $E'_s = E^{(1)}$ is the unique mouth module of the tube $\mathcal{T}^C_1$. Note that the modules $E'_1, \ldots, E'_s$ lie in $\mathcal{T}^C_\infty \cup \mathcal{T}^C_0 \cup \mathcal{T}^C_1$, and we are done.

Case 1.2°. Assume that $p = 1$ and $q \geq 2$. Then $\mathcal{T}^C_0$ is the unique tube of $\Gamma(\mathrm{mod}\, C)$ of rank at least 2, and there are precisely two modules in the set $\{E_1, \ldots, E_s\}$ lying in different tubes of the family $\{\mathcal{T}^C_\lambda\}_{\lambda \in K \setminus \{0\}}$ of stable tubes of $\mathcal{T}^C$ of rank 1. Without loss of generality, we may assume that $E_1 \in \mathcal{T}^C_{\lambda_1}$ and $E_s \in \mathcal{T}^C_{\lambda_2}$, for some $\lambda_1, \lambda_2 \in \mathbb{P}_1(K) \setminus \{0\}$ such that $\lambda_1 \neq \lambda_2$.

If $\lambda_1 = \infty$ or $\lambda_2 = \infty$, we proceed as in Case 1.1°. Indeed, if $\lambda_1 = \infty$ then $\sigma_1\alpha_1$ belongs to I_B and, as in case Case 1.1°, we show that $B \cong B' \cong KQ_B/I'_B$, where the ideal I'_B is obtained from the ideal I_B by interchanging the commutativity relation $\lambda_2\sigma_2\alpha_1 - \sigma_2\beta_q\ldots\beta_1$ with the commutativity relation $\sigma_2\alpha_1 - \sigma_2\beta_q\ldots\beta_1$.

Assume that $\lambda_1, \lambda_2 \in K \setminus \{0\}$. The algebra $B = C[E_1, \mathcal{L}^{(1)}, \ldots, E_s, \mathcal{L}^{(s)}]$ is then isomorphic to the bound quiver algebra KQ_B/I_B, where I_B is an admissible ideal of the path algebra KQ_B generated by the following elements:

- all the paths of length 2 that generate the ideals $I^{(1)}, \ldots, I^{(s)}$ of the branches $\mathcal{L}^{(1)}, \ldots, \mathcal{L}^{(s)}$,
- the paths $\sigma_k\beta_{j_k}$, for all $k \in \{2, \ldots, s-1\}$,
- $\lambda_1\sigma_1\alpha_1 - \sigma_1\beta_q\ldots\beta_1$,
- $\lambda_2\sigma_2\alpha_1 - \sigma_s\beta_q\ldots\beta_1$.

We define the surjective homomorphism of algebras

$$f : KQ_B \longrightarrow KQ_B/I_B$$

by setting $f(\alpha_1) = \alpha_1 - \lambda_1^{-1}\beta_q\ldots\beta_1 + I_B$, $f(\beta_q) = (\lambda_2^{-1} - \lambda_1^{-1})\beta_q + I_B$, and $f(\gamma) = \gamma + I_B$, for all arrows γ of Q_B that are different from the arrows α_1 and β_q. The definition is correct, because the inequality $\lambda_1 \neq \lambda_2$ yields $\lambda_2^{-1} - \lambda_1^{-1} \neq 0$.

Moreover, the following equalities hold in KQ_B/I_B

$$f(\sigma_1\alpha_1) = f(\sigma_1)f(\alpha_1) = (\sigma_1 + I_B) \cdot (\alpha_1 - \lambda_1^{-1}\beta_q\ldots\beta_1 + I_B)$$
$$= (\sigma_1\alpha_1 - \lambda_1^{-1}\sigma_1\beta_q\ldots\beta_1) + I_B$$
$$= \lambda_1^{-1}[(\lambda_1\sigma_1\alpha_1 - \sigma_1\beta_q\ldots\beta_1) + I_B] = 0 + I_B,$$

$$f(\sigma_s\alpha_1) = f(\sigma_s)f(\alpha_1) = (\sigma_s + I_B) \cdot (\alpha_1 - \lambda_1^{-1}\beta_q \ldots \beta_1 + I_B)$$
$$= (\sigma_s\alpha_1 - \lambda_1^{-1}\sigma_s\beta_q \ldots \beta_1) + I_B$$
$$= \lambda_2^{-1}(\lambda_2\sigma_s\alpha_1 + I_B) - (\lambda_1^{-1}\sigma_s\beta_q \ldots \beta_1 + I_B)$$
$$= \lambda_2^{-1}(\sigma_s\beta_q \ldots \beta_1 + I_B) - \lambda_1^{-1}(\sigma_s\beta_q \ldots \beta_1 + I_B)$$
$$= \sigma_s(\lambda_2^{-1} - \lambda_1^{-1})\beta_q \ldots \beta_1 + I_B$$
$$= (\sigma_s + I_B)(\lambda_2^{-1} - \lambda_1^{-1})(\beta_q + I_B) \ldots (\beta_1 + I_B)$$
$$= f(\sigma_s)f(\beta_q) \ldots f(\beta_1)$$
$$= f(\sigma_s\beta_q \ldots \beta_1).$$

This shows that $\sigma_1\alpha_1 \in \text{Ker } f$, $\sigma_s\alpha_1 - \sigma_s\beta_q \ldots \beta_1 \in \text{Ker } f$ and, consequently, the admissible ideal $I_B' = \text{Ker } f$ of KQ_B is obtained from the ideal I_B by interchanging

- the commutativity relation $\lambda_1\sigma_1\alpha_1 - \sigma_1\beta_q \ldots \beta_1$ with the zero relation $\sigma_1\alpha_1$,
- the commutativity relation $\lambda_2\sigma_s\alpha_1 - \sigma_s\beta_q \ldots \beta_1$ with the commutativity relation $\sigma_s\alpha_1 - \sigma_s\beta_q \ldots \beta_1$,

and keeping the remaining generators unchanged. Then f induces the isomorphism $KQ_B/I_B' \xrightarrow{\simeq} KQ_B/I_B$ and, consequently, B is isomorphic to the bound quiver algebra $B' = KQ_B/I_B'$ and B' is the branch extension

$$B' = C[E_1', \mathcal{L}^{(1)}, E_2', \mathcal{L}^{(2)}, \ldots, E_s', \mathcal{L}^{(s)}],$$

where $E_2' = E_2, \ldots, E_{s-1}' = E_{s-1}$ are modules in $\mathcal{T}_0^C$, $E_1' \in \mathcal{T}_\infty^C$, and $E_s' \in \mathcal{T}_1^C$. Hence, the modules $E_1', \ldots, E_s'$ lie in $\mathcal{T}_\infty^C \cup \mathcal{T}_0^C \cup \mathcal{T}_1^C$, and we are done.

Case 1.3°. Assume that $p = 1$ and $q = 1$, that is, $C = C(1,1)$ is the Kronecker algebra. It follows from (XI.4.6) that $s = 3$, all stable tubes of $\Gamma(\text{mod } C)$ are of rank 1, and there are pairwise different elements $\lambda_1, \lambda_2, \lambda_3 \in \mathbb{P}_1(K)$ such that $E_1 \in \mathcal{T}_{\lambda_1}^C$, $E_2 \in \mathcal{T}_{\lambda_2}^C$, and $E_3 \in \mathcal{T}_{\lambda_3}^C$.

1.3.1° Assume that the set $\{\lambda_1, \lambda_2, \lambda_3\}$ contains the elements 0 and ∞; say $\lambda_1 = \infty$ and $\lambda_2 = 0$. Then $\lambda_3 \in K \setminus \{0\}$, and by applying the arguments used in Case 1.1°, we show that the algebra

$$B = C[E_1, \mathcal{L}^{(1)}, E_2, \mathcal{L}^{(2)}, E_3, \mathcal{L}^{(3)}]$$

is isomorphic to an algebra

$$B' = C[E_1', \mathcal{L}^{(1)}, E_2', \mathcal{L}^{(2)}, E_3', \mathcal{L}^{(3)}],$$

where $E_1' = E_1 \in \mathcal{T}_\infty^C$, $E_2' = E_2 \in \mathcal{T}_0^C$, and $E_3' \in \mathcal{T}_1^C$.

1.3.2° Assume that the set $\{\lambda_1, \lambda_2, \lambda_3\}$ contains exactly one of the elements 0 and ∞; say $\lambda_1 = \infty$. Then, as in 1.2°, there is an isomorphism

$$B = C[E_1, \mathcal{L}^{(1)}, E_2, \mathcal{L}^{(2)}, E_3, \mathcal{L}^{(3)}] \cong B' = C[E_1', \mathcal{L}^{(1)}, E_2', \mathcal{L}^{(2)}, E_3', \mathcal{L}^{(3)}],$$

where $E_1' = E_1 \in \mathcal{T}_\infty^C$, $E_2' \in \mathcal{T}_0^C$, and $E_3' \in \mathcal{T}_1^C$. Similarly, if $\lambda_1 = 0$ then, as in 1.2°, there is an isomorphism

$$B = C[E_1, \mathcal{L}^{(1)}, E_2, \mathcal{L}^{(2)}, E_3, \mathcal{L}^{(3)}] \cong B' = C[E_1', \mathcal{L}^{(1)}, E_2', \mathcal{L}^{(2)}, E_3', \mathcal{L}^{(3)}],$$

where $E_1' \in \mathcal{T}_\infty^C$, $E_2' = E_2 \in \mathcal{T}_0^C$, and $E_3' \in \mathcal{T}_1^C$.

1.3.3°   Assume that $\lambda_1, \lambda_2, \lambda_3 \in K \setminus \{0\}$. Then B is isomorphic to the bound quiver algebra KQ_B/I_B, where I_B is an admissible ideal of the path algebra KQ_B generated by the following elements:

- all paths of length 2 that generate the ideals $I^{(1)}, I^{(2)}, I^{(3)}$ of the branches $\mathcal{L}^{(1)}, \mathcal{L}^{(2)}, \mathcal{L}^{(3)}$, and
- $\lambda_1 \sigma_1 \alpha_1 - \sigma_1 \beta_1$, $\lambda_2 \sigma_2 \alpha_1 - \sigma_2 \beta_1$, and $\lambda_3 \sigma_3 \alpha_1 - \sigma_3 \beta_1$.

We define the surjective homomorphism of algebras

$$f : KQ_B \longrightarrow KQ_B/I_B$$

by setting

$$f(\alpha_1) = \alpha_1 - \lambda_1^{-1}\beta_1 + I_B, \; f(\beta_1) = (\lambda_3 - \lambda_2)^{-1}\lambda_1^{-1}(\lambda_1 - \lambda_3)(\beta_1 - \lambda_2\alpha_1) + I_B,$$

and $f(\gamma) = \gamma + I_B$, for all arrows γ of Q_B that are different from the arrows α_1 and β_1. Observe that $\lambda_1 - \lambda_3 \neq 0$ and $\lambda_3 - \lambda_2 \neq 0$, because the elements $\lambda_1, \lambda_2, \lambda_3$ are pairwise different. Moreover, the following equalities hold in KQ_B/I_B

$$\begin{aligned}
f(\sigma_1\alpha_1) &= f(\sigma_1)f(\alpha_1) = (\sigma_1 + I_B) \cdot (\alpha_1 - \lambda_1^{-1}\beta_1 + I_B) \\
&= (\sigma_1\alpha_1 - \lambda_1^{-1}\sigma_1\beta_1) + I_B \\
&= \lambda_1^{-1}[(\lambda_1\sigma_1\alpha_1 - \sigma_1\beta_1) + I_B] = 0 + I_B, \\
f(\sigma_2\beta_1) &= f(\sigma_2)f(\beta_1) \\
&= (\sigma_2 + I_B)((\lambda_3 - \lambda_2)^{-1}\lambda_1^{-1}(\lambda_1 - \lambda_3)(\beta_1 - \lambda_2\alpha_1) + I_B) \\
&= (\lambda_3 - \lambda_2)^{-1}\lambda_1^{-1}(\lambda_1 - \lambda_3)((\sigma_2\beta_1 - \lambda_2\sigma_2\alpha_1) + I_B) = 0 + I_B, \\
f(\sigma_3\alpha_1) &= f(\sigma_3)f(\alpha_1) = (\sigma_3 + I_B) \cdot (\alpha_1 - \lambda_1^{-1}\beta_1 + I_B) \\
&= (\sigma_3\alpha_1 + I_B) - \lambda_1^{-1}(\sigma_3\beta_1 + I_B) \\
&= (\sigma_3\alpha_1 + I_B) - \lambda_3\lambda_1^{-1}(\sigma_3\alpha_1 + I_B) \\
&= (1 - \lambda_3\lambda_1^{-1})(\sigma_3\alpha_1 + I_B) = \lambda_1^{-1}(\lambda_1 - \lambda_3)(\sigma_3\alpha_1 + I_B) \\
&= (\lambda_3 - \lambda_2)^{-1}\lambda_1^{-1}(\lambda_1 - \lambda_3)((\lambda_3 - \lambda_2)\sigma_3\alpha_1 + I_B) \\
&= (\lambda_3 - \lambda_2)^{-1}\lambda_1^{-1}(\lambda_1 - \lambda_3)((\lambda_3\sigma_3\alpha_1 + I_B) - (\lambda_2\sigma_3\alpha_1 + I_B)) \\
&= (\lambda_3 - \lambda_2)^{-1}\lambda_1^{-1}(\lambda_1 - \lambda_3)((\sigma_3\beta_1 + I_B) - (\lambda_2\sigma_3\alpha_1 + I_B)) \\
&= (\sigma_3 + I_B)((\lambda_3 - \lambda_2)^{-1}\lambda_1^{-1}(\lambda_1 - \lambda_3))((\beta_1 - \lambda_2\alpha_1) + I_B) \\
&= f(\sigma_3)f(\beta_1) = f(\sigma_3\beta_1).
\end{aligned}$$

This shows that $\sigma_1\alpha_1 \in \operatorname{Ker} f$, $\sigma_2\beta_1 \in \operatorname{Ker} f$, $\sigma_3\alpha_1 - \sigma_3\beta_1 \in \operatorname{Ker} f$ and, consequently, the $I'_B = \operatorname{Ker} f$ of KQ_B is an admissible ideal of KQ_B generated by $\sigma_1\alpha_1$, $\sigma_2\beta_1$, $\sigma_3\alpha_1 - \sigma_3\beta_1$, and by all paths of length 2 generating the ideals $I^{(1)}$, $I^{(2)}$, and $I^{(3)}$ of the branches $\mathcal{L}^{(1)}$, $\mathcal{L}^{(2)}$, and $\mathcal{L}^{(s)}$. Then f induces the isomorphism $KQ_B/I'_B \xrightarrow{\simeq} KQ_B/I_B$ and, consequently, B is isomorphic to the branch extension

$$B' = C[E'_1, \mathcal{L}^{(1)}, E'_2, \mathcal{L}^{(2)}, E'_3, \mathcal{L}^{(3)}],$$

where $E'_1 \in \mathcal{T}^C_\infty$, $E'_2 \in \mathcal{T}^C_0$, and $E'_3 \in \mathcal{T}^C_1$. This finish the proof in Case 1°.

Case 2°. Assume that $r^B = \widehat{r}^B = (p', q')$, for some integers p' and q' such that $1 \leq p' < q'$. Because we assume that $C = C(p, q)$, where (p, q) is a pair of integers such that $1 \leq p \leq q$, then $r^C = \widehat{r}^C = (p, q)$, by (XII.2.8). We have three cases to consider.

Case 2.1°. Assume that $q \geq p \geq 2$. Then $\mathcal{T}^C_\infty$ and $\mathcal{T}^C_0$ are the unique tubes of $\Gamma(\operatorname{mod} C)$ of rank greater than or equal to 2. Then our assumption on r^B forces that the modules $E_1, \ldots, E_s$ belong to $\mathcal{T}^C_\infty \cup \mathcal{T}^C_0$, and we are done.

Case 2.2°. Assume that $p = 1$ and $q \geq 2$. Then $\mathcal{T}^C_0$ is the unique tube of $\Gamma(\operatorname{mod} C)$ of rank greater than or equal to 2. Assume that there exists a module $E \in \{E_1, \ldots, E_s\}$ such that E does not belong to $\mathcal{T}^C_\infty \cup \mathcal{T}^C_0$. By our assumption on r^B, E is a unique module in $\{E_1, \ldots, E_s\}$ with this property, and $E \in \mathcal{T}^C_\lambda$, for some $\lambda \in K \setminus \{0\}$. Without loss of generality, we may assume that $E = E_1$. Observe that then the modules $E_2, \ldots, E_s$ also belong to the tube $\mathcal{T}^C_0$. It follows that there is an isomorphism $B \cong KQ_B/I_B$, where I_B is an admissible ideal of the path algebra KQ_B generated by the following elements:

- all the paths of length 2 that generate the ideals $I^{(1)}, \ldots, I^{(s)}$ of the branches $\mathcal{L}^{(1)}, \ldots, \mathcal{L}^{(s)}$,
- the paths $\sigma_k\alpha_{j_k}$, for all $k \in \{2, \ldots, s\}$, and
- the commutativity relation $\lambda\sigma_1\alpha_1 - \sigma_1\beta_q \ldots \beta_1$.

Now we define the surjective homomorphism of algebras

$$f : KQ_B \longrightarrow KQ_B/I_B$$

by setting $f(\alpha_1) = (\alpha_1 - \lambda\beta_q \ldots \beta_1) + I_B$, and $f(\gamma) = \gamma + I_B$, for all arrows γ of Q_B that are different from the arrow α_1. As in Case 1.2°, we show that $f(\sigma_1\alpha_1) = 0$ and there is an isomorphism of algebras

$$B \cong KQ_B/I_B \cong KQ_B/I'_B,$$

where $I'_B = \operatorname{Ker} f$ is an admissible ideal of KQ_B obtained from the ideal I_B by interchanging the commutativity relation $\lambda\sigma_1\alpha_1 - \sigma_1\beta_q \ldots \beta_1$ with the zero relation $\sigma_1\alpha_1$, and keeping the remaining generators unchanged. Therefore, B is isomorphic to the bound quiver algebra $B' = KQ_B/I'_B$ and B' is the branch extension

$$B' = C[E'_1, \mathcal{L}^{(1)}, E'_2, \mathcal{L}^{(2)}, \ldots, E'_s, \mathcal{L}^{(s)}],$$

where $E'_1 \in \mathcal{T}^C_\infty$, the modules $E'_2, \ldots, E'_s$ belong to the tube $\mathcal{T}^C_0$.

Case 2.3°. Assume that $p = q = 1$, that is, $C = C(1,1)$ is the Kronecker algebra and $s = 2$, by (XI.4.6).

If one of the modules E_1 or E_2 belongs to $\mathcal{T}^C_\infty \cup \mathcal{T}^C_0$ then, as in Case 2.2°, the algebra

$$B = C[E_1, \mathcal{L}^{(1)}, E_2, \mathcal{L}^{(2)}]$$

is isomorphic to an algebra

$$B' = C[E'_1, \mathcal{L}^{(1)}, E'_2, \mathcal{L}^{(2)}],$$

the modules E'_1 and E'_2 belong $\mathcal{T}^C_\infty \cup \mathcal{T}^C_0$, and $E'_1 = E_1$, if $E_1 \in \mathcal{T}^C_\infty \cup \mathcal{T}^C_0$, and $E'_2 = E_2$, if $E_2 \in \mathcal{T}^C_\infty \cup \mathcal{T}^C_0$.

Assume that both modules E_1 and E_2 do not belong to $\mathcal{T}^C_\infty \cup \mathcal{T}^C_0$. Then there exist elements $\lambda_1, \lambda_2 \in K \setminus \{0\}$ such that $\lambda_1 \neq \lambda_2$, $E_1 \in \mathcal{T}^C_{\lambda_1}$, and $E_2 \in \mathcal{T}^C_{\lambda_2}$. Moreover, B is isomorphic to the bound quiver algebra KQ_B/I_B, where I_B is an admissible ideal of the path algebra KQ_B generated by the following elements:

- all the paths of length 2 that generate the ideals $I^{(1)}, I^{(2)}$ of the branches $\mathcal{L}^{(1)}, \mathcal{L}^{(2)}$, and
- $\lambda_1 \sigma_1 \alpha_1 - \sigma_1 \beta_1$ and $\lambda_2 \sigma_2 \alpha_1 - \sigma_2 \beta_1$.

We define the surjective homomorphism of algebras

$$f : KQ_B \longrightarrow KQ_B/I_B$$

by setting $f(\alpha_1) = (\alpha_1 - \lambda_1^{-1}\beta_1) + I_B$, $f(\beta_1) = (\beta_1 - \lambda_2\alpha_1) + I_B$, and $f(\gamma) = \gamma + I_B$, for all arrows γ of Q_B that are different from the arrows α_1 and β_1. It is easy to check that the paths $\sigma_1\alpha_1$ and $\sigma_2\beta_1$ belong to $\operatorname{Ker} f$ and, consequently, the ideal $I'_B = \operatorname{Ker} f$ of KQ_B is an admissible ideal of KQ_B obtained from the ideal I_B by interchanging the commutativity relations $\lambda_1\sigma_1\alpha_1 - \sigma_1\alpha_1$ and $\lambda_2\sigma_2\alpha_1 - \sigma_2\beta_1$ with the zero relations $\sigma_1\alpha_1$ and $\sigma_2\beta_1$, respectively, and keeping the remaining relations unchanged. Then f induces an isomorphism of algebras

$$KQ_B/I'_B \xrightarrow{\simeq} KQ_B/I_B$$

and, consequently, the algebra B is isomorphic to the branch extension $B' = C[E'_1, \mathcal{L}^{(1)}, E'_2, \mathcal{L}^{(2)}]$, where $E'_1 \in \mathcal{T}^C_\infty$ and $E'_2 \in \mathcal{T}^C_0$. This finishes the proof. $\square$

4.2. Proposition. *Let A be a concealed algebra of Euclidean type, B a domestic tubular extension of A, and C a canonical algebra of Euclidean type such that $r^B = r^C$. Then there exists a multiplicity-free tilting C-module T_C in the category* $\mathrm{add}\,(\mathcal{P}(C) \cup \boldsymbol{\mathcal{T}}^C)$ *and an isomorphism of algebras*

$$B \cong \mathrm{End}\,T_C.$$

Proof. Assume that A is a concealed algebra of Euclidean type. It follows from the structure theorem (XII.3.4) that the Auslander–Reiten quiver $\Gamma(\mathrm{mod}\,A)$ of A has a disjoint union decomposition

$$\Gamma(\mathrm{mod}\,A) = \mathcal{P}(A) \cup \mathcal{T}^A \cup \mathcal{Q}(A)$$

where $\mathcal{P}(A)$ is the unique postprojective component containing all the indecomposable projective A-modules, $\mathcal{Q}(A)$ is the unique preinjective component containing all the indecomposable injective A-modules, and

$$\boldsymbol{\mathcal{T}}^A = \{\mathcal{T}_\lambda^A\}_{\lambda \in \mathbb{P}_1(K)}$$

is a $\mathbb{P}_1(K)$-family of pairwise orthogonal standard stable tubes $\mathcal{T}_\lambda^A$. The family $\boldsymbol{\mathcal{T}}^A$ separates the component $\mathcal{P}(A)$ from $\mathcal{Q}(A)$ in the sense of (XII.3.3). Moreover, by (XII.2.8) and (XII.2.12), the mouth modules of the tubes of $\boldsymbol{\mathcal{T}}^A$ are described in Tables (XII.2.5) and (XII.2.9).

Assume that B is a domestic tubular extension of A. In view of (4.1) and (XV.3.9), without loss of generality, we may assume that B is a domestic branch extension

$$B = A[E_1, \mathcal{L}^{(1)}, E_2, \mathcal{L}^{(2)}, \dots, E_s, \mathcal{L}^{(s)}]$$

of A, where $E_1, E_2, \dots, E_s$ are modules in $\mathcal{T}_\infty^C \cup \mathcal{T}_0^C$, if B is of tubular type (p', q'), with $1 \leq p' \leq q'$, and $E_1, E_2, \dots, E_s$ are modules in $\mathcal{T}_\infty^C \cup \mathcal{T}_0^C \cup \mathcal{T}_1^C$, if B is of one of the tubular types $(2, 2, m-2)$, with $m \geq 4$, $(2, 3, 3)$, $(2, 3, 4)$, and $(2, 3, 5)$.

We assume that $A = C(p, q)$, with $1 \leq p \leq q$, or $A = C(p, q, r)$, where (p, q, r) is a triple of integers such that $r \geq q \geq p \geq 2$ and $\frac{1}{p} + \frac{1}{q} + \frac{1}{r} > 1$. It is easy to check that (p, q, r) is such a triple if and only if it is one of the following triples $(2, 2, m - 2)$, with $m \geq 4$, $(2, 3, 3)$, $(2, 3, 4)$, and $(2, 3, 5)$.

We recall from (XII.1.1) that the canonical algebra $C(p, q)$, with $1 \leq p \leq q$, is the path algebra $K\Delta(p, q)$ of the following acyclic Euclidean quiver

$$\Delta(p, q) = \Delta(\widetilde{\mathbb{A}}_{p,q}):$$

$$
\begin{array}{c}
a_1 \xleftarrow{\;\alpha_2\;} a_2 \longleftarrow \cdots \longleftarrow a_{p-1} \\
{\scriptstyle\alpha_1}\nearrow \qquad\qquad\qquad\qquad {\scriptstyle\alpha_p}\nwarrow \\
0 \qquad\qquad\qquad\qquad\qquad\qquad \omega \\
{\scriptstyle\beta_1}\nwarrow \qquad\qquad\qquad\qquad {\scriptstyle\beta_q}\swarrow \\
b_1 \xleftarrow[\;\beta_2\;]{} b_2 \longleftarrow \cdots \longleftarrow b_{q-1}
\end{array}
$$

while $C(p, q, r)$ is the bound quiver algebra $K\Delta(p, q, r)/I(p, q, r)$, where $\Delta(p, q, r)$ is the quiver

$$
\Delta(p, q, r): \quad 0
\begin{array}{c}
a_1 \xleftarrow{\alpha_2} a_2 \longleftarrow \cdots \xleftarrow{\alpha_{p-1}} a_{p-1} \\
\nwarrow \alpha_1 \qquad \qquad \qquad \qquad \qquad \qquad \nearrow \alpha_p \\
\xleftarrow{\beta_1} b_1 \xleftarrow{\beta_2} b_2 \longleftarrow \cdots \xleftarrow{\beta_{q-1}} b_{q-1} \xleftarrow{\beta_q} \omega \\
\nwarrow \gamma_1 \qquad \qquad \qquad \qquad \qquad \qquad \swarrow \gamma_r \\
c_1 \longleftarrow c_2 \longleftarrow \cdots \longleftarrow c_{r-1} \\
\gamma_2 \qquad \qquad \qquad \gamma_{r-1}
\end{array}
$$

and $I(p, q, r)$ is the two-sided ideal of the path K-algebra $K\Delta(p, q, r)$ generated by the element

$$\alpha_p \ldots \alpha_1 + \beta_q \ldots \beta_1 + \gamma_r \ldots \gamma_1.$$

Throughout this chapter, we set $a_p = b_q = c_r = \omega$.

Given $h \in \{1, \ldots, s\}$, we denote by

$$\mathcal{L}^{(h)} = (L^{(h)}, I^{(h)})$$

the branch, with the germ O_h. Note that O_h is the extension vertex of the one-point extension algebra $B^{(h)} = A[E_h]$ inside the algebra

$$B = A[E_1, \mathcal{L}^{(1)}, E_2, \mathcal{L}^{(2)}, \ldots, E_s, \mathcal{L}^{(s)}].$$

We denote by $Q^{(h)}$ the ordinary quiver of the algebra $B^{(h)} = A[E_h]$.

The description of the mouth modules of the tubes $\mathcal{T}_\infty^A$, $\mathcal{T}_0^A$ and $\mathcal{T}_1^A$ of $\Gamma(\mathrm{mod}\, C)$ yields the following statement, for each $h \in \{1, \ldots, s\}$.

- If the module E_h belongs to $\mathcal{T}_\infty^C$ then there is a vertex $a_{i(h)} \in \{a_1, \ldots, a_p\}$ and one arrow $O_h \xrightarrow{\sigma_h} a_{i(h)}$ in Q_B such that $B^{(h)} \cong KQ^{(h)}/I^{(h)}$, the quiver $Q^{(h)}$ is obtained from Q_A by adding the arrow $O_h \xrightarrow{\sigma_h} a_{i(h)}$, and $I^{(h)}$ is the ideal generated by I_A and the path $\sigma_h \alpha_{i(h)}$.

- If the module E_h belongs to $\mathcal{T}_0^C$ then there is a vertex $b_{j(h)} \in \{b_1, \ldots, b_p\}$ and one arrow $O_h \xrightarrow{\sigma_h} b_{j(h)}$ in Q_B such that $B^{(h)} \cong KQ^{(h)}/I^{(h)}$, the quiver $Q^{(h)}$ is obtained from Q_A by adding the arrow $O_h \xrightarrow{\sigma_h} b_{i(h)}$, and $I^{(h)}$ is the ideal generated by I_A and the path $\sigma_h \beta_{i(h)}$.

- If the module E_h belongs to $\mathcal{T}_1^C$ then there is a vertex $c_{k(h)} \in \{c_1, \ldots, c_r\}$ and one arrow $O_h \xrightarrow{\sigma_h} c_{k(h)}$ in Q_B such that $B^{(h)} \cong KQ^{(h)}/I^{(h)}$, the quiver $Q^{(h)}$ is obtained from Q_A by adding the arrow $O_h \xrightarrow{\sigma_h} c_{k(h)}$, and $I^{(h)}$ is the ideal generated by I_A and the path $\sigma_h \gamma_{k(h)}$.

It then follows that the algebra

$$B = A[E_1, \mathcal{L}^{(1)}, E_2, \mathcal{L}^{(2)}, \ldots, E_s, \mathcal{L}^{(s)}]$$

is isomorphic to the bound quiver algebra KQ_B/I_B, where the quiver Q_B is obtained from the disjoint union $L^{(1)} \cup L^{(2)} \cup \ldots \cup L^{(s)}$ of the quivers of the branches $\mathcal{L}^{(1)}, \mathcal{L}^{(2)}, \ldots, \mathcal{L}^{(s)}$ and the quiver Q_A of the algebra A, by adding the arrows $\sigma_1, \ldots, \sigma_s$. The admissible ideal I_B of the path algebra KQ_B is generated by the following elements:

- $\alpha_p \ldots \alpha_1 + \beta_q \ldots \beta_1 + \gamma_r \ldots \gamma_1$, if $A = C(p, q, r)$,
- all paths of length 2 that generate the ideals $I^{(1)}, I^{(2)}, \ldots, I^{(s)}$ of the branches $\mathcal{L}^{(1)}, \mathcal{L}^{(2)}, \ldots, \mathcal{L}^{(s)}$,
- the path $\sigma_h \alpha_{j(h)}$, for each $h \in \{1, \ldots, s\}$ such that $E_h \in \mathcal{T}_\infty^A$,
- the path $\sigma_h \beta_{j(h)}$, for each $h \in \{1, \ldots, s\}$ such that $E_h \in \mathcal{T}_0^A$,
- the path $\sigma_h \gamma_{k(h)}$, for each $h \in \{1, \ldots, s\}$ such that $E_h \in \mathcal{T}_1^A$,

Because $B = A[E_1, \mathcal{L}^{(1)}, E_2, \mathcal{L}^{(2)}, \ldots, E_s, \mathcal{L}^{(s)}]$ then, by (XV.4.3), the Auslander–Reiten quiver $\Gamma(\mathrm{mod}\, B)$ of B has a disjoint union decomposition

$$\Gamma(\mathrm{mod}\, B) = \mathcal{P}(B) \cup \boldsymbol{T}^B \cup \mathcal{Q}(B)$$

where

- $\mathcal{P}(B) = \mathcal{P}(A)$ is the unique postprojective component of $\Gamma(\mathrm{mod}\, B)$,
- $\mathcal{Q}(B)$ is a family of components of $\Gamma(\mathrm{mod}\, B)$ consisting of all indecomposable B-modules X such that the restriction $\mathrm{res}_A(X)$ of X to A is a module in the preinjective component $\mathcal{Q}(A)$ of $\Gamma(\mathrm{mod}\, A)$, and
- $\boldsymbol{T}^B = \{\mathcal{T}_\lambda^B\}_{\lambda \in \mathbb{P}_1(K)}$ is a $\mathbb{P}_1(K)$-family of pairwise orthogonal standard stable tubes $\mathcal{T}_\lambda^B$. It is obtained from the $\mathbb{P}_1(K)$-family $\boldsymbol{T}^A = \{\mathcal{T}_\lambda^A\}_{\lambda \in \mathbb{P}_1(K)}$ of pairwise orthogonal standard stable tubes $\mathcal{T}_\lambda^A$ of $\Gamma(\mathrm{mod}\, A)$ by iterated rectangle insertions,
- for each $\lambda \in \mathbb{P}_1(K)$, the rank r_λ^B of the ray tube $\mathcal{T}_\lambda^B$ is given by the formula

$$r_\lambda^B = r_\lambda^A + \sum_{h \in \mathcal{S}(\lambda)} \kappa_h,$$

where r_λ^A is the rank of the stable tube $\mathcal{T}_\lambda^A$, $\kappa_h = |\mathcal{L}^{(h)}|$ is the number of vertices of the branch $\mathcal{L}^{(h)}$, and the sum is taken over the subset $\mathcal{S}(\lambda)$ of $\{1, 2, \ldots, s\}$ consisting of all h such that the module E_h lies on $\mathcal{T}_\lambda^A$.

- The family $\boldsymbol{T}^A$ separates $\mathcal{P}(A)$ from $\mathcal{Q}(A)$ in the sense of (XII.3.3).

Because we assume that

$$E_1, E_2, \ldots, E_s \in \mathcal{T}_\infty^C \cup \mathcal{T}_0^C \cup \mathcal{T}_1^C$$

then $r_\lambda^B = r_\lambda^A$, for each $\lambda \in \mathbb{P}_1(K) \setminus \{0, 1, \infty\}$. Moreover, we have $r_1^B = r_1^A$, if B is of the tubular type (p', q'). The assumption that the algebra B is a domestic tubular extension of A yields that the tubular type $r^B = \hat{r}^B$ of B is one of the five domestic types (p', q'), with $1 \leq p' \leq q'$, $(2, 2, m' - 2)$, with $m' \geq 4$, $(2, 3, 3)$, $(2, 3, 4)$, and $(2, 3, 5)$.

Here we warn the reader that, in contrast to the tubular type $r^A = \hat{r}^A$ of A, the inequalities $r_0^B \geq r_\infty^B$ and $r_1^B \geq r_0^B \geq r_\infty^B$ are not necessarily satisfied.

Now we associate to our algebra B, that is a domestic tubular extension of a concealed algebra A of Euclidean type, a canonical algebra $C \cong KQ_C/I_C$ of Euclidean type such that $r^C = r^B$.

First we construct the quiver Q_C from the quiver Q_A of A as follows.

- for each $h \in \mathcal{S}(\infty)$, we replace the arrow

$$a_{i(h)-1} \xleftarrow{\quad \alpha_{i(h)} \quad} a_{i(h)}$$

of the quiver Q_A by the path

$$a_{i(h)-1} \xleftarrow{\alpha_0^{(h)}} a_0^{(h)} \xleftarrow{\alpha_1^{(h)}} a_1^{(h)} \xleftarrow{\quad} \cdots \xleftarrow{\alpha_{\kappa_h-1}^{(h)}} a_{\kappa_h-1}^{(h)} \xleftarrow{\alpha_{\kappa_h}^{(h)}} a_{\kappa_h}^{(h)} = a_{i(h)},$$

- for each $h \in \mathcal{S}(0)$, we replace the arrow

$$b_{j(h)-1} \xleftarrow{\quad \beta_{j(h)} \quad} b_{j(h)}$$

of the quiver Q_A by the path

$$b_{j(h)-1} \xleftarrow{\beta_0^{(h)}} b_0^{(h)} \xleftarrow{\beta_1^{(h)}} b_1^{(h)} \xleftarrow{\quad} \cdots \xleftarrow{\beta_{\kappa_h-1}^{(h)}} b_{\kappa_h-1}^{(h)} \xleftarrow{\beta_{\kappa_h}^{(h)}} b_{\kappa_h}^{(h)} = b_{j(h)},$$

- if $\hat{r}^A = (p, q, r)$, we replace the arrow

$$c_{k(h)-1} \xleftarrow{\quad \gamma_{k(h)} \quad} c_{j(h)}$$

of the quiver Q_A by the path

$$c_{j(h)-1} \xleftarrow{\gamma_0^{(h)}} c_0^{(h)} \xleftarrow{\gamma_1^{(h)}} c_1^{(h)} \xleftarrow{\quad} \cdots \xleftarrow{\gamma_{\kappa_h-1}^{(h)}} c_{\kappa_h-1}^{(h)} \xleftarrow{\gamma_{\kappa_h}^{(h)}} c_{\kappa_h}^{(h)} = c_{j(h)},$$

for any $h \in \mathcal{S}(1)$, and

- if $\hat{r}^A = (p, q)$, we insert into the quiver Q_A the path

$$0 \xleftarrow{\gamma_0^{(h)}} c_0^{(h)} \xleftarrow{\gamma_1^{(h)}} c_1^{(h)} \xleftarrow{\quad} \cdots \xleftarrow{\gamma_{\kappa_h-1}^{(h)}} c_{\kappa_h-1}^{(h)} \xleftarrow{\gamma_{\kappa_h}^{(h)}} c_{\kappa_h}^{(h)} = \omega$$

with the unique element $h \in \mathcal{S}(1)$. Note that in case $\hat{r}^A = (p, q)$, $\mathcal{T}_1^A$ is a stable tube of rank one and, hence, the set $\mathcal{S}(1)$ consists of one element. We recall that $\kappa_h = |\mathcal{L}^{(h)}|$ is the capacity of the branch $\mathcal{L}^{(h)}$.

It follows from the construction that the quiver Q_C can be described as follows.

- If $\hat{r}^A = (p', q')$, the quiver Q_C consists of three parallel paths u_∞, u_0, and u_1 with the common source ω and the common target 0. The path u_∞ is of length $r_\infty^B = p'$, and the path u_0 is of length $r_0^B = q'$.
- If $\hat{r}^A = (p', q', r')$, the quiver Q_C consists of three parallel paths u_∞, u_0, and u_1 with the common source ω and the common target 0. The path u_∞ is of length $r_\infty^B = p'$, the path u_0 is of length $r_0^B = q'$, and the path u_1 is of length $r_1^B = r'$.

Having constructed the quiver Q_C, we define the admissible ideal I_C of KQ_C by setting

- $I_C = (0)$, if $\hat{r}^B = (p', q')$, and
- $I_C = (u_\infty + u_0 + u_1)$ is the ideal generated by the element $u_\infty + u_0 + u_1$ of KQ_C, if $\hat{r}^B = (p', q', r')$.

We define the algebra C to be the bound quiver algebra $C = KQ_C/I_C$. It is easy to see that $C \cong C(p', q')$, if $\hat{r}^B = (p', q')$, and $C \cong C(p', q', r')$, if $\hat{r}^B = (p', q', r')$. It then follows that C is a canonical algebra of Euclidean type such that $r^C = r^B$. By the results of Chapter XII, the Auslander–Reiten quiver $\Gamma(\mathrm{mod}\, C)$ of C has a disjoint union decomposition

$$\Gamma(\mathrm{mod}\, C) = \mathcal{P}(C) \cup \boldsymbol{T}^C \cup \mathcal{Q}(C)$$

where $\mathcal{P}(C)$ is the unique postprojective component containing all the indecomposable projective C-modules, $\mathcal{Q}(C)$ is the unique preinjective component containing all the indecomposable injective C-modules, and

$$\boldsymbol{T}^C = \{\mathcal{T}_\lambda^C\}_{\lambda \in \mathbb{P}_1(K)}$$

is a $\mathbb{P}_1(K)$-family of pairwise orthogonal standard stable tubes $\mathcal{T}_\lambda^C$ separating $\mathcal{P}(C)$ from $\mathcal{Q}(C)$. Moreover, the tube $\mathcal{T}_\infty^C$ is of rank $r_\infty^C = r_\infty^B$, the tube $\mathcal{T}_0^C$ is of rank $r_0^C = r_0^B$, the tube $\mathcal{T}_1^C$ is of rank $r_1^C = r_1^B$, and $r_\lambda^C = r_\lambda^B = 1$, for all $\lambda \in \mathbb{P}_1(K) \setminus \{0, 1, \infty\}$. We should stress here that the tubes $\mathcal{T}_\infty^C$, $\mathcal{T}_0^C$, and $\mathcal{T}_1^C$ correspond to the paths u_∞, u_0, and u_1.

To finish the proof of the proposition, we construct now a partial tilting C-module T_C^{rg} such that the algebra $\mathrm{End}\, T_C^{rg}$ is isomorphic to the product $K\mathcal{L}^{(1)} \times \ldots \times K\mathcal{L}^{(s)}$ of the branch algebras $K\mathcal{L}^{(1)}, \ldots, K\mathcal{L}^{(s)}$.

Fix an element $h \in \{1, \ldots, s\}$. We define a stone cone $\mathcal{C}_h$ of $\boldsymbol{T}^C$, depending on the position of the module E_h in $\boldsymbol{T}^A$. We do it in three steps.

Step 1°. Assume that the module E_h lies on the tube $\mathcal{T}_\infty^A$. Then $\mathcal{T}_\infty^A$ admits a cone of the form

$$
\begin{array}{cccccccc}
E_{i(h)-1}^{(\infty)}[1] & & S(a_0^{(h)})[1] & & S(a_1^{(h)})[1] & \cdots & S(a_{\kappa_h-1}^{(h)})[1] & E_{i(h)}^{(\infty)}[1] \\
& \searrow & \nearrow & \searrow & \nearrow & \searrow & \nearrow & \searrow & \nearrow \\
& E_{i(h)-1}^{(\infty)}[2] & & S(a_0^{(h)})[2] & & \cdots & \cdots & S(a_{\kappa_h-1}^{(h)})[2] \\
& & \searrow & \nearrow & \searrow & & & \nearrow \\
& & E_{i(h)-1}^{(\infty)}[3] & & \cdots & & \cdots & \cdots \\
& & & \searrow & & \searrow & \nearrow & \\
& & & & S(a_0^{(h)})[\kappa_h] & & \cdots \\
& & & & \searrow & \nearrow & \searrow & \nearrow \\
& & & E_{i(h)-1}^{(\infty)}[\kappa_h+1] & & S(a_0^{(h)})[\kappa_h+1] \\
& & & & \searrow & \nearrow \\
& & & & E_{i(h)-1}^{(\infty)}[\kappa_h+2]
\end{array}
$$

where

- $S(a_t^{(h)})[1] = S(a_t^{(h)})$ is the simple C-module at the vertex $a_t^{(h)}$ of the path u_∞, for each $t \in \{0, 1, \ldots, \kappa_h - 1\}$,
- $E_{i(h)-1}^{(\infty)}[1]$ is the simple C-module $S(a_{i(h)-1})$, at the vertex $a_{i(h)-1}$, if $a_{i(h)-1} \neq 0$,
- $E_{i(h)}^{(\infty)}[1]$ is the simple C-module $S(a_{i(h)})$, at the vertex $a_{i(h)}$, if $a_{i(h)} \neq \omega$,
- $E_0^{(\infty)} = E_\omega^{(\infty)}$ is the unique non-simple C-module on the mouth of the tube $\mathcal{T}_\infty^C$.

Observe that the rank r_∞^C of the standard stable tube $\mathcal{T}_\infty^C$ is bigger than κ_h, because

$$
r_\infty^C = r_\infty^B = r_\lambda^A + \sum_{h \in \mathcal{S}(\infty)} \kappa_h.
$$

Therefore, by (1.6), the module $M_h = S(a_1^{(h)})[\kappa_h]$ is a stone and hence the cone $\mathcal{C}^{(h)} = \mathcal{C}(M_h)$ determined by M_h is a stone cone. Because $\kappa_h = \ell_\infty(M_h)$ is the depth of $\mathcal{C}(M_h)$ then there is an equivalence of categories

$$
F^{(h)} : \operatorname{add}\mathcal{C}^{(h)} \longrightarrow \operatorname{mod} H^{(h)},
$$

where $H^{(h)} = K\Delta(\mathbb{A}_{\kappa_h})$ is the path algebra of the equioriented quiver $\Delta(\mathbb{A}_{\kappa_h})$. Because $\mathcal{L}^{(h)} = (L^{(h)}, I^{(h)})$ is a branch of capacity κ_h then, by

(XVI.2.2), there exists a multiplicity-free tilting $H^{(h)}$-module $U_{H^{(h)}}^{(h)}$ such that the branch algebra $K\mathcal{L}^{(h)}$ is isomorphic to the tilted algebra $\operatorname{End} U_{H^{(h)}}^{(h)}$. It follows that there exists a multiplicity-free module $T_C^{(h)}$ in $\operatorname{add} \mathcal{C}^{(h)}$ such that $U_{H^{(h)}}^{(h)} \cong F^{(h)}(T_C^{(h)})$.

Note that $T_C^{(h)}$ is a partial tilting C-module, because the relation $T_C^{(h)}$ in $\operatorname{add} \mathcal{C}^{(h)}$ yields $\operatorname{pd} T_C^{(h)} \leq 1$, and we get the isomorphisms

$$\operatorname{Ext}_C^1(T^{(h)}, T^{(h)}) \cong D\operatorname{Hom}_C(T^{(h)}, \tau_C T^{(h)})$$

$$\cong D\operatorname{Hom}_{H^{(h)}}(U^{(h)}, \tau_{H^{(h)}} U^{(h)})$$

$$\cong \operatorname{Ext}_{H^{(h)}}^1(U^{(h)}, U^{(h)}) = 0.$$

The functor $F^{(h)}$ induces an isomorphism of the algebra $\operatorname{End} T_C^{(h)}$ with the algebra $\operatorname{End} U_{H^{(h)}}^{(h)} \cong K\mathcal{L}^{(h)}$. Note also that the indecomposable projective $K\mathcal{L}^{(h)}$-module at the germ O_h of the branch $\mathcal{L}^{(h)}$ corresponds to the direct summand $M_h = S(a_1^{(h)})[\kappa_h]$ of $T_C^{(h)}$, because the unique indecomposable projective-injective $H^{(h)}$-module is a direct summand of the tilting $H^{(h)}$-module $U_{H^{(h)}}^{(h)}$. Further, we observe that the cone

$$\tau_C \mathcal{C}^{(h)} = \mathcal{C}(\tau_C M_h)$$

is the cone $\mathcal{C}(S(a_0^{(h)})[\kappa_h])$ determined by the module $S(a_0^{(h)})[\kappa_h]$, and the simple composition factors of the modules in $\tau_C \mathcal{C}^{(h)}$ belong to the family

$$\left\{ S(a_0^{(h)}), S(a_1^{(h)}), \dots, S(a_{\kappa_h - 1}^{(h)}) \right\}.$$

Hence we conclude that $\operatorname{Hom}_C(P(d), \tau_C T^{(h)}) = 0$, for each indecomposable projective C-module $P(d)$ at any vertex d of Q_C, which is a vertex of Q_A.

Step $2°$. Assume that the module E_h lies on the tube $\mathcal{T}_0^A$. Then, similarly as in Step $1°$, we show that the tube $\mathcal{T}_0^A$ contains a stone cone $\mathcal{C}^{(h)} = \mathcal{C}(M_h)$ determined by the indecomposable module $M_h = S(b_1^{(h)})[\kappa_h]$. The mouth modules of the cone $\mathcal{C}^{(h)}$ are the modules

$$S(b_1^{(h)})[1] = S(b_1^{(h)}), S(b_2^{(h)})[1] = S(b_2^{(h)}), \dots, S(b_{\kappa_h - 1}^{(h)})[1] = S(b_{\kappa_h - 1}^{(h)}), E_{j(h)}^{(0)}[1],$$

where $E_{j(h)}^{(0)}[1]$ is the simple C-module $S(b_{j(h)})$ at $b_{j(h)}$, if $b_{j(h)} \neq \omega$, and $E_{j(h)}^{(0)}[1] = E_\omega^{(0)}[1]$ is the unique non-simple C-module on the mouth of the tube $\mathcal{T}_0^C$, if $b_{j(h)} = \omega$. One shows that there exists a multiplicity-free partial tilting C-module $T_C^{(h)}$ in $\operatorname{add} \mathcal{C}^{(h)}$ such that

• $T_C^{(h)}$ has κ_h pairwise non-isomorphic direct summands,

- the module $M_h = S(b_1^{(h)})[\kappa_h]$ is a direct summand of $T_C^{(h)}$,
- the algebra $\operatorname{End} T_C^{(h)}$ is isomorphic to the branch algebra $K\mathcal{L}^{(h)}$ and the germ O_h of the branch $\mathcal{L}^{(h)}$ corresponds to the direct summand M_h of $T_C^{(h)}$,
- the simple composition factors of the modules in the cone

$$\tau_C \mathcal{C}^{(h)} = \mathcal{C}(\tau_C M_h) = \mathcal{C}(S(b_0^{(h)})[\kappa_h])$$

belong to the family $\left\{ S(b_0^{(h)}), S(b_1^{(h)}), \dots, S(b_{\kappa_h-1}^{(h)}) \right\}$, and

- $\operatorname{Hom}_C(P(d), \tau_C T^{(h)}) = 0$, for each indecomposable projective C-module $P(d)$ at any vertex d of Q_C, which is a vertex of Q_A.

Step 3°. Assume that the module E_h lies on the tube $\mathcal{T}_1^A$. Then $\mathcal{T}_1^A$ is of rank $r_1^C \geq 2$ and, similarly as in Step 1°, we show that the tube $\mathcal{T}_1^A$ contains a stone cone $\mathcal{C}^{(h)} = \mathcal{C}(M_h)$ determined by the indecomposable module $M_h = S(c_1^{(h)})[\kappa_h]$. The mouth modules of the cone $\mathcal{C}^{(h)}$ are the modules

$$S(c_1^{(h)})[1] = S(c_1^{(h)}), S(c_2^{(h)})[1] = S(c_2^{(h)}), \dots, S(c_{\kappa_h-1}^{(h)})[1] = S(c_{\kappa_h-1}^{(h)}), E_{k(h)}^{(1)}[1],$$

where $E_{k(h)}^{(1)}[1]$ is the simple C-module $S(c_{k(h)})$ at $c_{k(h)}$, if $c_{k(h)} \neq \omega$, and $E_{k(h)}^{(1)}[1] = E_\omega^{(1)}[1]$ is the unique non-simple C-module on the mouth of the tube $\mathcal{T}_1^C$, if $c_{j(h)} = \omega$. One shows that there exists a multiplicity-free partial tilting C-module $T_C^{(h)}$ in $\operatorname{add} \mathcal{C}^{(h)}$ such that

- $T_C^{(h)}$ has κ_h pairwise non-isomorphic direct summands,
- the module $M_h = S(c_1^{(h)})[\kappa_h]$ is a direct summand of $T_C^{(h)}$,
- the algebra $\operatorname{End} T_C^{(h)}$ is isomorphic to the branch algebra $K\mathcal{L}^{(h)}$ and the germ O_h of the branch $\mathcal{L}^{(h)}$ corresponds to the direct summand M_h of $T_C^{(h)}$,
- the simple composition factors of the modules in the cone

$$\tau_C \mathcal{C}^{(h)} = \mathcal{C}(\tau_C M_h) = \mathcal{C}(S(c_0^{(h)})[\kappa_h])$$

belong to the family $\left\{ S(c_0^{(h)}), S(c_1^{(h)}), \dots, S(c_{\kappa_h-1}^{(h)}) \right\}$,

- $\operatorname{Hom}_C(P(d), \tau_C T^{(h)}) = 0$, for each indecomposable projective C-module $P(d)$ at any vertex d of Q_C, which is a vertex of Q_A.

Define the multiplicity-free C-module T_C to be the direct sum

$$T_C = T_C^{pp} \oplus T_C^{rg}, \quad \text{with} \quad T_C^{rg} = T_C^{(1)} \oplus \dots \oplus T_C^{(s)} \text{ and } T_C^{pp} = \bigoplus_{d \in (Q_A)_0^\bullet} P(d),$$

where $P(d) = e_d C$ is the indecomposable projective C-module at the vertex d, and d runs over the set $(Q_A)_0^\bullet$ of the vertices of the quiver Q_A that belong to Q_C.

First we prove that T_C^{rg} is a partial tilting C-module. Because $T_C^{rg} \in$ add $\boldsymbol{T}^C$ then $\mathrm{pd}\, T_C^{rg} \leq 1$. If two stone cones $\mathcal{C}^{(h)}$ and $\mathcal{C}^{(h')}$, with $h, h' \in \{1, \ldots, s\}$ and $h \neq h'$, are in the same tube of $\boldsymbol{T}^C$ then (1.7) yields

$$\mathcal{C}^{(h)} \cap \mathcal{C}^{(h')} = \emptyset, \quad \mathcal{C}^{(h)} \cap \tau_A \mathcal{C}^{(h')} = \emptyset, \text{ and } \mathcal{C}^{(h')} \cap \tau_A \mathcal{C}^{(h)} = \emptyset.$$

Hence we get the isomorphisms

$$\mathrm{Ext}_C^1(T_C^{rg}, T_C^{rg}) \cong D\mathrm{Hom}_C(T_C^{rg}, \tau_C T_C^{rg}) = 0,$$

because the tubes $\mathcal{T}_\infty^C$, $\mathcal{T}_0^C$, and $\mathcal{T}_1^C$ are pairwise orthogonal. Observe also that

- $\mathrm{End}\, T_C^{rg} \cong K\mathcal{L}^{(1)} \times \ldots \times K\mathcal{L}^{(s)}$, by our choice of the direct summands $T_C^{(1)}, \ldots, T_C^{(s)}$ of T_C^{rg},
- the module T_C^{pp} belongs to add $\mathcal{P}(C)$ and $\mathrm{pd}\, T_C^{pp} \leq 1$, because the postprojective component $\mathcal{P}(C)$ contains all the indecomposable projective C-modules and the module T_C^{pp} is projective,
- there is an isomorphism of algebras $A \cong \mathrm{End}\, T_C^{pp}$.

Now we show that the multiplicity-free C-module $T_C = T_C^{pp} \oplus T_C^{rg}$ is tilting. Let $n_A = \mathrm{rk}\, K_0(A)$, $n_B = \mathrm{rk}\, K_0(B)$, and $n_C = \mathrm{rk}\, K_0(C)$ be the ranks of the Grothendieck groups $K_0(A)$, $K_0(B)$, and $K_0(C)$ of the algebras A, B, and C, respectively. Then we have

$$n_B = n_A + \sum_{h=1}^s \kappa_h$$

and $n_C = n_B$, by the construction of the quiver Q_C. It follows that the module T_C is a direct sum of n_C pairwise non-isomorphic indecomposable C-modules. Moreover, $T_C \in \mathrm{add}\,(\mathcal{P}(C) \cup \boldsymbol{T}^C)$, $\mathrm{pd}\, T_C \leq 1$, and $\mathrm{Ext}_C^1(T_C^{pp}, T_C) = 0$, because the module T_C^{pp} is projective. We recall that $\mathrm{Ext}_C^1(T_C^{rg}, T_C^{rg}) = 0$. Finally, it follows that

$$\mathrm{Ext}_C^1(T_C^{rg}, T_C^{pp}) \cong D\mathrm{Hom}_C(T_C^{pp}, \tau_C T_C^{rg}) = 0,$$

because $T_C^{pp} \in \mathrm{add}\, \mathcal{P}(C)$, $T_C^{rg} \in \mathrm{add}\, \boldsymbol{T}^C$, and $T_C^{pi} \in \mathrm{add}\, \mathcal{Q}(C)$. This shows that T_C is a tilting module.

We note also that $\mathrm{Hom}_C(\boldsymbol{T}^C, T_C^{pp}) = 0$. Hence, by applying (3.5) and using the bound quiver presentation $B \cong KQ_B/I_B$ described earlier, we get an isomorphism of algebras $B = C[E_1, \mathcal{L}^{(1)}, E_2, \mathcal{L}^{(2)}, \ldots, E_s, \mathcal{L}^{(s)}] \cong \mathrm{End}\, T_C$. This completes the proof of the proposition. $\square$

Now we prove the sufficiency part of Theorem (3.5).

4.3. Theorem. *Let C be a concealed algebra of Euclidean type, B a domestic tubular extension of C with the tubular type $r^B = m_\Delta$, where Δ is one of the quivers $\Delta(\widetilde{\mathbb{A}}_{p,q})$, with $1 \leq p \leq q$, $\Delta(\widetilde{\mathbb{D}}_m)$, with $m \geq 4$, $\Delta(\widetilde{\mathbb{E}}_6)$, $\Delta(\widetilde{\mathbb{D}}_7)$, and $\Delta(\widetilde{\mathbb{D}}_8)$. Let $A = K\Delta$ be the path algebra of Δ. Then there exists a multiplicity-free tilting A-module T_A such that $T_A^{pi} = 0$ and $B \cong \operatorname{End} T_A$.*

Proof. Assume that B is a domestic tubular extension of C with the tubular type $r^B = m_\Delta$. Then, by (XV.3.9), B is a domestic branch extension

$$B = C[E_1, \mathcal{L}^{(1)}, E_2, \mathcal{L}^{(2)}, \ldots, E_s, \mathcal{L}^{(s)}]$$

of C. By the structure theorem (XII.3.4), the Auslander–Reiten quiver $\Gamma(\operatorname{mod} B)$ of B has a disjoint union decomposition

$$\Gamma(\operatorname{mod} B) = \mathcal{P}(B) \cup \boldsymbol{T}^B \cup \mathcal{Q}(B),$$

where $\mathcal{P}(B) = \mathcal{P}(C)$ is a unique postprojective component of $\Gamma(\operatorname{mod} B)$, $\mathcal{Q}(B)$ is a family of components of $\Gamma(\operatorname{mod} B)$ consisting of all modules X such that the restriction $\operatorname{res}_C(X)$ to C belongs to add $\mathcal{Q}(C)$, and the regular part $\boldsymbol{T}^B$ of $\Gamma(\operatorname{mod} B)$ is a $\mathbb{P}_1(K)$-family $\boldsymbol{T}^B = \{\mathcal{T}_\lambda^B\}_{\lambda \in \mathbb{P}_1(K)}$ of pairwise orthogonal standard ray tubes $\mathcal{T}_\lambda^B = \operatorname{Hom}_A(T, \mathcal{T}_\lambda^A)$, with $r_\lambda^B = r_\lambda^A$ separating $\mathcal{P}(B)$ from $\mathcal{Q}(B)$. Moreover, by our assumption, the tubular type r^B of B is domestic then it coincides with the tubular type $r^A = m_\Delta$ of the hereditary algebra $A = K\Delta$. It follows from (XII.3.1) that there are a uniquely determined quiver Δ' in the set

$$\left\{ \Delta(\widetilde{\mathbb{A}}_{p,q}), \text{ with } 1 \leq p \leq q, \Delta(\widetilde{\mathbb{D}}_m), \text{ with } m \geq 4, \Delta(\widetilde{\mathbb{E}}_6), \Delta(\widetilde{\mathbb{D}}_7), \Delta(\widetilde{\mathbb{D}}_8) \right\}$$

and a multiplicity-free postprojective tilting module $U_{A'}$ over the hereditary algebra $A' = K\Delta'$ such that $C \cong \operatorname{End} U_{A'}$. If we denote by Λ' the canonical algebra $C(\Delta')$ of the type Δ' then, according to (XII.1.5), (XII.1.8), and (XII.1.14), there are a multiplicity-free postprojective tilting Λ'-module $V_{A'}$ and an isomorphism of algebras

$$\Lambda' \cong \operatorname{End} V_{A'}.$$

Thus, by (VIII.4.5), there exist

   (i) a full translation subquiver $\mathcal{P}'(A') \cong (-\mathbb{N})(\Delta')^{\operatorname{op}}$ of $\mathcal{P}(A')$ closed under successors,

   (ii) a full translation subquiver $\mathcal{P}'(C) \cong (-\mathbb{N})(\Delta')^{\operatorname{op}}$ of $\mathcal{P}(C)$ closed under successors, and

   (iii) a full translation subquiver $\mathcal{P}'(\Lambda') \cong (-\mathbb{N})(\Delta')^{\operatorname{op}}$ of $\mathcal{P}(\Lambda')$ closed under successors,

such that the functors

$$\operatorname{Hom}_{A'}(U, -) : \operatorname{mod} A' \longrightarrow \operatorname{mod} C \text{ and } \operatorname{Hom}_{A'}(V, -) : \operatorname{mod} A' \longrightarrow \operatorname{mod} \Lambda'$$

induce equivalences of categories

$$\text{add}(\mathcal{P}'(A') \cup \boldsymbol{T}^{A'}) \xrightarrow{\simeq} \text{add}(\mathcal{P}'(C) \cup \boldsymbol{T}^{C}),$$
$$\text{add}(\mathcal{P}'(A') \cup \boldsymbol{T}^{A'}) \xrightarrow{\simeq} \text{add}(\mathcal{P}'(\Lambda') \cup \boldsymbol{T}^{\Lambda'}).$$

Then, by the Brenner-Butler theorem (VI.3.8), the functor $- \otimes_{\Lambda'} V$ induces an equivalence of categories

$$\text{add}\,(\mathcal{P}'(\Lambda') \cup \boldsymbol{T}^{\Lambda'}) \xrightarrow{\simeq} \text{add}\,(\mathcal{P}'(A') \cup \boldsymbol{T}^{A'}).$$

Therefore, the composition of functors $- \otimes_{\Lambda'} V$ and $\text{Hom}_{A'}(U, -)$ defines an equivalence

$$F : \text{add}\,(\mathcal{P}'(\Lambda') \cup \boldsymbol{T}^{\Lambda'}) \xrightarrow{\simeq} \text{add}\,(\mathcal{P}'(C) \cup \boldsymbol{T}^{C}).$$

of categories. Let B_B^{pp} be the direct sum of all indecomposable projective B-modules in the component $\mathcal{P}(B) = \mathcal{P}(C)$, and let B_B^{rg} be the direct sum of all indecomposable projective B-modules in $\boldsymbol{T}^B$. Then

$B_B = B_B^{pp} \oplus B_B^{rg}$, $C \cong \text{End}\, B_B^{pp}$, and $D = \text{End}\, B_B^{rg} \cong K\mathcal{L}^{(1)} \times \ldots \times K\mathcal{L}^{(s)}$.
Because $\text{Hom}_B(B_B^{rg}, B_B^{pp}) = 0$, then B has the lower triangular matrix form

$$B \cong \begin{bmatrix} \text{End}\, B_B^{pp} & 0 \\ \text{Hom}_B(B_B^{pp}, B_B^{rg}) & \text{End}\, B_B^{rg} \end{bmatrix} = \begin{bmatrix} C & 0 \\ {}_D M_C & D \end{bmatrix},$$

where ${}_D M_C = \text{Hom}_B(B_B^{pp}, B_B^{rg})$ is viewed as a D-C-bimodule in an obvious way. Note also that there is an isomorphism of C-modules

$$M_C \cong E_1^{n_1} \oplus E_2^{n_2} \oplus \ldots \oplus E_s^{n_s},$$

where n_k is the number of vertices on the maximal path in the quiver $L^{(k)}$ with target at the germ O_k of the branch $\mathcal{L}^{(k)}$, for each $k \in \{1, \ldots, s\}$. Because all but at most three tubes of the family $\boldsymbol{T}^C$ are of rank one, there is a common multiple m of the τ_C-periods of all indecomposable C-modules in the family $\boldsymbol{T}^C$. Note also that there is an integer $t > 0$ such that the module $\tau_C^{-tm} C_C$ belongs to $\text{add}\,\mathcal{P}'(C)$. Then, we get isomorphisms

$C \cong \text{End}\, C_C \cong \text{End}\, \tau_C^{-tm} C_C$, and
$M_C \cong \text{Hom}_C(C_C, M_C) \cong \text{Hom}_C(\tau_C^{-tm} C_C, \tau_C^{-tm} M_C) \cong \text{Hom}_C(\tau_C^{-tm} C_C, M_C)$,
because $\tau_C^{-tm} M_C \cong M_C$. Hence we conclude that there are algebra isomorphisms

$$B \cong \text{End}_B(\tau_C^{-tm} C_C \oplus B_B^{rg}) \cong \begin{bmatrix} \text{End}_B(\tau_C^{-tm} C_C) & 0 \\ \text{Hom}_B(\tau_C^{-tm} C_C, B_B^{rg}) & \text{End}\, B_B^{rg} \end{bmatrix}.$$

By applying the equivalence $F : \text{add}\,(\mathcal{P}'(\Lambda') \cup \boldsymbol{T}^{\Lambda'}) \xrightarrow{\simeq} \text{add}\,(\mathcal{P}'(C) \cup \boldsymbol{T}^{C})$, we find a module Z in $\text{add}\,\mathcal{P}(\Lambda')$ and indecomposable modules $E_1', \ldots, E_s'$ in $\boldsymbol{T}^{\Lambda'}$ such that

$$F(Z) \cong \tau_C^{-tm} C_C \text{ and } F(E_1') \cong E_1, \dots, F(E_s') \cong E_s.$$

Consider the branch extension

$$B' = \Lambda'[E_1', \mathcal{L}^{(1)}, E_2', \mathcal{L}^{(2)}, \dots, E_s', \mathcal{L}^{(s)}]$$

of Λ', and note that, by (XV.3.8), B' is a tubular extension of Λ', with $r^{B'} = r^B$, and hence, B' is a domestic tubular (branch) extension of Λ'. Decompose the B'-module B' as $B'_{B'} = (B')^{pp} \oplus (B')^{rg}$, where $(B')^{pp}$ is the direct sum of all indecomposable projective modules in $\mathcal{P}(B') = \mathcal{P}(\Lambda')$ and $(B')^{rg}$ is the direct sum of all indecomposable projective B'-modules lying in $\boldsymbol{\mathcal{T}}^{B'}$. Consider the algebra

$$B'' = \mathrm{End}_{B'}(\tau_C^{-tm} C_C \oplus (B')^{rg}).$$

Because the module $\tau_C^{-tm} C_C$ lies in the category $\mathrm{add}\,\mathcal{P}(B')$ then we have $\mathrm{Hom}_{B'}((B')^{rg}, \tau_C^{-tm} C_C) = 0$. Hence, there are isomorphisms of algebras

$$B'' \cong \begin{bmatrix} \mathrm{End}_{B'}(\tau_C^{-tm} C_C) & 0 \\ \mathrm{Hom}_{B'}(\tau_C^{-tm} C_C, (B')^{rg}) & \mathrm{End}_{B'}(B')^{rg} \end{bmatrix} \cong \begin{bmatrix} C & 0 \\ {}_D M_C & D \end{bmatrix} \cong B,$$

because it is easy to see that there are isomorphisms of algebras

$$\mathrm{End}_{B'}(\tau_C^{-tm} C_C) \cong C,$$
$$\mathrm{End}_{B'}(B')^{rg} \cong \mathrm{End}_C(E_1^{n_1} \oplus E_2^{n_2} \oplus \dots \oplus E_s^{n_s}) \cong \mathrm{End}\,M_C \cong D,$$

and, under these isomorphisms, there is an isomorphism of D-C-bimodules

$$\mathrm{Hom}_{B'}(\tau_C^{-tm} C_C, (B')^{rg}) \cong \mathrm{Hom}_B(\tau_C^{-tm} C_C, M_C) \cong {}_D M_C.$$

If $\Lambda = C(\Delta)$ is the canonical algebra $C(\Delta)$ of type Δ then, by (4.2), there exist a multiplicity-free tilting Λ-module $S_\Lambda \in \mathrm{add}\,(\mathcal{P}(\Lambda) \cup \boldsymbol{\mathcal{T}}^\Lambda)$ and an isomorphism of algebras $B' \cong \mathrm{End}\,S_\Lambda$. Note that $r^B = r^C = m_\Delta = r^A$, where $A = K\Delta$ is the path algebra of the quiver Δ. On the other hand, it follows from (XII.1.5), (XII.1.8), (XII.1.11), and (XII.1.14) that there exist a multiplicity-free postprojective tilting A-module W_A and an isomorphism of algebras $\Lambda \cong \mathrm{End}\,W_A$. Further, it follows from (VIII.4.5) that there exist

- a full translation subquiver $\mathcal{P}'(\Lambda) \cong (-\mathbb{N})\Delta^{\mathrm{op}}$ of $\mathcal{P}(\Lambda)$ closed under successors, and
- a full translation subquiver $\mathcal{P}'(A) \cong (-\mathbb{N})\Delta^{\mathrm{op}}$ of $\mathcal{P}(A)$ closed under successors,

such that the functor $\mathrm{Hom}_A(W, -) : \mathrm{mod}\,A \longrightarrow \mathrm{mod}\,\Lambda$ induces an equivalence

$$\mathrm{add}\,(\mathcal{P}'(A) \cup \boldsymbol{\mathcal{T}}^A) \xrightarrow{\;\approx\;} \mathrm{add}\,(\mathcal{P}'(\Lambda) \cup \boldsymbol{\mathcal{T}}^\Lambda)$$

of categories. Take now the common multiplicity r of ranks of the tubes in the family $\boldsymbol{\mathcal{T}}^A = \{\mathcal{T}_\lambda^A\}_{\lambda \in \mathbb{P}_1(K)}$ and an integer $\ell \geq 1$ such that the module $\tau_\Lambda^{-\ell r} S_\Lambda^{pp}$ lies in add $\mathcal{P}'(\Lambda)$. Then there are isomorphisms of Λ-modules

$$\tau_\Lambda^{-\ell r} S_\Lambda \cong \tau_\Lambda^{-\ell r} S_\Lambda^{pp} \oplus \tau_\Lambda^{-\ell r} S_\Lambda^{rg} \cong \tau_\Lambda^{-\ell r} S_\Lambda^{pp} \oplus S_\Lambda^{rg},$$

and hence we derive the isomorphisms of algebras

$$B' \cong \operatorname{End} S_\Lambda \cong \operatorname{End} \tau_\Lambda^{-\ell r} S_\Lambda \cong \operatorname{End} (\tau_\Lambda^{-\ell r} S_\Lambda^{pp} \oplus S_\Lambda^{rg})$$

Choose a module $N_A^{pp} \in \operatorname{add} \mathcal{P}'(A)$ and a module $N_A^{rg} \in \operatorname{add} \boldsymbol{\mathcal{T}}^A$ such that $\tau_\Lambda^{-\ell r} S_\Lambda \cong \operatorname{Hom}_A(W, N_A^{pp})$ and $S_\Lambda^{rg} \cong \operatorname{Hom}_A(W, N_A^{rg})$. Then the module $N_A = N_A^{pp} \oplus N_A^{rg}$ is multiplicity-free and lies in add $(\mathcal{P}(A) \cup \boldsymbol{\mathcal{T}}^A)$. Moreover, there are isomorphisms

$$\begin{aligned}
\operatorname{Ext}_A^1(N_A, N_A) &\cong \operatorname{Ext}_\Lambda^1(\operatorname{Hom}_A(W, N_A), \operatorname{Hom}_A(W, N_A)) \\
&\cong \operatorname{Ext}_\Lambda^1(\tau_\Lambda^{-\ell r} S_\Lambda, \tau_\Lambda^{-\ell r} S_\Lambda) \\
&\cong \operatorname{Ext}_\Lambda^1(S_\Lambda, S_\Lambda) = 0,
\end{aligned}$$

because S_Λ is a tilting Λ-module. This shows that N_A is a tilting A-module. Because there are isomorphisms of algebras

$$\operatorname{End} N_A \cong \operatorname{End} \operatorname{Hom}_A(W, N_A) \cong \operatorname{End} \tau_\Lambda^{-\ell r} S_\Lambda \cong B'$$

then B' is a tilted algebra for a tilting module N_A with $N_A^{pi} = 0$. Now, by applying (3.5), we get

- $\mathcal{P}(\Lambda') = \mathcal{P}(B') = \operatorname{Hom}_A(N, \mathcal{T}(N) \cap \mathcal{P}(A))$, and
- $\boldsymbol{\mathcal{T}}^{B'} = \operatorname{Hom}_A(N, \mathcal{T}(N) \cap \boldsymbol{\mathcal{T}}^A)$.

It follows that there exist A-modules

$$T_A^{pp} \in \operatorname{add} (\mathcal{T}(N) \cap \mathcal{P}(A)) \text{ and } T_A^{rg} \in \operatorname{add} (\mathcal{T}(N) \cap \boldsymbol{\mathcal{T}}^A)$$

such that $\operatorname{Hom}_A(N, T^{pp}) \cong \tau_C^{-tm} C_C$ and $\operatorname{Hom}_A(N, T^{rg}) \cong (B')^{rg}$.
    Consider the A-module

$$T_A = T_A^{pp} \oplus T_A^{rg},$$

and observe that T_A is multiplicity-free.
    Now we show that T_A is a tilting A-module. First, we note that, because the module $(B')^{rg}$ is the direct sum of all indecomposable projective modules lying in $\boldsymbol{\mathcal{T}}^{B'}$ and B' is isomorphic to the tilted algebra $\operatorname{End} N_A$, then the module T_A^{rg} is the direct sum of all indecomposable direct summands of N_A lying in $\boldsymbol{\mathcal{T}}^A$, that is, $T_A^{rg} \cong N_A^{rg}$. It follows that $\operatorname{Ext}_A^1(T_A^{rg}, T_A^{rg}) = 0$ and there are isomorphisms $\operatorname{Ext}_A^1(T_A^{pp}, T_A^{rg}) \cong D\operatorname{Hom}_A(T_A^{rg}, \tau_A T_A^{pp}) = 0$, because $\operatorname{Hom}_A(\boldsymbol{\mathcal{T}}^A, \mathcal{P}(A)) = 0$. By the Brenner-Butler theorem (VI.3.8), the functor $\operatorname{Hom}_A(N, -)$ induces an equivalence of categories $\mathcal{T}(N) \xrightarrow{\simeq} \mathcal{Y}(N)$, the category $\mathcal{T}(N)$ is closed under extensions in $\operatorname{mod} A$, and the category $\mathcal{Y}(N)$ is closed under extensions in $\operatorname{mod} B'$. Hence, we get isomorphisms

$$\mathrm{Ext}_A^1(T_A^{pp}, T_A^{pp}) \cong \mathrm{Ext}_{B'}^1(\tau_C^{-tm}C_C, \tau_C^{-tm}C_C) \cong \mathrm{Ext}_C^1(C_C, C_C) = 0,$$

and

$$\mathrm{Ext}_A^1(T_A^{rg}, T_A^{pp}) \cong \mathrm{Ext}_{B'}^1((B')^{rg}, \tau_C^{-tm}C_C) = 0,$$

because $(B')^{rg}$ is a projective B'-module.

Next, we note that the Grothendieck groups $K_0(B)$, $K_0(B')$, and $K_0(A)$ are isomorphic and, hence, the number of pairwise non-isomorphic indecomposable direct summands of T_A equals the rank of $K_0(A)$. This shows that T_A is a multiplicity-free tilting A-module with $T_A^{pi} = 0$, in the notation of (3.2). Moreover, B is a tilted algebra given by the module T_A over the hereditary algebra $A = K\Delta$, because of the isomorphisms of algebras

$$B \cong B' \cong \mathrm{End}_{B'}(\tau_C^{-tm}C_C \oplus (B')^{rg}) \cong \mathrm{End}_{B'}(\mathrm{Hom}_A(N, T_A)) \cong \mathrm{End}\, T_A.$$

Because the equalities $r^B = r^A = m_\Delta$ hold then the proof of the theorem is complete. □

We finish this section by a coextension analogue of (4.3) proving the sufficiency part of Theorem (3.6).

4.4. Theorem. *Let C be a concealed algebra of Euclidean type, B a domestic tubular coextension of C with the tubular type $r^B = m_\Delta$, where Δ is one of the quivers $\Delta(\widetilde{\mathbb{A}}_{p,q})$, with $1 \le p \le q$, $\Delta(\widetilde{\mathbb{D}}_m)$, with $m \ge 4$, $\Delta(\widetilde{\mathbb{E}}_6)$, $\Delta(\widetilde{\mathbb{D}}_7)$, and $\Delta(\widetilde{\mathbb{D}}_8)$, see Section XII.1. Let $A = K\Delta$ be the path algebra of Δ. Then there exists a multiplicity-free tilting A-module T_A such that $T_A^{pp} = 0$ and $B \cong \mathrm{End}\, T_A$.*

Proof. Apply (XV.4.4) and the arguments given in the proof of (4.3). The details are left to the reader. □

XVII.5. A classification of tilted algebras of Euclidean type

The aim of this section is to summarise the results of the preceeding sections concerning the structure of the representation-infinite algebras of Euclidean type. In particular, we prove that the number of the isomorphism classes of basic tilted algebras B of any fixed Euclidean type is finite.

The following theorem provides an important characterisation of representation-infinite algebras of Euclidean type, due to Ringel [525].

5.1. Theorem. *Let B be an arbitrary basic connected algebra. The following three statements are equivalent.*

(a) *B is a representation-infinite tilted algebra of Euclidean type.*

(b) *B is isomorphic to an algebra of one of the following two kinds:*

- *a domestic tubular extension of a concealed algebra of Euclidean type,*
- *a domestic tubular coextension of a concealed algebra of Euclidean type,*

(c) *The Auslander–Reiten quiver* $\Gamma(\operatorname{mod} B)$ *of* B *admits a connected component of the following two types:*
   - *an infinite postprojective component* $\mathcal{P}$ *containing all indecomposable projective* B-*modules and a section of Euclidean type,*
   - *an infinite preinjective component* $\mathcal{Q}$ *containing all indecomposable injective* B-*modules and a section of Euclidean type.*

Proof. The implication (a)$\Rightarrow$(b) follows from (3.5) and (3.6), while the converse implication (b)$\Rightarrow$(a) follows from (4.3) and (4.4). Further, the implication (a)$\Rightarrow$(c) is a consequence of (3.5) and (3.6).

It remains to prove that (c) implies (a). Assume that the Auslander–Reiten translation quiver $\Gamma(\operatorname{mod} B)$ of B admits an infinite postprojective component $\mathcal{P}$ containing all indecomposable projective B-modules and a section Σ of the Euclidean type. Let T_B be the direct sum of all indecomposable B-modules lying on Σ. Our assumptions on the component $\mathcal{P}$ yield:

- $\operatorname{Hom}_B(T, \tau_B T) = 0$, because the component $\mathcal{P}$ is directed and closed under predecessors in $\operatorname{mod} B$, see (VIII.2.5),
- every indecomposable projective B-module is a predecessor of Σ in $\mathcal{P}$,
- $\mathcal{P}$ consists of the τ_B-orbits of indecomposable projective B-modules and, hence, there is no indecomposable injective B-module that is a proper predecessor of Σ in $\mathcal{P}$,
- the injective envelope $f : B_B \hookrightarrow E(B)$ in $\operatorname{mod} B$ has a factorisation through a direct sum of modules lying on the section Σ, by (IV.5.1)(a), and hence
- the module B_B is cogenerated by T_B and T_B is a faithful B-module, see (VI.2.2).

Applying now the criterion (VIII.5.6), we infer that T_B is a multiplicity-free tilting B-module such that $A = \operatorname{End} T_B$ is a hereditary algebra isomorphic to the path algebra KQ of the Euclidean quiver $Q = \Sigma^{\mathrm{op}}$, and $\mathcal{P}$ is the connecting component $\mathcal{C}_{T^*}$ determined by the multiplicity-free tilting A-module $T^* = D(_A T)$. It follows that B is a representation-infinite tilted algebra of the Euclidean type Q, because the component $\mathcal{P}$ is infinite. Moreover, by (3.5) and (3.6), $\mathcal{P} = \mathcal{C}_{T^*}$ is a unique postprojective component of $\Gamma(\operatorname{mod} B)$ containing no indecomposable injective modules, and hence, the tilting A-module T_A^* has no postprojective direct summand, see (VIII.4.2).

Similarly, one shows that if $\Gamma(\mathrm{mod}\,B)$ admits an infinite preinjective component $\mathcal{Q}$ containing all indecomposable injective B-modules and a section
Σ of the Euclidean type then the direct sum T_B of all indecomposable B-
modules lying on Σ is a multiplicity-free tilting B-module, $A = \mathrm{End}\,T_B$
is a hereditary algebra isomorphic to the path algebra KQ of the Eu-
clidean quiver $Q = \Sigma^{\mathrm{op}}$, $T^* = D({}_A T)$ is a multiplicity-free tilting A-module
and without indecomposable preinjective direct summands, $B \cong \mathrm{End}\,T_A^*$
is a representation-infinite tilted algebra of the Euclidean type Q, and the
preinjective component $\mathcal{Q}$ is the connecting component $\mathcal{C}_{T^*}$ of mod B determined by the tilting A-module T^*. This finishes the proof of the implication
(c)$\Rightarrow$(a) and completes the proof of the theorem.   □

We complete Theorem (4.1) with a description of the Auslander–Reiten
quiver of representation-infinite tilted algebras of the Euclidean type.

5.2. Theorem. *Let Q be an acyclic quiver whose underlying unoriented
graph $\overline{Q}$ is Euclidean, and let B be a representation-infinite algebra of Euclidean type. The Auslander–Reiten quiver $\Gamma(\mathrm{mod}\,B)$ of B has a disjoint
union decomposition*

$$\Gamma(\mathrm{mod}\,B) = \mathcal{P}(B) \cup \boldsymbol{T}^B \cup \mathcal{Q}(B),$$

with the following properties.

- $\mathcal{P}(B)$ *is a unique postprojective component of* $\Gamma(\mathrm{mod}\,B)$.
- $\mathcal{Q}(B)$ *is a unique preinjective component of* $\Gamma(\mathrm{mod}\,B)$.
- $\boldsymbol{T}^B = \{\mathcal{T}_\lambda^B\}_{\lambda \in \mathbb{P}_1(K)}$ *is a $\mathbb{P}_1(K)$-family of pairwise orthogonal standard ray tubes, or $\boldsymbol{T}^B = \{\mathcal{T}_\lambda^B\}_{\lambda \in \mathbb{P}_1(K)}$ is a $\mathbb{P}_1(K)$-family of pairwise
  orthogonal standard coray tubes.*
- *Either $\mathcal{P}(B)$ contains all the indecomposable projective B-modules,
  or $\mathcal{Q}(B)$ contains all the indecomposable injective B-modules.*
- $\boldsymbol{T}^B$ *separates $\mathcal{P}(B)$ from $\mathcal{Q}(B)$ in the sense of* (XII.3.3).
- *The tubular type r^B of $\boldsymbol{T}^B$ depends only on the Euclidean graph $\overline{Q}$.*

Proof. Apply (3.5), (3.6), and (5.1).   □

5.3. Theorem. *Let Δ be a Euclidean graph. The number of the isomorphism classes of basic tilted algebras B of Euclidean types Q such that the
underlying graph $\overline{Q}$ equals Δ is finite.*

Proof. Assume that Δ is a Euclidean graph. First we note that the
number of quivers Q such that $\overline{Q} = \Delta$ is finite.

Fix a Euclidean quiver Q with $\overline{Q} = \Delta$. Let $A = KQ$ be the path
algebra of Q, T_A a multiplicity-free tilting A-module, and $B = \mathrm{End}\,T_A$
the associated tilted algebra. We recall from (XII.3.4) that the Auslander–
Reiten quiver $\Gamma(\mathrm{mod}\,A)$ of A has a disjoint union decomposition

$$\Gamma(\mathrm{mod}\,A) = \mathcal{P}(A) \cup \boldsymbol{T}^A \cup \mathcal{Q}(A),$$

where

- $\mathcal{P}(A) \cong (-\mathbb{N})Q^{\mathrm{op}}$ is a postprojective component having the starting section formed by all indecomposable projective A-modules,
- $\mathcal{Q}(A) \cong \mathbb{N}Q^{\mathrm{op}}$ is a preinjective component having the final section formed by all indecomposable injective A-modules,.
- $\boldsymbol{T}^A = \{\mathcal{T}_\lambda^A\}_{\lambda \in \mathbb{P}_1(K)}$ is a $\mathbb{P}_1(K)$-family of pairwise orthogonal standard stable tubes separating $\mathcal{P}(A)$ from $\mathcal{Q}(A)$.

For each $\lambda \in \mathbb{P}_1(K)$, we denote by r_λ^A the rank of the tube $\mathcal{T}_\lambda^A$, and we set

$$\Lambda(A) = \{\lambda \in \mathbb{P}_1(K);\ r_\lambda^A \geq 2\}.$$

It follows from (XII.3.4) that $|\Lambda(A)| \leq 3$.

For each $\lambda \in \mathbb{P}_1(K)$, we denote by $\mathcal{C}r(\mathcal{T}_\lambda^A)$ the stone crown of the tube $\mathcal{T}_\lambda^A$ consisting of all indecomposable A-modules X in $\mathcal{T}_\lambda^A$ of regular length at most $r_\lambda^A - 1$, and we define the **stone crown of A** to be the full translation subquiver

$$\mathcal{C}r(A) = \bigcup_{\lambda \in \Lambda(A)} \mathcal{C}r(\mathcal{T}_\lambda^A)$$

of $\boldsymbol{T}^A$. It follows from (1.6) that $\mathcal{C}r(A)$ consists of all stones in $\boldsymbol{T}^A$, that is, the indecomposable A-modules M of $\boldsymbol{T}^A$ such that $\mathrm{Ext}_A^1(M, M) = 0$.

In the notation of (3.2), the tilting A-module T_A has a canonical decomposition

$$T_A = T_A^{pp} \oplus T_A^{rg} \oplus T_A^{pi},$$

where T_A^{pp} is a postprojective A-module, T_A^{rg} is a regular A-module, and T_A^{pi} is a preinjective A-module. Observe that $T_A^{rg} \in \mathrm{add}\,\mathcal{C}r(A)$, because all indecomposable direct summands of T_A are stones.

It follows from (VII.5.10) and (XIV.2.4) that A is a minimal representation-infinite algebra, and hence the number of pairwise non-isomorphic non-sincere indecomposable A-modules is finite. In particular, there exists a minimal integer $p_A \geq 1$ such that $\mathrm{Hom}_A(P(a), \tau_A^{-i}P(b)) \neq 0$ for all $i \geq p_A$ and all vertices $a, b \in Q_0$. It is clear that, if $M = \tau_A^{-i}P(c)$ and $N = \tau_A^{-j}P(d)$ are two modules in $\mathcal{P}(A)$ such that $\mathrm{Hom}_A(M, \tau_A N) = 0$ then $j - i \leq p_A$, because the isomorphisms

$$0 = \mathrm{Hom}_A(M, \tau_A N) \cong \mathrm{Hom}_A(\tau_A^{-i}P(c), \tau_A^{-j+1}P(d))$$

$$\cong \mathrm{Hom}_A(P(c), \tau_A^{-(j-i)+1}P(d))$$

and the choice of p_A yield $j - i \leq p_A$.

Following the proof of (XIV.4.3), we define the **concealed domain** $\mathcal{D}\mathcal{P}(A)$ of $\mathcal{P}(A)$ to be the full translation subquiver of $\mathcal{P}(A)$ whose vertices are the modules $\tau_A^{-i}P(a)$, with $a \in Q_0$ and $i \in \{0, 1, 2, \dots, p_A\}$.

Similarly, we define a concealed domain of $\mathcal{Q}(A)$. Because A is a minimal representation-infinite algebra and the number of pairwise non-isomorphic non-sincere indecomposable A-modules is finite then there exists a minimal integer $q_A \geq 1$ such that $\mathrm{Hom}_A(I(a), \tau_A^i I(b)) \neq 0$ for all $i \geq q_A$ and all vertices $a, b \in Q_0$. It follows as above that if $U = \tau_A^i I(c)$ and $N = \tau_A^j I(d)$ are two modules in $\mathcal{Q}(A)$ such that $\mathrm{Hom}_A(U, \tau_A V) = 0$ then $i - j \leq q_A$. We define the **concealed domain** $\mathcal{DQ}(A)$ of $\mathcal{Q}(A)$ to be the full translation subquiver of $\mathcal{Q}(A)$ whose vertices are the modules $\tau_A^i I(a)$, with $a \in Q_0$ and $i \in \{0, 1, 2, \dots, q_A\}$.

Now we show that the tilted algebra $B = \mathrm{End}\, T_A$ is isomorphic to a tilted algebra $\mathrm{End}\, \widehat{T}_A$, where $\widehat{T}_A$ is a tilting A-module with a canonical decomposition (3.2)

$$\widehat{T}_A = \widehat{T}_A^{pp} \oplus \widehat{T}_A^{rg} \oplus \widehat{T}_A^{pi}$$

such that $\widehat{T}_A^{pp} \in \mathrm{add}\,\mathcal{DP}(A)$, $\widehat{T}_A^{rg} \in \mathrm{add}\,\mathcal{C}r(A)$, and $\widehat{T}_A^{pi} \in \mathrm{add}\,\mathcal{DQ}(A)$. Because $\mathcal{DP}(A) \cup \mathcal{C}r(A) \cup \mathcal{DQ}(A)$ has a finite number of indecomposable modules, this will imply that the number of the isomorphism classes of basic tilted algebras B of the Euclidean type Q is finite, and will finish the proof of the theorem. To construct such a tilting module $\widehat{T}_A$, we consider two cases.

<u>Case 1°</u>. Assume that the tilted algebra $B \cong \mathrm{End}\, T_A$ is representation-finite, and we show that the tilting module T_A has the required properties.

It follows from (VIII.4.3) that $T_A^{pp} \neq 0$ and $T_A^{pi} \neq 0$. Now we show that $T_A^{pp} \in \mathrm{add}\,\mathcal{DP}(A)$, and $T_A^{pi} \in \mathrm{add}\,\mathcal{DQ}(A)$. Assume, to the contrary, that $T_A^{pp} \notin \mathrm{add}\,\mathcal{DP}(A)$, that is, there exists an indecomposable direct summand X of T_A^{pp} such that $X \in \mathcal{P}(A) \setminus \mathcal{DP}(A)$. Let Y be an arbitrary module in $\mathcal{Q}(A)$. Then there exists a vertex $b \in Q_0$ and $s \geq 0$ such that $Y \cong \tau_A^s I(b)$. Because, by the construction of $\mathcal{DP}(A)$, the module $\tau_A^{-s} X$ is sincere then we get

$$\mathrm{Hom}_A(X, Y) \cong \mathrm{Hom}_A(X, \tau_A^s I(b)) \cong \mathrm{Hom}_A(\tau_A^{-s} X, I(b)) \neq 0.$$

On the other hand, because the algebra A is hereditary and the modules T_A^{pp}, T_A^{pi} are direct summands of the tilting module T_A then $\mathrm{Hom}_A(T_A^{pp}, \tau_A T_A^{pi}) \cong D\mathrm{Ext}_A^1(T_A^{pi}, T_A^{pp}) = 0$. It follows that $\mathrm{Hom}_A(X, Y) = 0$, for any indecomposable direct summand Y of $\tau_A T_A^{pi}$, and we get a contradiction with the choice of X. This shows that every indecomposable direct summand X of T_A^{pp} lies in $\mathcal{DP}(A)$, that is, $T_A^{pp} \in \mathrm{add}\,\mathcal{DP}(A)$. Because the relation $T_A^{pi} \in \mathrm{add}\,\mathcal{DQ}(A)$ follows in a similar way, then the tilting module $\widehat{T}_A = T_A$ has the required properties, and the proof is complete in this case.

<u>Case 2°</u>. Assume that the tilted algebra $B = \mathrm{End}\, T_A$ is representation-infinite. We know from (3.3) that $T_A^{pp} = 0$ or $T_A^{pi} = 0$.

First we consider the case when $T_A^{pi} = 0$. It follows from (3.5)(a) that $T_A^{pp} \neq 0$. Let

$$T_A = T_A^{pp} \oplus T_A^{rg} = T_1 \oplus \ldots \oplus T_m \oplus T_{m+1} \oplus \ldots \oplus T_n,$$

where $1 \leq m \leq n$, the modules $T_1, \ldots, T_m, T_{m+1}, \ldots, T_n$ are indecomposable, $T_A^{pp} = T_1 \oplus \ldots \oplus T_m$, and $T_A^{rg} = T_{m+1} \oplus \ldots \oplus T_n$. If $T_A^{rg} = 0$, we set $n = m$. Because the algebra A is hereditary and the modules $T_1, \ldots, T_m$ are in $\mathcal{P}(A)$ then, for each $i \in \{1, \ldots, m\}$, there exist a vertex $a_i \in Q_0$ and an integer $s_i \geq 0$ such that $T_i \cong \tau_A^{-s} P(a_i)$. Without loss of generality, we may assume that $s = s_1$ is the minimal integer in the set $\{s_1, \ldots, s_m\}$. Then the tilting vanishing condition yields $\operatorname{Hom}_A(T_1, \tau_A T_i) = 0$, for all $i \in \{1, \ldots, m\}$, and the choice of p_A yields $s_i - s \leq p_A$, for each $i \in \{1, \ldots, m\}$. Consider the A-module

$$\widehat{T}_A = \widehat{T}_1 \oplus \ldots \oplus \widehat{T}_m \oplus \widehat{T}_{m+1} \oplus \ldots \oplus \widehat{T}_n,$$

where $\widehat{T}_i = \tau_A^{-s} T_i$ is an indecomposable module, for each $i \in \{1, \ldots, n\}$. Note that

$$\operatorname{Hom}_A(\widehat{T}_A, \tau_A \widehat{T}_A) \cong \operatorname{Hom}_A(\tau_A^{-s} T_A, \tau_A^{-s+1} T_A) \cong \operatorname{Hom}_A(T_A, \tau_A T_A) = 0.$$

It follows that $\widehat{T}_A$ is a multiplicity-free tilting A-module such that

- $\widehat{T}_A^{pp} \cong \tau_A^{-s} T_A^{pp} \in \operatorname{add} \mathcal{DP}(A)$, $\widehat{T}_A^{rg} \cong \tau_A^{-s} T_A^{rg} \in \operatorname{add} \mathcal{C}r(A)$,
- $\widehat{T}_A \in \operatorname{add}(\mathcal{DP}(A) \cup \mathcal{C}r(A))$, and
- there are isomorphisms of algebras

$$\operatorname{End} \widehat{T}_A \cong \operatorname{End} \tau_A^{-s} T_A \cong \operatorname{End} T_A = B.$$

This finishes the proof in case $T_A^{pi} = 0$. Similarly, if $T_A^{pp} = 0$ then $T_A^{pi} \neq 0$, by (3.6), and we find a multiplicity-free tilting A-module

$$\widehat{T}_A \in \operatorname{add}(\mathcal{C}r(A) \cup \mathcal{DQ}(A))$$

such that $\operatorname{End} \widehat{T}_A \cong \operatorname{End} T_A = B$. This completes the proof of the theorem. $\square$

5.4. Corollary. *Given an integer $d \geq 1$, the number of the isomorphism classes of basic tilted algebras B of Euclidean type with $\dim_K B = d$ is finite.*

Proof. It is easy to see that the rank of the Grothendieck group $K_0(\Lambda)$ of any algebra Λ of K-dimension d is less than or equal to d. It follows that the number of Euclidean type quivers Q such that there is a tilted algebra B of the type Q is finite. Then the corollary is a consequence of (5.3). $\square$

We recall from (VIII.3.2) that every tilted algebra B satisfies the following distinguished homological conditions:

(a) $\operatorname{gl.dim} B \leq 2$, and
(b) $\operatorname{pd} X_B \leq 1$ or $\operatorname{id} X_B \leq 1$, for any indecomposable module in $\operatorname{mod} B$.

In particular, this is the case for any tilted algebra B of Euclidean type.

We end this section with an example showing that the class of algebras Λ satisfying the homological conditions (a) and (b) is larger than the class of tilted algebras of Euclidean type. However, for each algebra Λ constructed in our example, the structure of the module category $\text{mod}\,\Lambda$ is very similar to the structure of the module category $\text{mod}\,\Lambda$ for a representation-infinite tilted algebra B of Euclidean type. As a consequence, it shows that the finiteness of the number of the isomorphism classes of basic tilted algebras of Euclidean type of a fixed dimension $d \geq 1$, established in (5.4), is a very exclusive property of tilted algebras of Euclidean type.

5.5. Example. Let Q be the quiver

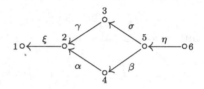

For each $a \in K \setminus \{0,1\}$, we consider the algebra
$$\Lambda^{(a)} = KQ/I^{(a)},$$
where $I^{(a)}$ is the ideal of the path algebra KQ of Q generated by the commutativity relations $\beta\alpha\xi - \sigma\gamma\xi$ and $\eta\beta\alpha - a\eta\sigma\gamma$. It is easy to see that

- $\dim_K \Lambda^{(a)} = 21$, for each $a \in K \setminus \{0,1\}$,
- $(1-a)\eta\sigma\gamma\xi \in I^{(a)}$, and hence
- $\eta\sigma\gamma\xi \in I^{(a)}$ and $\eta\beta\alpha\xi \in I^{(a)}$.

Now we show that
$$\Lambda^{(a)} \not\cong \Lambda^{(b)}, \quad \text{for all } a,b \in K \setminus \{0,1\} \text{ such that } a \neq b.$$

Assume, to the contrary, that there is an isomorphism of algebras $f : \Lambda^{(a)} \xrightarrow{\;\cong\;} \Lambda^{(b)}$, for some $a,b \in K \setminus \{0,1\}$. We show that $a = b$. First, we note that if there is an arrow $i \longrightarrow j$ in the quiver Q and e_i, e_j are the primitive idempotents of the algebra KQ corresponding to the vertices i, j of Q, then
$$e_i(\text{rad}\,KQ)e_j \cong K \quad \text{and} \quad e_i(\text{rad}^2\,KQ)e_j = 0.$$
Hence, there exist elements $c_\alpha, c_\beta, c_\gamma, c_\xi, c_\eta \in K \setminus \{0\}$ such that
$$f(\alpha+I^{(a)}) = c_\alpha\alpha+I^{(b)}, \; f(\beta+I^{(a)}) = c_\beta\beta+I^{(b)}, \; f(\gamma+I^{(a)}) = c_\gamma\gamma+I^{(b)},$$
$$f(\sigma+I^{(a)}) = c_\sigma\sigma+I^{(b)}, \; f(\xi+I^{(a)}) = c_\xi\xi+I^{(b)}, \; f(\eta+I^{(a)}) = c_\eta\eta+I^{(b)}.$$
It follows that $c_\beta c_\alpha = c_\sigma c_\gamma$, because of the following equalities

$$
\begin{aligned}
(c_\beta c_\alpha - c_\sigma c_\gamma)c_\xi \beta \alpha \xi + I^{(b)} &= (c_\beta c_\alpha c_\xi \beta \alpha \xi + I^{(b)}) - (c_\sigma c_\gamma c_\xi \beta \alpha \xi + I^{(b)}) \\
&= (c_\beta c_\alpha c_\xi \beta \alpha \xi + I^{(b)}) - (c_\sigma c_\gamma c_\xi \sigma \gamma \xi + I^{(b)}) \\
&= (c_\beta \beta + I^{(b)})(c_\alpha \alpha + I^{(b)})(c_\xi \xi + I^{(b)}) \\
&\quad - (c_\sigma \sigma + I^{(b)})(c_\gamma \gamma + I^{(b)})(c_\xi \xi + I^{(b)}) \\
&= f(\beta + I^{(a)})f(\alpha + I^{(a)})f(\xi + I^{(a)}) \\
&\quad - f(\sigma + I^{(a)})f(\gamma + I^{(a)})f(\xi + I^{(a)}) \\
&= f((\beta \alpha \xi + I^{(a)}) - (\sigma \gamma \xi + I^{(a)})) \\
&= f(0 + I^{(a)}) = 0.
\end{aligned}
$$

Hence, invoking the commutativity relation $\eta \beta \alpha = a \eta \sigma \gamma$ in $\Lambda^{(a)}$, we get

$$
\begin{aligned}
(b-a)c_\eta c_\sigma c_\gamma \eta \sigma \gamma + I^{(b)} &= (b c_\eta c_\sigma c_\gamma \eta \sigma \gamma + I^{(b)}) - (a c_\eta c_\sigma c_\gamma \eta \sigma \gamma + I^{(b)}) \\
&= c_\eta c_\sigma c_\gamma (b \eta \sigma \gamma + I^{(b)}) - (a c_\eta c_\sigma c_\gamma \eta \sigma \gamma + I^{(b)}) \\
&= c_\eta c_\beta c_\alpha (\eta \beta \alpha + I^{(b)}) - a c_\eta c_\sigma c_\gamma (\eta \sigma \gamma + I^{(b)}) \\
&= (c_\eta \eta + I^{(b)})(c_\beta \beta + I^{(b)})(c_\alpha \alpha + I^{(b)}) \\
&\quad - a(c_\eta \eta + I^{(b)})(c_\sigma \sigma + I^{(b)})(c_\gamma \gamma + I^{(b)}) \\
&= f(\eta + I^{(a)})f(\beta + I^{(a)})f(\alpha + I^{(a)}) \\
&\quad - af(\eta + I^{(a)})f(\sigma + I^{(a)})f(\gamma + I^{(a)}) \\
&= f((\eta \beta \alpha + I^{(a)}) - (a \eta \sigma \gamma + I^{(a)})) \\
&= f(0 + I^{(a)}) = 0.
\end{aligned}
$$

It follows that $b = a$, as we required.

Given a fixed element $a \in K \setminus \{0, 1\}$, we describe the Auslander–Reiten quiver $\Gamma(\operatorname{mod} \Lambda^{(a)})$ of the algebra $\Lambda^{(a)}$. Denote by C the path algebra $K\Delta$ of the quiver

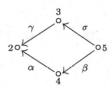

By (XII.1.1), C is the canonical algebra $C(2.2)$ and it follows from (XII.2.8) and (XII.3.4) that the Auslander–Reiten quiver $\Gamma(\operatorname{mod} C)$ of C has a disjoint union decomposition

$$
\Gamma(\operatorname{mod} C) = \mathcal{P}(C) \cup \boldsymbol{T}^C \cup \mathcal{Q}(C),
$$

where $\mathcal{P}(C)$ is a unique postprojective component containing all the indecomposable projective C-modules, $\mathcal{Q}(C)$ is a unique preinjective component containing all the indecomposable injective C-modules, and $\mathcal{T}^C$ is a $\mathbb{P}_1(K)$-family $\mathcal{T}^C = \{\mathcal{T}_\lambda^C\}_{\lambda \in \mathbb{P}_1(K)}$ of pairwise orthogonal standard stable tubes such that

- $\mathcal{T}_\infty^C$ is the stable tube of rank 2 with the mouth modules

$$E_1^{(\infty)} = S(3) : \quad 0 \diagup^{K}\diagdown_{0}^{0}, \qquad E_2^{(\infty)} = K \diagup^{0}\diagdown_{K}^{K}$$

- $\mathcal{T}_0^C$ is the stable tube of rank 2 with the mouth modules

$$E_1^{(0)} = S(4) : \quad 0 \diagup^{0}\diagdown_{0}^{K}, \qquad E_2^{(0)} : \quad K \diagup^{K}\diagdown_{0}^{K}$$

- for each $\lambda \in K \setminus \{0\}$, $\mathcal{T}_\lambda^C$ is the stable tube of rank 1 with the unique mouth module

$$E^{(\lambda)} : \quad K \diagup^{K}\diagdown_{K}^{K}.$$

Given $a \in K \setminus \{0, 1\}$, we consider the following two quotient algebras

$$\Lambda_- = KQ_-/I_- \quad \text{and} \quad \Lambda_+^{(a)} = KQ_+/I_+^{(a)}$$

of the algebra $\Lambda^{(a)}$, where Q^- and Q^+ are the following subquivers

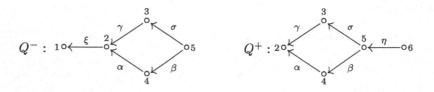

of the quiver Q of $\Lambda^{(a)}$, I_- is the ideal of the path algebra KQ_- of Q_- generated by the commutativity relation $\beta a \xi - \sigma \gamma \xi$, and $I_+^{(a)}$ is the ideal of the path algebra KQ_+ of Q_+ generated by the commutativity relation $\eta \beta a - a \eta \sigma \gamma$.

Observe that the algebra Λ_- is the domestic tubular (branch) coextension

$$\Lambda_- = [E^{(1)}, \mathcal{L}^{(1)}]C$$

of C of tubular type $r^{\Lambda_-} = \widehat{r}^{\Lambda_-} = (2, 2, 2)$, where $\mathcal{L}^{(1)}$ is the branch consisting of one vertex. Then, it follows from (3.6) and (4.4) that Λ_- is a representation-infinite tilted algebra of the Euclidean type $\widetilde{\mathbb{D}}_4$ and the Auslander–Reiten quiver $\Gamma(\operatorname{mod}\Lambda_-)$ of Λ_- has a disjoint union decomposition

$$\Gamma(\operatorname{mod}\Lambda_-) = \mathcal{P}(\Lambda_-) \cup \boldsymbol{\mathcal{T}}^{\Lambda_-} \cup \mathcal{Q}(\Lambda_-),$$

where

- $\mathcal{P}(\Lambda_-)$ is a unique postprojective component containing all the indecomposable projective Λ_--modules,
- $\mathcal{Q}(\Lambda_-) = \mathcal{Q}(C)$ is a unique preinjective component,
- $\boldsymbol{\mathcal{T}}^{\Lambda_-}$ is a $\mathbb{P}_1(K)$-family $\boldsymbol{\mathcal{T}}^{\Lambda_-} = \{\mathcal{T}_\lambda^{\Lambda_-}\}_{\lambda \in \mathbb{P}_1(K)}$ of pairwise orthogonal standard coray tubes such that $\mathcal{T}_\lambda^{\Lambda_-} = \mathcal{T}_\lambda^C$, for each $\lambda \in \mathbb{P}_1(K)\backslash\{1\}$, and $\mathcal{T}_1^{\Lambda_-}$ is a coray tube of rank 2 obtained from the stable tube $\mathcal{T}_1^C$ by insertion of one coray ending at the injective Λ_--module $I(1)_{\Lambda_-}$.
- $\boldsymbol{\mathcal{T}}^{\Lambda_-}$ separates $\mathcal{P}(\Lambda_-)$ from $\mathcal{Q}(\Lambda_-)$.

The algebra $\Lambda_+^{(a)}$ is the domestic tubular (branch) extension

$$\Lambda_+^{(a)} = C[E^{(a)}, \mathcal{L}^{(a)}]$$

of C of tubular type $r^{\Lambda_+} = \widehat{r}^{\Lambda_+} = (2, 2, 2)$, where $\mathcal{L}^{(a)}$ is the branch consisting of one vertex. Then, it follows from (3.5) and (4.3) that $\Lambda_+^{(a)}$ is a representation-infinite tilted algebra of the Euclidean type $\widetilde{\mathbb{D}}_4$ and the Auslander–Reiten quiver $\Gamma(\operatorname{mod}\Lambda_+^{(a)})$ of $\Lambda_+^{(a)}$ has a disjoint union decomposition

$$\Gamma(\operatorname{mod}\Lambda_+^{(a)}) = \mathcal{P}(\Lambda_+^{(a)}) \cup \boldsymbol{\mathcal{T}}^{\Lambda_+^{(a)}} \cup \mathcal{Q}(\Lambda_+^{(a)}),$$

where

- $\mathcal{P}(\Lambda_+^{(a)}) = \mathcal{P}(C)$ is a unique postprojective component,
- $\mathcal{Q}(\Lambda_+^{(a)})$ is a unique preinjective component containing all the indecomposable injective $\Lambda_+^{(a)}$-modules,
- $\boldsymbol{\mathcal{T}}^{\Lambda_+^{(a)}}$ is a $\mathbb{P}_1(K)$-family $\boldsymbol{\mathcal{T}}^{\Lambda_+^{(a)}} = \{\mathcal{T}_\lambda^{\Lambda_+^{(a)}}\}_{\lambda \in \mathbb{P}_1(K)}$ of pairwise orthogonal standard ray tubes such that $\mathcal{T}_\lambda^{\Lambda_+^{(a)}} = \mathcal{T}_\lambda^C$, for each $\lambda \in \mathbb{P}_1(K) \backslash \{a\}$, and $\mathcal{T}_a^{\Lambda_+^{(a)}}$ is a ray tube of rank 2 obtained from the stable tube $\mathcal{T}_a^C$ by insertion of one ray starting from the projective $\Lambda_+^{(a)}$-module $P(6)_{\Lambda_+^{(a)}}$,
- $\boldsymbol{\mathcal{T}}^{\Lambda_+^{(a)}}$ separates $\mathcal{P}(\Lambda_+^{(a)})$ from $\mathcal{Q}(\Lambda_+^{(a)})$.

It follows that the algebra $\Lambda^{(a)}$ is both

- the branch extension $\Lambda_-[E^{(a)}, \mathcal{L}^{(a)}]$ of Λ_- using the tube $\mathcal{T}_a^{\Lambda_-} = \mathcal{T}_a^C$, and

- the branch coextension $[E^{(1)}, \mathcal{L}^{(1)}]\Lambda_+^{(a)}$ of $\Lambda_+^{(a)}$ using the tube $\mathcal{T}_1^{\Lambda_+^{(a)}} = \mathcal{T}_1^C$.

Consequently, for each $a \in K \setminus \{0, 1\}$, the algebra $\Lambda^{(a)}$ has the branch coextension-extension form

$$\Lambda^{(a)} = [E^{(1)}, \mathcal{L}^{(1)}]C[E^{(a)}, \mathcal{L}^{(a)}].$$

By applying (XV.4.3) and (XV.4.4), we conclude that, for a fixed element $a \in K \setminus \{0, 1\}$, the Auslander–Reiten quiver $\Gamma(\operatorname{mod} \Lambda^{(a)})$ of the algebra $\Lambda^{(a)}$ has a disjoint union decomposition

$$\Gamma(\operatorname{mod} \Lambda^{(a)}) = \mathcal{P}(\Lambda^{(a)}) \cup \boldsymbol{\mathcal{T}}^{\Lambda^{(a)}} \cup \mathcal{Q}(\Lambda^{(a)}),$$

where

(i) $\mathcal{P}(\Lambda^{(a)}) = \mathcal{P}(\Lambda_-)$ is a unique postprojective component containing all the indecomposable projective Λ_--modules, except the module $P(6) = P(6)_{\Lambda^{(a)}}$, and the left hand part of $\mathcal{P}(\Lambda^{(a)})$ looks as follows

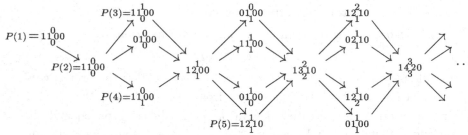

where we set $P(j) = P(j)_{\Lambda^{(a)}}$, for each $j \in \{1, 2, 3, 4, 5, 6\}$, and the indecomposable $\Lambda^{(a)}$-modules are represented by their dimension vectors,

(ii) $\mathcal{Q}(\Lambda^{(a)}) = \mathcal{Q}(\Lambda_+^{(a)})$ is a unique preinjective component containing all the indecomposable injective $\Lambda_+^{(a)}$-modules, except the module $I(6) = I(6)_{\Lambda^{(a)}}$, and the right hand part of $\mathcal{Q}(\Lambda^{(a)})$ looks as follows

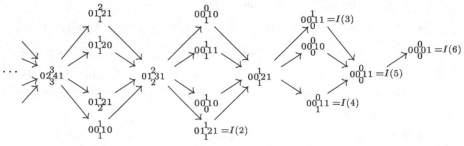

where we set $I(j) = I(j)_{\Lambda^{(a)}}$, for each $j \in \{1,2,3,4,5,6\}$, and the indecomposable $\Lambda^{(n)}$-modules are represented by their dimension vectors,

(iii) $\mathcal{T}^{\Lambda^{(a)}} = \{\mathcal{T}_\lambda^{\Lambda^{(a)}}\}_{\lambda \in \mathbb{P}_1(K)}$ is a $\mathbb{P}_1(K)$-family of pairwise orthogonal standard tubes separating $\mathcal{P}(\Lambda^{(a)})$ from $\mathcal{Q}(\Lambda^{(a)})$ and such that

- $\mathcal{T}_\infty^{\Lambda^{(a)}} = \mathcal{T}_\infty^C$ and $\mathcal{T}_0^{\Lambda^{(a)}} = \mathcal{T}_0^C$ are stable tubes of rank 2,
- $\mathcal{T}_\lambda^{\Lambda^{(a)}} = \mathcal{T}_\lambda^C$ is a stable tube of rank 1, for each $\lambda \in \mathbb{P}_1(K) \setminus \{0,1,a,\infty\}$,
- $\mathcal{T}_1^{\Lambda^{(a)}} = \mathcal{T}_1^{\Lambda_-}$ is a coray tube of rank 2, $\mathcal{T}_a^{\Lambda^{(a)}} = \mathcal{T}_a^{\Lambda_+^{(a)}}$ is a ray tube of rank 2, and the tubes $\mathcal{T}_1^{\Lambda^{(a)}}$ and $\mathcal{T}_a^{\Lambda^{(a)}}$ have the forms

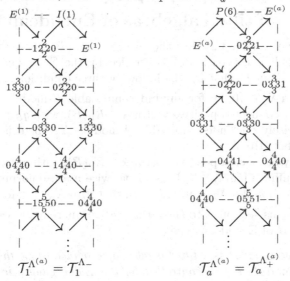

where $E^{(1)} = 01{\overset{1}{1}}10$, $I(1) = 11{\overset{1}{1}}10$, $P(6) = 01{\overset{1}{1}}11$, $E^{(a)} = 01{\overset{1}{1}}10$, and the indecomposable $\Lambda^{(a)}$-modules are represented by their dimension vectors,

(iv) $\mathrm{Hom}_{\Lambda^{(a)}}(D(\Lambda^{(a)}), \mathcal{P}(\Lambda^{(a)}) \cup (\mathcal{T}^{\Lambda^{(a)}} \setminus \{\mathcal{T}_1^{\Lambda^{(a)}}\})) = 0$,

(v) $\mathrm{Hom}_{\Lambda^{(a)}}(\Lambda_{\Lambda^{(a)}}^{(a)}, (\mathcal{T}^{\Lambda^{(a)}} \setminus \{\mathcal{T}_a^{\Lambda^{(a)}}\}) \cup \mathcal{Q}(\Lambda^{(a)})) = 0$,

(vi) $\mathrm{pd}\, X \leq 1$, for any module X in $\mathcal{P}(\Lambda^{(a)}) \cup (\mathcal{T}^{\Lambda^{(a)}} \setminus \{\mathcal{T}_1^{\Lambda^{(a)}}\})$,

(vii) $\mathrm{id}\, Y \leq 1$, for any module Y in $(\mathcal{T}^{\Lambda^{(a)}} \setminus \{\mathcal{T}_a^{\Lambda^{(a)}}\}) \cup \mathcal{Q}(\Lambda^{(a)})$,

(viii) $\mathrm{pd}\, X \leq 1$ or $\mathrm{id}\, X \leq 1$, for any indecomposable module X in $\mathrm{mod}\, \Lambda^{(a)}$, and

(ix) $\mathrm{gl.dim}\, \Lambda^{(a)} \leq 2$.

The statements (vi) and (vii) follow from (IV.2.7), because we assume that $a \neq 1$. The statement (ix) follows from (vi), (vii), and the fact that the radical of any indecomposable projective $\Lambda^{(a)}$-module lies in

$$\mathcal{P}(\Lambda^{(a)}) \cup (\mathcal{T}^{\Lambda^{(a)}} \setminus \{\mathcal{T}_1^{\Lambda^{(a)}}\}),$$

see (XV.4.3).

Finally, we note that the algebra $\Lambda^{(a)}$ is representation-infinite, but it is not a tilted algebra of the Euclidean type, because $\mathcal{T}_1^{\Lambda^{(a)}}$ is a coray tube, $\mathcal{T}_a^{\Lambda^{(a)}}$ is a ray tube, and (5.2) applies. This finishes the example.

XVII.6. A controlled property of the Euler form of tilted algebras of Euclidean type

It was shown in (XII.4.2) that the category $\mathrm{mod}\, B$ over a concealed algebra B of Euclidean type is link controlled by the Euler quadratic form $q_B : K_0(B) \longrightarrow \mathbb{Z}$ of B, that is, the following three conditions are satisfied.

- $q_A(\mathbf{dim}\, X) \in \{0, 1\}$, for any indecomposable B-module X.
- For any connected positive vector $\mathbf{x} \in K_0(B)$, with $q_B(\mathbf{x}) = 1$, there is precisely one indecomposable A-module X, up to isomorphism, such that $\mathbf{dim}\, X = \mathbf{x}$.
- For any connected positive vector $\mathbf{x} \in K_0(B)$, with $q_B(\mathbf{x}) = 0$, there is an infinite family $\{X_\lambda\}_{\lambda \in \Lambda}$ of pairwise non-isomorphic indecomposable modules X_λ in $\mathrm{mod}\, B$ such that $\mathbf{dim}\, X_\lambda = \mathbf{x}$, for any $\lambda \in \Lambda$.

The aim of this section is to show that the result remains valid, for any tilted algebra B of Euclidean type.

6.1. Theorem. *If B is a tilted algebra of Euclidean type then the module category $\mathrm{mod}\, B$ is link controlled by the Euler quadratic form $q_B : K_0(B) \longrightarrow \mathbb{Z}$ of B.*

Proof. Let Q be an acyclic Euclidean quiver and let $A = KQ$ be the path algebra of Q. Assume that B is a tilted algebra of the Euclidean type Q. Then $B \cong \mathrm{End}\, T_A$, where T_A is a multiplicity-free tilting A-module. It follows from (VI.3.5) that the algebra B is connected. We also recall from (VIII.3.2) that $\mathrm{gl.dim}\, B \leq 2$. Now we split the proof into two cases.

Case $1°$. Assume that the tilted algebra B is representation-finite. By (IV.5.4) and (VIII.4.3), the Auslander–Reiten quiver $\Gamma(\mathrm{mod}\, B)$ of B is both postprojective and preinjective, and consequently, B is a representation-directed algebra. Because $\mathrm{gl.dim}\, B \leq 2$ then (IX.3.3) applies. Hence, the Euler quadratic form $q_B : K_0(B) \longrightarrow \mathbb{Z}$ of B is weakly positive and the correspondence $X \mapsto \mathbf{dim}\, X$ defines a bijection between the indecomposable modules in $\mathrm{mod}\, B$ and the positive roots of q_B. Observe also that the Grothendieck group $K_0(B) \cong K_0(A)$ has no positive vector $\mathbf{x}$ such that $q_B(\mathbf{x}) - 0$, although q_B is $\mathbb{Z}$-congruent to the positive semidefinite Euler quadratic form $q_A : K_0(A) \longrightarrow \mathbb{Z}$ of A, see (VI.4.7) and (VII.4.2). It follows that $q_B(\dim X) = 1$, for any indecomposable module X in $\operatorname{mod} B$, and hence, the category $\operatorname{mod} B$ is link controlled by the Euler quadratic form q_B.

Case 2°. Assume that the tilted algebra $B \cong \operatorname{End} T_A$ is representation-infinite. In the notation of (3.2), the tilting A-module T_A has a canonical decomposition

$$T_A = T_A^{pp} \oplus T_A^{rg} \oplus T_A^{pi},$$

where T_A^{pp} is a postprojective A-module, T_A^{rg} is a regular A-module, and T_A^{pi} is a preinjective A-module. Because B is representation-infinite then $T_A^{pp} = 0$ or $T_A^{pi} = 0$, by (VIII.4.3) and (VIII.4.4). Without loss of generality, we may assume that $T_A^{pi} = 0$. Then, by (3.5), $C = \operatorname{End} T_A^{pp}$ is a concealed algebra of Euclidean type and B is a domestic tubular extension of C. In case $T_A^{rg} = 0$, we have $B = C$ and (XII.4.2) applies. It follows that the category $\operatorname{mod} B = \operatorname{mod} C$ is link controlled (and is even controlled) by the Euler quadratic form q_B.

Assume that $T_A^{rg} \neq 0$. Then B is a proper domestic tubular extension of C and, by (XV.3.9), B is a domestic branch extension

$$B = C[F_1, \mathcal{L}^{(1)}, F_2, \mathcal{L}^{(2)}, \ldots, F_s, \mathcal{L}^{(s)}]$$

of C and the ordinary quiver Q_B of B has the form

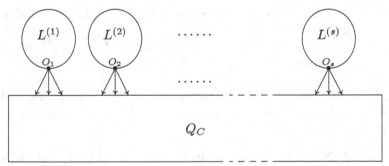

where

- $F_1, F_2, \ldots, F_s$ are pairwise different C-modules lying on the mouths of the tubes of the $\mathbb{P}_1(K)$-family $\mathcal{T}^C = \{\mathcal{T}_\lambda^C\}_{\lambda \in \mathbb{P}_1(K)}$ of pairwise orthogonal standard stable tubes of $\Gamma(\operatorname{mod} C)$,

- $\mathcal{L}^{(1)} = (L^{(1)}, I^{(1)}), \mathcal{L}^{(2)} = (L^{(2)}, I^{(2)}), \ldots, \mathcal{L}^{(s)} = (L^{(s)}, I^{(s)})$ are branches with the germs $O_1, O_2, \ldots, O_s$,
- for each $j \subset \{1, \ldots, s\}$, the vertex O_j is the extension vertex of the one-point extension algebra $C[F_j]$, and
- the quivers $Q_C, L^{(1)}, L^{(2)}, \ldots, L^{(s)}$ are full convex subquivers of Q_B.

On the other hand, the algebra B has the lower triangular matrix form

$$B \cong \begin{bmatrix} C & 0 \\ {}_D M_C & D \end{bmatrix},$$

where $C = \operatorname{End} T_A^{pp}$, $D = \operatorname{End} T_A^{rg}$, and ${}_D M_C = \operatorname{Hom}_A(T_A^{pp}, T_A^{rg})$ is viewed as a D-C-bimodule in an obvious way. Moreover, the algebra D is the product

$$D \cong D_1 \times D_2 \times \ldots \times D_s$$

of the branch algebras $D_1 \cong K\mathcal{L}^{(1)}, D_2 \cong K\mathcal{L}^{(2)}, \ldots, D_s \cong K\mathcal{L}^{(s)}$. Hence we easily conclude that the Cartan matrix $\mathbf{C}_B$ of B has the upper triangular matrix form

$$\mathbf{C}_B \cong \begin{bmatrix} \mathbf{C}_C & * \\ 0 & \mathbf{C}_D \end{bmatrix},$$

there is a canonical embedding

$$K_0(C) = \mathbb{Z}^{(Q_C)_0} \hookrightarrow \mathbb{Z}^{(Q_B)_0} = K_0(B)$$

of the Grothendieck groups of C and B, and the Euler quadratic form $q_C : K_0(C) \longrightarrow \mathbb{Z}$ is the restriction of $q_B : K_0(B) \longrightarrow \mathbb{Z}$ to the subgroup $K_0(C)$ of $K_0(B)$.

By (3.5), the Auslander–Reiten quiver $\Gamma(\operatorname{mod} B)$ of B has a disjoint union decomposition $\Gamma(\operatorname{mod} B) = \mathcal{P}(B) \cup \boldsymbol{\mathcal{T}}^B \cup \mathcal{Q}(B)$, where

- $\mathcal{P}(B) = \operatorname{Hom}_A(T, \mathcal{T}(T) \cap \mathcal{P}(A)) = \mathcal{P}(C)$ is a unique postprojective component of $\Gamma(\operatorname{mod} B)$,
- $\mathcal{Q}(B)$ is a unique preinjective component of $\Gamma(\operatorname{mod} B)$ containing all the indecomposable injective B-modules,
- $\boldsymbol{\mathcal{T}}^B = \operatorname{Hom}_A(T, \mathcal{T}(T) \cap \boldsymbol{\mathcal{T}}^A)$ is a $\mathbb{P}_1(K)$-family $\boldsymbol{\mathcal{T}}^B = \{\mathcal{T}_\lambda^B\}_{\lambda \in \mathbb{P}_1(K)}$ of pairwise orthogonal standard ray tubes $\mathcal{T}_\lambda^B = \operatorname{Hom}_A(T, \mathcal{T}_\lambda^A)$, with $r_\lambda^B = r_\lambda^A$ obtained from the $\mathbb{P}_1(K)$-family $\boldsymbol{\mathcal{T}}^C = \{\mathcal{T}_\lambda^C\}_{\lambda \in \mathbb{P}_1(K)}$ of pairwise orthogonal standard stable tubes of $\Gamma(\operatorname{mod} C)$ by iterated rectangle insertions, and
- $\boldsymbol{\mathcal{T}}^B$ is of the tubular type $\mathbf{m}_Q$ (see (XVI.4.7)), and separates $\mathcal{P}(B)$ from $\mathcal{Q}(B)$.

Because $B \cong \operatorname{End} T_A$ is a tilted algebra then, by (VI.4.3) and (VI.4.5), there exists a group isomorphism $f : K_0(A) \overset{\cong}{\longrightarrow} K_0(B)$ of the Grothendieck groups of A and B such that, for any A-module M, the following two equalities hold

- $f(\mathbf{dim}\, M) = \mathbf{dim}\, \operatorname{Hom}_A(T, M) - \mathbf{dim}\, \operatorname{Ext}_A^1(T, M)$, and
- $q_B(f(\mathbf{dim}\, M)) = q_A(\mathbf{dim}\, M)$.

Now we show that $q_B(\dim X) \in \{0,1\}$, for any indecomposable B-module X. If X is an indecomposable module in $\mathcal{P}(B) \cup \mathcal{Q}(B)$ then X is directing and, according to (IX.1.5), we have $q_B(\dim X) = 1$. Assume that X is an indecomposable module in $\mathcal{T}^B$. Then X has the form $X \cong \mathrm{Hom}_A(T,M)$, where M is an indecomposable module in $\mathcal{T}(T) \cap \mathcal{T}^A$. Because $A = KQ$ is a concealed algebra of Euclidean type then, by (XII.4.2), the category $\mathrm{mod}\,A$ is link controlled by the Euler quadratic form q_A of A and, hence, $q_A(\dim M) \in \{0,1\}$. Thus, we get

$$\begin{aligned} q_B(\dim X) &= q_B(\dim \mathrm{Hom}_A(T,M)) \\ &= q_B(f(\dim M)) \\ &= q_A(\dim M) \in \{0,1\}. \end{aligned}$$

We recall from (VI.4.7) and (VII.4.2) that the form q_B is semidefinite of corank one and

$$\mathrm{rad}\, q_B = \mathbb{Z} \cdot \mathbf{h}_B,$$

for a vector $\mathbf{h}_B \in K_0(B)$. On the other hand, because C is a concealed algebra of Euclidean type then, by (XI.3.7), there exists a unique positive vector $\mathbf{h}_C \in K_0(C)$ such that $\mathrm{rad}\, q_C = \mathbb{Z} \cdot \mathbf{h}_C$ and all coordinates of $\mathbf{h}_C$ are positive. It follows that $\mathrm{rad}\, q_B = \mathrm{rad}\, q_C$ and $\mathbf{h}_B = \mathbf{h}_C$, because q_C is the restriction of q_B to $K_0(C)$, under the canonical embedding $K_0(C) \hookrightarrow K_0(B)$.

We recall from (XII.4.2) that if X is an indecomposable module in a tube $\mathcal{T}_\lambda^C$ of rank $r_\lambda^C \geq 1$ then $\dim X = m \cdot \mathbf{h}_C$, for some $m \geq 1$, if and only if $r\ell(X) = m \cdot r_\lambda^C$. Hence we conclude that any positive vector $\mathbf{x} \in K_0(B)$, with $q_B(\mathbf{x}) = 0$, belongs to the subgroup $K_0(C) \hookrightarrow K_0(B)$ of $K_0(B)$ and there exists a $\mathbb{P}_1(K)$-family $\{X_\lambda\}_{\lambda \in \mathbb{P}_1(K)}$ of pairwise non-isomorphic indecomposable C-modules (hence, B-modules) such that $\dim X_\lambda = \mathbf{x}$, for each $\lambda \in \mathbb{P}_1(K)$. Moreover, each indecomposable B-module X such that $q_B(\dim X) = 0$ is a C-module.

Consequently, to prove that the module category $\mathrm{mod}\,B$ is link controlled by the Euler quadratic form $q_B : K_0(B) \longrightarrow \mathbb{Z}$, it is sufficient to show that, for every connected positive vector $\mathbf{x} \in K_0(B)$ with $q_B(\mathbf{x}) = 1$, there is precisely one indecomposable B-module X, up to isomorphism, such that $\dim X = \mathbf{x}$.

To show this, assume that $\mathbf{x} \in K_0(B)$ is a connected positive vector such that $q_B(\mathbf{x}) = 1$. We make the identification $K_0(B) = \mathbb{Z}^{(Q_A)_0}$ and we consider the support

$$\mathrm{supp}_A(\mathbf{x}) = \{j \in (Q_A)_0; \; x_j \neq 0\}$$

of the vector $\mathbf{x} \in K_0(B)$ in the group $K_0(A)$. Assume that $\mathrm{supp}_A(\mathbf{x})$ is contained in $(Q_C)_0$. Then, by (XII.4.2), there exists a unique indecomposable

C-module X, up to isomorphism, such that $\dim X = \mathbf{x}$. It follows that any indecomposable B-module Y such that $\dim Y = \mathbf{x}$ is a C-module and there is an isomorphism $Y \cong X$.

Assume now that $\operatorname{supp}_A(\mathbf{x})$ is not contained in $(Q_{D_j})_0$, for some $j \in \{1, \ldots, s\}$. Because the algebra $D = \operatorname{End} T_A^{rg}$ is the product

$$D \cong D_1 \times D_2 \times \ldots \times D_s$$

of the branch algebras

$$D_1 \cong K\mathcal{L}^{(1)}, D_2 \cong K\mathcal{L}^{(2)}, \ldots, D_s \cong K\mathcal{L}^{(s)},$$

then the Cartan matrix $\mathbf{C}_D$ of D has the diagonal form

$$\mathbf{C}_D \cong \begin{bmatrix} \mathbf{C}_{D_1} & 0 & 0 & \cdots & 0 \\ 0 & \mathbf{C}_{D_2} & 0 & \cdots & 0 \\ \vdots & & \ddots & & \vdots \\ 0 & 0 & \cdots & \cdots & \mathbf{C}_{D_s} \end{bmatrix},$$

and, for each $j \in \{1, \ldots, s\}$,

• the Euler quadratic form $q_{D_j} : K_0(D_j) \longrightarrow \mathbb{Z}$ is the restriction of the form $q_B : K_0(B) \longrightarrow \mathbb{Z}$ to the subgroup $K_0(D_j)$ of $K_0(B)$ under a canonical embedding

$$K_0(D_j) = \mathbb{Z}^{(Q_{D_j})_0} \hookrightarrow \mathbb{Z}^{(Q_B)_0} = K_0(B)$$

of the Grothendieck groups of D_j and B, and hence

• the restriction $\mathbf{x}^{(j)}$ of the vector $\mathbf{x} \in K_0(B)$ to the subgroup $K_0(D_j)$ of $K_0(B)$ is a positive vector of $K_0(D_j)$ and the equality $q_{D_j}(\mathbf{x}^{(j)}) = 1$ holds.

It follows from (XVI.2.2) that, for each $i \in \{1, \ldots, s\}$, D_i is a tilted algebra of the equioriented Dynkin type $\Delta(\mathbb{A}_{n_i})$, where n_i is the capacity of the branch $\mathcal{L}^{(i)}$, and hence, D_i is representation-directed. Applying (IX.3.3) again, we conclude there exists a unique indecomposable D_j-module X, up to isomorphism, such that $\dim X = \mathbf{x}$. Clearly, then X is a unique indecomposable B-module, up to isomorphism, such that $\dim X = \mathbf{x}$.

Assume that $\operatorname{supp}_A(\mathbf{x})$ is contained neither in $(Q_C)_0$ nor in $(Q_B)_0$. Let X be a B-module with $\dim X = \mathbf{x}$ such that $\dim_K \operatorname{End}_B(X)$ is minimal, and $X = X_1 \oplus \ldots \oplus X_t$ a decomposition of X into a direct sum of indecomposable B-modules.

If $t = 1$ then $\mathbf{x} = \dim X = \dim X_1$, and there is nothing to show. Assume that $t \geq 2$. Then $\operatorname{Ext}_A^1(X_i, X_j) \neq 0$, for all $i \neq j$, by (XIV.2.3). Because gl.dim $B \leq 2$ then we have the following equalities

$$
\begin{aligned}
1 &= q_B(\mathbf{x}) \\
&= q_B(\mathbf{dim}\, X) \\
&= \chi_B(X, X) \\
&= \dim_K \mathrm{End}_B(X) - \dim_K \mathrm{Ext}^1_B(X, X) + \dim_K \mathrm{Ext}^2_B(X, X) \\
&= \sum_{i \neq j} \left[\dim_K \mathrm{Hom}_B(X_i, X_j) + \dim_K \mathrm{Ext}^2_B(X_i, X_j) \right] \\
&\quad + \sum_{i=1}^{t} \left[\dim_K \mathrm{End}_B(X_i) - \dim_K \mathrm{Ext}^1_B(X_i, X_i) + \dim_K \mathrm{Ext}^2_B(X_i, X_i) \right] \\
&= \sum_{i \neq j} \left[\dim_K \mathrm{Hom}_B(X_i, X_j) + \dim_K \mathrm{Ext}^2_B(X_i, X_j) \right] + \sum_{i=1}^{t} \chi_B(X_i, X_i) \\
&= \sum_{i \neq j} \left[\dim_K \mathrm{Hom}_B(X_i, X_j) + \dim_K \mathrm{Ext}^2_B(X_i, X_j) \right] + \sum_{i=1}^{t} q_B(\mathbf{dim}\, X_i).
\end{aligned}
$$

Now we show that there exists $i \in \{1, \dots, t\}$ such that $q_B(\mathbf{dim}\, X_i) = 1$. Suppose, to the contrary, that $q_B(\mathbf{dim}\, X_i) = 0$, for any $i \in \{1, \dots, t\}$. Because $\mathrm{rad}\, q_B = \mathrm{rad}\, q_C = \mathbb{Z} \cdot \mathbf{h}_C$ then the B-modules $X_1, \dots, X_t$ are C-modules. Hence, the equalities $X = X_1 \oplus \dots \oplus X_t$ and $\mathbf{x} = \mathbf{dim}\, X$ imply that the support $\mathrm{supp}_A(\mathbf{x})$ is contained in $(Q_C)_0$ and we get a contradiction. This shows that there exists $i \in \{1, \dots, t\}$ such that $q_B(\mathbf{dim}\, X_i) = 1$.

Without loss of generality, we may suppose that $q_B(\mathbf{dim}\, X_1) = 1$. Then

- $q_B(\mathbf{dim}\, X_i) = 0$,   for $i \in \{2, \dots, t\}$, and
- $\sum_{i \neq j} \left[\dim_K \mathrm{Hom}_B(X_i, X_j) + \dim_K \mathrm{Ext}^2_B(X_i, X_j) \right] = 0$.

In particular, if $2 \leq i \leq t$, then $X_i \in \mathcal{T}^C_{\lambda_i} = \mathcal{T}^B_{\lambda_i}$, for some $\lambda_i \in \mathbb{P}_1(K)$, and (XII.4.2) yields $\mathbf{dim}\, X_i = m_i \cdot \mathbf{h}_C$, where $m_i \geq 1$ is an integer such that $m_i \cdot r^C_{\lambda_i}$ is the regular length $r\ell(X_i)$ of X_i in $\mathcal{T}^C_{\lambda_i}$. On the other hand, it follows from our assumption on $\mathbf{x}$ that the indecomposable B-module X_1 is not a C-module and, hence, $\mathrm{supp}_A(\mathbf{dim}\, X_1)$ contains a vertex of a branch $\mathcal{L}^{(k)}$, for some $k \in \{1, \dots, s\}$. Because

$$ \mathbf{x} = \mathbf{dim}\, X = \mathbf{dim}\, X_1 + \dots + \mathbf{dim}\, X_t $$

and the vectors $\mathbf{dim}\, X_2, \dots, \mathbf{dim}\, X_t$ are contained in $(Q_C)_0$ then the set $\mathrm{supp}_A(\mathbf{dim}\, X_1)$ contains the germ O_k of the branch $\mathcal{L}^{(k)}$, because the vector $\mathbf{x}$ is connected. Note also that X_1 is not a postprojective B-module, because $\mathcal{P}(B) = \mathcal{P}(C)$ consists of C-modules.

Assume that the module X_1 belongs to the preinjective component $\mathcal{Q}(B)$ of the Auslander–Reiten quiver $\Gamma(\mathrm{mod}\, B)$ of B. We prove that the restriction $\mathrm{res}_C X_1$ of X_1 to C is zero. Assume, to the contrary, that the module $Y_1 = \mathrm{res}_C X_1$ is non-zero. Then (XV.4.3) yields $Y_1 \in \mathrm{add}\, \mathcal{Q}(C)$ and, applying (XII.3.6), we conclude that $\mathrm{Hom}_B(X_2, Y_1) \neq 0$, because X_2 is an indecomposable C-module of regular length

$$r\ell(X_2) = m_2 \cdot r^C_{\lambda_2} \geq r^C_{\lambda_2}$$

in $\mathcal{T}^C_{\lambda_2}$. Hence we get the contradiction

$$0 = \operatorname{Hom}_B(X_2, X_1) \cong \operatorname{Hom}_B(X_2, Y_1) \neq 0.$$

It follows that if X_1 belongs to $\mathcal{Q}(B)$ then X_1 is an indecomposable module over the branch algebra D_k of the branch $\mathcal{L}^{(k)}$ with the germ O_k that belongs to the support of $\dim X_1$.

Let $\mathcal{T}^B_{\lambda_1}$ be the ray tube of $\Gamma(\operatorname{mod} B)$ containing the indecomposable projective B-module $P(O_k)_B = e_{O_k}B$ at the vertex O_k. We know from (XVI.2.2) and (XVI.2.3) that the branch algebra D_k is a tilted algebra $D_k \cong \operatorname{End}_{H_{n_k}}(R_k)$, where H_{n_k} is the path algebra of the equioriented quiver

$$\Delta(\mathbb{A}_{n_k}) : \overset{1}{\circ} \longleftarrow \overset{2}{\circ} \longleftarrow \ldots \longleftarrow \overset{n_k}{\circ},$$

R_k is a multiplicity-free tilting H_{n_k}-module, and n_k is the capacity of the branch $\mathcal{L}^{(k)}$. The finite Auslander–Reiten quiver $\Gamma(\operatorname{mod} H_{n_k})$ admits precisely one section Σ_k isomorphic to the equioriented linear quiver $\Delta(\mathbb{A}_{n_k})$, the source of Σ_k is the indecomposable projective D_k-module

$$P(O_k)_{B_k} = e_{O_k}B_k$$

at the germ O_k of $\mathcal{L}^{(k)}$, and the sink of Σ_k is the indecomposable injective D_k-module $I(O_k)_{B_k}$ at the germ O_k of $\mathcal{L}^{(k)}$. It follows that section Σ_k consists of all indecomposable D_k-modules M such that the support of $\dim M$ contains the germ O_k.

On the other hand, the ray tube $\mathcal{T}^B_{\lambda_1}$ is the image of $\operatorname{Hom}_A(T, \mathcal{T}(T) \cap \mathcal{T}^A_{\lambda_1})$ of the torsion part $\mathcal{T}(T) \cap \mathcal{T}^A_{\lambda_1}$ of the stable tube $\mathcal{T}^A_{\lambda_1}$ of $\Gamma(\operatorname{mod} A)$ containing the indecomposable summand V_k of T^{rg}_A creating the indecomposable projective B-module $P(O_k)_B$ at the vertex O_k. Moreover, the cone $\mathcal{C}(V_k)$ of the tube $\mathcal{T}^A_{\lambda_1}$ determined by V_k is of depth n_k and there is an equivalence of categories $G_k : \operatorname{mod} H_{n_k} \longrightarrow \operatorname{add}\mathcal{C}(V_k)$ such that the image $G_k(R_k)$ of the tilting H_{n_k}-module R_k via the functor G_k is the direct sum of all indecomposable direct summands of T^{rg}_A contained in the cone $\mathcal{C}(V_k)$. It follows that the tube $\mathcal{T}^A_{\lambda_1}$ of $\Gamma(\operatorname{mod} A)$ admits a coray containing a sectional path

$$[n_k+1]E_k \longrightarrow [n_k]E_k \longrightarrow [n_k-1]E_k \longrightarrow \ldots \longrightarrow [2]E_k \longrightarrow [1]E_k$$

entirely contained in $\mathcal{T}(T)$ such that $V_k = [n_k]E_k$, E_k lies on the mouth of $\mathcal{T}_{\lambda_1}^A$, and the image of this path via the functor G_k is a sectional path

$$(*) \qquad Z_{n_k+1} \longrightarrow Z_{n_k} \longrightarrow Z_{n_k-1} \longrightarrow \cdots \longrightarrow Z_2 \longrightarrow Z_1$$

of the ray tube $\mathcal{T}_{\lambda_1}^B$ of $\Gamma(\mathrm{mod}\, B)$ such that Z_{n_k+1} is the B-module F_k and the direct summand of the radical of the indecomposable projective B-module

$$Z_{n_k} = \mathrm{Hom}_A(T, V_k) \cong P(O_k)_B$$

at the vertex O_k, by the construction of the ray tube

$$\mathcal{T}_{\lambda_1}^B = \mathrm{Hom}_A(T, \mathcal{T}(T) \cap \mathcal{T}_{\lambda_1}^A)$$

given in the proof of (2.3).

On the other hand, the ray tube $\mathcal{T}_{\lambda_1}^B$ is obtained from the stable tube $\mathcal{T}_{\lambda_1}^C$ of $\Gamma(\mathrm{mod}\, C)$ by iterated rectangle insertions. Hence we conclude that by applying the restriction functor

$$\mathrm{res}_{D_k} : \mathrm{mod}\, B \longrightarrow \mathrm{mod}\, D_k$$

to the sectional path $(*)$ we get the unique section

$$\Sigma_k : \qquad P(O_k)_{D_k} = N_{n_k} \longrightarrow N_{n_k-1} \longrightarrow \cdots \longrightarrow N_2 \longrightarrow N_1 = I(O_k)_{D_k},$$

of type $\Delta(\mathbb{A}_{n_k})$ in $\Gamma(\mathrm{mod}\, D_k)$, where

$$N_{n_k} = \mathrm{res}_{D_k}(Z_{n_k}), \ldots, N_1 = \mathrm{res}_{D_k}(Z_1).$$

Hence we conclude that every indecomposable D_k-module M such that $\mathrm{Hom}_{D_k}(P(O_k)_{D_k}, M) \neq 0$ is isomorphic to a module N_ℓ, for some $\ell \in \{1, \ldots, n_k\}$. Because $\mathcal{T}_{\lambda_1}^B$ is a ray tube then, for each $\ell \in \{1, \ldots, n_k, n_k + 1\}$, there exists a unique infinite sectional path

$$(S_\ell) \qquad Z_\ell = Z_\ell[1] \longrightarrow Z_\ell[2] \longrightarrow \cdots \longrightarrow Z_\ell[j] \longrightarrow Z_\ell[j+1] \longrightarrow \cdots$$

in the tube $\mathcal{T}_{\lambda_1}^B$. We note also that the infinite path (S_{n_k+1}) starting from the C-module $F_k = Z_{n_k+1}$ consists entirely of C-modules and (S_{n_k+1}) is the ray

$$F_k = F_k[1] \longrightarrow F_k[2] \longrightarrow \cdots \longrightarrow F_k[j] \longrightarrow F_k[j+1] \longrightarrow \cdots$$

of the stable tube $\mathcal{T}_{\lambda_1}^C$ of $\Gamma(\mathrm{mod}\, C)$ starting from the mouth module F_k. It follows from the rectangle insertion procedure described in (XV.2.5) that every indecomposable B-module Y in $\mathbf{T}^B$ such that $\mathrm{Hom}_B(P(O_k)_B, Y) \neq 0$ is isomorphic to a module $Z_\ell[j]$, for some $\ell \in \{1, \ldots, n_k\}$ and some $j \geq 1$. Moreover, for each $t \in \{1, \ldots, n_k\}$ and $j \geq 1$, we have

$$\mathbf{dim}\, Z_t[j] = \mathbf{dim}\, N_t + \mathbf{dim}\, F_k[j].$$

Now we define an indecomposable B-module M in $\mathcal{T}^B_{\lambda_1}$ such that $\dim M = \mathbf{x}$. We recall that $\dim X_i = m_i \cdot \mathbf{h}_C$, for each $i \in \{2, \dots, t\}$, where $m_i \cdot \mathbf{h}_C = r\ell(X_i)$ is the regular length of X_i in $\mathcal{T}^B_{\lambda_1}$. Let $r = r^C_{\lambda_1}$ be the rank of the tube $\mathcal{T}^C_{\lambda_1}$, and we set $m = m_2 + \dots + m_t$. Then (XII.4.2) yields

$$\dim F_k[m \cdot r] = m \cdot \mathbf{h}_C = \dim X_2 + \dots + \dim X_t.$$

Because the support of the vector $\dim X_1$ contains the vertex O_k then $\operatorname{Hom}_B(P(O_k)_B, X_1) \neq 0$ and, consequently, the D_k-module $\operatorname{res}_{D_k}(X_1)$ is isomorphic to a module $N_\ell = \operatorname{res}_{D_k}(Z_\ell)$, for some $\ell \in \{1, \dots, n_k\}$. To construct the indecomposable B-module M, we have two cases to consider.

If $\operatorname{res}_{D_k}(X_1) = 0$ then we set $M = Z_\ell[m \cdot r]$. Obviously, M is an indecomposable B-module in the ray tube $\mathcal{T}^B_{\lambda_1}$ and

$$\dim M = \dim N_\ell + \dim F_k[m \cdot r] = \dim X_1 + \dim X_2 + \dots + \dim X_t = \mathbf{x}.$$

This shows that M is a required indecomposable B-module such that $\dim M = \mathbf{x}$.

If $\operatorname{res}_{D_k}(X_1) \neq 0$ then $X_1 \cong Z_\ell[j]$, for some $j \geq 1$, and we set $M = Z_\ell[j + m \cdot r]$. Then

$$\begin{aligned} \dim M &= \dim N_\ell + \dim F_k[j + m \cdot r] \\ &= (\dim N_\ell + \dim F_k[j]) + \dim F_k[m \cdot r] \\ &= \dim X_1 + \dim X_2 + \dots + \dim X_t = \mathbf{x}. \end{aligned}$$

It follows that M is a required indecomposable B-module such that $\dim M = \mathbf{x}$.

To finish the proof, it suffices to show that if X and Y are indecomposable B-modules such that

$$\mathbf{x} = \dim X = \dim Y, \quad q_B(\mathbf{x}) = 1,$$

and $\operatorname{supp}_B(\mathbf{x})$ is contained neither in $(Q_C)_0$ nor in $(Q_D)_0$, then $X \cong Y$.

Assume that X and Y satisfy the preceding conditions. If X belongs to the preinjective component $\mathcal{Q}(B)$ then X is directing and (IX.3.1) yields $X \cong Y$.

Because $\mathcal{P}(B) = \mathcal{P}(C)$ consists of C-modules then, without loss of generality, we may suppose that X belongs to $\boldsymbol{\mathcal{T}}^B$. We recall that the $\mathbb{P}_1(K)$-family

$$\boldsymbol{\mathcal{T}}^B = \{\mathcal{T}^B_\lambda\}_{\lambda \in \mathbb{P}_1(K)}$$

of pairwise orthogonal standard ray tubes of the Auslander–Reiten quiver $\Gamma(\operatorname{mod} B)$ of B is obtained from the $\mathbb{P}_1(K)$-family

$$\boldsymbol{\mathcal{T}}^C = \{\mathcal{T}^C_\lambda\}_{\lambda \in \mathbb{P}_1(K)}$$

of pairwise orthogonal standard stable tubes of $\Gamma(\mathrm{mod}\, C)$ by iterated ray insertions created by the branch extension

$$B = C[F_1, \mathcal{L}^{(1)}, F_2, \mathcal{L}^{(2)}, \ldots, F_s, \mathcal{L}^{(s)}]$$

of C. By our assumption on $\mathbf{x}$, there is exactly one branch $\mathcal{L}^{(k)}$ such that the support $\mathrm{supp}_B(\mathbf{x})$ of $\mathbf{x}$ contains the germ O_k of $\mathcal{L}^{(k)}$. It follows that $\mathrm{Hom}_B(P(O_k)_B, X) \neq 0$ and $\mathrm{Hom}_B(P(O_k)_B, Y) \neq 0$ and, in the notation introduced earlier, there are isomorphisms $X \cong Z_t[j]$ and $Y \cong Z_u[\ell]$, for some $j, \ell \geq 1$ and $t, u \in \{1, \ldots, n_k\}$. Hence, there exist isomorphisms $\mathrm{res}_{D_k} X \cong N_t$ and $\mathrm{res}_{D_k} Y \cong N_u$ of D_k-modules and the dimension vectors $\mathbf{dim}\, N_t$ and $\mathbf{dim}\, N_u$ coincide, because they are restrictions of the vector $\mathbf{x} = \mathbf{dim}\, X = \mathbf{dim}\, Y$ to $(Q_{D_k})_0$. Thus (IX.3.3) yields $N_t \cong N_u$, because the algebra D_k is representation-directed. It follows that $t = u$. On the other hand, the equalities

$$\mathbf{dim}\, X = \mathbf{dim}\, N_t + \mathbf{dim}\, F_k[j] \quad \text{and} \quad \mathbf{dim}\, Y = \mathbf{dim}\, N_u + \mathbf{dim}\, F_k[\ell]$$

yield $\mathbf{dim}\, F_k[j] = \mathbf{dim}\, F_k[\ell]$, and hence, there is an isomorphism

$$F_k[j] \cong F_k[\ell],$$

because the modules $F_k[j]$ and $F_k[\ell]$ lie on a common ray of a stable tube of the family $\mathcal{T}^C$. It follows that $j = \ell$ and, consequently, there is an isomorphism $X \cong Y$. This finishes the proof of the theorem. □

Now we give an example of a tilted algebra B of Euclidean type such that the module category $\mathrm{mod}\, B$ is link controlled by the Euler quadratic form $q_B : K_0(B) \longrightarrow \mathbb{Z}$, but the category $\mathrm{mod}\, B$ is not controlled by q_B in the sense of (4.1).

6.2. Example. Let B be the path algebra given by the quiver

$$
\begin{array}{ccccccc}
1 & \underset{\beta}{\overset{\alpha}{\rightleftarrows}} & 2 & \overset{\gamma}{\longleftarrow} & 3 & \overset{\sigma}{\longrightarrow} & 4 \\
\circ & & \circ & & \circ & & \circ
\end{array}
$$

bound by one zero relation $\gamma\alpha = 0$. Let A be the path algebra of the Kronecker quiver

$$
\begin{array}{ccc}
1 & \underset{\beta}{\overset{\alpha}{\rightleftarrows}} & 2 \\
\circ & & \circ
\end{array}.
$$

In the notation of (XI.4.3), B is a tubular extension $B = A[E_\infty, \mathcal{L}^{(\infty)}]$ of A of domestic tubular type $r^B = \hat{r}^B = (r_0^B, r_\infty^B) = (1, 3)$, where

$$E_\infty = \left(K \underset{1}{\overset{0}{\longleftarrow}} K \right)$$

is the mouth A-module of the tube $\mathcal{T}_\infty^A$ and $\mathcal{L}^{(\infty)}$ is the branch $3 \circ \longrightarrow \circ 4$ of capacity 2, with the germ vertex 3. The Cartan matrix $\mathbf{C}_B$ of B and its inverse $\mathbf{C}_B^{-1}$ are of the forms

$$\mathbf{C}_B = \begin{bmatrix} 1 & 2 & 1 & 0 \\ 0 & 1 & 1 & 0 \\ 0 & 0 & 1 & 0 \\ 0 & 0 & 1 & 1 \end{bmatrix} \quad \text{and} \quad \mathbf{C}_B^{-1} = \begin{bmatrix} 1 & -2 & 1 & 0 \\ 0 & 1 & -1 & 0 \\ 0 & 0 & 1 & 0 \\ 0 & 0 & -1 & 1 \end{bmatrix},$$

with respect to the canonical ordering $P(1)$, $P(2)$, $P(3)$, and $P(4)$ of the indecomposable projective B-modules. Then $K_0(B) = \mathbb{Z}^4$ and the Euler quadratic form $q_B : \mathbb{Z}^4 \longrightarrow \mathbb{Z}$ is defined by the formula

$$q_B(\mathbf{x}) = \mathbf{x}^t \cdot (\mathbf{C}_B^{-1})^t \cdot \mathbf{x} = x_1^2 + x_2^2 + x_3^2 + x_4^2 - 2x_1 x_2 - x_2 x_3 - x_3 x_4 + x_1 x_3,$$

for $\mathbf{x} = [x_1 \; x_2 \; x_3 \; x_4]^t \in \mathbb{Z}^4 = K_0(B)$.

Then, for the vector $\mathbf{y} = [1 \; 1 \; 0 \; 1]^t \in \mathbb{Z}^4$, we have $q_B(\mathbf{y}) = 1$, that is, $\mathbf{y}$ is a non-connected root of q_B, and obviously, $\mathbf{y}$ is not the dimension vector of any indecomposable B-module. This shows that the category $\operatorname{mod} B$ is not controlled by q_B, whereas $\operatorname{mod} B$ is link controlled by q_B, by (6.1).

We end this section with an interesting example of a representation-finite tilted algebra of Euclidean type.

6.3. Example. Let $A = K\Delta$ be the path algebra given by the Euclidean quiver

$$\Delta = \Delta(\widetilde{\mathbb{D}}_4) :$$
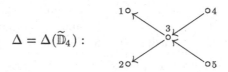

The standard calculation technique shows that the left hand part of the component $\mathcal{P}(A)$ of $\Gamma(\operatorname{mod} A)$ looks as follows

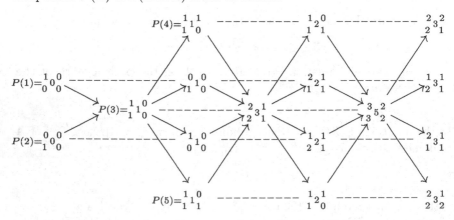

and the right hand part of the component $\mathcal{Q}(A)$ of $\Gamma(\operatorname{mod} A)$ looks as follows

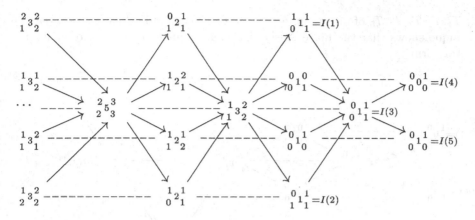

where the indecomposable modules are represented by their dimension vectors.

It follows from (XII.2.6) and (XII.2.9) that $\Gamma(\mathrm{mod}\,A)$ admits a stable tube $\mathcal{T}_1^A$ of rank 2 with the mouth formed by the modules

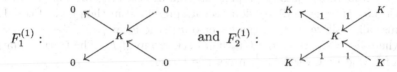

Consider the indecomposable A-modules

$$T_1 = P(1),\ T_2 = P(2),\ T_3 = F_2^{(1)},\ T_4 = I(4),\ \text{and}\ \ T_5 = I(5),$$

A simple checking shows that the tilting vanishing condition $\mathrm{Hom}_A(T_i, \tau_A T_j) = 0$ is satisfied, for all $i, j \in \{1, 2, 3, 4, 5\}$, and consequently

$$T_A = T_1 \oplus T_2 \oplus T_3 \oplus T_4 \oplus T_5,$$

is a multiplicity-free tilting A-module such that the summand $T_A^{pp} = T_1 \oplus T_2$ of T_A is postprojective, the summand $T_A^{pi} = T_4 \oplus T_5$ is preinjective, and the summand $T_A^{rg} = T_3$ is a regular module. Hence, by (VIII.4.3), $B = \mathrm{End}\,T_A$ is a representation-finite tilted algebra of the Euclidean type $\Delta(\widetilde{\mathbb{D}}_4)$. It is easy to see that B is given by the quiver

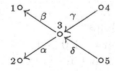

bound by the zero relations $\gamma\alpha = 0$, $\gamma\beta = 0$, $\delta\alpha = 0$, and $\delta\beta = 0$, where the ordering $1, 2, 3, 4, 5$ of the vertices of Q_B corresponds to the ordering

T_1, T_2, T_3, T_4, T_5 of direct summands of T_A. The standard calculation technique shows that the finite Auslander–Reiten quiver $\Gamma(\mathrm{mod}\,B)$ of B is of the form

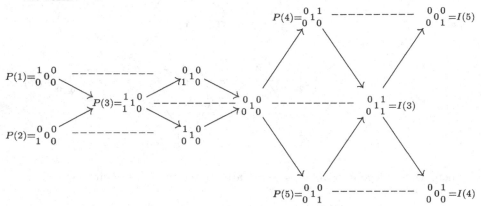

where the indecomposable modules are represented by their dimension vectors. It follows that B is a representation-directed algebra and, according to (IX.3.3), the eleven dimension vectors presented in the quiver $\Gamma(\mathrm{mod}\,B)$ are just all positive roots of the Euler quadratic form $q_B : \mathbb{Z}^5 \longrightarrow \mathbb{Z}$ of B, where $\mathbb{Z}^5 = K_0(B)$ is the Grothendieck group of B. The Cartan matrix $\mathbf{C}_B$ of B and its inverse $\mathbf{C}_B^{-1}$ are of the forms

$$\mathbf{C}_B = \begin{bmatrix} 1 & 0 & 1 & 0 & 0 \\ 0 & 1 & 1 & 0 & 0 \\ 0 & 0 & 1 & 1 & 1 \\ 0 & 0 & 0 & 1 & 0 \\ 0 & 0 & 0 & 0 & 1 \end{bmatrix} \quad \text{and} \quad \mathbf{C}_B^{-1} = \begin{bmatrix} 1 & 0 & -1 & 1 & 1 \\ 0 & 1 & -1 & 1 & 1 \\ 0 & 0 & 1 & -1 & -1 \\ 0 & 0 & 0 & 1 & 0 \\ 0 & 0 & 0 & 0 & 1 \end{bmatrix}.$$

Then the Euler quadratic form $q_B : \mathbb{Z}^4 \longrightarrow \mathbb{Z}$ is defined by the formula

$$\begin{aligned} q_B(\mathbf{x}) &= \mathbf{x}^t \cdot (\mathbf{C}_B^{-1})^t \cdot \mathbf{x} \\ &= x_1^2 + x_2^2 + x_3^2 + x_4^2 + x_5^2 - x_1 x_3 - x_2 x_3 - x_3 x_4 - x_3 x_5 \\ &\quad + x_1 x_4 + x_2 x_4 + x_1 x_5 + x_2 x_4, \end{aligned}$$

for $\mathbf{x} = [x_1\ x_2\ x_3\ x_4\ x_5]^t \in \mathbb{Z}^5 = K_0(B)$. Note that

$$4q_B(\mathbf{x}) = (2x_1 - x_3 + x_4 + x_5)^2 + (2x_2 - x_3 + x_4 + x_5)^2 + 2x_3^2 + 2(x_4 - x_5)^2.$$

It follows that

- q_B is positive semidefinite,
- $\mathrm{rad}\, q_B = \mathbb{Z} \cdot \mathbf{h}$, where $\mathbf{h} = [1\,1\ 0\,1 - 1]$, and
- q_B is weakly positive, that is, $q_B(\mathbf{x}) > 0$, for any positive vector $\mathbf{x} \in K_0(B)$.

This finishes the example.

XVII.7. Exercises

1. Assume that A is an algebra and $\mathcal{T}_\lambda$ is a hereditary standard stable tube of $\Gamma(\mathrm{mod}\,A)$ of rank $r_\lambda \geq 1$, and M is an indecomposable A-module in $\mathcal{T}_\lambda$ such that $\ell_\lambda(M) = r_\lambda$. Prove that

   (a) the module M is a brick,

   (b) M is not a stone, and

   (c) there is an isomorphism $\mathrm{Ext}_A^1(M, M) \cong K$.

2. Let $B = KQ$ be the algebra given by the quiver

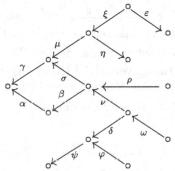

bound by the relations $\rho\sigma\gamma = \rho\beta\alpha$, $\mu\gamma = 0$, $\xi\eta = 0$, $\nu\sigma = 0$, $\omega\delta = 0$, and $\varphi\psi = 0$.

   (a) Prove that there exists a tilting module T_A over the path algebra $A = K\Delta(\widetilde{\mathbb{D}}_{13})$ of the quiver $\Delta(\widetilde{\mathbb{D}}_{13})$ (XII.1.5) such that $B \cong \mathrm{End}\,T_A$.

   (b) Describe the Auslander–Reiten quiver $\Gamma(\mathrm{mod}\,B)$ of B.

3. Let $B = KQ$ be the algebra given by the quiver

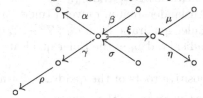

bound by the zero relations $\sigma\gamma\rho = 0$, $\beta\xi = 0$, $\sigma\xi = 0$, and $\mu\eta = 0$.

   (a) Prove that there exists a tilting module T_A over the path algebra $A = K\Delta(\widetilde{\mathbb{E}}_8)$ of the quiver $\Delta(\widetilde{\mathbb{E}}_8)$ (XII.1.14) such that $B \cong \mathrm{End}\,T_A$.

   (b) Describe the Auslander–Reiten quiver $\Gamma(\mathrm{mod}\,B)$ of B.

4. Let $B = KQ$ be the algebra given by the quiver

bound by three commutativity relations $\rho\alpha = \omega\beta$, $\xi\beta = \eta\gamma$, and $\mu\gamma = \varepsilon\sigma$.

  (a) Prove that there exists a tilting module T_A over the path algebra $A = K\Delta(\widetilde{\mathbb{E}}_7)$ of the quiver $\Delta(\widetilde{\mathbb{E}}_7)$ (XII.1.8) such that $B \cong \operatorname{End} T_A$.

  (b) Describe the Auslander–Reiten quiver $\Gamma(\operatorname{mod} B)$ of B.

5. Let B be the algebra given by the quiver

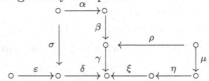

bound by the commutativity relations $\sigma\rho = \alpha\beta\gamma$ and $\rho\gamma = \mu\eta\xi$.

  (a) Prove that there exists a tilting module T_A over the path algebra $A = K\Delta(\widetilde{\mathbb{E}}_8)$ of the quiver $\Delta(\widetilde{\mathbb{E}}_8)$ (XII.1.14) such that $B \cong \operatorname{End} T_A$.

  (b) Describe the Auslander–Reiten quiver $\Gamma(\operatorname{mod} B)$ of B.

6. Let B be the algebra given by the quiver

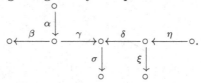

bound by the zero relations $\alpha\beta = 0$, $\gamma\sigma = 0$, $\delta\sigma = 0$, and $\eta\xi = 0$.

  (a) Prove that B is a representation-finite tilted algebra of the Euclidean type $\Delta(\widetilde{\mathbb{E}}_7)$ (XII.1.11).

  (b) Find a tilting module T_A over the path algebra $A = K\Delta(\widetilde{\mathbb{E}}_7)$ of the quiver $\Delta(\widetilde{\mathbb{E}}_7)$ such that $B \cong \operatorname{End} T_A$.

  (c) Describe the Auslander–Reiten quiver $\Gamma(\operatorname{mod} B)$ of B.

  (d) Describe the Euler quadratic form $q_B : \mathbb{Z}^8 \longrightarrow \mathbb{Z}$ of the algebra B.

  (e) Describe the radical $\operatorname{rad} q_B$ of the Euler quadratic form $q_B : \mathbb{Z}^8 \longrightarrow \mathbb{Z}$ of B.

  (f) Describe the positive roots of the quadratic form $q_B : \mathbb{Z}^8 \longrightarrow \mathbb{Z}$.

7. Let B be the algebra given by the quiver

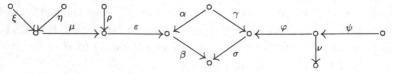

bound by the relations $\gamma\sigma = \alpha\beta$, $\xi\mu = 0$, $\eta\mu = 0$, $\rho\varepsilon = 0$, $\varphi\sigma = 0$, and $\psi\nu = 0$.

  (a) Prove that B is a representation-finite tilted algebra of the Euclidean type $\Delta(\widetilde{\mathbb{D}}_{11})$ (XII.1.5).

(b) Find a tilting module T_A over the path algebra $A = K\Delta(\widetilde{\mathbb{D}}_{11})$ of the quiver $\Delta(\widetilde{\mathbb{D}}_{11})$ such that $B \cong \operatorname{End} T_A$.

(c) Describe the Auslander–Reiten quiver $\Gamma(\operatorname{mod} B)$ of B.

(d) Describe the Euler quadratic form $q_B : \mathbb{Z}^{12} \longrightarrow \mathbb{Z}$ of B.

(e) Describe the radical $\operatorname{rad} q_B$ of the quadratic form $q_B : \mathbb{Z}^{12} \longrightarrow \mathbb{Z}$.

(f) Describe the positive roots of the quadratic form $q_B : \mathbb{Z}^{12} \longrightarrow \mathbb{Z}$.

8. Let B be the algebra given by the quiver

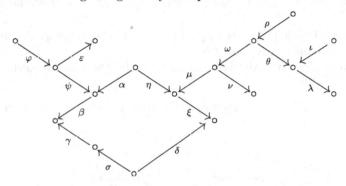

bound by the zero relations $\alpha\beta = 0$, $\eta\xi = 0$, $\varphi\varepsilon = 0$, $\rho\theta = 0$, $\iota\lambda = 0$, and $\omega\nu = 0$.

(a) Prove that B is a representation-finite tilted algebra of the Euclidean type $\Delta(\widetilde{\mathbb{A}}_{6,11})$ (XII.1.1).

(b) Find a tilting module T_A over the path algebra $A = K\Delta(\widetilde{\mathbb{A}}_{6,11})$ of the quiver $\Delta(\widetilde{\mathbb{A}}_{6,11})$ such that $B \cong \operatorname{End} T_A$.

(c) Describe the Auslander–Reiten quiver $\Gamma(\operatorname{mod} B)$ of B.

(d) Describe the Euler quadratic form $q_B : \mathbb{Z}^{18} \longrightarrow \mathbb{Z}$ of B.

(e) Describe the radical $\operatorname{rad} q_B$ of the quadratic form $q_B : \mathbb{Z}^{18} \longrightarrow \mathbb{Z}$.

(f) Describe the positive roots of the quadratic form $q_B : \mathbb{Z}^{18} \longrightarrow \mathbb{Z}$.

9. Let $C = B/(\sigma\gamma)$ be the quotient algebra of the algebra B of Exercise 8 by the ideal $(\sigma\gamma)$ of B generated by the path $\sigma\gamma$. Prove that C is not a tilted algebra.

10. Let $p \geq 1$ and $q \geq 1$ be integers such that $1 \leq p \leq q$. Describe all representation-infinite tilted algebras of the Euclidean type $\Delta(\widetilde{\mathbb{A}}_{p,q})$ (XII.1.1).

11. Classify all tilted algebras of the Euclidean type $\Delta(\widetilde{\mathbb{D}}_4)$ (XII.1.5).

12. Classify all tilted algebras of the Euclidean type $\Delta(\widetilde{\mathbb{D}}_5)$ (XII.1.5).

13. Classify all tilted algebras of the Euclidean type $\Delta(\widetilde{\mathbb{E}}_6)$ (XII.1.8).

14. Let $\Lambda^{(a)}$, with $a \in K \setminus \{0, 1\}$, be the algebra given by the quiver

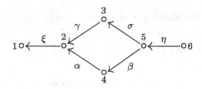

bound by the commutativity relations $\beta\alpha\xi = \sigma\gamma\xi$ and $\eta\beta\alpha = a\eta\sigma\gamma$, see (5.5).

(a) Describe the Euler quadratic form $q_{\Lambda^{(a)}} : \mathbb{Z}^6 \longrightarrow \mathbb{Z}$ of the algebra $\Lambda^{(a)}$.

(b) Describe the radical $\operatorname{rad} q_{\Lambda^{(a)}}$ of $q_{\Lambda^{(a)}}$ and show that $q_{\Lambda^{(a)}}$ is positive semidefinite of corank two.

(c) Show that $\operatorname{mod} \Lambda^{(a)}$ is link controlled by the Euler quadratic form $q_{\Lambda^{(a)}} : \mathbb{Z}^6 \longrightarrow \mathbb{Z}$.

15. Let $\Lambda = \Lambda^{(1)}$ be the algebra of Exercise 14, with $a = 1$, given by the quiver

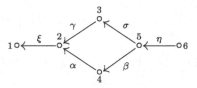

bound by the commutativity relations $\beta\alpha\xi = \sigma\gamma\xi$ and $\eta\beta\alpha = \eta\sigma\gamma$.

(i) Prove that the algebra Λ is representation-infinite and describe the Auslander–Reiten quiver $\Gamma(\operatorname{mod}\Lambda)$ of Λ.

(ii) Prove that $\operatorname{gl.dim}\Lambda = 3$.

(iii) Prove that $\operatorname{pd} X_\Lambda \le 2$ and $\operatorname{id} X_\Lambda \le 2$, for any indecomposable Λ-module X_Λ in $\operatorname{mod}\Lambda$.

(iv Describe the Euler quadratic form $q_\Lambda : \mathbb{Z}^6 \longrightarrow \mathbb{Z}$ of Λ.

(v) Prove that q_Λ is positive semidefinite of corank one, describe the radical $\operatorname{rad} q_\Lambda$ of q_Λ, and show that the group $\operatorname{rad} q_\Lambda$ is generated by a positive vector.

(vi) Prove that the quadratic form $q_{\Lambda^{(a)}} : \mathbb{Z}^6 \longrightarrow \mathbb{Z}$ is $\mathbb{Z}$-congruent to the Euler quadratic form $q_A : \mathbb{Z}^6 \longrightarrow \mathbb{Z}$ of the path algebra $A = K\Delta(\widetilde{\mathbb{D}}_5)$ of the quiver $\Delta(\widetilde{\mathbb{D}}_5)$ (XII.1.5).

Chapter XVIII

Wild hereditary algebras and tilted algebras of wild type

Throughout, we let Q be a connected and acyclic quiver Q with n vertices, that is, $n = |Q_0|$. We assume that K is an algebraically closed field and we denote by A the hereditary path K-algebra KQ of Q.

We have seen in Chapters VII and XIII that, for any hereditary algebra $A = KQ$ such that the underlying graph $\overline{Q}$ of Q is Dynkin or Euclidean, there exists an explicit description of the isomorphism classes of the indecomposable A-modules and a description of the components of the Auslander–Reiten quiver $\Gamma(\text{mod}\,A)$ of A. Moreover, the structure of the quiver $\Gamma(\text{mod}\,A)$ is presented.

One of our objectives in the present chapter is to describe the quiver $\Gamma(\text{mod}\,A)$ and its components in case $A = KQ$ is the path algebra of a wild quiver Q in the following sense.

Definition. (a) A **wild quiver** (or a **representation-wild quiver**) is a finite, connected and acyclic quiver Q whose underlying graph $\overline{Q}$ is neither Dynkin nor Euclidean.

(b) A hereditary K-algebra A is **wild** (or a **representation-wild algebra**) if A is isomorphic to the path K-algebra KQ of a representation-wild quiver Q.

We show in Section 4 that any representation-wild quiver Q is wild in the sense of Drozd [201] and [202], that is, the classification of indecomposable K-linear representations of Q 'contains' the classification of all indecomposable modules over any finite dimensional algebra Λ. Because there is a lot of finite dimensional algebras Λ with very complicated module categories mod Λ, this justifies why we call such a quiver Q representation-wild.

In the first half of the chapter we study the components in the regular part $\mathcal{R}(A)$ of $\Gamma(\text{mod}\,A)$ in case $A = KQ$ is the path algebra of a wild quiver Q.

We recall from Chapter VIII that, for any representation-wild algebra $A = KQ$, the Auslander–Reiten quiver $\Gamma(\operatorname{mod} A)$ of A has the shape

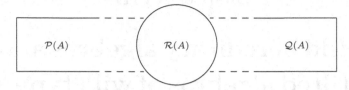

where $\mathcal{P}(A)$ is the unique postprojective component containing all the indecomposable projective A-modules, $\mathcal{Q}(A)$ is the unique preinjective component containing all the indecomposable injective A-modules, and $\mathcal{R}(A)$ is the regular part consisting of the remaining components.

In Section 1, we show that the regular A-modules are not τ-periodic, the regular part $\mathcal{R}(A)$ contains no tube and any component $\mathcal{C}$ of $\mathcal{R}(A)$ is of the following type

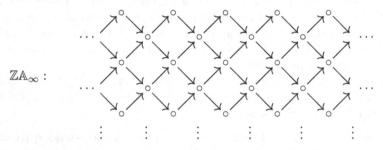

Moreover, each component $\mathcal{C}$ in $\mathcal{R}(A)$ contains at most finitely many non-sincere indecomposable modules, any indecomposable module M in a regular component $\mathcal{C}$ is uniquely determined in $\mathcal{C}$ by the dimension vector $\mathbf{dim}\,M$ of M, and the set of components in $\mathcal{R}(A)$ is of cardinality $\operatorname{card}(K)$, the cardinality of the field K.

In Section 2, we describe the basic properties of homomorphisms between regular A-modules over wild hereditary algebras $A = KQ$.

In Section 3, we associate to a given algebra A and a given module X in $\operatorname{mod} A$ a pair of subcategories $X^{\perp}$ and $^{\perp}X$ of $\operatorname{mod} A$, called the right and the left perpendicular category of X. This gives a new reduction technique, that is converse to the one-point extension procedure $X \mapsto A[X]$ and the one-point coextension procedure $X \mapsto [X]A$ studied in Section XV.1. We prove that if $A = KQ$ is a hereditary algebra and T is a partial tilting A-module then the right perpendicular category $T^{\perp}$ of T is equivalent to a module category $\operatorname{mod} B$ of a hereditary algebra B and

$$\operatorname{rk} K_0(A) = r + \operatorname{rk} K_0(B),$$

where $\operatorname{rk} K_0(A)$ and $\operatorname{rk} K_0(B)$ are the ranks of the Grothendieck groups $K_0(A)$ and $K_0(B)$ of A and B, and $r \geq 1$ is the number of pairwise non-isomorphic indecomposable direct summands of the module T. This is applied in Section 4, where the wild behaviour of the module category $\operatorname{mod} A$ of a wild hereditary algebra $A = KQ$ is discussed.

In the final section of the chapter, we describe basic properties of the module category $\operatorname{mod} B$ of any concealed algebra B of wild type Q, that is, the algebra of the form $B = \operatorname{End} T_A$, where T_A is a postprojective tilting module over a connected representation-wild hereditary algebra $A = KQ$. We also exhibit other classes of tilted algebras of wild type and we discuss the structure of their module categories. In particular, we prove a theorem of Ringel [527] asserting that, given a finite connected quiver Q, there exists a regular tilting A-module over the path algebra $A = KQ$ if and only if Q is a wild quiver with at least three vertices. We also present an effective procedure of Baer [35] and [36], for constructing regular tilting modules over wild hereditary algebras $A = KQ$.

XVIII.1. Regular components

Throughout this section, we let A be the path algebra KQ of a wild quiver Q with n vertices and $A = KQ$ is a wild hereditary algebra.

Without loss of generality we may assume that $Q_0 = \{1, \ldots, n\}$. It follows from (III.3.5) that the Grothendieck group $K_0(A)$ of A is isomorphic to $\mathbb{Z}^{|Q_0|} = \mathbb{Z}^n$. As usual, we denote by $\mathbf{e}_1, \ldots, \mathbf{e}_n$ the canonical basis of the free abelian group $\mathbb{Z}^n$.

It follows from (VII.4.5) that the algebra $A = KQ$ is representation-wild if and only if the Euler quadratic form $q_A : \mathbb{Z}^n \longrightarrow \mathbb{Z}$ of A is not positive semidefinite or, equivalently, there exists a non-zero vector $\mathbf{y} \in \mathbb{Z}^n$ such that $q_A(\mathbf{y}) < 0$.

We recall from Section VII.1 that, for any pair of vectors $\mathbf{x}, \mathbf{y} \in \mathbb{Z}^n$, we have $q_A(\mathbf{y}) = q_Q(\mathbf{y}) = \langle \mathbf{y}, \mathbf{y} \rangle_Q$, where $\langle -, - \rangle_Q : \mathbb{Z}^n \times \mathbb{Z}^n \longrightarrow \mathbb{Z}$ is the bilinear (non-symmetric) form of the quiver Q defined by the formula

$$\langle \mathbf{x}, \mathbf{y} \rangle_Q = \sum_{i \in Q_0} x_i y_i - \sum_{\alpha \in Q_1} x_{s(\alpha)} y_{t(\alpha)}.$$

We denote by $(-, -)_Q : \mathbb{Z}^n \times \mathbb{Z}^n \longrightarrow \frac{1}{2} \cdot \mathbb{Z}$ the symmetric bilinear form associated to $\langle -, - \rangle_Q$, that is,

$$(\mathbf{x}, \mathbf{y})_Q = \frac{1}{2}(\langle \mathbf{x}, \mathbf{y} \rangle_Q + \langle \mathbf{y}, \mathbf{x} \rangle_Q),$$

for all $\mathbf{x}, \mathbf{y} \in \mathbb{Z}^n$. The bilinear form $\langle -, - \rangle_Q$ equals the Euler bilinear form $\langle -, - \rangle_A$ of A defined in terms of the Cartan matrix $\mathbf{C}_A$ of A by the formula

$\langle \mathbf{x}, \mathbf{y} \rangle_A = \mathbf{x}^t (\mathbf{C}_A^{-1})^t \mathbf{y}$, for $\mathbf{x}, \mathbf{y} \in \mathbb{Z}^n$. We recall from (III.3.14) that the Coxeter matrix $\mathbf{\Phi}_A$ of A is the matrix

$$\mathbf{\Phi}_A = -\mathbf{C}_A^t \mathbf{C}_A^{-1}.$$

The Coxeter transformation of A is the group homomorphism $\mathbf{\Phi}_A : \mathbb{Z}^n \longrightarrow \mathbb{Z}^n$ defined by the formula $\mathbf{\Phi}_A(\mathbf{x}) = \mathbf{\Phi}_A \cdot \mathbf{x}$, for all $\mathbf{x} = [x_1 \ldots x_n]^t \in \mathbb{Z}^n$. It follows from (IV.2.9) that, for any regular A-module M, we have

$$\dim \tau M = \mathbf{\Phi}_A(\dim M).$$

We recall from (VII.1.9) that, because A is hereditary, the Auslander–Reiten translation $\tau = D\mathrm{Tr}$ is isomorphic to $D\mathrm{Ext}_A^1(-, A)$, hence the functor

$$\tau \cong D\mathrm{Ext}_A^1(-, A) : \mathrm{mod}\, A \longrightarrow \mathrm{mod}\, A$$

is a left exact and defines an equivalence from the full subcategory of $\mathrm{mod}\, A$ formed by the modules without non-zero projective direct summands to the full subcategory of $\mathrm{mod}\, A$ formed by the modules without non-zero injective direct summands. Dually, there is a functorial isomorphism

$$\tau^{-1} = \mathrm{Tr}\, D \cong \mathrm{Ext}_A^1(D(-), A)$$

and therefore τ^{-1} is a right exact functor. Moreover, the functors τ and τ^{-1} restrict to the mutually inverse exact self-equivalences

$$\mathrm{add}\,\mathcal{R}(A) \underset{\tau_A^{-1}}{\overset{\tau_A}{\rightleftarrows}} \mathrm{add}\,\mathcal{R}(A)$$

of the category $\mathrm{add}\,\mathcal{R}(A)$ of all regular A-modules. The equivalences preserve the irreducibility and carry almost split sequences to almost split sequences.

The main objective of this section is to describe the shape of all regular components of $\Gamma(\mathrm{mod}\, A)$. We start with the following three technical results on regular modules.

1.1. Proposition. *Let $A = KQ$ be a wild hereditary algebra, X an indecomposable regular A-module and $m \geq 1$ an integer. Then $\dim X \neq \dim \tau^m X$.*

Proof. Suppose, to the contrary, that $\dim X = \dim \tau^m X$. We may assume that $m > 0$. Consider the vector $\mathbf{d} = \sum_{i=0}^{m-1} \dim \tau^i X$. It follows from (IV.2.9) that $\dim \tau M = \mathbf{\Phi}_A(\dim M)$, for any regular A-module M, where $\mathbf{\Phi}_A : \mathbb{Z}^n \longrightarrow \mathbb{Z}^n$ is the Coxeter transformation of A. Therefore, we get the equalities

$$\mathbf{d} = \sum_{i=1}^{m} \dim \tau^i X = \sum_{i=1}^{m} \mathbf{\Phi}_A^i(\dim X)$$

$$= \mathbf{\Phi}_A \left(\sum_{i=1}^{m} \mathbf{\Phi}_A^{i-1}(\dim X) \right) = \mathbf{\Phi}_A \left(\sum_{i=0}^{m-1} \dim \tau^i X \right) = \mathbf{\Phi}_A(\mathbf{d}).$$

Applying now (III.3.16) we conclude that $\langle \mathbf{d}, -\rangle_Q + \langle -, \mathbf{d} \rangle_Q = 0$. We recall from (VII.4) that the symmetric bilinear form $(-, -)_Q : \mathbb{Z}^n \times \mathbb{Z}^n \longrightarrow \frac{1}{2} \cdot \mathbb{Z}$ associated to $\langle -, - \rangle_A = \langle -, - \rangle_Q$ is given by the formula

$$(\mathbf{x}, \mathbf{y})_Q = \sum_{i \in Q_0} x_i y_i - \frac{1}{2} \sum_{\alpha \in Q_1} (x_{s(\alpha)} y_{t(\alpha)} + x_{t(\alpha)} y_{s(\alpha)}),$$

for any $\mathbf{x}, \mathbf{y} \in \mathbb{Z}^n$. Then $0 = \langle \mathbf{d}, -\rangle_Q + \langle -, \mathbf{d} \rangle_Q = 2(\mathbf{d}, -)_Q$ leads to the equalities

$$0 = 2(\mathbf{d}, \mathbf{e}_i)_Q = 2d_i - \sum_{\alpha \in Q_1, t(\alpha) = i} d_{s(\alpha)} - \sum_{\alpha \in Q_1, s(\alpha) = i} d_{t(\alpha)},$$

for all $i \in Q_0$. We claim that $d_i > 0$, for any $i \in Q_0$. Suppose that it is not the case. Because Q is connected, then there exist two vertices i and j connected by an arrow such that $d_i = 0$ and $d_j > 0$, and clearly then $(\mathbf{d}, \mathbf{e}_i)_Q < 0$, a contradiction.

The above equalities can be written as $\mathbf{F} \cdot \mathbf{d} = 0$, where $\mathbf{F} = \mathbf{F}_Q = (f_{ij})$ is the symmetric $n \times n$-matrix defined as follows:

- $f_{ii} = 2$, for any $i \in Q_0$ and,
- $-f_{ij}$ is the number of arrows in Q between the vertices i and j, for each pair $i, j \in Q_0$ such that $i \neq j$.

Because Q is a wild quiver, it contains a proper subquiver Q' whose underlying graph is Euclidean. Denote by $\mathbf{d}' \in \mathbb{Z}^{|Q_0'|}$ the restriction of $\mathbf{d} \in \mathbb{Z}^{|Q_0|} = \mathbb{Z}^n$ to Q_0', and by $\mathbf{F}' = \mathbf{F}_{Q'} = (f'_{ij})$ the corresponding symmetric matrix associated to Q'. Then, for each $i \in Q_0'$, we have

$$(\mathbf{F}' \cdot \mathbf{d}')_i = \sum_{j \in Q_0'} f'_{ij} d'_j = 2d_i - \sum_{\alpha \in Q_1', t(\alpha) = i} d'_{s(\alpha)} - \sum_{\alpha \in Q_1', s(\alpha) = i} d'_{t(\alpha)}$$

$$\geq 2d_i - \sum_{\alpha \in Q_1, t(\alpha) = i} d_{s(\alpha)} - \sum_{\alpha \in Q_1, s(\alpha) = i} d_{t(\alpha)} = 0,$$

and hence $\mathbf{F}' \cdot \mathbf{d}' \in \mathbb{N}^{|Q_0'|}$. Moreover, because Q' is a proper subquiver of Q, we conclude that either there exist $i \in Q_0'$ and $j \in Q_0$ such that $f_{ij} \neq 0$ and j is not in Q_0', or there exist $i, j \in Q_0'$ such that $|f'_{ij}| < |f_{ij}|$. Because $d_i > 0$, for any $i \in Q_0$, we then conclude that $(\mathbf{F}' \cdot \mathbf{d}')_i > 0$, for some $i \in Q_0$, and consequently, $\mathbf{F}' \cdot \mathbf{d}' \neq 0$.

Let $A' = KQ'$. Because the underlying graph of Q' is Euclidean then (VII.4.2) yields $\operatorname{rad} q_{A'} = \mathbb{Z} \cdot \mathbf{h}'$, for a vector $\mathbf{h}' \in \mathbb{Z}^{|Q_0'|}$ whose coordinates are all positive. Then $\mathbf{\Phi}_{A'} \cdot \mathbf{h}' = \mathbf{h}'$, and consequently

$$2(\mathbf{h}', -)_{Q'} = \langle \mathbf{h}', - \rangle_{Q'} + \langle -, \mathbf{h}' \rangle_{Q'} = 0.$$

But then $(\mathbf{h}', \mathbf{e}_i)_Q = 0$, for any $i \in Q_0'$, and hence $\mathbf{F}' \cdot \mathbf{h}' = 0$. Then

$$(\mathbf{F}' \cdot \mathbf{d}')^t \cdot \mathbf{h}' = (\mathbf{d}')^t \cdot (\mathbf{F}')^t \cdot \mathbf{h}' = (\mathbf{d}')^t \cdot (\mathbf{F}' \cdot \mathbf{h}') = 0,$$

because the matrix $\mathbf{F}'$ is symmetric. On the other hand, $\mathbf{F}' \cdot \mathbf{d}'$ is a non-zero vector from $\mathbb{N}^{|Q_0'|}$ and $\mathbf{h}'$ has all coordinates positive. Therefore, the obtained contradiction shows that in fact $\dim X \neq \dim \tau^m X$. $\square$

1.2. Lemma. *Let $A = KQ$ be a wild hereditary algebra, M a non-zero regular A-module, m a positive integer, and $f : M \longrightarrow \tau^{-m} M$ a homomorphism. Then f is neither a monomorphism nor an epimorphism.*

Proof. Observe first that f is not an isomorphism. Indeed, if f is an isomorphism, then there is an indecomposable direct summand X of M such that $X \cong \tau^{-sm} X$, for some $s \geq 1$; a contradiction with (1.1). Suppose now that f is a proper monomorphism. Because the functor τ is left exact, we then get a sequence of proper monomorphisms

$$\cdots \longrightarrow \tau^{(r+1)m} M \xrightarrow{\tau^{(r+1)m} f} \tau^{rm} \longrightarrow \cdots \longrightarrow \tau^{2m} M \xrightarrow{\tau^{2m} f} \tau^m M \xrightarrow{\tau^m f} M,$$

a contradiction, because M is finite dimensional. Similarly, if f is a proper epimorphism, invoking the fact that τ^{-1} is right exact, we obtain a sequence of proper epimorphisms

$$M \xrightarrow{f} \tau^{-m} M \xrightarrow{\tau^{-m} f} \tau^{-2m} M \longrightarrow \cdots \longrightarrow \tau^{-rm} M \xrightarrow{\tau^{-rm} f} \tau^{-(r+1)m} M \longrightarrow \cdots,$$

again a contradiction. $\square$

1.3. Theorem. *Let $A = KQ$ be a wild hereditary algebra, X be an indecomposable regular A-module and*

$$0 \longrightarrow X \longrightarrow \bigoplus_{i=1}^{r} Y_i \longrightarrow Z \longrightarrow 0$$

an almost split sequence, with $Y_1, \ldots, Y_r$ indecomposable. Then

(a) *$r \leq 2$, and*
(b) *if $r = 2$ and $\dim_K Y_1 \leq \dim_K Y_2$ then $\dim_K Y_1 < \dim_K X < \dim_K Y_2$ and $\dim_K Y_1 < \dim_K Z < \dim_K Y_2$.*

Proof. (a) We divide the proof into several steps.

Step 1° First we show that if $\bigoplus_{i=1}^{r} Y_i = Y' \oplus Y''$ is a decomposition such that $\dim_K X \leq \dim_K Y'$, then $\dim_K X > \dim_K Y''$. Suppose, to the contrary, that $\dim_K X \leq \dim_K Y'$ and $\dim_K X \leq \dim_K Y''$. It follows that $Y' \neq 0$, $Y'' \neq 0$ and the induced irreducible morphisms $X \longrightarrow Y'$, $X \longrightarrow Y''$ are monomorphisms. Because X is finite dimensional then there exists a positive integer m such that $\dim_K \tau^m X \leq \dim_K \tau^{m+1} X$. Moreover, there is an almost split sequence

$$0 \longrightarrow \tau^{m+1} X \longrightarrow \tau^{m+1} Y' \oplus \tau^{m+1} Y'' \longrightarrow \tau^m X \longrightarrow 0,$$

where the induced irreducible morphisms $\tau^{m+1}X \longrightarrow \tau^{m+1}Y'$ and $\tau^{m+1}X \longrightarrow \tau^{m+1}Y''$ are non-bijective monomorphisms, because the functor τ is left exact. Then

$$\dim_K \tau^{m+1}Y' + \dim_K \tau^{m+1}Y'' = \dim_K \tau^{m+1}X + \dim_K \tau^m X$$
$$\leq 2\dim_K \tau^{m+1}X$$
$$< \dim_K \tau^{m+1}Y' + \dim_K \tau^{m+1}Y'',$$

and we get a contradiction.

Step 2° Next we show that $\dim_K X < \dim_K Y_i + \dim_K Y_j$, for each pair $i,j \in \{1,\dots,r\}$ such that $i \neq j$. Suppose, to the contrary, that there exists a pair $i,j \in \{1,\dots,r\}$ such that $i \neq j$ and $\dim_K X \geq \dim_K Y_i + \dim_K Y_j$. Then the induced irreducible morphism $X \longrightarrow Y_i \oplus Y_j$ is an epimorphism.

Choose a positive integer $s \geq 1$ such that $\dim_K \tau^{-s}X \leq \dim_K \tau^{-s-1}X$. Because the functor τ^{-1} carries epimorphisms to epimorphisms then we derive two irreducible epimorphisms

$$\tau^{-s}X \longrightarrow \tau^{-s}Y_i \bigoplus \tau^{-s}Y_j \quad \text{and} \quad \tau^{-s-1}X \longrightarrow \tau^{-s-1}Y_i \bigoplus \tau^{-s-1}Y_j.$$

Consider the almost split sequences

$$0 \longrightarrow \tau Y_i \longrightarrow X \oplus V_i \longrightarrow, Y_i \longrightarrow 0 \quad \text{and} \quad 0 \longrightarrow \tau Y_j, \longrightarrow X \oplus V_j \longrightarrow Y_j \longrightarrow 0$$

ending at Y_i and Y_j. By applying the functor τ^{-s-1}, we get the almost split sequences

$$0 \longrightarrow \tau^{-s}Y_i \longrightarrow \tau^{-s-1}X \bigoplus \tau^{-s-1}V_i \longrightarrow \tau^{-s-1}Y_i \longrightarrow 0,$$
$$0 \longrightarrow \tau^{-s}Y_j \longrightarrow \tau^{-s-1}X \bigoplus \tau^{-s-1}V_j \longrightarrow \tau^{-s-1}Y_j \longrightarrow 0.$$

Altogether this yields the inequalities

$$2\dim_K \tau^{-s-1}X \leq \dim_K \tau^{-s}Y_i + \dim_K \tau^{-s-1}Y_i$$
$$+ \dim_K \tau^{-s}Y_j + \dim_K \tau^{-s-1}Y_j$$
$$< \dim_K \tau^{-s}X + \dim_K \tau^{-s-1}X$$
$$\leq 2\dim_K \tau^{-s-1}X,$$

and we get a contradiction.

Step 3° The inequality $r \leq 3$ holds. Indeed, otherwise $r \geq 4$ and it follows from Step 2° that

$$\dim_K X < \dim_K(Y_1 \oplus Y_2) \quad \text{and} \quad \dim_K X < \dim_K(Y_3 \oplus Y_4).$$

But this is a contradiction with Step 1°.

Step 4° The inequality $r \leq 2$ holds. Assume, to the contrary, that $r = 3$. We claim that $\dim_K Y_i < \dim_K X$, for all $i \in \{1,2,3\}$. Suppose, to the contrary, that there exists an $i \in \{1, 2, 3\}$ such that $\dim_K X \leq \dim_K Y_i$. We may assume that $i = 1$. It follows from Step 2° that

$$\dim_K X < \dim_K (Y_2 \oplus Y_3).$$

But this contradicts again Step 1°. Therefore, for each $i \in \{1, 2, 3\}$, there exists an irreducible epimorphism $X \longrightarrow Y_i$. Hence, the composite homomorphism

$$\tau Z \oplus \tau Z \oplus \tau Z \cong X \oplus X \oplus X \longrightarrow Y_1 \oplus Y_2 \oplus Y_3 \longrightarrow Z$$

is surjective. By the same type of arguments applied to the regular module $\tau^t Z$, there exists an epimorphism $\tau^{t+1} Z \oplus \tau^{t+1} Z \oplus \tau^{t+1} Z \longrightarrow \tau^t Z$, for any $t \geq 1$.

Now we define inductively a chain of homomorphisms

$$\cdots \longrightarrow \tau^{t+1} Z \xrightarrow{f_{t+1}} \tau^t Z \xrightarrow{f_t} \cdots \longrightarrow \tau^2 Z \xrightarrow{f_2} \tau Z \xrightarrow{f_1} Z$$

such that $g_t = f_1 \ldots f_t \neq 0$, for any $t \geq 1$. We take for $f_1 : \tau Z \longrightarrow Z$ a non-zero restriction of the epimorphism $\tau Z \oplus \tau Z \oplus \tau Z \longrightarrow Z$ to a direct summand τZ.

Assume that, for each $p \in \{1, \ldots, t\}$, we have defined a homomorphism $f_p : \tau^p Z \longrightarrow \tau^{p-1} Z$ such that $g_t = f_1 \ldots f_t \neq 0$. To define the required homomorphism $f_{t+1} : \tau^{t+1} Z \longrightarrow \tau^t Z$, we consider an epimorphism

$$h = (h_1, h_2, h_3) : \tau^{t+1} Z \oplus \tau^{t+1} Z \oplus \tau^{t+1} Z \longrightarrow \tau^t Z.$$

It follows that $g_t h \neq 0$ and, hence, there exists $i \in \{1, 2, 3\}$ such that $g_t h_i \neq 0$. Letting $f_{t+1} = h_i$ we have $g_{t+1} = f_1 \ldots f_t f_{t+1} = g_t h_i \neq 0$. If, for each $t \geq 1$, we denote by L_t the image of g_t, then we get a decreasing chain

$$L_1 \supseteq L_2 \supseteq \ldots \supseteq L_t \supseteq L_{t+1} \supseteq \ldots$$

of submodules of Z. Because $\dim_K L_1$ is finite, then there exists a positive integer m_0 such that $L_s = L_t$, for all $s, t \geq m_0$. Let $L = L_{m_0}$. Thus, $g_t : \tau^t Z \longrightarrow L$ is an epimorphism, for any $t \geq m_0$. We note also that L is a regular module, because the category add $\mathcal{R}(A)$ of regular modules is closed under images. Applying now the right exact functor τ^{-t}, for each $t \geq m_0$, we obtain an epimorphism $\tau^{-t} g_t : Z \longrightarrow \tau^{-t} L$. In particular, for each $t \geq m_0$, we get $\dim \tau^{-t} L \leq \dim Z$.

If U is an indecomposable direct summand of L then $\dim \tau^{-t} U \leq \dim Z$, for all $t \geq m_0$, and consequently, $\dim \tau^{-s} U = \dim \tau^{-t} U$, for some $s > t \geq m_0$. It follows that, for the indecomposable regular module $M = \tau^{-s} U$ and $m = s - t$, we have $\dim M = \dim \tau^m M$, contrary to (1.1). This shows that $r \leq 2$, and proves the statement (a) of the theorem.

To prove (b), we assume that $r = 2$ and that $\dim_K Y_1 \leq \dim_K Y_2$. Because each of the irreducible morphisms $X \longrightarrow Y_1$ and $X \longrightarrow Y_2$ is a proper monomorphism or a proper epimorphism then, by Step 1°,

- either $\dim_K Y_1 \leq \dim_K Y_2 < \dim_K X$, or
- $\dim_K Y_1 < \dim_K X < \dim_K Y_2$.

Dually, we have

- either $\dim_K Y_1 \leq \dim_K Y_2 < \dim_K Z$, or
- $\dim_K Y_1 < \dim_K Z < \dim_K Y_2$.

Moreover, in view of the equality

$$\dim_K X + \dim_K Z = \dim_K Y_1 + \dim_K Y_2,$$

the inequalities

- $\dim_K Y_1 \leq \dim_K Y_2 < \dim_K X$ and
- $\dim_K Y_1 \leq \dim_K Y_2 < \dim_K Z$

do not hold simultaneously. It follows also that the inequalities

$$\dim_K Y_1 < \dim_K X < \dim_K Y_2$$

hold if and only if the inequalities

$$\dim_K Y_1 < \dim_K Z < \dim_K Y_2$$

hold. This finishes the proof.   □

1.4. Definition. Let $A = KQ$ be a wild hereditary algebra. An indecomposable regular A-module X is said to be **quasi-simple** if the middle term E in the almost split sequence $0 \longrightarrow X \longrightarrow E \longrightarrow \tau^{-1}X \longrightarrow 0$ is indecomposable.

1.5. Corollary. *Assume that $A = KQ$ is a connected wild hereditary algebra and let X be a quasi-simple regular A-module.*

(a) *For any integer $r \in \mathbb{Z}$, the regular module $\tau^r X$ is quasi-simple.*

(b) *There exist an infinite chain of irreducible monomorphisms*

$$X = X[1] \longrightarrow X[2] \longrightarrow X[3] \longrightarrow \cdots \longrightarrow X[r] \longrightarrow \cdots$$

and an infinite chain of irreducible epimorphisms

$$\cdots \longrightarrow [r]X \longrightarrow \cdots \longrightarrow [3]X \longrightarrow [1]X = X,$$

where $X[i]$ and $[r]X$ are indecomposable regular modules, for all $i, r \geq 1$. The chains are uniquely determined by X, up to isomorphism.

(c) *For each integer $m \in \mathbb{Z}$, there exist isomorphisms $\tau^m X[i] \cong (\tau^m X)[i]$ and $\tau^m [i]X \cong [i](\tau^m X)$ of A-modules.*

Proof. We recall that the functors $\tau, \tau^{-1} : \operatorname{add} \mathcal{R}(A) \longrightarrow \operatorname{add} \mathcal{R}(A)$ are equivalences of categories inverse to each other. Because the functors preserve the irreducibility and carry almost split sequences to almost split sequences then the corollary is an immediate consequence of (1.3) and the Definition 1.4.   □

Let X be a quasi-simple regular A-module and let $Y = X[i]$, $Z = [i]X$ be as in (1.5). Following Ringel [515], we call the number $i \geq 1$ the **quasi-length** of the regular modules Y and Z, respectively.

We usually consider the irreducible monomorphisms $X[n] \longrightarrow X[n+1]$ in (1.5) as inclusions. If we think of the infinite chains in (1.5) as a part of the Auslander–Reiten quiver $\Gamma(\operatorname{mod} A)$ of A, we view them as the upper peak angle

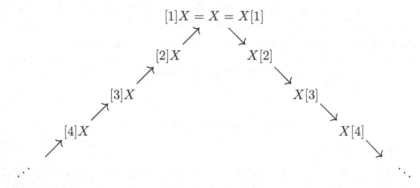

We show in the next section that, in contrast to the situation for regular modules over hereditary algebras of Euclidean type, the quasi-simple modules over wild hereditary algebras may contain proper regular submodules.

We may now describe the shape of the regular components of $\Gamma(\operatorname{mod} A)$. The result was independently established by Ringel in [515] and Auslander, Bautista, Platzeck, Reiten, and Smalø in [28].

1.6. Theorem. *Let $A = KQ$ be a wild hereditary algebra and $\mathcal{C}$ be a regular component of $\Gamma(\operatorname{mod} A)$. Then $\mathcal{C}$ is of type*

$\mathbb{Z}A_\infty$:

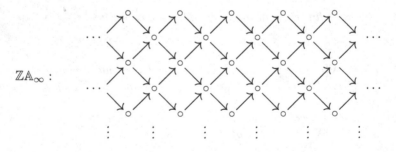

(see (X.1)) and consists of the modules $\tau^m X[i]$ defined in (1.5), where X is a quasi-simple regular A-module in $\mathcal{C}$, $m \in \mathbb{Z}$ and $i \geq 1$.

Proof. Let X be an indecomposable module in $\mathcal{C}$ of minimal K-dimension among all indecomposable modules from $\mathcal{C}$. It follows from (1.3) that X is quasi-simple and, according to (1.5), all modules $\tau^m X$, with $m \in \mathbb{Z}$, are quasi-simple. Moreover, applying (1.1), we conclude that the modules $\tau^m X$, with $m \in \mathbb{Z}$, are pairwise non-isomorphic. Further, it follows from (1.3), that every indecomposable module Z in $\mathcal{C}$ has exactly two direct predecessors and two direct successors, or it is quasi-simple. It follows that Z is quasi-simple, or Z contains a quasi-simple module X connected with Z by a finite chain $X \longrightarrow \ldots \longrightarrow Z$ of irreducible monomorphisms. Hence, in view of (1.5), if

$$X = X[1] \longrightarrow X[2] \longrightarrow \cdots \longrightarrow X[r] \longrightarrow \cdots$$

is a chain of irreducible monomorphisms shown in (1.5), then $Z = X[r]$, for some $r \geq 1$, and we derive the chain

$$\tau^m X = \tau^m X[1] \longrightarrow \tau^m X[2] \longrightarrow \cdots \longrightarrow \tau^m X[r] \longrightarrow \cdots$$

of irreducible monomorphisms, for any $m \in \mathbb{Z}$. Similarly, for a chain

$$\cdots \longrightarrow [r]X \longrightarrow \cdots \longrightarrow [2]X \longrightarrow [1]X = X$$

of irreducible epimorphisms shown in (1.5), we derive the chain

$$\cdots \longrightarrow \tau^m [r]X \longrightarrow \cdots \longrightarrow \tau^m [2]X \longrightarrow \tau^m [1]X = \tau^m X$$

of irreducible epimorphisms, for any $m \in \mathbb{Z}$. Therefore, the component $\mathcal{C}$ is of type $\mathbb{Z}A_\infty$, consists of the modules $\tau^m X[i]$, where $m \in \mathbb{Z}$ and $i \geq 1$, and is formed by the τ^m-shifts of the upper peak angle shown above, for all $m \in \mathbb{Z}$. $\square$

We prove now the following interesting fact, due to Y. Zhang [688], asserting that, for a hereditary wild algebra $A = KQ$, there is no pair of non-isomorphic indecomposable modules M and N, lying in the same regular component, with $\mathbf{dim}\, M = \mathbf{dim}\, N$.

1.7. Theorem. *Let $A = KQ$ be a wild hereditary algebra and $\mathcal{C}$ be a regular component of $\Gamma(\mathrm{mod}\, A)$. If M and N are two non-isomorphic indecomposable modules in $\mathcal{C}$, then $\mathbf{dim}\, M \neq \mathbf{dim}\, N$.*

Proof. Suppose, to the contrary, that M and N are non-isomorphic indecomposable modules in $\mathcal{C}$ such that $\mathbf{dim}\, M = \mathbf{dim}\, N$. Let X be the quasi-simple module in $\mathcal{C}$ such that $M \cong X[r]$, for some $r \geq 1$. Then $N \cong \tau^m X[r + i]$, for some integer m and $-r < i$. Clearly, $m \neq 0$ and, without loss of generality, we may assume that $m > 0$. Moreover, it follows from (1.1) that $i \neq 0$. We have two cases to consider.

Case 1° Assume that $i > 0$. Let $\mathbf{\Phi}_A : \mathbb{Z}^n \longrightarrow \mathbb{Z}^n$ be the Coxeter transformation of A (see (III.3.14)). In view of (IV.2.9), the equality $\mathbf{dim}\, X[r] = \mathbf{dim}\,\tau^m X[r+i]$ yields the equalities

$$\mathbf{dim}\,\tau^{sm} X[r] = \mathbf{\Phi}_A^{sm}(\mathbf{dim}\, X[r])$$
$$= \mathbf{\Phi}_A^{sm}(\mathbf{dim}\,\tau^m X[r+i])$$
$$= \mathbf{dim}\,\tau^{(s+1)m} X[r+i],$$

for all $s \geq 0$. Because $\mathbf{dim}\,\tau^{sm} X[r] < \mathbf{dim}\,\tau^{sm} X[r+i]$, for any $s \geq 0$, we get a strictly decreasing infinite chain

$$\mathbf{dim}\, X[r] > \mathbf{dim}\,\tau^m X[r] > \mathbf{dim}\,\tau^{2m} X[r] > \mathbf{dim}\,\tau^{3m} X[r] > \dots,$$

and we get a contradiction.

Case 2° Assume that $i < 0$. By (IV.2.9), the equality $\mathbf{dim}\, X[r] = \mathbf{dim}\,\tau^m X[r+i]$ yields the equalities

$$\mathbf{dim}\,\tau^{-sm} X[r+i] = \mathbf{\Phi}_A^{-sm}(\mathbf{dim}\, X[r+i]) = \mathbf{\Phi}^{-(s+1)m}\mathbf{\Phi}_A^m(\mathbf{dim}\, X[r+i])$$
$$= \mathbf{\Phi}_A^{-(s+1)m}(\mathbf{dim}\,\tau^m X[r+i]) = \mathbf{\Phi}_A^{-(s+1)m}(\mathbf{dim}\, X[r])$$
$$= \mathbf{dim}\,\tau^{-(s+1)m} X[r],$$

for all $s \geq 0$. Together with the inequality $\mathbf{dim}\,\tau^{-sm} X[r+i] < \mathbf{dim}\,\tau^{-sm} X[r]$, for any $s \geq 0$, this yields a strictly decreasing infinite chain

$$\mathbf{dim}\, X[r+i] > \mathbf{dim}\,\tau^{-m} X[r+i]$$
$$> \mathbf{dim}\,\tau^{-2m} X[r+i]$$
$$> \mathbf{dim}\,\tau^{-3m} X[r+i] > \dots,$$

and we get a contradiction. Consequently, $\mathbf{dim}\, M \neq \mathbf{dim}\, N$, and we are done. $\square$

1.8. Corollary. *If $A = KQ$ is a connected wild hereditary K-algebra, then the Auslander–Reiten quiver $\Gamma(\mathrm{mod}\, A)$ of A has exactly* $\mathrm{card}(K)$ *regular components, where* $\mathrm{card}(K)$ *is the cardinality of the field K.*

Proof. By our hypothesis, A is the path algebra KQ of a connected, acyclic and wild quiver Q. It follows that there exists a two-sided ideal $\mathcal{I}$ in KQ generated by some arrows of Q and some idempotents of KQ such that the quotient algebra $A' = KQ/\mathcal{I}$ is the path algebra KQ' of a quiver Q', whose underlying graph $\overline{Q'}$ is Euclidean. By (XI.3.5), (XII.3.4), and (XII.4.2), there exists a $\mathbb{P}_1(K)$-family $\{R_\lambda\}_{\lambda \in \mathbb{P}_1(K)}$, of pairwise non-isomorphic indecomposable regular A'-modules R_λ such that

- $\mathrm{End}_{A'}(R_\lambda) \cong K$,
- $\mathrm{Ext}^1_{A'}(R_\lambda, R_\lambda) \cong K$, and
- $\mathbf{dim}\, R_\lambda = \mathbf{dim}\, R_\mu$, for all $\lambda, \mu \in \mathbb{P}_1(K)$.

Clearly, each R_λ is an indecomposable A-module and there are isomorphisms
$$\mathrm{Ext}_A^1(R_\lambda, R_\lambda) \cong \mathrm{Ext}_{A'}^1(R_\lambda, R_\lambda) \cong K.$$
Hence, the module R_λ is neither in the postprojective component $\mathcal{P}(A)$ nor in the preinjective component $\mathcal{Q}(A)$, because both consist of modules without self-extensions. Consequently, R_λ is a regular module. It follows now from (1.7) that, for $\lambda \neq \mu$, the modules R_λ and R_μ lie in different regular components of $\Gamma(\mathrm{mod}\,A)$, because $\mathbf{dim}\,R_\lambda = \mathbf{dim}\,R_\mu$ and $R_\lambda \not\cong R_\mu$. This shows that the cardinality of the set of components in $\mathcal{R}(A)$ is greater than or equal to the cardinality of $\mathbb{P}_1(K) = K \cup \{\infty\}$.

On the other hand, because there are only countably many dimension vectors of indecomposable A-modules then $\Gamma(\mathrm{mod}\,A)$ has exactly $\mathrm{card}(K)$ regular components. $\qquad \square$

1.9. Example. Let A be the path algebra of the enlarged Kronecker quiver

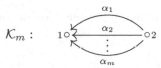

$$\mathcal{K}_m: \qquad 1 \circ \begin{smallmatrix} \alpha_1 \\ \alpha_2 \\ \vdots \\ \alpha_m \end{smallmatrix} \circ 2$$

with $m \geq 3$ arrows. Then A is isomorphic to the triangular matrix algebra $\begin{bmatrix} K & 0 \\ K^m & K \end{bmatrix}$, called an enlarged Kronecker algebra, the Grothendieck group $K_0(A)$ of A is isomorphic to $\mathbb{Z}^2$, the Cartan matrix $C_A \in \mathbb{M}_2(\mathbb{Z})$ of A and its inverse $C_A^{-1} \in \mathbb{M}_2(\mathbb{Z})$ are of the forms
$$C_A = \begin{bmatrix} 1 & m \\ 0 & 1 \end{bmatrix}, \qquad C_A^{-1} = \begin{bmatrix} 1 & -m \\ 0 & 1 \end{bmatrix},$$
and the Coxeter matrix $\Phi_A = -C_A^t C_A^{-1} \in \mathbb{M}_2(\mathbb{Z})$ of A and its inverse are the matrices
$$\Phi_A = \begin{bmatrix} -1 & m \\ -m & m^2 - 1 \end{bmatrix}, \qquad \Phi_A^{-1} = \begin{bmatrix} m^2 - 1 & -m \\ m & -1 \end{bmatrix}.$$
Moreover,

- $\mathbf{dim}\,P(1) = \begin{bmatrix} 1 \\ 0 \end{bmatrix}$, $\mathbf{dim}\,P(2) = \begin{bmatrix} m \\ 1 \end{bmatrix}$, $\mathbf{dim}\,I(1) = \begin{bmatrix} 1 \\ m \end{bmatrix}$, $\mathbf{dim}\,I(2) = \begin{bmatrix} 0 \\ 1 \end{bmatrix}$,
- the postprojective component $\mathcal{P}(A)$ of A is of the form

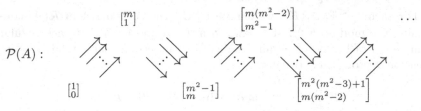

- the preinjective component $\mathcal{Q}(A)$ of A is of the form

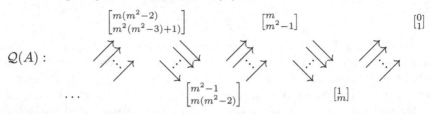

$$\mathcal{Q}(A):$$

where the indecomposable modules are represented by their dimension vectors.

We recall from (IV.2.9) that $\dim \tau^m X = \mathbf{\Phi}_A^m(\dim X)$, for any non-postprojective indecomposable module X, and $\dim \tau^{-m}Y = \mathbf{\Phi}_A^{-m}(\dim Y)$, for any non-preinjective indecomposable module Y and any $m \geq 0$. For each m-tuple

$$\lambda = (\lambda_1, \dots, \lambda_n) \in K^m \setminus \{(0, \dots, 0)\},$$

we denote by $E^{(\lambda)}$ the K-linear representation of the quiver $\mathcal{K}_m$ given by

$$E^{(\lambda)}: \quad K \overset{\overset{\lambda_1}{\underset{\lambda_2}{\underset{\vdots}{\longleftarrow}}}}{\underset{\lambda_m}{\longleftarrow}} K$$

Clearly, each $E^{(\lambda)}$ is a brick and, therefore, it is indecomposable.

By (III.2.12), (III.3.13), and (VII.4.1), the Euler quadratic form $q_A : \mathbb{Z}^2 \longrightarrow \mathbb{Z}$ of A is given by the formula $q_A(\mathbf{x}) = x_1^2 + x_2^2 - mx_1x_2$, for any $\mathbf{x} = \begin{bmatrix} x_1 \\ x_2 \end{bmatrix} \in \mathbb{Z}^2$.

It follows that $\dim E^{(\lambda)} = \begin{bmatrix} 1 \\ 1 \end{bmatrix}$, and $q_A(\dim E^{(\lambda)}) = -m + 2 \leq -1$, because we assume that $m \geq 3$. Hence, according to (VIII.2.7), the A-module $E^{(\lambda)}$ is neither postprojective nor preinjective. Thus, $E^{(\lambda)}$ is regular. Note also that the A-module $E^{(\lambda)}$ is quasi-simple. Moreover, for $\lambda = (\lambda_1, \dots, \lambda_m)$ and $\mu = (\mu_1, \dots, \mu_m)$ in $K^m \setminus \{(0, \dots, 0)\}$, we have

- $E^{(\lambda)} \cong E^{(\mu)}$ if and only if there exists $a \in K \setminus \{0\}$ such that $\mu = a\lambda$, or equivalently,
- $E^{(\lambda)} \cong E^{(\mu)}$ if and only if λ and μ define the same point of the projective space $\mathbb{P}_{m-1}(K)$.

For each $\lambda \in \mathbb{P}_{m-1}(K)$, we denote by $\mathcal{C}_\lambda^A$ the component in $\mathcal{R}(A)$ containing the module $E^{(\lambda)}$. It follows from (1.6) and (1.7) that each regular component of $\Gamma(\text{mod } A)$ admits at most one indecomposable module of dimension vector $(1, 1)$. Because

$$E^{(\lambda)} \not\cong E^{(\mu)} \quad \text{and} \quad \dim E^{(\lambda)} = \dim E^{(\mu)},$$

for $\lambda \neq \mu$ in $\mathbb{P}_{m-1}(K)$, then the $\mathbb{P}_{m-1}(K)$-family

$$\mathbf{\mathcal{C}}^A = \{\mathcal{C}_\lambda^A\}_{\lambda \in \mathbb{P}_{m-1}(K)}$$

consists of pairwise different regular components. By applying the Coxeter transformations Φ_A, $\Phi_A^{-1} : \mathbb{Z}^2 \longrightarrow \mathbb{Z}^2$, we show that the regular component $\mathcal{C}_\lambda^A$ is of the form

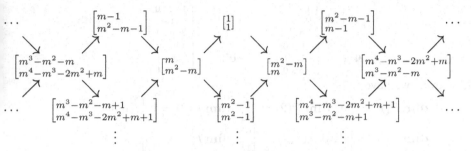

where the indecomposable modules are represented by their dimension vectors.

Observe also that

$$\dim_K \operatorname{Ext}_A^1(E^{(\lambda)}, E^{(\lambda)}) = m - 1 \geq 2,$$

for any $\lambda \in \mathbb{P}_{m-1}(K)$. Indeed, because $q_A(\mathbf{x}) = x_1^2 + x_2^2 - mx_1x_2$, for any vector $\mathbf{x} = [\begin{smallmatrix} x_1 \\ x_2 \end{smallmatrix}] \in \mathbb{Z}^2$, then we get the equalities

$$\begin{aligned} 2 - m &= q_A(\mathbf{dim}\, E^{(\lambda)}) \\ &= \langle \mathbf{dim}\, E^{(\lambda)}, \mathbf{dim}\, E^{(\lambda)} \rangle_A \\ &= \chi_A(E^{(\lambda)}, E^{(\lambda)}) \\ &= \dim_K \operatorname{End}_A(E^{(\lambda)}) - \dim_K \operatorname{Ext}_A^1(E^{(\lambda)}, E^{(\lambda)}) \\ &= 1 - \dim_K \operatorname{Ext}_A^1(E^{(\lambda)}, E^{(\lambda)}), \end{aligned}$$

and, consequently, $\dim_K \operatorname{Ext}_A^1(E^{(\lambda)}, E^{(\lambda)}) = m - 1$.

Finally, we note that, by the Auslander–Reiten formulae (IV.2.14), there are isomorphisms of K-vector spaces

$$\operatorname{Ext}_A^1(E^{(\lambda)}, E^{(\lambda)}) \cong \operatorname{Hom}_A(E^{(\lambda)}, \tau E^{(\lambda)}) \cong \operatorname{Hom}_A(\tau^{-1} E^{(\lambda)}, E^{(\lambda)}).$$

It follows that there is a lot of A-homomorphisms from $\tau^{-1} E^{(\lambda)}$ to $E^{(\lambda)}$, and from $E^{(\lambda)}$ to $\tau E^{(\lambda)}$.

1.10. Example. Let $A = KQ$ be the path algebra of the quiver

$$Q: \quad 1 \underset{\beta}{\overset{\alpha}{\longleftarrow\!\!\!\longleftarrow}} 2 \overset{\gamma}{\longleftarrow} 3.$$

Then A is isomorphic to the triangular matrix algebra $\begin{bmatrix} K & 0 & 0 \\ K^2 & K & 0 \\ K^2 & K & K \end{bmatrix}$ with the obvious multiplication, $K_0(A) \cong \mathbb{Z}^3$, and

$$C_A = \begin{bmatrix} 1 & 2 & 2 \\ 0 & 1 & 1 \\ 0 & 0 & 1 \end{bmatrix}, \qquad C_A^{-1} = \begin{bmatrix} 1 & -2 & 0 \\ 0 & 1 & -1 \\ 0 & 0 & 1 \end{bmatrix},$$

$$\Phi_A = \begin{bmatrix} -1 & 2 & 0 \\ -2 & 3 & 1 \\ -2 & 3 & 0 \end{bmatrix}, \qquad \Phi_A^{-1} = \begin{bmatrix} 3 & 0 & -2 \\ 2 & 0 & -1 \\ 0 & 1 & -1 \end{bmatrix}.$$

Moreover, we have

(i) $\mathbf{dim}\, P(1) = \begin{bmatrix} 1 \\ 0 \\ 0 \end{bmatrix}$, $\mathbf{dim}\, P(2) = \begin{bmatrix} 2 \\ 1 \\ 0 \end{bmatrix}$, $\mathbf{dim}\, P(3) = \begin{bmatrix} 2 \\ 1 \\ 1 \end{bmatrix}$, and

(ii) $\mathbf{dim}\, I(1) = \begin{bmatrix} 1 \\ 2 \\ 2 \end{bmatrix}$, $\mathbf{dim}\, I(2) = \begin{bmatrix} 0 \\ 1 \\ 1 \end{bmatrix}$, $\mathbf{dim}\, I(3) = \begin{bmatrix} 0 \\ 0 \\ 1 \end{bmatrix}$.

Therefore, the postprojective component $\mathcal{P}(A)$ and the preinjective component $\mathcal{Q}(A)$ of $\Gamma(\mathrm{mod}\, A)$ are of the forms

respectively, where the indecomposable modules are represented by their dimension vectors.

Consider the $\mathbb{P}_1(K)$-family

$$\mathcal{E}^A = \{E^{(\lambda)}\}_{\lambda \in \mathbb{P}_1(K)}$$

of the indecomposable A-modules $E^{(\lambda)}$ defined as follows

$$E^{(\lambda)}: \quad K \overset{1}{\underset{\lambda}{\Longleftarrow}} K \longleftarrow 0, \qquad \text{for} \quad \lambda \in K, \quad \text{and}$$

$$E^{(\infty)}: \quad K \overset{0}{\underset{1}{\Longleftarrow}} K \longleftarrow 0, \qquad \text{for} \quad \lambda = \infty,$$

where $\mathbb{P}_1(K) = K \cup \{\infty\}$ is the projective line with $\infty = (0 : 1)$. It is clear that the family $\mathcal{E}^A = \{E^{(\lambda)}\}_{\lambda \in \mathbb{P}_1(K)}$ is induced from the $\mathbb{P}_1(K)$-family of pairwise non-isomorphic simple homogeneous modules over the Kronecker algebra described in Section XI.4, that is, the path algebra of the Kronecker quiver $\circ \overset{\alpha}{\underset{\beta}{\Longleftarrow}} \circ$.

We note that, for each $\lambda \in \mathbb{P}_1(K)$, the A-module $E^{(\lambda)}$ is indecomposable and regular. To see this, we apply the explicit description of $\mathcal{P}(A)$ and $\mathcal{Q}(A)$ presented above, or we use (VIII.2.7) and the isomorphism

$$\text{Ext}_A^1(E^{(\lambda)}, E^{(\lambda)}) \cong K$$

of vector spaces. It is easy to check that, for each $\lambda \in \mathbb{P}_1(K)$, the A-module $E^{(\lambda)}$ is quasi-simple.

For each $\lambda \in \mathbb{P}_1(K)$, we denote by $\mathcal{C}_\lambda^A$ the connected component in $\mathcal{R}(A)$ containing the module $E^{(\lambda)}$. It follows from (1.6) and (1.7) that the $\mathbb{P}_1(K)$-family

$$\boldsymbol{\mathcal{C}}^A = \{\mathcal{C}_\lambda^A\}_{\lambda \in \mathbb{P}_1(K)}$$

consists of pairwise different regular components.

By applying the Coxeter transformations $\boldsymbol{\Phi}_A : \mathbb{Z}^3 \longrightarrow \mathbb{Z}^3$ and its inverse $\boldsymbol{\Phi}_A^{-1} : \mathbb{Z}^3 \longrightarrow \mathbb{Z}^3$, we show that the regular component $\mathcal{C}_\lambda^A$ is of the form

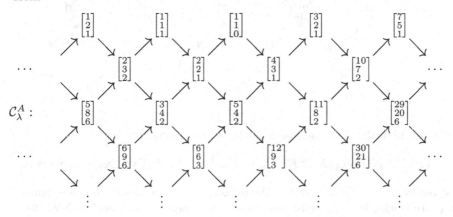

where the indecomposable modules are represented by their dimension vectors.

We also note that there are isomorphisms

$$\operatorname{Ext}_A^1(\tau^m E^{(\lambda)}, \tau^m E^{(\lambda)}) \cong \operatorname{Ext}_A^1(E^{(\lambda)}, E^{(\lambda)}) \cong K,$$

for any $m \in \mathbb{Z}$ and $\lambda \in \mathbb{P}_1(K)$; hence all quasi-simple modules in the component $\mathcal{C}_\lambda^A$ have self-extensions.

Consider now the indecomposable A-module X of the form

$$K \underset{(1,0)}{\overset{(0,1)}{\longleftarrow}} K^2 \longleftarrow 0,$$

induced by the indecomposable injective module over the Kronecker algebra. It is clear that

$$\mathbf{dim}\, X = [1,2,0]^t = \begin{bmatrix} 1 \\ 2 \\ 0 \end{bmatrix},$$

and X is the unique indecomposable A-module of dimension vector $[1,2,0]^t$.

It follows also from the above description of the postprojective component $\mathcal{P}(A)$ of $\Gamma(\operatorname{mod} A)$ and the preinjective component $\mathcal{Q}(A)$ of $\Gamma(\operatorname{mod} A)$ that X is a regular A-module. Hence, by applying (1.6), (1.7), and the Coxeter transformations $\Phi_A, \Phi_A^{-1} : \mathbb{Z}^3 \longrightarrow \mathbb{Z}^3$, we conclude that the regular component $\mathcal{C}_X^A$ of $\Gamma(\operatorname{mod} A)$ containing the module X is of the form

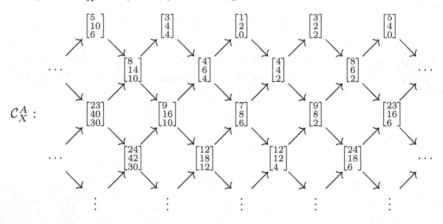

We also note that, for each $m \in \mathbb{Z}$, there exist isomorphisms

$$\operatorname{End}_A(\tau^m X) \cong \operatorname{End}_A(X) \cong K \quad \text{and} \quad \operatorname{Ext}_A^1(\tau^m X, \tau^m X) \cong \operatorname{Ext}_A^1(X, X) = 0,$$

of vector spaces. Consequently, the quasi-simple modules in the component $\mathcal{C}_X^A$ are bricks, have no self-extensions and, hence, are stones, by (XV.1.2).

XVIII.2. Homomorphisms between regular modules

Throughout this section, Q is a connected wild quiver and $A = KQ$ the associated connected wild hereditary K-algebra. The main objective of this section is to describe basic properties of homomorphisms between regular A-modules. We start with the following lemma showing that the category add $\mathcal{R}(A)$ of regular A-modules is not closed under kernels and cokernels, and consequently is not abelian (in contrast to the Euclidean case).

2.1. Lemma. *Let $A = KQ$ be a connected wild hereditary K-algebra.*

(i) *Let X be a non-zero regular A-module without non-trivial regular quotient modules. Then there exist a positive integer $m \geq 1$ and a short exact sequence*

$$0 \longrightarrow X \longrightarrow \tau^m X \longrightarrow I \longrightarrow 0$$

in mod A, *with a preinjective module I.*

(ii) *Let Y be a non-zero regular A-module without non-trivial regular submodules. Then there exist a positive integer $m \geq 1$ and a short exact sequence*

$$0 \longrightarrow P \longrightarrow \tau^{-m} Y \longrightarrow Y \longrightarrow 0$$

in mod A, *with a postprojective module P.*

Proof. We prove only (i), because the proof of (ii) is dual.

Let $n \geq 1$ be the rank of the Grothendieck group $K_0(A)$ of A. Consider the module $Z = \bigoplus_{i=0}^{n} \tau^{2i} X$. It follows from our assumption that X is an indecomposable regular module. By (1.6), X lies in a component of type $\mathbb{Z}\mathbb{A}_\infty$. Hence, in view of (1.7) we conclude that Z is a direct sum of $n + 1$ pairwise non-isomorphic indecomposable regular modules, and therefore $\mathrm{Hom}_A(Z, \tau Z) \neq 0$, by (VIII.5.3). Hence we conclude that

$$\mathrm{Hom}_A(X, \tau^{2(j-i)+1} X) \cong \mathrm{Hom}_A(\tau^{2i} X, \tau^{2j+1} X) \neq 0,$$

for some $i \neq j$. Let $m = 2(j - i) + 1$ and let $f : X \longrightarrow \tau^m X$ be a non-zero homomorphism. Observe that f has to be injective, because Im f is regular and X has no non-trivial regular quotient modules. Hence, by applying (1.2), we conclude that $m \geq 1$. We then get a short exact sequence

$$0 \longrightarrow X \xrightarrow{f} \tau^m X \xrightarrow{g} I \longrightarrow 0,$$

where I is the cokernel of f. Clearly, $I \neq 0$, because f is not an isomorphism, by (1.7). Because the functor τ^{-m} is right exact, we infer that $\tau^m X$ has no non-trivial regular quotient modules. Therefore, I is preinjective, and the proof is complete.   □

We note that if R is an arbitrary non-zero regular A-module and X is a non-zero regular quotient module of R of minimal dimension, then X has no non-trivial regular quotient modules. Dually, a non-zero regular submodule Y of R of minimal dimension, has no non-trivial regular submodules.

2.2. Lemma. *Let $A = KQ$ be a wild hereditary algebra and X a non-zero regular A-module. Then there exists a natural number m_0 such that, for any regular A-module R and any $m \geq m_0$, all homomorphisms $f : \tau^m X \longrightarrow R$ have regular kernel and all homomorphisms $g : R \longrightarrow \tau^{-m} X$ have regular cokernel.*

Proof. We prove only the first claim, because the proof of the second one is similar. We know from (IX.5.4) that there exists a natural number m_0 such that $\dim_K \tau^{-m} P > \dim_K X$, for all non-zero postprojective A-modules P and all $m \geq m_0$. Let R be a regular A-module and $f : \tau^m X \longrightarrow R$ a homomorphism. Denote by I the image of f and by L the kernel of f. We have a short exact sequence

$$0 \longrightarrow L \longrightarrow \tau^m X \longrightarrow I \longrightarrow 0.$$

Observe that I is regular and L has no preinjective direct summands. Thus, applying the functor τ^{-m}, we get a short exact sequence

$$0 \longrightarrow \tau^{-m} L \longrightarrow X \longrightarrow \tau^{-m} I \longrightarrow 0,$$

and hence $\dim_K \tau^{-m} L \leq \dim_K X$. Then, by our choice of m, L has no non-zero postprojective direct summands, and so L is regular.   □

We know from (IX.5.6) and its dual that all but finitely many modules from the postprojective (respectively, preinjective) component of $\Gamma(\mathrm{mod}\, A)$ are sincere. We prove now that all regular components of $\Gamma(\mathrm{mod}\, A)$ have the same property. First we prove the following technical fact.

2.3. Lemma. *Let $A = KQ$ be a wild hereditary algebra and X an indecomposable regular A-module. Then all but finitely many modules $\tau^i X$, with $i \in \mathbb{Z}$, are sincere.*

Proof. Assume first that X has no non-trivial regular quotient modules. Then, by (2.1), there exists a short exact sequence

$$0 \longrightarrow X \longrightarrow \tau^m X \longrightarrow I \longrightarrow 0,$$

with $m \geq 1$ and I non-zero preinjective. Because the modules in the sequence have no non-zero postprojective direct summands then, for each $t \geq 0$, we have the induced exact sequence

$$0 \longrightarrow \tau^t X \longrightarrow \tau^{m+t} X \longrightarrow \tau^t I \longrightarrow 0.$$

On the other hand, by (IX.5.6), there is a natural number $n_0 \geq 0$ such that all modules $\tau^t I$, with $t \geq n_0$, are sincere. Clearly, then all modules $\tau^{m+t} X$, with $t \geq n_0$, are also sincere.

Now, let X be an arbitrary indecomposable regular A-module. By applying (2.2), we conclude that there is a natural number $n \geq 1$ such that all homomorphisms $f : \tau^n X \longrightarrow R$, with R regular, have regular kernel.

Let Y be a regular quotient module of $\tau^n X$ having no non-trivial regular quotient modules. Then we have a short exact sequence

$$0 \longrightarrow L \longrightarrow \tau^n X \longrightarrow Y \longrightarrow 0,$$

with L regular. Then, for each $s \geq 0$, we have the induced exact sequence

$$0 \longrightarrow \tau^s L \longrightarrow \tau^{n+s} X \longrightarrow \tau^s Y \longrightarrow 0.$$

It follows from the first part of our proof that all but finitely many modules $\tau^s Y$, with $s \geq 0$, are sincere. Therefore, we have proved that all but finitely many modules $\tau^i X$, with $i \geq 0$, are sincere. Invoking the functor τ^{-1} we prove similarly that all but finitely many modules $\tau^i X$, with $i \leq 0$, are also sincere. □

2.4. Corollary. Let $A = KQ$ be a wild hereditary algebra and $\mathcal{C}$ be a regular component of $\Gamma(\mathrm{mod}\, A)$. Then all but finitely many modules in $\mathcal{C}$ are sincere.

Proof. Let X be a quasi-simple module from $\mathcal{C}$. It follows from (1.6) that, for any $m \in \mathbb{Z}$, there exist a chain of irreducible monomorphisms

$$\tau^m X = \tau^m X[1] \longrightarrow \tau^m X[2] \longrightarrow \cdots \longrightarrow \tau^m X[r] \longrightarrow \cdots$$

and a chain of irreducible epimorphisms

$$\cdots \longrightarrow \tau^m[r]X \longrightarrow \cdots \longrightarrow \tau^m[2]X \longrightarrow \tau^m[1]X = \tau^m X,$$

such that $\mathcal{C}$ consists of the modules $\tau^m X[i] = \tau^{m-i}[i]X$, with $m \in \mathbb{Z}$ and $i \geq 1$. From (2.3) we know that all but finitely many modules $\tau^m X$, with $m \in \mathbb{Z}$, are sincere. This clearly implies that all but finitely many modules from $\mathcal{C}$ are also sincere. □

We have also the following interesting fact.

2.5. Corollary. *Let $A = KQ$ be a wild hereditary algebra, $\mathcal{C}$ be a regular component of $\Gamma(\operatorname{mod} A)$, P an indecomposable postprojective A-module, and I an indecomposable preinjective A-module. Then, for all but finitely many modules X from $\mathcal{C}$, we have $\operatorname{Hom}_A(P, X) \neq 0$ and $\operatorname{Hom}_A(X, I) \neq 0$.*

Proof. Because P is an indecomposable postprojective module, then there exist an $m \geq 0$ and an indecomposable projective A-module $P(i)$, with $i \in Q_0$, such that $P \cong \tau^{-m} P(i)$. Then, for each module X from $\mathcal{C}$, we have $\operatorname{Hom}_A(P, X) \cong \operatorname{Hom}_A(P(i), \tau^m X)$. It follows from (2.4) that $\operatorname{Hom}_A(P(i), Z) \neq 0$, for all but finitely many modules Z from $\mathcal{C}$, and consequently, $\operatorname{Hom}_A(P, X) \neq 0$, for all but finitely many modules X from $\mathcal{C}$. The second statement follows from the first one by applying the standard duality functor $D : \operatorname{mod} A \longrightarrow \operatorname{mod} A^{\mathrm{op}}$, or by the dual arguments. $\square$

We are now in position to prove the following important facts on homomorphisms between regular modules, due to Baer [35] and Kerner [343].

2.6. Theorem. *Let $A = KQ$ be a wild hereditary algebra and X, Y be non-zero regular A-modules. Then there exists a natural number m_0 such that*

$$\operatorname{Hom}_A(\tau^m X, Y) = 0 \quad and \quad \operatorname{Hom}_A(X, \tau^m Y) \neq 0,$$

for all $m \geq m_0$.

Proof. We recall from (IV.2.9) that

$$\dim \tau^{-m} Z = \mathbf{\Phi}_A^{-m} \dim Z,$$

for each regular A-module Z and any natural number $m \in \mathbb{N}$, where $\mathbf{\Phi}_A : K_0(A) \longrightarrow K_0(A)$ is the Coxeter transformation of A.

Let X and Y be non-zero regular A-modules. Let $\{\mathbf{x}^{(1)}, \dots, \mathbf{x}^{(r)}\}$ be the set of the dimension vectors $\mathbf{x}^{(i)}$ of all regular A-modules such that $||\mathbf{x}^{(i)}|| \leq \dim_K Y$, where $||\mathbf{z}|| = \sum_{i \in Q_0} z_i$, for any vector $\mathbf{z} \in K_0(A) = \mathbb{Z}^{|Q_0|}$. We recall from (1.7) that two non-isomorphic modules from the same regular component of $\Gamma(\operatorname{mod} A)$ have different dimension vectors. Altogether this implies that there exists an integer $m_1 \geq 1$ such that $||\mathbf{\Phi}^{-m}(\mathbf{x}^{(i)})|| > \dim_K X$, for all $m \geq m_1$ and $i = 1, \dots, r$.

Suppose now that there exists a non-zero homomorphism $f : \tau^m X \longrightarrow Y$, for some $m \geq m_1$. Consider the exact sequence

$$0 \longrightarrow L \longrightarrow \tau^m X \longrightarrow Z \longrightarrow 0,$$

where L is the kernel of f and Z is the image of f. Because Z is a regular submodule of Y then $\dim Z = \mathbf{x}^{(j)}$, for some $j \in \{1, \dots, r\}$. Hence, by applying the functor τ^{-m}, we get the induced short exact sequence

$$0 \longrightarrow \tau^{-m} L \longrightarrow X \longrightarrow \tau^{-m} Z \longrightarrow 0.$$

Because, on the other hand, we have

$$\dim_K \tau^{-m}Z = ||\mathbf{dim}\,\tau^{-m}Z|| = ||\mathbf{\Phi}_A^{-m}\mathbf{dim}\,Z|| = ||\mathbf{\Phi}_A^{-m}\mathbf{x}^{(j)}|| > \dim_K X,$$

we get a contradiction. It follows that $\mathrm{Hom}_A(\tau^m X, Y) = 0$, for all $m \geq m_1$.

To prove the second claim, we may assume without loss of generality that the module X has no non-trivial regular quotient modules. Indeed, there is an epimorphism $X \longrightarrow X'$, where X' is a regular module without non-trivial regular quotient modules, and clearly $\mathrm{Hom}_A(X', Z) \neq 0$ implies $\mathrm{Hom}_A(X, Z) \neq 0$. By (2.1), there exists an exact sequence $0 \longrightarrow X \longrightarrow \tau^m X \longrightarrow I \longrightarrow 0$, with $m > 0$ and I non-zero preinjective. Because Y is a regular A-module then, by applying (2.5) and the Auslander–Reiten formulae (IV.2.14), we conclude that there is a natural number m_2 such that

$$\mathrm{Ext}_A^1(I, \tau^s Y) \cong D\mathrm{Hom}_A(\tau^{s-1}Y, I) \neq 0,$$

for all $s \geq m_2$. Further, it follows from the first part of the proof that there exists a natural number m_3 such that

$$\mathrm{Ext}_A^1(\tau^m X, \tau^s Y) \cong D\mathrm{Hom}_A(\tau^s Y, \tau^{m+1}X) = 0,$$

for all $s \geq m_3$. Let $m_4 = \max\{m_2, m_3\}$. By applying $\mathrm{Hom}_A(-, \tau^s Y)$, with $s \geq m_4$, to the exact sequence $0 \longrightarrow X \longrightarrow \tau^m X \longrightarrow I \longrightarrow 0$ we get the induced long exact sequence

$$\cdots \longrightarrow \mathrm{Hom}_A(X, \tau^s Y) \longrightarrow \mathrm{Ext}_A^1(I, \tau^s Y) \longrightarrow \mathrm{Ext}_A^1(\tau^m X, \tau^s Y) = 0.$$

It follows that $\mathrm{Hom}_A(X, \tau^s Y) \neq 0$. Therefore, if $m_0 = \max\{m_1, m_4\}$ then $\mathrm{Hom}_A(\tau^m X, Y) = 0$ and $\mathrm{Hom}_A(X, \tau^m Y) \neq 0$, for all $m \geq m_0$. $\square$

We have also the following interesting fact.

2.7. Lemma. *Let $A = KQ$ be a wild hereditary algebra. Let X be an indecomposable regular A-module and assume that $\mathrm{Hom}_A(X, \tau^{-m}X) \neq 0$, for some $m \geq 1$. Then $\mathrm{Hom}_A(X, \tau^{-i}X) \neq 0$, for all $i \in \{1, \dots, m\}$.*

Proof. Suppose, to the contrary, that $\mathrm{Hom}_A(X, \tau^{-i}X) = 0$, for some index $i \in \{1, \dots, m\}$. Then $m \geq 2$ and there exists a $j \in \{1, \dots, m\}$ such that $\mathrm{Hom}_A(X, \tau^{-(j-1)}X) = 0$ and $\mathrm{Hom}_A(X, \tau^{-s}X) \neq 0$, for any $s \in \{j, \dots, m\}$. Then the Auslander–Reiten formulae (IV.2.13) yields

$$\mathrm{Ext}_A^1(\tau^{-j}X, X) \cong D\mathrm{Hom}_A(X, \tau^{-(j-1)}X) = 0.$$

Then, it follows from (VIII.3.3) that any non-zero homomorphism from X to $\tau^{-j}X$ is a monomorphism or an epimorphism. But this contradicts (1.2).

$\square$

It follows from (XI.2.8) and (XII.3.4) that the regular part $\mathcal{R}(A)$ of the Auslander–Reiten quiver $\Gamma(\mathrm{mod}\,A)$ of a hereditary algebra $A = KQ$ of Euclidean type is a disjoint union of a $\mathbb{P}_1(K)$-family

$$\mathcal{T}^A = \{\mathcal{T}_\lambda^A\}_{\lambda \in \mathbb{P}_1(K)}$$

of pairwise orthogonal standard stable tubes $\mathcal{T}_\lambda^A$, and hence consists entirely of nondirecting indecomposable A-modules.

2.8. Lemma. *Let $A = KQ$ be a wild hereditary algebra.*

(a) *For every pair of indecomposable regular A-modules M and N, there exists a path $M = Z_0 \longrightarrow Z_1 \longrightarrow Z_2 \longrightarrow \cdots \longrightarrow Z_{m-1} \longrightarrow Z_m = N$ in $\mathrm{mod}\,A$.*

(b) *Every indecomposable regular A-module is nondirecting.*

Proof. (a) Assume that M and N are indecomposable regular A-modules. It follows from (1.6) that N belongs to a component $\mathcal{C}$ of $\Gamma(\mathrm{mod}\,A)$ of type $\mathbb{Z}\mathbb{A}_\infty$. Then there exists a quasi-simple module X in $\mathcal{C}$ and an integer $n \geq 1$ such that $N \cong X[n]$. On the other hand, by (2.6), there exists an integer $m \geq 1$ such that $\mathrm{Hom}_A(M, \tau^m X) \neq 0$. Hence (a) follows, because there exists a path of irreducible morphisms

$$\tau^m X = \tau^m X[1] \longrightarrow \cdots \longrightarrow X[1] \longrightarrow \cdots \longrightarrow X[n] = N$$

in $\mathrm{mod}\,A$.

The statement (b) is a direct consequence of (a). $\square$

We recall that a **brick** in $\mathrm{mod}\,A$ is an indecomposable A-module X such that $\mathrm{End}_A(X) \cong K$. A **stone** is an indecomposable A-module X such that $\mathrm{Ext}_A^1(X, X) = 0$, see (XV.1.2).

We know from (VIII.3.3) that every stone in $\mathrm{mod}\,A$ is a brick. On the other hand, we have seen in (1.9) that for enlarged Kronecker algebras there are bricks (even quasi-simple regular) which are not stones. We also recall from (VIII.2.7) that all indecomposable postprojective modules and all indecomposable preinjective modules are stones.

The following proposition characterises the quasi-simple regular bricks.

2.9. Proposition. *Let $A = KQ$ be a wild hereditary algebra and X a regular brick in $\mathrm{mod}\,A$. Then X is quasi-simple if and only if $\mathrm{Hom}_A(X, \tau^{-1}X) = 0$.*

Proof. Because X is a regular brick then $\mathrm{Ext}_A^1(\tau^{-1}X, X) \cong D\mathrm{End}_A(X)$ is a one-dimensional vector space over K generated by an almost split sequence

$$0 \longrightarrow X \longrightarrow Y \longrightarrow \tau^{-1}X \longrightarrow 0.$$

Moreover, Y is indecomposable if and only if X is quasi-simple, because X lies in a component of the form $\mathbb{Z}\mathbb{A}_\infty$, by (1.6).

Assume that $\operatorname{Hom}_A(X, \tau^{-1}X) \neq 0$, and let $f : X \longrightarrow \tau^{-1}X$ be a non-zero homomorphism. We know from (1.2), that f is neither a monomorphism nor an epimorphism. Letting $M = \operatorname{Im} f$ we get a short exact sequence

$$0 \longrightarrow \operatorname{Ker} f \longrightarrow X \overset{g}{\longrightarrow} M \longrightarrow 0,$$

where $\operatorname{Ker} f \neq 0$ and M is a proper submodule of $\tau^{-1}X$. Applying the functor $\operatorname{Hom}_A(\tau^{-1}X/M, -)$ we get an epimorphism

$$\operatorname{Ext}^1_A(\tau^{-1}X/M, g) : \operatorname{Ext}^1_A(\tau^{-1}X/M, X) \longrightarrow \operatorname{Ext}^1_A(\tau^{-1}X/M, M),$$

because A is hereditary. Therefore there is a commutative diagram

$$
\begin{array}{ccccccccc}
0 & \longrightarrow & X & \overset{h'}{\longrightarrow} & N & \longrightarrow & \tau^{-1}X/M & \longrightarrow & 0 \\
& & {\scriptstyle g}\downarrow & & {\scriptstyle g'}\downarrow & & {\scriptstyle 1}\downarrow & & \\
0 & \longrightarrow & M & \overset{h}{\longrightarrow} & \tau^{-1}X & \longrightarrow & \tau^{-1}X/M & \longrightarrow & 0
\end{array}
$$

with exact rows. It follows that the exact sequence

$$0 \longrightarrow X \overset{\left[\begin{smallmatrix} g \\ -h' \end{smallmatrix}\right]}{\longrightarrow} M \oplus N \overset{[h \; g']}{\longrightarrow} \tau^{-1}X \longrightarrow 0$$

is non-split, because $\dim_K M < \dim_K X$ and $\dim_K M < \dim_K \tau^{-1}X$. Hence there exists a non-split exact sequence

$$0 \longrightarrow X \longrightarrow E \longrightarrow \tau^{-1}X \longrightarrow 0,$$

with E decomposable. But then $Y \cong E$ is decomposable, and consequently X is not quasi-simple. On the other hand, if X is not quasi-simple then there is a non-zero homomorphism from X to $\tau^{-1}X$, which is the composition of an irreducible epimorphism and an irreducible monomorphism, because X lies in a component of type $\mathbb{Z}\mathbb{A}_\infty$. $\square$

2.10. Corollary. *Let $A = KQ$ be a wild hereditary algebra and X be a quasi-simple brick in $\operatorname{mod} A$. Then $\operatorname{Hom}_A(X, \tau^{-m}X) = 0$, for all $m \geq 1$.*

Proof. Apply (2.7) and (2.9). $\square$

2.11. Lemma. *Let $A = KQ$ be a wild hereditary algebra and X an indecomposable regular A-module which is not a stone. Then $\operatorname{Hom}_A(X, \tau^m X) \neq 0$, for all $m \geq 1$.*

Proof. Because $\operatorname{Ext}^1_A(X, X) \neq 0$, we have

$$\operatorname{Hom}_A(X, \tau X) \neq 0 \quad \text{and} \quad \operatorname{Hom}_A(\tau^{-1}X, X) \neq 0.$$

To show the claim, suppose to the contrary, that there is an integer $m \geq 1$ such that $\mathrm{Hom}_A(X, \tau^{m+1}X) = 0$ and $\mathrm{Hom}_A(X, \tau^m X) \neq 0$. Then

$$\mathrm{Ext}_A^1(\tau^m X, X) \cong D\mathrm{Hom}_A(X, \tau^{m+1}X) = 0.$$

Take a non-zero homomorphism $f : X \longrightarrow \tau^m X$. It follows from (VIII.3.3) that f is a monomorphism or an epimorphism. If f is a monomorphism, then composing it with a non-zero homomorphism $g \in \mathrm{Hom}_A(\tau^{-1}X, X)$ we get a non-zero homomorphism $fg : \tau^{-1}X \longrightarrow \tau^m X$ and, consequently,

$$\mathrm{Hom}_A(X, \tau^{m+1}X) \cong \mathrm{Hom}_A(\tau^{-1}X, \tau^m X) \neq 0.$$

Similarly, if f is an epimorphism then, for a non-zero homomorphism

$$h \in \mathrm{Hom}_A(\tau^m X, \tau^{m+1}X) \cong \mathrm{Hom}_A(X, \tau X),$$

the composite homomorphism $hf \in \mathrm{Hom}_A(X, \tau^{m+1}X)$ is non-zero, and hence $\mathrm{Hom}_A(X, \tau^{m+1}X) \neq 0$. This finishes the proof. $\qquad \square$

To establish a further relationship between regular bricks and stones over wild hereditary algebras $A = KQ$ we need the following lemma.

2.12. Lemma. *Let Y be an indecomposable A-module, X be a quasi-simple regular A-module,*

$$X = X[1] \longrightarrow X[2] \longrightarrow X[3] \longrightarrow \cdots \longrightarrow X[r] \longrightarrow \cdots$$

an infinite sequence of irreducible monomorphisms (1.5), and $i \geq 1$ a natural number.

(a) *If $Y \not\cong X[i]/X[j]$, for $1 \leq j < i$, then each homomorphism $f : X[i] \longrightarrow Y$ factors through the canonical irreducible monomorphism $\varepsilon : X[i] \hookrightarrow X[i+1]$.*

(b) *If $Y \not\cong X[j]$, for $1 \leq j < i$, then each homomorphism $g : Y \longrightarrow X[i]$ factors through the canonical irreducible epimorphism $\pi : (\tau X)[i+1] \longrightarrow X[i]$.*

Proof. We prove only (a), because the proof of (b) is dual. If $i = 1$, then the irreducible monomorphism $\varepsilon : X \longrightarrow X[2]$ is a minimal left almost split morphism, and consequently any homomorphism $f : X \longrightarrow Y$ with $X \not\cong Y$ factors through ε.

Assume that $i \geq 2$. Consider the canonical almost split sequence

$$0 \longrightarrow X[i] \xrightarrow{[\varepsilon, \pi]^t} X[i+1] \oplus \tau^{-1}X[i-1] \xrightarrow{[\pi', \varepsilon']} \tau^{-1}X[i] \longrightarrow 0.$$

Then each homomorphism $f : X[i] \longrightarrow Y$, with $Y \not\cong X[i]/X[j]$, $1 \leq j < i$, factors through $[\varepsilon, \pi]^t$, that is, $f = g\varepsilon + h\pi$, for some homomorphisms

$$g : X[i{+}1] \longrightarrow Y \quad \text{and} \quad h : \tau^{-1}X[i{-}1] \longrightarrow Y.$$

By induction, we have $h = h'\varepsilon'$, for some homomorphism $h' : \tau^{-1}X[i] \longrightarrow Y$. Hence

$$f = g\varepsilon + h\pi = g\varepsilon + h'\varepsilon'\pi = g\varepsilon - h'\pi'\varepsilon = (g - h'\pi')\varepsilon,$$

as required. □

Throughout this chapter we freely use the terminology and notation introduced in Section XVI.1. In particular, given an indecomposable A-module M in a component $\mathcal{C}$ of the Auslander–Reiten quiver $\Gamma(\mathrm{mod}\,A)$ of an algebra Λ, by a cone determined by M we mean a full translation subquiver $\mathcal{C}(M)$ of $\mathcal{C}$ of the form

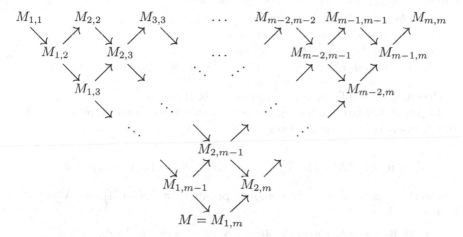

The module $M = M_{1,m}$ is called the **germ** of the cone $\mathcal{C}(M)$, see (XVI.1.4).

It is easy to see that, for a wild hereditary algebra $A = KQ$, the definition of cone coincides with the following one.

2.13. Definition. Let $A = KQ$ be a wild hereditary algebra.

(a) A full translation subquiver $\mathcal{C}$ of $\Gamma(\mathrm{mod}\,A)$ is said to be **mesh-closed** if any mesh of $\Gamma(\mathrm{mod}\,A)$ with source and target in $\mathcal{C}$ is contained in $\mathcal{C}$.

(b) Let $A = KQ$ be a wild hereditary algebra, let $\mathcal{C}$ be a regular component of $\Gamma(\mathrm{mod}\,A)$, X a quasi-simple module in $\mathcal{C}$, and $m \geq 1$ a natural number.

(b1) The mesh-closed full subquiver of $\mathcal{C}$ formed by the modules $\tau^{-i}X[j]$ (1.5), with $1 \leq i + j \leq m$, is called the **cone** with germ $X[m]$, and denoted by $\mathcal{C}(X[m])$.

(b2) The cone $C([X[m])$ is called **standard** if the full subcategory of mod A formed by the modules from $C(X[m])$ is equivalent to the mesh subcategory $KC(X[m])$ of $C(X[m])$.

We recall that a pair of A-modules X and Y in mod A is said to be **orthogonal** if $\operatorname{Hom}_A(X,Y) = 0$ and $\operatorname{Hom}_A(Y,X) = 0$.

2.14. Proposition. *Let $A = KQ$ be a wild hereditary algebra, X a quasi-simple regular A-module,*

$$X = X[1] \longrightarrow X[2] \longrightarrow \cdots \longrightarrow X[r] \longrightarrow \cdots$$

be a chain of irreducible monomorphisms (1.5) and $m \geq 2$ be a natural number. The following conditions are equivalent.

(a) *The cone $C(X[m])$ is standard.*
(b) *The module $X[m]$ is a brick.*
(c) *The module $X[m-1]$ is a stone.*
(d) *The modules $X, \tau^{-1}X, \ldots, \tau^{-(m-1)}X$ are pairwise orthogonal stones.*

Proof. The implication (a) implies (b) is obvious.

To prove that (b) implies (c), we assume that the module $X[m]$ is a brick and $X[m-1]$ is not a stone. Then

$$D\operatorname{Hom}_A(\tau^{-1}X[m-1], X[m-1]) \cong \operatorname{Ext}^1_A(X[m-1], X[m-1]) \neq 0,$$

and so there exists a non-zero homomorphism $f : \tau^{-1}X[m-1] \longrightarrow X[m-1]$. Moreover, there are

- an irreducible epimorphism $\pi : X[m] \longrightarrow \tau^{-1}X[m-1]$, and
- an irreducible monomorphism $\varepsilon : X[m-1] \longrightarrow X[m]$.

Hence $\varepsilon f \pi$ is a non-zero homomorphism in $\operatorname{rad}\operatorname{End}_A(X[m])$. This is a contradiction, because $X[m]$ is a brick.

To prove that (c) implies (d), we recall from (VIII.3.3) that every stone in mod A is a brick. Assume that $X[m-1]$ is a stone. It follows that $X[m-1]$ is a brick, and, by the implication (b)$\Rightarrow$(c) applied to the brick $X[m-1]$, the module $X[m-2]$ is a stone. An easy induction shows that the modules

$$X[m-1], \ X[m-2], \ \ldots, \ X[2], \ X[1] = X$$

are stones and bricks. In particular, it follows from (2.10) that, for each $t \geq 1$, we have $\operatorname{Hom}_A(X, \tau^{-t}X) = 0$ and, hence, $\operatorname{Hom}_A(\tau^{-i}X, \tau^{-j}X) = 0$, for all $0 \leq i < j \leq m - 1$. Moreover, the modules

$$X, \tau^{-1}X, \tau^{-2}X, \ldots, \tau^{-(m-1)}X$$

are stones. It remains to show that $\mathrm{Hom}_A(\tau^{-i}X, \tau^{-j}X) = 0$, for all $0 \leq j < i \leq m - 1$. Assume, to the contrary, that this is not the case, that is, $\mathrm{Hom}_A(\tau^{-i}X, \tau^{-j}X) \neq 0$, for some $j < i$ such that $0 \leq j, i \leq m - 1$. Then, for $r = (m-1) - i$ and $s = j + r$, we have

$$\mathrm{Hom}_A(\tau^{-(m-1)}X, \tau^{-s}X) \cong \mathrm{Hom}_A(\tau^{-i}X, \tau^{-j}X) \neq 0.$$

Let $h : \tau^{-(m-1)}X \longrightarrow \tau^{-s}X$ be a non-zero homomorphism. Applying (2.12)(b) to h, we conclude that h has a factorisation $h = h_2 h_1$, where

$$h_1 : \tau^{-(m-1)}X \longrightarrow X[s+1] \text{ and } h_2 : X[s+1] \longrightarrow \tau^{-s}X.$$

Note that $s + 1 \leq m - 1$ and, hence, there exists a monomorphism $w : X[s+1] \longrightarrow X[m-1]$. Clearly, there exists also an epimorphism $p : \tau^{-1}X[m-1] \longrightarrow \tau^{-(m-1)}X$. Consequently, the composite homomorphism $wh_1 p : \tau^{-1}X[m-1] \longrightarrow X[m-1]$ is non-zero and, hence,

$$\mathrm{Ext}_A^1(X[m-1], X[m-1]) \cong D\mathrm{Hom}_A(\tau^{-1}X[m-1], X[m-1]) \neq 0,$$

which is a contradiction. Hence, the modules $X, \tau^{-1}X, \ldots, \tau^{-(m-1)}X$ are pairwise orthogonal.

To prove that (d) implies (a), we assume, to the contrary, that (d) holds and there exists a non-zero homomorphism $f : U \longrightarrow V$, for some

$$U = \tau^{-r}X[l] \quad \text{and} \quad V = \tau^{-s}X[t],$$

with $r + l \leq m$ and $s + t \leq m$.

We show, by applying the induction on $l \geq 1$, that f is either an isomorphism and a scalar multiple of the identity map 1_U on U, or a composition of irreducible morphisms corresponding to arrows of the cone $\mathcal{C}(X[m])$.

Step 1° We assume that $l = 1$. Because the module $U = \tau^{-r}X[1]$ is a brick, we may assume that f is not an isomorphism. Then the assumption (d) forces $t \geq 2$. For each $i \in \{0, 1, \ldots, t-2\}$, we look at the canonical short exact sequences

$$0 \longrightarrow \tau^{-(s+i)}X \longrightarrow \tau^{-(s+i)}X[t-i] \xrightarrow{\pi_{i+1}} \tau^{-(s+i+1)}X[t-i-1] \longrightarrow 0,$$

with $\pi_1, \ldots, \pi_{t-1}$ irreducible morphisms. Observe that

$$\mathrm{rad}_A(U, \tau^{-(s+t-2)}X) = \mathrm{rad}_A(\tau^{-r}X, \tau^{-(s+t-2)}X) = 0,$$

because the modules $X, \tau^{-1}X, \ldots, \tau^{-(m-1)}X$ are pairwise orthogonal stones and, hence, bricks. In particular, we have $\pi_{t-1} \cdot \ldots \cdot \pi_2 \cdot \pi_1 \cdot f = 0$.

By considering the composite homomorphism

$$\pi_{i+1}\pi_i \ldots \pi_2 \pi_1 f : U \longrightarrow \tau^{-(s+i+1)}X[t-i-1],$$

for each $i \in \{0, 1, \ldots, t-2\}$, we conclude that there exists exactly one index $i \in \{0, 1, \ldots, t-2\}$ such that $\pi_{i+1} \ldots \pi_1 f = 0$ but $\pi_i \ldots \pi_1 f \neq 0$ (we put $\pi_0 = \text{id}$). Then $\text{Hom}_A(\tau^{-r} X, \tau^{-(s+i)} X) \neq 0$, or equivalently $r = s + i$. Hence $s \leq r$ and $t - (r - s) \geq 2$, because $i \in \{0, 1, \ldots, t-2\}$.

We claim that $s = r$. Suppose, to the contrary, that $s < r$. Consider the composite epimorphism

$$\pi = \pi_{r-s} \ldots \pi_1 : \tau^{-s} X[t] \longrightarrow \tau^{-r} X[t-(r-s)] = \tau^{-(s+r-s)} X[t-(r-s)]$$

and the canonical exact sequence

$$0 \longrightarrow \tau^{-r} X \longrightarrow \tau^{-r} X[t-(r-s)] \xrightarrow{\pi_{r-s+1}} \tau^{-(r+1)} X[t-(r-s)-1] \longrightarrow 0.$$

Because $\pi f = \pi_i \ldots \pi_2 \pi_1 f \neq 0$ and $\pi_{r-s+1}(\pi f) = \pi_{i+1} \pi_i \ldots \pi_2 \pi_1 f = 0$, then the image of

$$\pi f : \tau^{-r} X \longrightarrow \tau^{-r} X[t-(r-s)]$$

is contained in the submodule $\tau^{-r} X$ of $\tau^{-r} X[t-(r-s)]$ and, consequently, $\text{rad} \, \text{End}_A(\tau^{-r} X) \neq 0$. This is a contradiction, because $\tau^{-r} X$ is a brick. Therefore $r = s$, and f is a composition of irreducible monomorphisms corresponding to the arrows of a path from $\tau^{-r} X$ to $\tau^{-r} X[t]$ in $\mathcal{C}(X[m])$.

Step 2° Now we assume that $l \geq 2$, and we consider the restriction of f to $\tau^{-r} X$. If this restriction is non-zero, it follows from the assumption (d) and the earlier considerations that $r = s$, $l \leq t$, and $f : \tau^{-r} X[l] \longrightarrow \tau^{-r} X[t]$ is either an isomorphism and a scalar multiple of the identity map 1_U on U (if $l = t$), or is a composition of irreducible morphisms corresponding to the arrows of a path from $\tau^{-r} X[l]$ to $\tau^{-r} X[t]$ in $\mathcal{C}(X[m])$. Assume that the restriction of f to $\tau^{-r} X$ is zero. Then $f = g\pi$, where

- $g : \tau^{-r-1} X[l-1] \longrightarrow \tau^{-s} X[t]$ is a homomorphism , and
- $\pi : \tau^{-r} X[l] \longrightarrow \tau^{-r-1} X[l-1]$ is the canonical irreducible epimorphism.

By the inductive hypothesis, g is either an isomorphism and a scalar multiple of the identity map on $X[l-1]$, or a composition of irreducible morphisms corresponding to some arrows of the cone $\mathcal{C}(X[m])$. Hence, the homomorphism f has also this property. Therefore we have proved that the cone $\mathcal{C}(X[m])$ is standard. This finishes the proof.   □

2.15. Proposition. *Let $A = KQ$ be a wild hereditary algebra and let X be a quasi-simple regular A-module such that the chain*

$$X = X[1] \longrightarrow X[2] \longrightarrow \cdots \longrightarrow X[r] \longrightarrow \cdots,$$

see (1.5), *of irreducible monomorphisms contains a brick* $X[m]$, *for some* $m \geq 2$.

   (a) *The module* $T = X[1] \oplus \ldots \oplus X[m{-}1]$ *is a partial tilting A-module,*
   (b) $m \leq n - 1$, *where n is the rank of $K_0(A)$,*
   (c) *The cone $\mathcal{C}(X[m])$ is standard and consists of bricks,*
   (d) *All modules in the cone $\mathcal{C}(X[m])$, except the module $X[m]$, are stones.*

Proof. (a) By (2.14), the cone $\mathcal{C}(X[m])$ is standard. Because the modules $X[i]$ and $\tau^{-1}X[i]$, for $1 \leq i \leq m - 1$, lie in the standard cone $\mathcal{C}(X[m])$, we have $\mathrm{Ext}_A^1(T, T) \cong D\mathrm{Hom}_A(\tau^{-1}T, T) = 0$, and so T is a partial tilting A-module.

(b) Let $n = \mathrm{rk}\, K_0(A) = |Q_0|$ be the rank of the Grothendieck group $K_0(A)$ of A. Because, by (a), $T_A = X[1] \oplus \ldots \oplus X[m{-}1]$ is a multiplicity-free partial tilting A-module then applying (VI.2.4) and (VI.4.4) yields $m - 1 \leq n$, that is, $m \leq n + 1$. To prove that $m \leq n - 1$, we need to exclude the equalities $m = n + 1$ and $m = n$.

First, assume to the contrary, that $m - 1 = n$. Then T is a direct sum of n pairwise non-isomorphic indecomposable A-modules and, hence, is a tilting A-module, by (VI.4.4). Because the cone $\mathcal{C}(X[m])$ is standard then the tilted algebra $B = \mathrm{End}\, T_A$ is isomorphic to the hereditary path algebra $K\Delta(\mathbb{A}_{m-1})$ of the equioriented quiver

$$\Delta(\mathbb{A}_{m-1}) : \overset{1}{\circ} \longleftarrow \overset{2}{\circ} \longleftarrow \ldots \longleftarrow \overset{m-1}{\circ} .$$

Then, by (VI.5.6), T_A is a separating tilting module. Because B is representation-finite, the algebra $A = KQ$ is representation-finite. This is a contradiction, because we assume that Q is not a Dynkin quiver, see (VII.5.10).

Finally, assume to the contrary, that $m = n$. Because T_A is a multiplicity-free partial tilting A-module and T_A is a direct sum of $n-1 = m-1$ pairwise non-isomorphic indecomposable A-modules then, by (VI.2.4) and (VI.4.4), there exists an indecomposable A-module N such that $T'_A = T_A \oplus N$ is a tilting A-module and the tilting vanishing condition yields $\mathrm{Hom}_A(T'_A, \tau_A T'_A) \cong D\mathrm{Ext}_A^1(T'_A, T'_A) = 0$. Moreover, we have the torsion pair $(\mathcal{T}(T'), \mathcal{F}(T'))$ in $\mathrm{mod}\, A$, where

$$\mathcal{T}(T') = \{X_A;\ \mathrm{Ext}_A^1(T', X) = 0\} = \{X_A;\ \mathrm{Hom}_A(X, \tau_A T') = 0\},\ \text{and}$$
$$\mathcal{F}(T') = \{Y_A;\ \mathrm{Hom}_A(T', Y) = 0\}.$$

We set $M = X[m{-}1]$. We prove that the module N does not belong to the cone $\mathcal{C}(M)$ determined by M. Assume, to the contrary, that N belongs to $\mathcal{C}(M)$. Because $N \not\cong X[i]$, for any $i \in \{1, \ldots, m - 1\}$, then $m \geq 3$ and N belongs to the cone $\mathcal{C}(\tau_A^{-1}X[m{-}2])$ determined by $\tau_A^{-1}X[m{-}2]$. But then there is an epimorphism $X[j] \longrightarrow \tau_A N$, for some $j \in \{1, \ldots, m - 2\}$.

Hence, $\operatorname{Hom}_A(T'_A, \tau_A T'_A) \neq 0$ and we get a contradiction. This shows that $N \notin \mathcal{C}(M)$.

Now we note that, to finish the proof of (b), it is sufficient to show that

    (i) the tilted algebra $B' = \operatorname{End} T'_A$ is hereditary,

    (ii) T'_A is a separating tilting A-module, and hence

    (iii) any indecomposable module in $\operatorname{mod} A$ belongs to $\mathcal{T}(T')$ or to $\mathcal{F}(T')$.

Indeed, if (i) holds then (VI.5.6) yields (ii) and, hence, we get (iii). It follows that any successor of an indecomposable A-module from $\mathcal{T}(T')$ in $\operatorname{mod} A$ lies in $\mathcal{T}(T')$. Because, by (2.8), there is a path in $\operatorname{mod} A$ from the module $X[1]$ to $\tau_A X[1]$ then $\tau_A X[1]$ lies in $\mathcal{T}(T')$. On the other hand, $\operatorname{Hom}_A(T'_A, \tau_A T'_A) = 0$ implies that $\tau_A X[1]$ lies in $\mathcal{F}(T')$ and we get a contradiction. This shows that $m \leq n - 1$ and finishes the proof of (b).

Then, to complete the proof, it remains to show that (i) holds. We recall from the first part of the proof that the algebra $B = \operatorname{End} T_A$ is isomorphic to the hereditary path algebra $H_{m-1} = K\Delta(\mathbb{A}_{m-1})$ of the equioriented quiver $\Delta(\mathbb{A}_{m-1}) : \overset{1}{\circ} \leftarrow \overset{2}{\circ} \leftarrow \ldots \leftarrow \overset{m-1}{\circ}$. Further, by (VIII.3.3), the equality $\operatorname{Ext}^1_A(N, N) = 0$ yields $\operatorname{End} N_A \cong K$.

Now we show that $\operatorname{Hom}_A(N, M) = 0$ or $\operatorname{Hom}_A(M, N) = 0$. Indeed, because the tilting vanishing condition yields

$$\operatorname{Ext}^1_A(M, N) = 0 \text{ and } \operatorname{Ext}^1_A(N, M) = 0$$

then (VIII.3.3) yields

- any non-zero homomorphism $N \to M$ is a monomorphism or an epimorphism, and
- any non-zero homomorphism $M \to N$ is a monomorphism or an epimorphism,

because the algebra $A = KQ$ is hereditary. It follows that $\operatorname{Hom}_A(N, M) = 0$ or $\operatorname{Hom}_A(M, N) = 0$, because the modules M and N are non-isomorphic bricks.

Assume that $m \geq 3$. We prove that $\operatorname{Hom}_A(N, Y) = 0$, for any indecomposable module Y in the cone $\mathcal{C}(X[m-2])$. Assume, to the contrary, that there is a non-zero homomorphism $f : N \longrightarrow Y$ in $\operatorname{mod} A$, with $Y \in \mathcal{C}(X[m-2])$. Because $\mathcal{C}(X[m-2]) \subseteq \mathcal{C}(X[m-1]) = \mathcal{C}(M)$ and $N \notin \mathcal{C}(M)$ then, by (IV.5.1), there exist $i \in \{1, \ldots, m-1\}$ and a chain of irreducible morphisms

$$\tau_A X[i] = Y_t \overset{g_t}{\longrightarrow} Y_{t-1} \overset{g_{t-1}}{\longrightarrow} Y_{t-1} \longrightarrow \ldots \longrightarrow Y_1 \overset{g_1}{\longrightarrow} Y_0 = Y.$$

and a homomorphism $h : N \longrightarrow Y_t$ in $\operatorname{mod} A$ such that $g_1 \cdot \ldots \cdot g_t \cdot h \neq 0$. This shows that $D\operatorname{Ext}^1_A(X[i], N) \cong \operatorname{Hom}_A(N, \tau_A X[i]) \neq 0$ and we get a contradiction. Consequently, $\operatorname{Hom}_A(N, Y) = 0$, for any indecomposable module $Y \in \mathcal{C}(X[m-2])$.

Because the cone $\mathcal{C}(M)$ is standard then the additive category $\operatorname{add}\mathcal{C}(M)$ is equivalent to the module category $\operatorname{mod} H_{m-1}$, where $H_{m-1} = K\Delta(\mathbb{A}_{m-1})$. Hence, for each $j \in \{2, \ldots, m-1\}$, there exists a short exact sequence

$$0 \longrightarrow X[j-1] \xrightarrow{u_j} X[j] \longrightarrow E_j \longrightarrow 0,$$

where u_j is a canonical irreducible monomorphism and $E_j = \tau_A^{-(j-1)}X[1]$. By applying the functor $\operatorname{Hom}_A(-, N)$, we derive an exact sequence

$$0 \longrightarrow \operatorname{Hom}_A(E_j, N) \longrightarrow \operatorname{Hom}_A(X[j], N) \longrightarrow \operatorname{Hom}_A(X[j-1], N)$$
$$\longrightarrow \operatorname{Ext}_A^1(E_j, N) \longrightarrow 0.$$

Because $\tau_A^{-(j-2)}X[1] \in \mathcal{C}(X[m-2])$ then

$$\operatorname{Ext}_A^1(E_j, N) \cong D\operatorname{Hom}_A(N, \tau_A E_j) \cong D\operatorname{Hom}_A(N, \tau_A^{-(j-2)}X[1]) = 0.$$

Moreover, because there exists an epimorphism $\tau_A^{-1}X[j-1] \longrightarrow E_j$ then the isomorphisms

$$\operatorname{Hom}_A(\tau_A^{-1}X[j-1], N) \cong D\operatorname{Ext}_A^1(N, X[j-1]) = 0$$

force $\operatorname{Hom}_A(E_j, N) = 0$ and, for each $j \in \{2, \ldots, m-1\}$, we derive an isomorphism

$$\operatorname{Hom}_A(u_j, N) : \operatorname{Hom}_A(X[j], N) \xrightarrow{\ \cong\ } \operatorname{Hom}_A(X[j-1], N).$$

Assume that $\operatorname{Hom}_A(N, M) = 0$. Then $\operatorname{Hom}_A(N, T) = 0$. Moreover, there is an isomorphism $\operatorname{Hom}_A(X[i], N) \cong \operatorname{Hom}_A(M, N)$, for each $i \in \{1, \ldots, m-1\}$. Further, by (VI.3.5), the tilted algebra $B' = \operatorname{End} T_A' = \operatorname{End}_A(T \oplus N)$ is connected. It follows that $\operatorname{Hom}_A(M, N) \neq 0$ and B' is isomorphic to the path algebra of the quiver

where $r = \dim_K \operatorname{Hom}_A(M, N)$.

Now assume that $\operatorname{Hom}_A(M, N) = 0$. Then $\operatorname{Hom}_A(T, N) = 0$, because there is an isomorphism

$$\operatorname{Hom}_A(X[i], N) \cong \operatorname{Hom}_A(M, N),$$

for any $i \in \{1, \ldots, m-1\}$. Moreover there is an isomorphism $\operatorname{Hom}_A(N, T) \cong \operatorname{Hom}_A(N, M)$, for each $i \in \{1, \ldots, m-1\}$. Further, by (VI.3.5), the tilted algebra $B' = \operatorname{End} T'_A = \operatorname{End}_A(T \oplus N)$ is connected. It follows that $\operatorname{Hom}_A(N, M) \neq 0$ and B' is isomorphic to the path algebra of the quiver

$$\Delta': \qquad \underset{1}{\circ} \longleftarrow \underset{2}{\circ} \longleftarrow \cdots \longleftarrow \underset{m-2}{\circ} \longleftarrow \underset{m-1}{\circ} \overset{\alpha_1}{\underset{\alpha_{r'}}{\overset{\alpha_2}{\longrightarrow}}} \circ \, \omega'$$

where $r' = \dim_K \operatorname{Hom}_A(N, M)$. It follows that, in each of the two cases, the tilted algebra B' is hereditary. This finishes the proof of (b).

Because the statements (c) and (d) are direct consequences of the standardness of $\mathcal{C}(X[m])$, the proof of the proposition is complete. □

2.16. Corollary. *Let $A = KQ$ be a wild hereditary algebra and X be a quasi-simple regular A-module such that the chain*

$$X = X[1] \longrightarrow X[2] \longrightarrow \cdots \longrightarrow X[r] \longrightarrow \cdots,$$

see (1.5), of irreducible monomorphisms contains a stone $X[r]$, for some $r \geq 1$. If n is the rank of $K_0(A)$, then $r \leq n - 2$.

Proof. Apply (2.14) and (2.15). □

It follows from (2.16) that if A is a wild hereditary algebra with two simple modules (that is, A is an enlarged Kronecker algebra) then every stone is either postprojective or preinjective. The same holds for the tame hereditary algebra with two simple modules (that is, for the Kronecker algebra).

The following example shows that, for $n \geq 3$, the upper bound $n - 2$ on the positions of regular stones given in (2.14) is the best possible.

2.17. Example. Let $n \geq 3$ be a positive integer. Consider the wild quiver

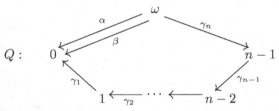

Denote by $\Delta = \Delta(1, n)$ the subquiver of Q obtained from Q by deleting the arrow β, and by Σ the subquiver

$$\Delta(\mathbb{A}_n): \qquad \underset{\gamma_1}{\overset{0}{\circ} \longleftarrow} \underset{\gamma_2}{\overset{1}{\circ} \longleftarrow} \overset{3}{\circ} \longleftarrow \cdots \longleftarrow \underset{\gamma_{n-1}}{\overset{n-2}{\circ} \longleftarrow} \overset{n-1}{\circ}.$$

of Q obtained from Q by removing the vertex ω and the arrows α, β, γ_n. The path algebras of the quivers Q, Δ and Σ have the forms

$$A = KQ \cong \begin{bmatrix} K & 0 & \dots & 0 \\ K & K & \dots & 0 \\ \vdots & \vdots & \ddots & \vdots \\ K^3 & K & \dots & K \end{bmatrix}$$

$$\Lambda = K\Delta \cong \begin{bmatrix} K & 0 & \dots & 0 \\ K & K & \dots & 0 \\ \vdots & \vdots & \ddots & \vdots \\ K^2 & K & \dots & K \end{bmatrix}$$

$$H = K\Sigma \cong \begin{bmatrix} K & 0 & \dots & 0 \\ K & K & \dots & 0 \\ \vdots & \vdots & \ddots & \vdots \\ K & K & \dots & K \end{bmatrix}.$$

Clearly, there are canonical algebra surjections $A \longrightarrow \Lambda \longrightarrow H$, Λ is the hereditary canonical algebra $C(1, n)$ of the Euclidean type $\widetilde{\mathbb{A}}_n$, and H is a hereditary algebra of the Dynkin type $\mathbb{A}_n$. Note also that A is the one-point extension $A = H[M]$ of H by the projective H-module

$$M = P(0) \oplus P(0) \oplus P(n{-}1).$$

We know from (XII.2.8) that $\Gamma(\operatorname{mod}\Lambda)$ contains a standard stable tube $\mathcal{T}$ of rank $n \geq 1$ such that the simple modules $S(1), \dots, S(n{-}1)$ are mouth modules in $\mathcal{T}$. Moreover, it follows from the standardness of $\mathcal{T}$ that, for each $i \in \{1, \dots, n-1\}$, the module $S(i)[n{-}1]$ (respectively, $S(i)[n]$) is a stone (respectively, brick), and clearly $n - 1 = (n + 1) - 2$, where $n + 1$ is the rank of the Grothendieck group $K_0(\Lambda)$ of Λ.

Observe now that the simple modules $S(1), \dots, S(n{-}1)$ are also regular A-modules. Indeed, these modules are non-directing modules in $\operatorname{mod}\Lambda$, hence in $\operatorname{mod} A$ and, consequently, they are neither postprojective nor preinjective.

Further, the Auslander–Reiten quiver $\Gamma(\operatorname{mod} H)$ of H is of the form

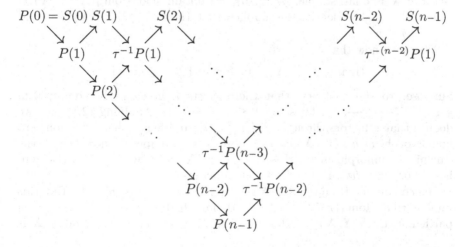

Let $X = S(1)$, $Y = \tau^{-1}P(n-2)$, and denote by $\mathcal{C}$ the full translation subquiver of $\Gamma(\operatorname{mod} H)$ formed by all modules except $P(0), \dots, P(n-1)$, that is, by all non-projective modules.

Observe that $\operatorname{Hom}_H(M, Z) = 0$, for any indecomposable H-module Z in the cone $\mathcal{C}(\tau^{-1}P(n-3))$. Then, by (XV.1.7), $\mathcal{C}$ is a mesh-closed translation subquiver of $\Gamma(\operatorname{mod} A)$. Moreover, $X = S(1)$ is a quasi-simple regular A-module, because $S(1)$ is simple. Therefore, $\mathcal{C}$ is a mesh-closed translation subquiver of the cone $\mathcal{C}(X[n])$ of a component of type $\mathbb{Z}\mathbb{A}_\infty$ in $\Gamma(\operatorname{mod} A)$, and $Y = X[n-1]$. Clearly, there are isomorphisms $\operatorname{Ext}^1_A(X[n-1], X[n-1]) \cong \operatorname{Ext}^1_H(Y, Y) = 0$, and so $X[n-1]$ is a stone in $\operatorname{mod} A$ and $n-1 = (n+1) - 2$, with $n+1$, being the rank of $K_0(A)$. Moreover, by (2.14), $X[n]$ is a brick.

Finally, we note that, although $X[n-1]$ is not a sincere A-module, by (2.3), all but finitely many modules from the family $\tau^r X[n-1] = (\tau^r X)[n]$, with $r \in \mathbb{Z}$, are sincere stones.

We have also the following consequence of (2.14).

2.18. Lemma. *Let $A = KQ$ be a wild hereditary algebra, X be a quasi-simple regular A-module and $Y = X[m]$, for some $m \geq 2$. Then Y is a brick if and only if $\operatorname{Hom}_A(Y, X) = 0$.*

Proof. If Y is a brick, then it follows from (2.14) that the cone $\mathcal{C}(Y) = \mathcal{C}(X[m])$ is standard, and consequently $\operatorname{Hom}_A(Y, X) = 0$.

To prove the inverse implication, we assume that $\operatorname{Hom}_A(Y, X) = 0$. We prove that the A-modules $X, \tau^{-1}X, \dots, \tau^{-(m-1)}X$ are pairwise orthogonal stones. This implies, by (2.14), that $Y = X[m]$ is a brick.

It follows from (2.12)(a) that any non-zero non-isomorphism $f : X \longrightarrow X$ factors through $Y = X[m]$, and so our assumption implies that X is a brick. Because X is quasi-simple, by (2.10), we obtain that $\operatorname{Hom}_A(X, \tau^{-r}X) = 0$, for all $r \geq 1$. Clearly, this implies that $\operatorname{Hom}_A(\tau^{-i}X, \tau^{-j}X) = 0$, for $0 \leq i < j \leq m - 1$.

We prove now that

$$\operatorname{Hom}_A(\tau^{-s}X, X) = 0, \text{ for all } 1 \leq s \leq m - 1.$$

Suppose, to the contrary, that there exists a non-zero homomorphism $g : \tau^{-t}X \longrightarrow X$, for some $1 \leq t \leq m - 1$. Applying (2.13)(a), we deduce that g factors through $(\tau^{-t}X)[m-t]$, and hence there is a non-zero homomorphism $h : (\tau^{-t}X)[m-t] \longrightarrow X$. Composing now h with the canonical epimorphism $\pi : Y \longrightarrow (\tau^{-t}X)[m-t]$ we obtain a non-zero homomorphism $h\pi : Y \longrightarrow X$, a contradiction.

Hence, we get $\operatorname{Hom}_A(\tau^{-s}X, X) = 0$, for all $1 \leq s \leq m - 1$. But this implies that $\operatorname{Hom}_A(\tau^{-i}X, \tau^{-j}X) = 0$, for all $0 \leq j < i \leq m - 1$. In particular, $\operatorname{Ext}^1_A(X, X) = D\operatorname{Hom}_A(\tau^{-1}X, X) = 0$, that is, the module X is a stone. Consequently, the modules $X, \tau^{-1}X, \ldots, \tau^{-(m-1)}X$ are pairwise orthogonal stones. This finishes the proof. □

We end our discussion on the distribution of regular stones with the following proposition.

2.19. Proposition. *Assume that $A = KQ$, where Q is an acyclic connected wild quiver with at least three vertices. Then $\Gamma(\operatorname{mod} A)$ admits infinitely many pairwise different regular components containing sincere quasi-simple stones.*

Proof. Observe first that, if an indecomposable regular A-module X is a stone then its whole τ-orbit $\tau^m X$, with $m \in \mathbb{Z}$, consists of stones and, by (2.3), infinitely many of them are sincere. Moreover, it follows from (2.15) that if a regular component $\mathcal{C}$ of $\Gamma(\operatorname{mod} A)$ contains a stone then $\mathcal{C}$ contains a quasi-simple stone.

Therefore, it remains to show that $\Gamma(\operatorname{mod} A)$ admits infinitely many pairwise different regular components containing stones. Because Q has at least three vertices and the underlying graph $\overline{Q}$ of Q is neither Dynkin nor Euclidean, there exists a full connected proper subquiver Q' of Q such that $A' = KQ'$ is representation-infinite. Consider the postprojective component $\mathcal{P}(A')$ of $\Gamma(\operatorname{mod} A')$. We know that $\mathcal{P}(A')$ consists of directing modules, and hence of stones. Obviously, these indecomposable A'-modules are non-sincere stones in the category of A-modules.

On the other hand, it follows from (IX.5.6) and its dual that all but finitely many indecomposable postprojective A-modules are sincere and all but finitely many indecomposable preinjective A-modules are sincere. Hence, all but finitely many modules from $\mathcal{P}(A')$ are regular stones.

Finally, note that, by (2.4), all but finitely many modules of any given regular component of $\Gamma(\operatorname{mod} A)$ are sincere. Clearly, then the required claim follows. □

XVIII.3. Perpendicular categories

In Chapter XV we have associated to a given K-algebra A and an A-module X the one-point extension algebra $A[X]$ and the one-point coextension algebra $[X]A$ having one more simple module than the algebra A has. Here, we associate to X two subcategories $X^{\perp}$ and $^{\perp}X$ of $\operatorname{mod} A$, and we show that under some assumptions they are again module categories of some algebras having less simple modules than the algebra A has.

3.1. Definition. Let A be an arbitrary (not necessarily hereditary) K-algebra and X an A-module. Then the full subcategory of $\operatorname{mod} A$

$$X^{\perp} = \{M_A \mid \operatorname{Hom}_A(X, M) = 0 = \operatorname{Ext}^1_A(X, M)\}$$

is said to be the **right perpendicular subcategory** of X, and the full
subcategory of $\operatorname{mod} A$

$$X^\perp = \{M_A \mid \operatorname{Hom}_A(M, X) = 0 = \operatorname{Ext}^1(M, X)\}$$

is said to be **the left perpendicular category** of X.

3.2. Example. Let A be an algebra, e an idempotent of A, $P = eA$
the corresponding projective A-module and $I = D(Ae)$ the corresponding
injective A-module. Then clearly we have $P^\perp \cong \operatorname{mod} A/AeA \cong {}^\perp I$. In
particular, if P is the new indecomposable projective module over the one-
point extension algebra $B[X]$ then $P^\perp = \operatorname{mod} B$. Similarly, if I is the new
indecomposable injective module over the one-point coextension algebra
$[X]B$ then ${}^\perp I = \operatorname{mod} B$.

3.3. Example. Let A be a hereditary algebra and X an A-module
without non-zero projective direct summands. Then $X^\perp = {}^\perp(\tau X)$. This
follows from the functorial isomorphisms

$$\operatorname{Hom}_A(M, \tau X) \cong D\operatorname{Ext}^1_A(X, M) \quad \text{and} \quad \operatorname{Ext}^1_A(M, \tau X) \cong D\operatorname{Hom}_A(X, M)$$

established in (IV.2.14).

3.4. Example. Let $A = KQ$ be the path algebra of the quiver

$$Q: \quad \begin{array}{c} 1 \\ \nwarrow \\ \\ \searrow \\ 2 \end{array} \nearrow\!\!\!\searrow \; 3 \longleftarrow 4 \longrightarrow 5 \longrightarrow 6$$

of Dynkin type $\mathbb{D}_6$. Then $\Gamma(\operatorname{mod} A)$ is of the form

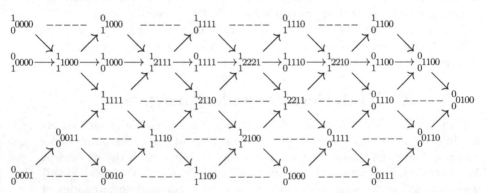

where the indecomposable modules are replaced by their dimension vectors.

Now we describe the categories $X^\perp$ and ${}^\perp X$, for three particular choices
of the module X.

Case 1° We take for X the indecomposable A-module with $\mathbf{dim}\, X = {}^1_1 2221$.
Because the algebra A is hereditary, we have

$$\mathrm{Ext}^1_A(X,-) \cong D\mathrm{Hom}_A(-,\tau X) \quad \text{and} \quad \mathrm{Ext}^1_A(-,X) \cong D\mathrm{Hom}_A(\tau^{-1}X,-).$$

Then a direct calculation shows that $X^\perp$ consists of direct sums of the following indecomposable modules

$$\,^1_0 1111$$

$$\,^0_1 1111$$

$$\,^1_1 2110$$

$$\,^1_1 1110 \qquad\qquad \,^1_1 2100$$

$$\,^0_0 0010 \qquad\qquad \,^1_1 1100 \qquad\qquad \,^0_0 1000$$

while $\,^\perp X$ consists of direct sums of indecomposable modules

$$\,^0_1 1110$$

$$\,^1_0 1110$$

$$\,^1_1 2211$$

$$\,^1_1 2100 \qquad\qquad \,^0_0 1111$$

$$\,^1_1 1100 \qquad\qquad \,^0_0 1000 \qquad\qquad \,^0_0 0111.$$

Observe that there exist equivalences of categories $X^\perp \cong \mathrm{mod}\, KQ' \cong \,^\perp X$, where Q' is the (non-connected) quiver

$$1 \qquad 2 \qquad 3 \longleftarrow 4 \longleftarrow 5.$$

<u>Case 2°</u> Now take for X the simple A-module $S(3)$ at the vertex 3. Then the category $S(3)^\perp$ consists of direct sums of all indecomposable A-modules except those of dimension vectors

$$\,^1_0 0000, \quad \,^0_1 0000, \quad \,^1_1 1000, \quad \,^1_1 1111, \quad \,^1_1 1110, \quad \,^1_1 1100, \quad \,^0_0 1000, \quad \,^0_0 1111, \quad \,^0_0 1110, \quad \,^0_0 1100.$$

Hence, we get an equivalence of categories $S(3)^\perp \cong \mathrm{rep}_K(\Delta) \cong \mathrm{mod}\, K\Delta$, where Δ is the quiver

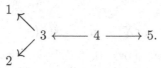

Similarly, the category ${}^{\perp}S(3)$ consists of direct sums of all indecomposable A-modules except those of dimension vectors

$$
{}^{1}_{1}1000,\quad {}^{0}_{1}1000,\quad {}^{1}_{0}1000,\quad {}^{1}_{1}2111,\quad {}^{1}_{1}2110,\quad {}^{1}_{1}2100,\quad {}^{0}_{0}1000,\quad {}^{0}_{0}0111,\quad {}^{0}_{0}0110,\quad {}^{0}_{0}0100,
$$

and again we have an equivalence of categories ${}^{\perp}S(3) \cong \operatorname{rep}_K(\Delta) \cong \operatorname{mod} K\Delta$.

Case 3° Finally, take for X the injective A-module $I = I(1) \oplus I(2) \oplus I(3)$. We already know that
$$
{}^{\perp}I \cong \operatorname{mod} KQ'',
$$
where Q'' is the full subquiver of Q formed by the vertices 4, 5 and 6.

On the other hand, a direct calculation shows that the category $I^{\perp}$ consists of direct sums of indecomposable modules of dimension vectors

$$
{}^{0}_{0}0001,\quad {}^{0}_{0}1000,\quad {}^{0}_{0}1111,\quad {}^{0}_{0}1110,\quad {}^{0}_{0}0111,\quad {}^{0}_{0}0110,
$$

and that $I^{\perp} \cong \operatorname{rep}_K(\Gamma)$, where Γ is the quiver $1 \longleftarrow 2 \longrightarrow 3$.

Note that, in this case, the category $I^{\perp}$ is not equivalent to ${}^{\perp}I$, and the category $I^{\perp}$ is not the dual category of ${}^{\perp}I$.

3.5. Example. Let A be the path algebra KQ of the four subspace quiver

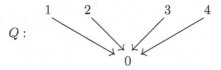

The indecomposable A-modules have been described in Chapter XIII. In particular, it was shown that $\Gamma(\operatorname{mod} A)$ consists of the following components:

- the postprojective component $\mathcal{P}(A)$ is of the form

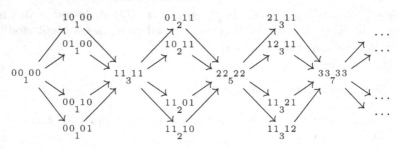

- the preinjective component $\mathcal{Q}(A)$ is of the form

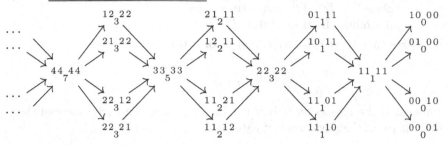

- the stable standard tubes $\mathcal{T}_0^A$, $\mathcal{T}_1^A$, $\mathcal{T}_\infty^A$ of rank 2 are of the forms, see (XIII.3.17),

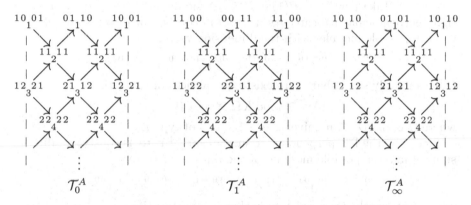

where the indecomposable modules are represented by their dimension vectors.

- For each $\lambda \in K \setminus \{0,1\}$, the tube $\mathcal{T}_\lambda^A$ of rank one whose simple regular module has the dimension vector $h_Q = {}^{11}{}_2{}^{11}$.

Now we describe the categories $X^\perp$ and $^\perp X$, for two particular choices of the module X.

Case 1° We take for X the direct sum of three indecomposable modules U, V, W of dimension vectors ${}^{10}{}_1{}^{01}$, ${}^{11}{}_1{}^{00}$, and ${}^{10}{}_1{}^{10}$, respectively, lying in different tubes of rank 2. Then the right perpendicular subcategory

$$X^\perp = \{M \mid \operatorname{Hom}_A(X, M) = 0 = \operatorname{Ext}_A^1(X, M)\}$$
$$= \{M \mid \operatorname{Hom}_A(X, M) = 0 = \operatorname{Hom}_A(M, \tau X)\}$$

to X consists of direct sums of the following indecomposable modules:

- the postprojective modules $\tau^{-2m} P(1)$, with $m \geq 0$,
- the preinjective modules $\tau^{2m+1} I(1)$, with $m \geq 0$,

- the modules $(\tau U)[2m]$, $(\tau V)[2m]$, $(\tau W)[2m]$, with $m \geq 1$, from the tubes $\mathcal{T}_0^A$, $\mathcal{T}_1^A$, $\mathcal{T}_\infty^A$, respectively, and
- all modules from the tubes $\mathcal{T}_\lambda^A$, with $\lambda \in K \setminus \{0,1\}$.

It follows that there exist equivalences of categories

$$X^\perp \cong \operatorname{mod} \begin{bmatrix} K & 0 \\ K^2 & K \end{bmatrix} \cong \operatorname{rep}_K(\Delta),$$

where Δ is the Kronecker quiver $1 \circ \overset{\longleftarrow}{\underset{\longleftarrow}{\rule{0pt}{0pt}}} \circ 2$. Similarly, one can show that there exist equivalences of categories

$$^\perp X \cong \operatorname{mod} \begin{bmatrix} K & 0 \\ K^2 & K \end{bmatrix} \cong \operatorname{rep}_K(\Delta).$$

<u>Case 2°</u> Take now $Y = P(1) \oplus I(2)$. A simple analysis shows that the right perpendicular subcategory $Y^\perp$ to Y consists of direct sums of six indecomposable modules with the dimension vectors

$$01_1 00, \quad 00_0 10, \quad 00_0 01, \quad 01_1 11, \quad 01_1 10, \quad 01_1 01.$$

One can easily show that there exist equivalences of categories

$$Y^\perp \cong \operatorname{mod} K\Omega \cong \operatorname{rep}_K(\Omega),$$

where Ω is the Dynkin quiver $2 \longrightarrow 1 \longleftarrow 3$ of type $\mathbb{A}_3$.

Similarly, the left perpendicular subcategory $^\perp Y$ to Y consists of direct sums of indecomposable modules of the dimension vectors

$$00_1 10, \quad 00_1 01, \quad 10_2 11, \quad 10_1 01, \quad 10_1 10, \quad 10_0 00,$$

and, consequently, there exist equivalences of categories

$$^\perp Y \cong \operatorname{mod} K\Omega^{\mathrm{op}} \cong \operatorname{rep}_K(\Omega^{\mathrm{op}}).$$

3.6. Lemma. *Let A be an arbitrary (not necessarily hereditary) algebra, X an A-module and $0 \longrightarrow M' \longrightarrow M \longrightarrow M'' \longrightarrow 0$ an exact sequence in $\operatorname{mod} A$.*

(a) *If $\operatorname{pd}_A X \leq 1$ then $X^\perp$ is an abelian subcategory of $\operatorname{mod} A$ which is closed under extensions. Moreover, if two of the modules M', M, M'' belong to $X^\perp$, the third one also belongs to $X^\perp$.*

(b) *If $\operatorname{id}_A X \leq 1$ then $^\perp X$ is an abelian subcategory of $\operatorname{mod} A$ which is closed under extensions. Moreover, if two of the modules M', M, M'' belong to $^\perp X$, the third one also belongs to $^\perp X$.*

Proof. We prove only (a), because the proof of (b) is dual. Assume that $\operatorname{pd}_A X \leq 1$. Then, for any short exact sequence

$$0 \longrightarrow M' \longrightarrow M \longrightarrow M'' \longrightarrow 0$$

in mod A, we get the exact sequence

$$0 \longrightarrow \operatorname{Hom}_A(X, M') \longrightarrow \operatorname{Hom}_A(X, M) \longrightarrow \operatorname{Hom}_A(X, M'')$$
$$\longrightarrow \operatorname{Ext}^1_A(X, M') \longrightarrow \operatorname{Ext}^1_A(X, M)$$
$$\longrightarrow \operatorname{Ext}^1_A(X, M'') \longrightarrow \operatorname{Ext}^2_A(X, M') = 0.$$

Hence, if two of the modules M', M, M'' belong to $X^\perp$, the third one also belongs to $X^\perp$. Let $f : M \longrightarrow N$ be a homomorphism in mod A with M and N from $X^\perp$. Then we have two exact sequences

$$0 \longrightarrow \operatorname{Ker} f \longrightarrow M \longrightarrow \operatorname{Im} f \longrightarrow 0$$
$$0 \longrightarrow \operatorname{Im} f \longrightarrow N \longrightarrow \operatorname{Coker} f \longrightarrow 0.$$

The equalities $\operatorname{Hom}_A(X, M) = 0 = \operatorname{Hom}_A(X, N)$ yield $\operatorname{Hom}_A(X, \operatorname{Ker} f) = 0$ and $\operatorname{Hom}_A(X, \operatorname{Im} f) = 0$. But then from the above long exact sequence applied to $M' = \operatorname{Ker} f$ and $M'' = \operatorname{Im} f$ we get $\operatorname{Ext}^1_A(X, \operatorname{Ker} f) = 0 = \operatorname{Ext}^1_A(X, \operatorname{Im} f)$, because $\operatorname{Ext}^1_A(X, M) = 0$. Hence, we conclude that $\operatorname{Ker} f$, $\operatorname{Im} f$, and hence also $\operatorname{Coker} f$, belong to $X^\perp$. This finishes the proof. □

3.7. Proposition. *Let A be an arbitrary (not necessarily hereditary) algebra and T a partial tilting A-module. Then the inclusion functor $T^\perp \hookrightarrow \operatorname{mod} A$ admits a canonical left adjoint functor $p_T : \operatorname{mod} A \longrightarrow T^\perp$, that is, for each module $N \in T^\perp$ and each module M in $\operatorname{mod} A$, there is a functorial isomorphism of K-vector spaces*

$$\operatorname{Hom}_A(p_T(M), N) \xrightarrow{\simeq} \operatorname{Hom}_A(M, N).$$

In particular, $p_T(N) \cong N$, for each module N in $T^\perp$.

Proof. Consider the torsion class $\operatorname{Gen}(T)$ in mod A consisting of all A-modules generated by T (VI.2.3), and denote by t the associated idempotent radical such that $\operatorname{Gen}(T) = \{M \mid tM = M\}$, see (VI.1.4). Let M be an A-module and $\dim_K \operatorname{Ext}^1_A(T, M) = d$. We show that there exists a universal exact sequence

$$0 \longrightarrow M \longrightarrow U_T(M) \longrightarrow T^d \longrightarrow 0$$

with $\operatorname{Ext}^1_A(T, U_T(M)) = 0$. If $d = 0$ we put $U_T(M) = M$. For $d \geq 1$ we get the sequence by a slight generalisation of the Bongartz lemma (VI.2.4) as follows. Let $\varepsilon_1, \ldots, \varepsilon_d$ be a basis of the K-vector space $\operatorname{Ext}^1_A(T, M)$, and represent each ε_i by a short exact sequence

$$0 \longrightarrow M \xrightarrow{f_i} E_i \xrightarrow{g_i} T \longrightarrow 0.$$

Consider the commutative diagram

$$
\begin{array}{ccccccccc}
0 & \longrightarrow & M^d & \xrightarrow{\ f\ } & \overset{d}{\underset{i=1}{\bigoplus}} E_i & \xrightarrow{\ g\ } & T^d & \longrightarrow & 0 \\
& & {\scriptstyle k}\downarrow & & {\scriptstyle h}\downarrow & & {\scriptstyle 1}\downarrow & & \\
(*) & & 0 \ \longrightarrow \ M & \xrightarrow{\ v\ } & U_T(M) & \xrightarrow{\ w\ } & T^d & \longrightarrow & 0,
\end{array}
$$

with exact rows, where

$$
f = \begin{bmatrix} f_1 & & 0 \\ & \ddots & \\ 0 & & f_d \end{bmatrix}, \qquad g = \begin{bmatrix} g_1 & & 0 \\ & \ddots & \\ 0 & & g_d \end{bmatrix},
$$

and $k = [1, \dots, 1]$ is the codiagonal homomorphism. Denote by ε the element of $\mathrm{Ext}_A^1(T^d, M)$ represented by the lower sequence $(*)$. Moreover, for each $i \in \{1, \dots, d\}$, let $u_i : T \longrightarrow T^d$ be the inclusion homomorphism in the ith coordinate. Then $\varepsilon_i = \varepsilon u_i$, for each $i \in \{1, \dots, d\}$. Applying $\mathrm{Hom}_A(T, -)$ to $(*)$ we derive an exact sequence

$$
\cdots \to \mathrm{Hom}_A(T, T^d) \xrightarrow{\ \delta\ } \mathrm{Ext}_A^1(T, M) \to \mathrm{Ext}_A^1(T, U_T(M)) \to \mathrm{Ext}_A^1(T, T^d) = 0.
$$

Because $\varepsilon_i = \varepsilon u_i = \delta(u_i)$, each basis element of $\mathrm{Ext}_A^1(T, M)$ lies in the image of the connecting homomorphism δ, which is therefore surjective. Hence, we get $\mathrm{Ext}_A^1(T, U_T(M)) = 0$, as required. For each M in $\mathrm{mod}\,A$, we put

$$
p_T(M) = U_T(M)/t(U_T(M)).
$$

Because $t(p_T(M)) = 0$, then we get $\mathrm{Hom}_A(T, p_T(M)) = 0$. Moreover,

$$
\{X \mid tX = X\} = \mathrm{Gen}(T) \subseteq \mathcal{T}(T) = \{X \mid \mathrm{Ext}_A^1(T, X) = 0\},
$$

by (VI.2.3), and $\mathrm{Ext}_A^1(T, t(U_T(M))) = 0$. Hence, $\mathrm{Ext}_A^1(T, U_T(M)) = 0$ forces $\mathrm{Ext}_A^1(T, p_T(M)) = 0$, and consequently $p_T(M) \in T^\perp$.

Denote by $f_M : M \longrightarrow p_T(M)$ the composite homomorphism

$$
M \xrightarrow{\ v\ } U_T(M) \xrightarrow{\ \pi\ } p_T(M),
$$

where π is the canonical epimorphism. Let X be a module in $T^\perp$. Observe that the canonical exact sequence

$$
0 \longrightarrow t(U_T(M)) \longrightarrow U_T(M) \xrightarrow{\ \pi\ } p_T(M) \longrightarrow 0
$$

induces an exact sequence

$$
0 \longrightarrow \mathrm{Hom}_A(p_T(M), X) \longrightarrow \mathrm{Hom}_A(U_T(M), X) \longrightarrow \mathrm{Hom}_A(t(U_T(M)), X) = 0.
$$

Applying $\mathrm{Hom}_A(-, X)$ to the exact sequence $(*)$ we get an exact sequence

$$0 = \mathrm{Hom}_A(T^d, X) \to \mathrm{Hom}_A(U_T(M), X) \to \mathrm{Hom}_A(M, X) \to \mathrm{Ext}^1_A(T^d, X) = 0.$$

Therefore, the homomorphism $f_M : M \longrightarrow p_T(M)$ induces an isomorphism

$$\mathrm{Hom}_A(f_M, X) : \mathrm{Hom}_A(p_T(M), X) \xrightarrow{\sim} \mathrm{Hom}_A(M, X).$$

In particular, for each A-module N, we have a commutative diagram

$$\begin{array}{ccc}
\mathrm{Hom}_A(p_T(M), N) & \xrightarrow{\mathrm{Hom}_A(f_M, N)} & \mathrm{Hom}_A(M, N) \\[4pt]
{\scriptstyle \mathrm{Hom}_A(p_T(M), f_N)}\Big\downarrow \cong & & {\scriptstyle \mathrm{Hom}_A(M, f_N)}\Big\downarrow \cong \\[6pt]
\mathrm{Hom}_A(p_T(M), p_T(N)) & \xrightarrow{\mathrm{Hom}_A(f_M, p_T(N))} & \mathrm{Hom}_A(M, p_T(N)),
\end{array}$$

where the lower horizontal map is an isomorphism. Then we may assign to each homomorphism $g : M \longrightarrow N$ in $\mathrm{mod}\, A$ the unique homomorphism $p_T(g) : p_T(M) \longrightarrow p_T(N)$ such that $p_T(g)f_M = f_N g$. Clearly, then p_T becomes a K-linear functor from $\mathrm{mod}\, A$ to $T^\perp$. Moreover, for each $N \in T^\perp$, we have a natural isomorphism of K-vector spaces

$$\mathrm{Hom}_A(p_T(M), N) \xrightarrow{\;\sim\;} \mathrm{Hom}_A(M, N),$$

and consequently p_T is left adjoint to the inclusion functor $T^\perp \hookrightarrow \mathrm{mod}\, A$. $\quad\square$

3.8. Corollary. *Let A be an arbitrary (not necessarily hereditary) algebra, T be a partial tilting A-module, and let $p_T : \mathrm{mod}\, A \longrightarrow T^\perp$ be as in (3.7). Then the functor*

$$\mathrm{Hom}_A(p_T(A), \ ) : T^\perp \longrightarrow \mathrm{mod}\,\mathrm{End}_A(p_T(A))$$

is an equivalence of categories.

Proof. For any A-module N, the vector space $\mathrm{Hom}_A(p_T(A), N)$ has a natural structure of a right module over $\mathrm{End}_A(p_T(A))$. We know from (3.6) that $T^\perp$ is an abelian subcategory of $\mathrm{mod}\, A$ which is closed under extensions. Further, by (3.7), the functor $\mathrm{Hom}_A(p_T(A), -) : T^\perp \longrightarrow \mathrm{mod}\, K$ is equivalent to the restriction of $\mathrm{Hom}_A(A, -)$ to $T^\perp$. It follows that $p_T(A)$ is a projective object of the category $T^\perp$.

We claim that, for any module N in $T^\perp$, there is an epimorphism

$$p_T(A)^r \longrightarrow N,$$

with $r \geq 0$. To see this, we choose an integer $r \geq 0$ and an epimorphism $f : A^r \longrightarrow N$. Because the module N lies in $T^\perp$, it follows from (3.7) that $p_T(N) \cong N$. Moreover, the functor $p_T : \mathrm{mod}\, A \longrightarrow T^\perp$ is right exact and, hence, the homomorphism $p_T(f) : p_T(A^r) \longrightarrow p_T(N)$ is surjective. This proves our claim, because $p_T(A^r) \cong p_T(A)^r$ and $p_T(N) \cong N$.

Now, by applying the next result to the category $\mathcal{A} = T^\perp$, we conclude that the functor $\mathrm{Hom}_A(p_T(A), -) : T^\perp \longrightarrow \mathrm{mod}\,\mathrm{End}_A(p_T(A))$ is an equivalence of categories. $\quad\square$

3.9. Proposition. *Let A be a K-algebra and let $\mathcal{A}$ be an abelian full subcategory of* $\operatorname{mod} A$ *containing a projective object P such that, for any module N of $\mathcal{A}$, there exist an $s \geq 0$ and an epimorphism $P^s \longrightarrow N$ in* $\operatorname{mod} A$. *Then the K-linear exact functor*

$$\operatorname{Hom}_A(P, -) : \mathcal{A} \longrightarrow \operatorname{mod} \operatorname{End}_A(P)$$

is an equivalence of categories.

Proof. Let $B = \operatorname{End}_A(P)$. For any right A-module M, the K-vector space $\operatorname{Hom}_A(P, M)$ has a structure of right B-module defined by the formula $(f \cdot \varphi)(p) = f(\varphi(p))$, for $f \in \operatorname{Hom}_A(P, M)$, $\varphi \in \operatorname{End}_A(P)$, and $p \in P$. In view of (A.2.5) in [11, Appendix], it suffices to prove that the functor $\operatorname{Hom}_A(P, -)$ is full, faithful, and dense.

To show that the functor $\operatorname{Hom}_A(P, -)$ is faithful, assume that $\varphi : M \to N$ is a homomorphism in $\mathcal{A}$ such that the induced B-module homomorphism $\operatorname{Hom}_A(P, \varphi) : \operatorname{Hom}_A(P, M) \longrightarrow \operatorname{Hom}_A(P, N)$ is zero. By our assumption, there is an epimorphism $h : P^s \longrightarrow \operatorname{Im} \varphi$ of A-modules, where $s \geq 1$. By the projectivity of P, there is an A-module homomorphism $h' : P^s \longrightarrow M$ such that $h = \varphi h'$. This shows that $h = \varphi h' = \operatorname{Hom}_A(P, \varphi)(h') = 0$. It follows that $\operatorname{Im} \varphi = 0$ and therefore $\varphi = 0$. Consequently, the functor $\operatorname{Hom}_A(P, -)$ is faithful.

To show that the functor $\operatorname{Hom}_A(P, -)$ is full, we prove that the K-linear map

$$H_{M,N} : \operatorname{Hom}_A(M, N) \longrightarrow \operatorname{Hom}_B(\operatorname{Hom}_A(P, M), \operatorname{Hom}_A(P, N)),$$

given by $\varphi \mapsto \operatorname{Hom}_A(P, \varphi)$, is surjective, for all modules M and N in $\mathcal{A}$. We note that the map $H_{M,N}$ is injective, because $\operatorname{Hom}_A(P, -)$ is faithful.

The surjectivity of $H_{M,N}$ is obvious when M is a direct sum of finitely many copies of P, because then the dimensions of both sides are equal.

Assume that M and N are arbitrary modules in $\mathcal{A}$ and let f be an element of $\operatorname{Hom}_B(\operatorname{Hom}_A(P, M), \operatorname{Hom}_A(P, N))$. By our assumption, there are exact sequences

$$P^r \xrightarrow{h_1} P^s \xrightarrow{h_0} M \longrightarrow 0 \quad \text{and} \quad P^u \xrightarrow{h'_1} P^v \xrightarrow{h'_0} N \longrightarrow 0$$

in $\operatorname{mod} A$ (and in $\mathcal{A}$). It follows that there is a commutative diagram

$$
\begin{array}{ccccccc}
\operatorname{Hom}_A(P, P^r) & \xrightarrow{\operatorname{Hom}_A(P,h_1)} & \operatorname{Hom}_A(P, P^s) & \xrightarrow{\operatorname{Hom}_A(P,h_0)} & \operatorname{Hom}_A(P, M) & \longrightarrow & 0 \\
\downarrow{\scriptstyle f_1} & & \downarrow{\scriptstyle f_0} & & \downarrow{\scriptstyle f} & & \\
\operatorname{Hom}_A(P, P^u) & \xrightarrow{\operatorname{Hom}_A(P,h'_1)} & \operatorname{Hom}_A(P, P^v) & \xrightarrow{\operatorname{Hom}_A(P,h'_0)} & \operatorname{Hom}_A(P, N) & \longrightarrow & 0
\end{array}
$$

with exact rows in mod B. The B-module homomorphisms f_0 and f_1 do exist, because the B-modules of the form $\mathrm{Hom}_A(P, P^i)$ on the left of the diagram are free. Because the map $H_{M,N}$ is bijective, for any A-module M in add P, then there exist A-homomorphisms $\varphi_1 : P^r \longrightarrow P^u$ and $\varphi_0 : P^s \longrightarrow P^v$ such that $f_1 = \mathrm{Hom}_A(P, \varphi_1)$ and $f_0 = \mathrm{Hom}_A(P, \varphi_0)$. Because the above diagram is commutative and the functor $\mathrm{Hom}_A(P, -)$ is faithful, then the diagram

$$
\begin{array}{ccccccc}
P^r & \xrightarrow{\ h_1\ } & P^s & \xrightarrow{\ h_0\ } & M & \longrightarrow & 0 \\
\downarrow{\varphi_1} & & \downarrow{\varphi_0} & & & & \\
P^u & \xrightarrow{\ h_1'\ } & P^v & \xrightarrow{\ h_0'\ } & N & \longrightarrow & 0
\end{array}
$$

with exact rows is commutative, and there is an A-module homomorphism $\varphi : M \longrightarrow N$ such that $\varphi h_0 = h_0' \varphi_0$. Hence it easily follows that $f = \mathrm{Hom}_A(P, \varphi)$ and consequently that the functor $\mathrm{Hom}_A(P, -)$ is full.

Finally, we show that the functor $\mathrm{Hom}_A(P, -)$ is dense. Assume that X is a module in mod B and fix an exact sequence

$$
B^r \xrightarrow{\ h\ } B^s \longrightarrow X_B \longrightarrow 0
$$

in mod B. Because the map $H_{M,N}$ is bijective, for any pair of A-modules M and N in $\mathcal{A}$, then the composite B-module homomorphism

$$
\mathrm{Hom}_A(P, P^r) \cong B^r \xrightarrow{\ h\ } B^s \cong \mathrm{Hom}_A(P, P^s)
$$

is of the form $\mathrm{Hom}_A(P, d')$, where $d' \in \mathrm{Hom}_A(P^r, P^s)$. By our assumption on $\mathcal{A}$, the image $\mathrm{Im}\, d'$ of d' belongs to $\mathcal{A}$ and, hence, the module $M = P^s/\mathrm{Im}\, d'$ also belongs to $\mathcal{A}$. Moreover, there is a commutative diagram

$$
\begin{array}{ccccccc}
B^r & \xrightarrow{\ \ h\ \ } & B^s & \longrightarrow & X_B & \longrightarrow & 0 \\
\cong\downarrow & & \cong\downarrow & & & & \\
\mathrm{Hom}_A(P, P^r) & \xrightarrow{\mathrm{Hom}_A(P,d')} & \mathrm{Hom}_A(P, P^s) & \longrightarrow & \mathrm{Hom}_A(P, M) & \longrightarrow & 0
\end{array}
$$

in mod B, with exact rows and bijective vertical homomorphisms, because P is projective in $\mathcal{A}$. It follows that there is a B-module isomorphism

$$
g : X_B \xrightarrow{\ \sim\ } \mathrm{Hom}_A(P, M)
$$

completing the above diagram to a commutative right hand square. This finishes the proof. $\qquad\square$

For hereditary algebras we have even more precise information on the right perpendicular categories induced by partial tilting modules.

3.10. Theorem. *Let A be a hereditary algebra and T be a partial tilting module.*

  (i) *There exist a hereditary algebra B and a K-linear equivalence of categories $T^{\perp} \xrightarrow{\;\cong\;} \operatorname{mod} B$.*

  (ii) *$\operatorname{rk} K_0(A) = \operatorname{rk} K_0(B) + r$, where $r \geq 1$ is the number of pairwise non-isomorphic indecomposable direct summands of T and $\operatorname{rk} K_0(A)$, $\operatorname{rk} K_0(B)$ are the ranks of the Grothendieck groups $K_0(A)$ and $K_0(B)$ of A and B.*

Proof. It follows from (3.8) that there exists an equivalence of categories $T^{\perp} \xrightarrow{\;\cong\;} \operatorname{mod} B$, where $B = \operatorname{End}_A(p_T(A))$. We claim that B is hereditary. Because $T^{\perp}$ is an abelian category, the notions of an Ext-projective object in $T^{\perp}$ and a relative projective object in $T^{\perp}$ coincide. Let X be a submodule of $p_T(A)$. Then, for each module N from $T^{\perp}$, we have an exact sequence

$$0 = \operatorname{Ext}_A^1(p_T(A), N) \longrightarrow \operatorname{Ext}_A^1(X, N) \longrightarrow \operatorname{Ext}_A^2(p_T(A)/X, N) = 0,$$

because A is hereditary and $p_T(A)$ is projective in $T^{\perp}$. Hence X is Ext-projective in $T^{\perp}$. Therefore, any submodule of $p_T(A)$ is a projective object of $T^{\perp}$, and so $B = \operatorname{End}_A(p_T(A))$ is hereditary.

To prove the second statement, we may assume that $T = T_1 \oplus \ldots \oplus T_r$, where $T_i \not\cong T_j$, for $i \neq j$. Moreover, because $\operatorname{Ext}_A^1(T, T) = 0$ and A is hereditary, we may also assume that $T_1, \ldots, T_r$ are ordered in such a way that $\operatorname{Hom}_A(T_i, T_j) \neq 0$ implies $i \leq j$, see (VIII.3.3) and (VIII.3.4). We apply induction on $r \geq 1$.

First we assume that $r = 1$, that is, T is indecomposable. If $T = eA$, for some primitive idempotent e of A, then there is an equivalence of categories

$$T^{\perp} \cong \operatorname{mod} A/AeA$$

and our claim follows.

Next we assume that T is not projective and consider the universal exact sequence

$$0 \longrightarrow A \longrightarrow U_T(A) \longrightarrow T^d \longrightarrow 0,$$

where $d = \dim_K \operatorname{Ext}_A^1(T, A)$. We know from (VI.2.4) that $T \oplus U_T(A)$ is a tilting A-module and there exists an exact sequence

$$0 \to \operatorname{Hom}_A(T, A) \to \operatorname{Hom}_A(T, U_T(A)) \to \operatorname{Hom}_A(T, T^d) \xrightarrow{\;\delta\;} \operatorname{Ext}_A^1(T, A) \to 0.$$

Because T is an indecomposable module and $\operatorname{Ext}_A^1(T, T) = 0$, then (VIII.3.3) yields a K-algebra isomorphism $\operatorname{End}_A(T) \cong K$, and so δ is an isomorphism. Moreover, we have $\operatorname{Hom}_A(T, A) = 0$, because A is hereditary and T is not projective. Therefore, we get $\operatorname{Hom}_A(T, U_T(A)) = 0$, and so $U_T(A) = p_T(A) \in T^\perp$. Clearly, then $K_0(B)$ has rank $n - 1$, where $n = \operatorname{rk} K_0(A)$.

Now we assume that $r \geq 2$. From the first part of the proof we know that there is an equivalence of categories $T_r^\perp \cong \operatorname{mod} H$, where H is a hereditary algebra such that $K_0(H) \cong \mathbb{Z}^{n-1}$. If we take $V = T_1 \oplus \ldots \oplus T^{r-1}$, then V is a partial tilting module belonging to $T_r^\perp$. Observe also that

$$\{X \in T_r^\perp; \ \operatorname{Hom}_A(V, X) = 0 = \operatorname{Ext}_A^1(V, X)\} = 0\}$$
$$= \{X \in \operatorname{mod} A \mid \operatorname{Hom}_A(T, X) = 0 = \operatorname{Ext}_A^1(T, X)\}$$
$$= T^\perp \cong \operatorname{mod} B.$$

Therefore, by induction, the rank $\operatorname{rk} K_0(B)$ of the Grothendieck group $K_0(B)$ equals

$$\operatorname{rk} K_0(B) = (n - 1) - (r - 1) = n - r = \operatorname{rk} K_0(A) - r,$$

and (ii) follows. $\square$

We have seen in (3.4) that the right perpendicular category $X^\perp$ of a stone X over a hereditary algebra is a module category of not necessarily connected (hereditary) algebra. Further, as the example (3.5) shows, the right perpendicular category of a partial tilting module over a hereditary algebra of Euclidean type can be a module category of a representation-finite algebra.

We show later that this cannot happen for the perpendicular categories of quasi-simple regular stones over connected representation-infinite hereditary algebras.

From now on we assume that A is a representation-infinite connected hereditary algebra, n is the rank of $K_0(A)$, and X is a quasi-simple regular stone in $\operatorname{mod} A$.

Clearly, then $n \geq 3$, by (2.16). Our main objective is to prove that $X^\perp$ has the same representation type as $\operatorname{mod} A$. We know from (3.10) that $X^\perp$ is equivalent to a module category $\operatorname{mod} C$, where C is a hereditary algebra with $K_0(C)$ of rank $n - 1$. Moreover, there are pairwise non-isomorphic indecomposable modules $Y_1, \ldots, Y_{n-1}$ such that

$$Y = Y_1 \oplus \ldots \oplus Y_{n-1}$$

is a projective object of $X^\perp$, $C = \operatorname{End}_A(Y)$, and the functor $\operatorname{Hom}_A(Y, -)$ induces an equivalence of categories

$$\operatorname{Hom}_A(Y, -) : X^\perp \xrightarrow{\ \cong\ } \operatorname{mod} C.$$

We identify $X^\perp$ with $\operatorname{mod} C$. Then the Auslander–Reiten translations τ_C and τ_C^{-1} induce functors $\tau_C, \tau_C^{-1} : X^\perp \longrightarrow X^\perp$.

We have the following lemma.

3.11. Lemma. *Let A be a representation-infinite connected hereditary algebra, X be a quasi-simple regular stone in $\operatorname{mod} A$,*

$$0 \longrightarrow \tau_A X \longrightarrow Z \longrightarrow X \longrightarrow 0$$

an almost split sequence in $\operatorname{mod} A$, and Y the module defined above. Then

(a) *the module Z is an object of $X^{\perp}$, and*
(b) *the module $T_A = X \oplus Y$ is a tilting A-module.*

Proof. (a) Applying $\operatorname{Hom}_A(X, -)$ to the given almost split sequence ending at X, we obtain the exact sequence

$$0 \longrightarrow \operatorname{Hom}_A(X, \tau_A X) \longrightarrow \operatorname{Hom}_A(X, Z) \longrightarrow \operatorname{Hom}_A(X, X)$$
$$\xrightarrow{\cong} \operatorname{Ext}_A^1(X, \tau_A X) \longrightarrow \operatorname{Ext}_A^1(X, Z) \longrightarrow \operatorname{Ext}_A^1(X, X) = 0,$$

where $\operatorname{Hom}_A(X, \tau_A X) \cong D\operatorname{Ext}_A^1(X, X) = 0$ and $\operatorname{Ext}_A^1(X, \tau_A X) \cong D\operatorname{End}(X) \cong K$, because X is a stone and, hence, a brick. Therefore, we have $\operatorname{Hom}_A(X, Z) = 0$ and $\operatorname{Ext}_A^1(X, Z) = 0$, and so $Z \in X^{\perp}$.

(b) First observe that the module $T = X \oplus Y$ is a direct sum of n pairwise non-isomorphic indecomposable A-modules. By our assumption $\operatorname{Ext}_A^1(X, X) = 0$. Next we note that $\operatorname{Ext}_A^1(Y, Y) = 0$ and $\operatorname{Ext}_A^1(Y, Z) = 0$, because Y is a projective object of $X^{\perp}$ and $Z \in X^{\perp}$.

Because $Y \in X^{\perp}$, then $\operatorname{Ext}_A^1(X, Y) = 0$. Finally, the given almost split sequence $0 \longrightarrow \tau_A X \longrightarrow Z \longrightarrow X \longrightarrow 0$ induces an exact sequence

$$0 = \operatorname{Ext}_A^1(Y, Z) \longrightarrow \operatorname{Ext}_A^1(Y, X) \longrightarrow \operatorname{Ext}_A^2(Y, \tau_A X) = 0$$

and, therefore, $\operatorname{Ext}_A^1(Y, X) = 0$. It follows that $\operatorname{Ext}_A^1(T, T) = 0$ and $T = X \oplus Y$ is a tilting A-module. $\qquad \square$

The tilting A-module $T = X \oplus Y$ in (3.11) induces the torsion pair $(\mathcal{F}, \mathcal{T})$ in $\operatorname{mod} A$ with
$$\mathcal{F} = \mathcal{F}(T) = \{M; \ \operatorname{Hom}_A(T, M) = 0\} = \operatorname{Cogen}(\tau T) \text{ and}$$
$$\mathcal{T} = \mathcal{T}(T) = \{M; \ \operatorname{Ext}_A^1(T, M) = 0\} = \operatorname{Gen}(T),$$
see (VI.2.5). Denote by g_T the torsion radical such that $\mathcal{T} = \{M \mid g_T M = M\}$. We may then define a functor $\tau_{\mathcal{T}} : \mathcal{T} \longrightarrow \mathcal{T}$ which assigns to each module M in $\mathcal{T}$ the module $g_T \tau_A M$. The following lemma shows that $X^{\perp} \subseteq \mathcal{T}$ and so we may investigate the relationship between τ_C, $\tau_{\mathcal{T}}$ and τ_A.

3.12. Lemma. *Let A be a representation-infinite connected hereditary algebra, M be an A-module, $T = X \oplus Y$ the tilting module in (3.11) and $(\mathcal{F}, \mathcal{T})$ the torsion pair in $\operatorname{mod} A$ defined above. Then $M \in \mathcal{T}$ if and only if $\operatorname{Ext}_A^1(X, M) = 0$. In particular, $X^{\perp}$ is a full subcategory of $\mathcal{T}$.*

Proof. If M is a module in $\mathcal{T} = \{N \mid \operatorname{Ext}^1_A(T,N) = 0\}$ then obviously $\operatorname{Ext}^1_A(X,M) = 0$. Conversely, assume that $\operatorname{Ext}^1_A(X,M) = 0$. Denote by $g_X M$ the largest submodule of M generated by X. Because $\operatorname{pd}_A X \leq 1$ then, by applying $\operatorname{Hom}_A(X,-)$ to the exact sequence

$$0 \longrightarrow g_X M \longrightarrow M \longrightarrow M/g_X M \longrightarrow 0,$$

we derive an exact sequence

$$0 \to \operatorname{Hom}_A(X, g_X M) \overset{\alpha}{\longrightarrow} \operatorname{Hom}_A(X,M) \longrightarrow \operatorname{Hom}_A(X, M/g_X M)$$
$$\longrightarrow \operatorname{Ext}^1_A(X, g_X M) \longrightarrow \operatorname{Ext}^1_A(X,M) \longrightarrow \operatorname{Ext}^1_A(X, M/g_X M) \to 0,$$

where α is an isomorphism, $\operatorname{Ext}^1_A(X, g_X M) = 0$, by (VI.2.3) (because $g_X M$ is in $\operatorname{Gen}(X)$) and $\operatorname{Ext}^1_A(X,M) = 0$, by our assumption. Hence we obtain

$$\operatorname{Hom}_A(X, M/g_X M) = 0 = \operatorname{Ext}^1_A(X, M/g_X M),$$

and so $M/g_X M \in X^\perp$. Because Y is a projective object in $X^\perp$ we then have $\operatorname{Ext}^1_A(Y, M/g_X M) = \operatorname{Ext}^1_C(Y, M/g_X M) = 0$. Moreover, there exists an epimorphism $X^r \longrightarrow g_X M$, for some $r \geq 1$, and hence we get an exact sequence

$$\operatorname{Ext}^1_A(Y, X^r) \longrightarrow \operatorname{Ext}^1_A(Y, g_X M) \longrightarrow 0.$$

Because $X \oplus Y$ is a tilting module then $\operatorname{Ext}^1_A(Y,X) = 0$ and we get $\operatorname{Ext}^1_A(Y, g_X M) = 0$. This, together with $\operatorname{Ext}^1_A(Y, M/g_X M) = 0$, implies $\operatorname{Ext}^1_A(Y,M) = 0$. Therefore, $\operatorname{Ext}^1_A(T,M) = \operatorname{Ext}^1_A(X \oplus Y, M) = 0$, and the module M belongs to $\mathcal{T} = \mathcal{T}(T)$. $\qquad\square$

For M and N from $X^\perp$, we have the following chain of natural isomorphisms
$$\operatorname{Hom}_C(N, \tau_C M) \overset{\simeq}{\longrightarrow} D\operatorname{Ext}^1_C(M,N) \overset{\simeq}{\longrightarrow} D\operatorname{Ext}^1_A(M,N) \overset{\simeq}{\longrightarrow} \operatorname{Hom}_A(N, \tau_A M)$$
of K-vector spaces. Hence there exists a natural homomorphism of A-modules

$$\Theta_M : \tau_C M \longrightarrow \tau_A M$$

inducing an isomorphism $\operatorname{Hom}_C(N, \tau_C M) \overset{\simeq}{\longrightarrow} \operatorname{Hom}_A(N, \tau_A M)$ of K-vector spaces, for any A-module $N \subset X^\perp$.

3.13. Lemma. *Let $T = X \oplus Y$ and $(\mathcal{F}, \mathcal{T})$ be as in (3.11). If $M \in X^\perp$ then the image of the natural homomorphism $\Theta_M : \tau_C M \longrightarrow \tau_A M$ is the module $\tau_\mathcal{T} M$, and the kernel of Θ_M is isomorphic to*

$$\tau_A X \otimes_K D\operatorname{Ext}^1_A(M,X) \cong (\tau_A X)^r,$$

where $r = \dim_K \operatorname{Ext}^1_A(M,X)$.

Proof. Because $\tau_C M \in X^\perp \subseteq \mathcal{T}$ then we have $\operatorname{Im} \Theta_M \subseteq g_\mathcal{T} \tau_A M = \tau_\mathcal{T} M$. Conversely, for each $N \in X^\perp$, we have an isomorphism

$$\operatorname{Hom}_C(N, \tau_C M) \xrightarrow{\simeq} \operatorname{Hom}_A(N, \tau_A M)$$

induced by Θ_M, and so any homomorphism from N to $\tau_A M$ factors through $\operatorname{Im} \Theta_M$. Because $T = X \oplus Y$ is an epimorphic image of the module $Z \oplus Y$ in $X^\perp$, we easily conclude that $\operatorname{Im} \Theta_M = g_\mathcal{T} \tau_A M = \tau_\mathcal{T} M$.

To prove the second claim, consider the universal exact sequence

$$\eta_M : 0 \longrightarrow (\tau_A X)^r \xrightarrow{\xi_M} E \xrightarrow{\varrho_M} \tau_\mathcal{T} M \longrightarrow 0,$$

where $r = \dim_K \operatorname{Ext}^1_A(\tau_\mathcal{T} M, \tau_A X) = \dim_K D\operatorname{Ext}^1_A(\tau_\mathcal{T} M, \tau_A X)$.

We recall that if the exact sequences

$$\eta_i : 0 \longrightarrow \tau_A X \xrightarrow{f_i} E_i \xrightarrow{g_i} \tau_\mathcal{T} M \longrightarrow 0,$$

viewed as elements of $\operatorname{Ext}^1_A(\tau_\mathcal{T} M, \tau_A X)$, form a K-basis of $\operatorname{Ext}^1_A(\tau_\mathcal{T} M, \tau_A X)$ then the sequence η_M is the upper sequence in the following commutative diagram

$$
\begin{array}{ccccccccc}
0 & \longrightarrow & (\tau_A X)^r & \longrightarrow & E & \longrightarrow & \tau_\mathcal{T} M & \longrightarrow & 0 \\
& & \downarrow{\scriptstyle \mathrm{id}} & & \downarrow{\scriptstyle u} & & \downarrow{\scriptstyle v} & & \\
0 & \longrightarrow & (\tau_A X)^r & \xrightarrow{f} & \overset{r}{\underset{i=1}{\bigoplus}} E & \xrightarrow{g} & (\tau_\mathcal{T} M)^r & \longrightarrow & 0,
\end{array}
$$

with exact rows, where

$$f = \begin{bmatrix} f_1 & & 0 \\ & \ddots & \\ 0 & & f_d \end{bmatrix}, \qquad g = \begin{bmatrix} g_1 & & 0 \\ & \ddots & \\ 0 & & g_d \end{bmatrix},$$

and $v = [1, \dots, 1]^t$ is the diagonal homomorphism. Applying $\operatorname{Hom}_A(X, -)$, we obtain the exact sequence

$$0 \longrightarrow \operatorname{Hom}_A(X, (\tau_A X)^r) \longrightarrow \operatorname{Hom}_A(X, E) \longrightarrow \operatorname{Hom}_A(X, \tau_\mathcal{T} M)$$
$$\xrightarrow{\partial} \operatorname{Ext}^1_A(X, (\tau_A X)^r) \longrightarrow \operatorname{Ext}^1_A(X, E) \longrightarrow \operatorname{Ext}^1_A(X, \tau_\mathcal{T} M) \longrightarrow 0,$$

where

$$\operatorname{Hom}_A(X, (\tau_A X)^r) \xrightarrow{\simeq} D\operatorname{Ext}^1_A(X, X)^r = 0,$$
$$\dim_K \operatorname{Hom}_A(X, \tau_\mathcal{T} M) = \dim_K D\operatorname{Ext}^1_A(\tau_\mathcal{T} M, \tau_A X) = r,$$
$$\dim_K \operatorname{Ext}^1_A(X, (\tau_A X)^r) = \dim_K D\operatorname{Hom}_A(X, X^r) = r,$$

because X is a brick (being a stone), and $\mathrm{Ext}^1_A(X, \tau_{\mathcal{T}}M) = 0$, because $\tau_{\mathcal{T}}M$ is in $\mathcal{T} = \mathcal{T}(X \oplus Y)$. Moreover, it follows from the construction of η_M that ∂ is an isomorphism. Therefore, we obtain $\mathrm{Hom}_A(X, E) = 0$ and $\mathrm{Ext}^1_A(X, E) = 0$, that is, E is an object of $X^\perp$. Let $N \in X^\perp$. Applying $\mathrm{Hom}_A(N, -)$ to the sequence η_M yields the exact sequence

$$0 = \mathrm{Hom}_A(N, (\tau_A X)^r) \longrightarrow \mathrm{Hom}_A(N, E)$$
$$\longrightarrow \mathrm{Hom}_A(N, \tau_{\mathcal{T}}M) \longrightarrow \mathrm{Ext}^1_A(N, (\tau_A X)^r) = 0,$$

because $\mathrm{Ext}^1_A(N, \tau_A X) \cong D\mathrm{Hom}_A(X, N) = 0$. This gives a sequence

$$\mathrm{Hom}_A(N, E) \overset{\simeq}{\longrightarrow} \mathrm{Hom}_A(N, \tau_{\mathcal{T}}M) \overset{\simeq}{\longrightarrow} \mathrm{Hom}_A(N, \tau_A M) \overset{\simeq}{\longrightarrow} \mathrm{Hom}_C(N, \tau_C M),$$

of isomorphisms, which is moreover functorial in $N \in X^\perp$. Consequently, there is an isomorphism $E \cong \tau_C M$ in $X^\perp = \mathrm{mod}\, C$ whose composition with Θ_M is the homomorphism $\varrho_M : E \longrightarrow \tau_{\mathcal{T}}M$ occurring in the exact sequence η_M. Hence we get an isomorphism $\mathrm{Ker}\, \Theta_M \cong (\tau_A X)^r$. To finish the proof we observe that there are isomorphisms

$$D\mathrm{Ext}^1_A(\tau_{\mathcal{T}}M, \tau_A X) \cong \mathrm{Hom}_A(X, \tau_{\mathcal{T}}M) \cong \mathrm{Hom}_A(X, \tau_A M) \cong D\mathrm{Ext}^1_A(M, X)$$

and, consequently, $\mathrm{Ker}\, \Theta_M \cong (\tau_A X)^r \cong \tau_A X \otimes_K D\mathrm{Ext}^1_A(M, X)$. $\square$

3.14. Lemma. *Let $T = X \oplus Y$ and $(\mathcal{F}, \mathcal{T})$ be as in (3.11).*

(a) *For any module $M \in \mathcal{T}$ with $\mathrm{Hom}_A(M, X) = 0$, we have $\tau_{\mathcal{T}}M = \tau_A M$.*

(b) *For any module $M \in X^\perp$ with $\mathrm{Ext}^1_A(M, Z) = 0$, the epimorphism $\Theta'_M : \tau_C M \longrightarrow \tau_{\mathcal{T}}M$ induced by $\Theta_M : \tau_C M \longrightarrow \tau_A M$ is an isomorphism.*

Proof. (a) By applying the Auslander–Reiten formulae (IV.2.14), we get the isomorphisms

$$\mathrm{Ext}^1_A(X, \tau_A M) \cong D\mathrm{Hom}_A(\tau_A M, \tau_A X) \cong D\mathrm{Hom}_A(M, X) = 0$$

and, hence, $\tau_A M \in \mathcal{T}$, by (3.12). Hence $\tau_{\mathcal{T}}M = g_{\mathcal{T}}\tau_A M = \tau_A M$.

(b) By (3.13), there is an exact sequence

$$0 \longrightarrow (\tau_A X)^r \overset{u}{\longrightarrow} \tau_C M \overset{\Theta'_M}{\longrightarrow} \tau_{\mathcal{T}}M \longrightarrow 0.$$

We show that $u = 0$. By applying the functor $\mathrm{Hom}_A(-, \tau_C M)$ to the almost split sequence $0 \longrightarrow \tau_A X \longrightarrow Z \longrightarrow X \longrightarrow 0$, we derive the exact sequence

$$0 \longrightarrow \mathrm{Hom}_A(X, \tau_C M) \longrightarrow \mathrm{Hom}_A(Z, \tau_C M)$$
$$\longrightarrow \mathrm{Hom}_A(\tau_A X, \tau_C M) \longrightarrow \mathrm{Ext}^1_A(X, \tau_C M).$$

Because $M \in X^{\perp} = \operatorname{mod} C$ then $\tau_C M \in X^{\perp}$ and, hence, $\operatorname{Hom}_A(X, \tau_C M) = 0$ and $\operatorname{Ext}^1_A(X, \tau_C M) = 0$. Thus the above exact sequence yields

$$\operatorname{Hom}_A(\tau_A X, \tau_C M) \cong \operatorname{Hom}_A(Z, \tau_C M) = \operatorname{Hom}_C(Z, \tau_C M)$$
$$\cong D\operatorname{Ext}^1_C(M, Z) = D\operatorname{Ext}^1_A(M, Z) = 0.$$

It follows that the homomorphism u is zero, and therefore the epimorphism Θ'_M is an isomorphism. □

We are now in position to prove the following important fact on the right perpendicular categories of quasi-simple regular stones.

3.15. Theorem. *Let A be a representation-infinite connected hereditary algebra, X a quasi-simple regular stone in $\operatorname{mod} A$, and C a basic hereditary algebra such that $X^{\perp} \cong \operatorname{mod} C$. Then*

(a) *the algebra C is connected and representation-infinite,*
(b) *C is of Euclidean type if and only if A is of Euclidean type,*
(c) *the module $[2]X$ is a quasi-simple regular brick in $\operatorname{mod} C$.*

Proof. (a) It follows from (XI.3.5) and (2.15) that there exists $r \geq 2$ such that $X[r-1]$ is a stone, but $X[r]$ is not a stone in $\operatorname{mod} A$. Then $X[r]$ is a brick and the cone $\mathcal{C}(X[r])$ in the regular component of $\Gamma(\operatorname{mod} A)$ containing X is standard, see (XI.3.4), (XVII.1.6) and (2.14). Then the module $[r-1]X = \tau_A^{r-2} X[r-1]$ is a stone and the module $[r]X = \tau_A^{r-1} X[r]$ is a brick but not a stone. There is a chain

$$[r]X \longrightarrow [r-1]X \longrightarrow \cdots \longrightarrow [1]X = X$$

of irreducible epimorphisms in $\operatorname{mod} A$, and these modules, as well as $\tau_A X$, belong to the standard cone $\mathcal{C}([r]X) = \mathcal{C}((\tau^{r-1}X)[r])$. Hence, we conclude that

$$\operatorname{Hom}_A(X, [t]X) = 0 \quad \text{and} \quad \operatorname{Ext}^1_A(X, [t]X) \cong D\operatorname{Hom}_A([t]X, \tau X) = 0,$$

for any $t \in \{2, \ldots, r\}$, and consequently, we have the chain

$$[r]X \longrightarrow [r-1]X \longrightarrow \cdots \longrightarrow [2]X$$

of irreducible epimorphisms in the category $X^{\perp} = \operatorname{mod} C$.

Further, because $\operatorname{Ext}^1_C([r]X, [r]X) = \operatorname{Ext}^1_A([r]X, [r]X) \neq 0$, we deduce that C is representation-infinite and $[2]X, [3]X, \ldots, [r-1]X, [r]X$ are indecomposable regular C-modules.

We prove now that the algebra C is connected. Without loss of generality, we may assume that $C = \operatorname{End}_A(Y)$, where Y is a projective object of the category $X^\perp$ and Y is a direct sum of pairwise non-isomorphic indecomposable modules.

Suppose, to the contrary, that C has a non-trivial K-algebra decomposition $C = C' \times C''$. It follows that there exists a decomposition $Y = Y' \oplus Y''$ of Y into a direct sum of two orthogonal non-zero submodules Y' and Y'' such that

$$C' \cong \operatorname{End}_A(Y') \text{ and } C'' \cong \operatorname{End}_A(Y'').$$

Because the module $[2]X$ is indecomposable, we may assume that $[2]XC'' = 0$. We know from (3.11) that $T = X \oplus Y$ is a tilting A-module. In view of (VI.3.5), the assumption that A is connected implies that the associated tilted algebra $B = \operatorname{End}_A(T)$ is also connected.

On the other hand, by applying the functor $\operatorname{Hom}_A(Y'', -)$ to the almost split sequence

$$0 \longrightarrow \tau_A X \longrightarrow [2]X \longrightarrow X \longrightarrow 0,$$

we derive an exact sequence

$$\operatorname{Hom}_A(Y'', [2]X) \longrightarrow \operatorname{Hom}_A(Y'', X) \longrightarrow \operatorname{Ext}^1_A(Y'', \tau_A X).$$

Because $[2]XC'' = 0$ and X is an epimorphic image of $[2]X$ then also $XC'' = 0$, and consequently

$$\operatorname{Hom}_A(Y'', [2]X) = 0 \quad \text{and} \quad \operatorname{Ext}^1_A(Y'', \tau_A X) \cong D\operatorname{Hom}_A(X, Y'') = 0.$$

It follows that the middle term of the preceding sequence is also zero, that is, $\operatorname{Hom}_A(Y'', X) = 0$.

On the other hand, $Y \in X^\perp$ yields $Y'' \in X^\perp$ and therefore $\operatorname{Hom}_A(X, Y'') = 0$. Consequently, the decomposition $Y = Y' \oplus Y''$ induces a non-trivial K-algebra direct product decomposition

$$B = \operatorname{End}_A(T) = \operatorname{End}_A(X \oplus Y) \cong \operatorname{End}_A(X \oplus Y') \times \operatorname{End}_A(Y'')$$

of a connected algebra B, a contradiction. This proves that the algebra C is connected.

(b) First we observe that if A is of Euclidean type then $\Gamma(\operatorname{mod} A)$ admits a stable tube $\mathcal{T}$ which is orthogonal to the tube containing the module X. Clearly, then all modules from the tube $\mathcal{T}$ belong to $X^\perp$, and consequently $\mathcal{T}$ is a connected component of $\Gamma(\operatorname{mod} C)$. Because C is hereditary and connected, we then deduce that C is of Euclidean type.

Conversely, assume that A is not of Euclidean type, that is, A is wild hereditary. We claim that C is also wild hereditary. Suppose, to the contrary, that C is of Euclidean type. Take a module M from a stable tube of rank one in $\Gamma(\operatorname{mod} C)$ which does not contain the module $[2]X$. Then we get

- $\operatorname{Hom}_A(M, [2]X) = \operatorname{Hom}_C(M, [2]X) = 0$, and
- $\operatorname{Ext}_A^1(M, [2]X) = \operatorname{Ext}_C^1(M, [2]X) \cong D\operatorname{Hom}_C([2]X, \tau_C M) = 0$,

because different tubes in $\Gamma(\operatorname{mod} C)$ are orthogonal. Note that the equality $\operatorname{Hom}_A(M, [2]X) = 0$ implies $\operatorname{Hom}_A(M, X) = 0$, because $[2]X$ is the unique immediate predecessor of X in $\Gamma(\operatorname{mod} A)$. By applying (3.14), we then conclude that

$$M = \tau_C M \cong \tau_T M \cong \tau_A M$$

and we get a contradiction, because A is wild hereditary, see (1.6), (VIII.2.3), and (VIII.2.7). This finishes the proof of the statement (b).

(c) Assume first that A is of Euclidean type. Then X lies in a standard stable tube $\mathcal{C}$ of $\Gamma(\operatorname{mod} A)$ and, in view of (XI.3.5), it follows from the choice of r that $\mathcal{C}$ is of rank r.

By passing from mod A to $X^\perp$, we remove from $\mathcal{C}$ all modules lying on the ray of $\mathcal{C}$ starting from X and all modules lying on the coray of $\mathcal{C}$ ending at τX, and, by shrinking the corresponding paths of length 2 to single arrows, we obtain a tube $\mathcal{C}'$ of rank $r-1$ in $\Gamma(\operatorname{mod} C)$ whose quasi-simple modules are

$$[2]X, \tau^2 X, \dots, \tau^{r-1} X,$$

see the construction of the stable tube $\mathcal{T}_\lambda''$ in (XVII.2.3), with V, $\mathcal{T}_\lambda$ and X, $\mathcal{C}$ interchanged. In particular, the module $[2]X$ is a quasi-simple regular brick in mod C.

Assume now that A is a wild hereditary algebra. We know that $[2]X$ is a regular C-module and also a brick, because $r \geq 2$. Hence, to show that $[2]X$ is quasi-simple in mod C, it is sufficient to prove that

$$\operatorname{Hom}_C(\tau_C[2]X, [2]X) = 0,$$

by (2.9). Because X is a quasi-simple brick in mod A, by applying (2.9) and (2.10), we conclude that $\operatorname{Hom}_A(\tau_A^n X, X) = 0$, for all $n \geq 1$. In particular,

$$\operatorname{Ext}_A^1(X, \tau_A^2 X) \cong D\operatorname{Hom}_A(\tau_A X, X) = 0,$$

and hence $\tau_A^2 X \in \mathcal{T} = \mathcal{T}(T)$, by (3.12). Moreover, because $T_A = X \oplus Y$ is a tilting A-module then

$$\operatorname{Hom}_A(T, \tau_A X) \cong D\operatorname{Ext}_A^1(T, X) = 0,$$

and, consequently, the module $\tau_A X$ belongs to the torsion-free part $\mathcal{F} = \mathcal{F}(T)$. Because there exists an almost split sequence

$$0 \longrightarrow \tau_A^2 X \longrightarrow \tau_A[2]X \longrightarrow \tau_A X \longrightarrow 0,$$

we get $\tau_T[2]X = g_T \tau_A[2]X \cong \tau_A^2 X$. Thus, in view of (3.13), there exists a short exact sequence

$$0 \longrightarrow (\tau_A X)^s \longrightarrow \tau_C[2]X \longrightarrow \tau_A^2 X \longrightarrow 0.$$

It follows that $\operatorname{Hom}_A((\tau_A X)^s, X) = 0$. Hence, $\operatorname{Hom}_A(\tau_A^2 X, X) = 0$ yields $\operatorname{Hom}_A(\tau_C[2]X, X) = 0$.

We also recall that $\operatorname{Hom}_A(\tau_C[2]X, \tau_A X) \cong D\operatorname{Ext}_A^1(X, \tau_C[2]X) = 0$, because $\tau_C([2]X) \in X^\perp$. Thus, by applying $\operatorname{Hom}_A(\tau_C[2]X, -)$ to the exact sequence

$$0 \longrightarrow \tau_A X \longrightarrow [2]X \longrightarrow X \longrightarrow 0,$$

we get $\operatorname{Hom}_C(\tau_C[2]X, [2]X) \cong \operatorname{Hom}_A(\tau_C[2]X, [2]X) = 0$. It follows that the module $[2]X$ is a quasi-simple regular brick in $\operatorname{mod} C$. This finishes the proof. $\square$

XVIII.4. Wild behaviour of the module category

We have seen in Chapters VII and XIII that for any hereditary algebra of Dynkin or Euclidean type there exists an explicit description of the isomorphism classes of indecomposable modules, and consequently of all modules. One of our objectives in this section is to show that, if Q is a connected acyclic quiver whose underlying graph $\overline{Q}$ of Q is neither Dynkin nor Euclidean and $A = KQ$, then a classification of indecomposable A-modules 'contains' the classification of all indecomposable modules over any finite dimensional algebra Λ. Because there is a lot of finite dimensional algebras Λ with very complicated module categories $\operatorname{mod} \Lambda$, this justifies why we call such hereditary algebras $A = KQ$ (respectively, quivers Q) wild or, equivalently, representation-wild. More precisely, we prove the following theorem (see [104] and [235]).

We recall that a connected acyclic quiver Q is called wild if the underlying graph $\overline{Q}$ of Q is neither Dynkin nor Euclidean.

4.1. Theorem. *Let Q be a finite connected acyclic quiver with $n = |Q_0|$ points, $q_Q : \mathbb{Z}^n \longrightarrow \mathbb{Z}$ be the quadratic form of Q, and let $A = KQ$. The following four conditions are equivalent.*

  (a) *Q is a wild quiver.*

(b) *There exists a full, faithful, exact, K-linear functor*

$$F : \operatorname{mod} \Lambda_0 \longrightarrow \operatorname{mod} A,$$

where Λ_0 is the local algebra $K[t_1, t_2]/(t_1, t_2)^2$ of dimension three.

(c) *For any (finite dimensional) K-algebra Λ, there exists a full, faithful, exact, and K-linear functor $F : \operatorname{mod} \Lambda \longrightarrow \operatorname{mod} A$.*

(d) *For any (finite dimensional) K-algebra Λ, there exists a full, faithful, exact, and K-linear functor $F : \operatorname{mod} \Lambda \longrightarrow \operatorname{mod} A$ such that*

$$q_Q(\operatorname{\mathbf{dim}} F(M)) \le -1,$$

for all non-zero modules M in $\operatorname{mod} \Lambda$.

Proof. We recall from (VII.4.1) that the quadratic form $q_Q : \mathbb{Z}^n \longrightarrow \mathbb{Z}$ of Q coincides with the Euler quadratic form q_A of the algebra $A = KQ$.

It is obvious that (d) implies (c), and (c) implies (b). To prove that (b) implies (a), we assume that there exists a full, faithful, exact, and a K-linear functor $F : \operatorname{mod} \Lambda_0 \longrightarrow \operatorname{mod} A$, where Λ_0 is the local algebra $K[t_1, t_2]/(t_1, t_2)^2$. We claim that then Q is a wild quiver. Indeed, denote by $S = \Lambda_0/\operatorname{rad} \Lambda_0$ the unique simple Λ_0-module, up to isomorphism. Then the Grothendieck group $K_0(\Lambda_0)$ of Λ_0 is infinite cyclic, (that is $K_0(\Lambda_0 \cong \mathbb{Z})$, $\dim_K \operatorname{End}_{\Lambda_0} S = 1$ and, according to (III.2.12), we get $\dim_K \operatorname{Ext}^1_{\Lambda_0}(S, S) = 2$, because

$$\Lambda_0 = KQ'/\mathcal{I},$$

where Q' is the quiver

consisting of one vertex and two loops α and β, and $\mathcal{I}$ is the ideal of KQ' generated by the four paths $\alpha\beta$, $\beta\alpha$, α^2, and β^2 of Q'.

Because the functor F is full, faithful, exact, and K-linear, we then obtain

$$\dim_K \operatorname{End}_A(F(S)) = 1 \quad \text{and} \quad \dim_K \operatorname{Ext}^1_A(F(S), F(S)) \ge 2.$$

Therefore, in view of (III.3.13), we get

$$q_A(\operatorname{\mathbf{dim}} F(S)) = \dim_K \operatorname{End}_A(F(S)) - \dim_K \operatorname{Ext}^1_A(F(S), F(S)) < 0,$$

that is, the Euler quadratic form $q_A : K_0(A) \longrightarrow \mathbb{Z}$ of A is indefinite. Because $q_A = q_Q$, by (VII.4.1), then, according to (VII.4.5), the quiver Q is wild.

To prove that (a) implies (d), we assume that Q is a wild quiver and $A = KQ$. Let Λ be an arbitrary finite dimensional K-algebra. We show by induction on the number $|Q_0|$ of vertices of the quiver Q that there exists a full, faithful, exact and K-linear functor $F : \operatorname{mod}\Lambda \longrightarrow \operatorname{mod} A$ such that $q_A(\operatorname{\mathbf{dim}} F(M)) < 0$, for all non-zero modules M in $\operatorname{mod}\Lambda$.

Because Q is wild and acyclic then $|Q_0| \geq 2$. Consider first the case when Q has only two vertices, that is, Q is the enlarged Kronecker quiver

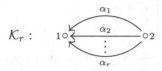

with $r \geq 3$ arrows. Let $m = \dim_K \Lambda$ and $b_1, \ldots, b_m$ be a K-basis of Λ. We identify $\operatorname{mod} A$ with $\operatorname{rep}_K(\mathcal{K}_r)$, according to (III.1.6). We define a K-linear functor $F : \operatorname{mod}\Lambda \longrightarrow \operatorname{mod} A$ by associating to each module M from $\operatorname{mod}\Lambda$ the representation

$$F(M) = (M^{m+2}, M^{m+2}; f_{\alpha_1}, \ldots, f_{\alpha_r}),$$

where $f_{\alpha_1}, f_{\alpha_2} : M^{m+2} \longrightarrow M^{m+2}$ are K-linear maps defined by the matrices

$$f_{\alpha_1} = \begin{bmatrix} 0 & 1 & 0 & \ldots & 0 \\ 0 & 0 & 1 & \ldots & 0 \\ \vdots & \vdots & \vdots & \ddots & \vdots \\ 0 & 0 & 0 & \ldots & 1 \\ 0 & 0 & 0 & \ldots & 0 \end{bmatrix}, \quad f_{\alpha_2} = \begin{bmatrix} 0 & 0 & 0 & \ldots & 0 & 0 & 0 \\ 1 & 0 & 0 & \ldots & 0 & 0 & 0 \\ \tilde{b}_1 & 1 & 0 & \ldots & 0 & 0 & 0 \\ 0 & \tilde{b}_2 & 1 & \ldots & 0 & 0 & 0 \\ \vdots & \vdots & \vdots & \ddots & \ddots & \vdots & \vdots \\ 0 & 0 & 0 & \ddots & 1 & 0 & 0 \\ 0 & 0 & 0 & \ldots & \tilde{b}_m & 1 & 0 \end{bmatrix},$$

and $f_{\alpha_3}, \ldots, f_{\alpha_r} : M^{m+2} \longrightarrow M^{m+2}$ are the identity maps. Here $\tilde{b}_j : M \longrightarrow M$ is the K-linear map defined by $m \mapsto m \cdot b_j$. For any homomorphism $g : M \longrightarrow N$ in $\operatorname{mod}\Lambda$, we define the homomorphism $F(g) : F(M) \longrightarrow F(N)$ to be the diagonal K-linear map

$$F(g) = \begin{bmatrix} g & 0 & \ldots & 0 \\ 0 & g & \ldots & 0 \\ \vdots & \vdots & \ddots & \vdots \\ 0 & 0 & \ldots & g \end{bmatrix}$$

from M^{m+2} to M^{m+2}. It is clear that F is a K-linear, exact and faithful functor.

We claim that F is also full. Let M, N be Λ-modules. Assume that $\varphi = (\varphi_1, \varphi_2)$ is a homomorphism from $F(M)$ to $F(N)$ in the category $\operatorname{mod} A \cong \operatorname{rep}_K(\mathcal{K}_r)$. Then $\varphi_1, \varphi_2 : M^{m+2} \longrightarrow N^{m+2}$ are K-linear maps such that $\varphi_1 f_{\alpha_i} = f_{\alpha_i} \varphi_2$, for any $i \in \{1, \ldots, r\}$. Because $r \geq 3$ and $f_{\alpha_3} = \operatorname{id}$, we get $\varphi_1 = \varphi_2$. From the equalities $\varphi_1 f_{\alpha_1} = f_{\alpha_1} \varphi_2 = f_{\alpha_1} \varphi_1$ we easily conclude that $\varphi_1 : M^{m+2} \longrightarrow N^{m+2}$ is given by a matrix

$$
h = \begin{bmatrix}
h_1 & h_2 & h_3 & \cdots & h_m & h_{m+1} & h_{m+2} \\
0 & h_1 & h_2 & \cdots & h_{m-1} & h_m & h_{m+1} \\
0 & 0 & h_1 & \cdots & h_{m-2} & h_{m-1} & h_m \\
\vdots & \vdots & \vdots & \ddots & \vdots & \vdots & \vdots \\
0 & 0 & 0 & \cdots & h_1 & h_2 & h_3 \\
0 & 0 & 0 & \cdots & 0 & h_1 & h_2 \\
0 & 0 & 0 & \cdots & 0 & 0 & h_1
\end{bmatrix},
$$

where $h_1, \ldots, h_{m+2} : M \longrightarrow N$ are K-linear maps.

The equalities $\varphi_1 f_{\alpha_2} = f_{\alpha_2}\varphi_2 = f_{\alpha_2}\varphi_1$ yield

- $h_2 = h_3 = \ldots = h_{m+1} = h_{m+2} = 0$ and
- $\widetilde{b}_j h_1 = h_1 \widetilde{b}_j$, for any $j \in \{1, \ldots, m\}$.

Because $b_1, \ldots, b_m$ is a K-basis of Λ, we conclude that $g = h_1 : M \longrightarrow N$ is a Λ-homomorphism such that $\varphi = F(g)$, that is, the functor F is full, see also (XIX.1.7).

By (III.3.13) and (VII.4.1), the Euler form $q_A : \mathbb{Z}^2 \longrightarrow \mathbb{Z}$ of the algebra A is given by the formula

$$q_A(\mathbf{x}) = x_1^2 + x_2^2 - rx_1 x_2, \quad \text{for any } \mathbf{x} = \begin{bmatrix} x_1 \\ x_2 \end{bmatrix} \in \mathbb{Z}^2.$$

Because $\mathbf{dim}\, F(M) = (\dim_K M) \cdot \begin{bmatrix} m+2 \\ m+2 \end{bmatrix}$ and $r \geq 3$ then

$$q_A(\mathbf{dim}\, F(M)) = (\dim_K M)^2 (m+2)^2 (2-r) < 0,$$

for any non-zero module M in $\mathrm{mod}\,\Lambda$. This finishes the proof in case $|Q_0| = 2$.

Assume now that $n \geq 2$, $A = KQ$, $|Q_0| = n + 1$ and, for each connected wild hereditary algebra $A' = KQ'$, with $2 \leq |Q'_0| \leq n$, there exists a full, faithful, exact, and K-linear functor $F' : \mathrm{mod}\,\Lambda \longrightarrow \mathrm{mod}\,A'$ such that $q_{A'}(\mathbf{dim}\, F'(M)) < 0$, for all non-zero modules M in $\mathrm{mod}\,\Lambda$.

It follows from (2.19) that $\mathrm{mod}\,A$ contains a quasi-simple regular stone X. Then, by (3.10) and (3.15), the right perpendicular category $X^\perp$ is equivalent to a module category $\mathrm{mod}\,C$, where C is a connected wild hereditary algebra whose quiver has n vertices. For simplicity, we identify $X^\perp$ with $\mathrm{mod}\,C$. Hence, by our inductive assumption applied to $A' = C$, there exists a full, faithful, exact and K-linear functor

$$G : \mathrm{mod}\,\Lambda \longrightarrow X^\perp = \mathrm{mod}\,C$$

such that $q_C(\mathbf{dim}\, G(M)) < 0$, for all non-zero modules M in $\mathrm{mod}\,\Lambda$. Then the composition of G with the canonical embedding

$$\mathrm{mod}\,C = X^\perp \hookrightarrow \mathrm{mod}\,A$$

gives the required full, faithful, exact and K-linear functor

$$F : \operatorname{mod} \Lambda \longrightarrow \operatorname{mod} A.$$

By (3.6), the category $X^{\perp}$ is closed under extensions. Then the canonical embedding $X^{\perp} \hookrightarrow \operatorname{mod} A$ induces a K-algebra isomorphism

$$\operatorname{End}_C(G(M)) \cong \operatorname{End}_A(F(M))$$

and a K-linear isomorphism

$$\operatorname{Ext}^1_C(G(M), G(M)) \cong \operatorname{Ext}^1_A(F(M), F(M)),$$

for any non-zero module M in $\operatorname{mod} \Lambda$. Hence, in view of (III.3.13) and (VII.4.1), for any non-zero module M in $\operatorname{mod} \Lambda$, we get

$$
\begin{aligned}
q_A(\operatorname{\mathbf{dim}} F(M)) &= \dim_K \operatorname{End}_A(F(M)) - \dim_K \operatorname{Ext}^1_A(F(M), F(M)) \\
&= \dim_K \operatorname{End}_C(G(M)) - \dim_K \operatorname{Ext}^1_C(G(M), G(M)) \\
&= q_C(\operatorname{\mathbf{dim}} G(M)) < 0.
\end{aligned}
$$

This finishes the proof. □

Following Drozd [201] and [202], a finite dimensional K-algebra R is called **strictly representation-wild** if, for every finite dimensional K-algebra Λ, there exists a full, faithful, exact, and K-linear functor $F : \operatorname{mod} \Lambda \longrightarrow \operatorname{mod} R$.

It follows from the definition that the functor F induces a K-linear isomorphism

$$\operatorname{Hom}_\Lambda(M, N) \xrightarrow{\simeq} \operatorname{Hom}_R(F(M), F(N)),$$

given by $f \mapsto F(f)$, for any M and N in $\operatorname{mod} \Lambda$. In particular, F reflects isomorphisms and carries indecomposable Λ-modules to indecomposable R-modules, because the algebra homomorphism

$$\operatorname{End}_\Lambda(M) \xrightarrow{\simeq} \operatorname{End}_R(F(M)),$$

given by $f \mapsto F(f)$, is an isomorphism, see (I.4.8). This shows that a classification of indecomposable A-modules contains a corresponding classification of indecomposable Λ-modules (via the functor F), where Λ is an arbitrary finite dimensional K-algebra.

It follows from (4.1) that the path K-algebra $A = KQ$ of a finite, acyclic and connected wild quiver Q is strictly representation-wild. We also have the following important improvement of (4.1).

4.2. Corollary. *Let Q be a finite connected acyclic wild quiver and let $A = KQ$. Then, for any finite dimensional K-algebra Λ, there exists a full, faithful, exact and a K-linear functor*

$$F : \operatorname{mod} \Lambda \longrightarrow \operatorname{add} \mathcal{R}(A),$$

where $\operatorname{add} \mathcal{R}(A)$ *is the category of all regular A-modules.*

Proof. Let Λ be an arbitrary finite dimensional K-algebra. By (4.1), there exists a full, faithful, exact and K-linear functor $F : \operatorname{mod} \Lambda \longrightarrow \operatorname{mod} A$ such that $q_A(\operatorname{\mathbf{dim}} F(M)) < 0$, for any indecomposable Λ-module M. Hence, according to (VIII.2.7), the indecomposable module $F(M)$ is regular. It follows that the image of F is contained in the category $\operatorname{add} \mathcal{R}(A)$ of all regular A-modules. □

Recall that if A is a hereditary algebra of Dynkin type then every finite dimensional indecomposable A-module is a brick. Recall also from (VIII.2.7) that all indecomposable postprojective modules and all indecomposable preinjective modules over any algebra are bricks. Moreover, if A is a hereditary algebra of Euclidean type then according to (X.2.7) (see also (XI.3.5)) the endomorphism algebra of any finite dimensional indecomposable A-module is isomorphic to a truncated polynomial algebra $K[t]/(t^r)$, for some $r \geq 1$. It follows that not all finite dimensional local K-algebras are of the form $\operatorname{End}_A(M)$, where M is a finite dimensional indecomposable A-module.

Our second objective in this section is to prove that, if A is a wild hereditary algebra then, for any finite dimensional local K-algebra Λ, there exists a quasi-simple regular A-module M such that $\operatorname{End}_A(M)$ is isomorphic to Λ. We need some preliminary facts.

Let A be a connected wild hereditary algebra and $\mathcal{X}$ a family of pairwise orthogonal bricks which are not stones. Consider the full subcategory

$$\mathcal{E}_A(\mathcal{X}) = \mathcal{EXT}_A(\mathcal{X})$$

of $\operatorname{mod} A$ formed by all modules Y in $\operatorname{mod} A$ having a chain of submodules

$$0 = Y_0 \subset Y_1 \subset \ldots \subset Y_{m-1} \subset Y_m = Y,$$

for some $m \geq 1$, such that $Y_i/Y_{i-1} \in \mathcal{X}$, for each $i \in \{1, \ldots, m\}$.

It follows from (X.2.1) that $\mathcal{E}_A(\mathcal{X})$ is an exact abelian subcategory of $\operatorname{mod} A$ which is closed under extensions, and $\mathcal{X}$ is the family of all simple objects in $\mathcal{E}_A(\mathcal{X})$. We then have the following useful fact.

4.3. Lemma. *Let A be a connected wild hereditary algebra and $\mathcal{X}$ a family of pairwise orthogonal bricks which are not stones.*

(a) *For each non-zero module Y in the category $\mathcal{E}_A(\mathcal{X}) = \mathcal{E}\mathcal{X}\mathcal{T}_A(\mathcal{X})$, we have $\operatorname{Ext}^1_A(Y,Y) \neq 0$. In particular, the category $\mathcal{E}_A(\mathcal{X})$ consists of regular A-modules.*

(b) *Every τ-orbit in $\operatorname{mod} A$ of an indecomposable regular module contains at most one module from $\mathcal{E}_A(\mathcal{X})$, up to isomorphism.*

Proof. (a) Let Y be a non-zero module from $\mathcal{E}_A(\mathcal{X}) = \mathcal{E}\mathcal{X}\mathcal{T}_A(\mathcal{X})$. Then there exist modules $X_1, \ldots, X_r$ in $\mathcal{X}$ and positive integers $m_1, \ldots, m_r$ such that $\mathbf{dim}\, Y = \sum_{i=1}^{r} m_i \mathbf{dim}\, X_i$. Because $X_1, \ldots, X_r$ are bricks and they are not stones, then $\dim_K \operatorname{End}_A(X_i) = 1$, $\dim_K \operatorname{Ext}^1_A(X_i, X_i) \geq 1$ and, hence,

$$q_A(\mathbf{dim}\, X_i) = \dim_K \operatorname{End}_A(X_i) - \dim_K \operatorname{Ext}^1_A(X_i, X_i) \leq 0,$$

for any $i \in \{1, \ldots, r\}$. Then, in view of (III.3.13) and (VII.4.1), we get

$$\dim_K \operatorname{End}_A(Y) - \dim_K \operatorname{Ext}^1_A(Y,Y) = q_A(\mathbf{dim}\, Y) = \langle \mathbf{dim}\, Y, \mathbf{dim}\, Y \rangle_Q$$

$$= \sum_{i=1}^{r} m_i^2 \langle \mathbf{dim}\, X_i, \mathbf{dim}\, X_i \rangle_Q + \sum_{i \neq j} m_i m_j \langle \mathbf{dim}\, X_i, \mathbf{dim}\, X_j \rangle_Q$$

$$= \sum_{i=1}^{r} m_i^2 q_A(\mathbf{dim}\, X_i) + \sum_{i \neq j} m_i m_j \dim_K \operatorname{Hom}_A(X_i, X_j)$$

$$- \sum_{i \neq j} m_i m_j \dim_K \operatorname{Ext}^1_A(X_i, X_j)$$

$$= \sum_{i=1}^{r} m_i^2 q_A(\mathbf{dim}\, X_i) - \sum_{i \neq j} m_i m_j \dim_K \operatorname{Ext}^1_A(X_i, X_j) \leq 0,$$

because the bricks $X_1, \ldots, X_r$ are pairwise orthogonal. It follows that $\operatorname{Ext}^1_A(Y,Y) \neq 0$. Hence, in view of (VIII.2.7), every indecomposable module in $\mathcal{E}_A(\mathcal{X})$ is neither postprojective nor preinjective and, therefore, is regular. Consequently, the category $\mathcal{E}_A(\mathcal{X})$ consists of regular modules.

(b) Suppose, to the contrary, that there exist an indecomposable regular A-module Y in $\mathcal{E}_A(\mathcal{X})$ and an integer $n \geq 1$ such that the module $\tau^n Y$ also belongs to $\mathcal{E}_A(\mathcal{X})$. Because Y is a non-zero module from $\mathcal{E}_A(\mathcal{X})$, then there exists a monomorphism $X \longrightarrow Y$, for some $X \in \mathcal{X}$, which induces a monomorphism $f : \tau^n X \longrightarrow \tau^n Y$ by applying the left exact functor τ^n.

Further, because X is not a stone then

$$\operatorname{Ext}^1_A(X, X) \neq 0 \quad \text{and} \quad \operatorname{Hom}_A(X, \tau^n X) \neq 0,$$

by (2.11). Clearly, then the composition $fg : X \longrightarrow \tau^n Y$ of f with any non-zero homomorphism $g : X \longrightarrow \tau^n X$ is non-zero. Because X is a simple object of $\mathcal{E}_A(\mathcal{X})$ and $\tau^n Y$ belongs to $\mathcal{E}_A(\mathcal{X})$, we then conclude that all non-zero homomorphisms from X to $\tau^n X$ are proper monomorphisms, by (1.1). Take a non-zero homomorphism $h : X \longrightarrow \tau^n X$. Then h is a proper monomorphism and we obtain a commutative diagram

$$
\begin{array}{ccccccccc}
0 & \longrightarrow & X & \xrightarrow{h} & \tau^n X & \xrightarrow{u} & M & \longrightarrow & 0 \\
 & & {\scriptstyle 1}\big\downarrow & & {\scriptstyle f}\big\downarrow & & {\scriptstyle w}\big\downarrow & & \\
0 & \longrightarrow & X & \xrightarrow{fh} & \tau^n Y & \longrightarrow & N & \longrightarrow & 0
\end{array}
$$

with exact rows, where $M = \operatorname{Coker} h$ and $N = \operatorname{Coker} fh$. Observe that w is a monomorphism, because so is f. Because $\mathcal{E}_A(\mathcal{X})$ is an abelian category we conclude that N belongs to $\mathcal{E}_A(\mathcal{X})$, and consequently N is a regular module. This implies that M is also a regular module. We know also that the restriction of the functor τ^{-1} to the category of regular A-modules is an exact functor. Thus, applying the exact functors $\tau^{-n}, \tau^{-2n}, \tau^{-3n}, \dots, \tau^{-sn}, \dots$ to the proper monomorphism h, we derive a chain of proper monomorphisms

$$
\cdots \longrightarrow \tau^{-sn} X \longrightarrow \cdots \longrightarrow \tau^{-2n} X \longrightarrow \tau^{-n} X \longrightarrow X,
$$

a contradiction. This finishes the proof. $\qquad\square$

4.4. Lemma. *Let A be a connected wild hereditary algebra and $\mathcal{X}$ a family of pairwise orthogonal bricks in* $\operatorname{mod} A$ *which are not stones. Let Z be a quasi-simple regular A-module and X a module in $\mathcal{X}$ such that $\operatorname{Hom}_A(X, Z) \neq 0$ and the module $Y = Z[r]$ belongs to $\mathcal{E}_A(\mathcal{X}) = \mathcal{E}\mathcal{X}\mathcal{T}_A(\mathcal{X})$, for some $r \geq 2$. Then the modules Z and $\tau^{-1} Z[r-1]$ also belong to $\mathcal{E}_A(\mathcal{X})$.*

Proof. It follows from (1.5) that there exists an infinite chain of irreducible monomorphisms

$$
Z = Z[1] \longrightarrow Z[2] \longrightarrow Z[3] \longrightarrow \cdots \longrightarrow Z[r] \longrightarrow \cdots
$$

between indecomposable regular modules in $\operatorname{mod} A$. Let $\varepsilon : Z \hookrightarrow Y = Z[r]$ be the composed embedding and $f : X \longrightarrow Z$ a non-zero homomorphism. Clearly, $\varepsilon f : X \longrightarrow Y$ is non-zero. Because X and Y belong to the abelian category $\mathcal{E}_A(\mathcal{X}) = \mathcal{E}\mathcal{X}\mathcal{T}_A(\mathcal{X})$ and X is simple in $\mathcal{E}_A(\mathcal{X})$, we conclude that in fact εf is a monomorphism. If f is an isomorphism then Z belongs to $\mathcal{X} \subset \mathcal{E}_A(\mathcal{X})$, and hence $\tau^{-1} Z[r-1] \cong Y/Z$ also belongs to $\mathcal{E}_A(\mathcal{X})$.

Assume that f is not an isomorphism. Denote by $\pi : Y \longrightarrow Y/Z$ the irreducible epimorphism $Y = Z[r] \longrightarrow \tau^{-1} Z[r-1] \cong Y/Z$. Then π has the factorisation $\pi = hg$ with

$$
Y \xrightarrow{g} Y/f(X) \xrightarrow{h} Y/Z.
$$

The irreducibility of π implies that h is a split epimorphism, and so

$$
Y/f(X) \cong Y/Z \oplus Z/f(X).
$$

Because $\mathcal{E}_A(\mathcal{X})$ is an abelian full subcategory of mod A and $\varepsilon f : X \longrightarrow Y$ is a morphism between two objects of $\mathcal{E}_A(\mathcal{X})$, we conclude that the module $Y/f(X) \cong Y/\varepsilon f(X)$ is an object of $\mathcal{E}_A(\mathcal{X})$, and consequently Y/Z and $Z/f(X)$ also belong to $\mathcal{E}_A(\mathcal{X})$. Finally, Z is an object of $\mathcal{E}_A(\mathcal{X})$, because it is the kernel of the epimorphism $\pi : Y \longrightarrow Y/Z$ between two objects of $\mathcal{E}_A(\mathcal{X})$. Consequently, Z and $\tau^{-1}Z[r-1]$ belong to $\mathcal{E}_A(\mathcal{X})$. $\square$

We may present now the following important properties of the category $\mathcal{E}_A(\mathcal{X})$.

4.5. Proposition. *Let A be a connected wild hereditary algebra and $\mathcal{X}$ a family of pairwise orthogonal bricks in mod A which are not stones.*

(a) *Every indecomposable non-brick in $\mathcal{E}_A(\mathcal{X}) = \mathcal{E}\mathcal{X}\mathcal{T}_A(\mathcal{X})$ is a quasi-simple regular A-module.*

(b) *Every regular component in $\Gamma(\text{mod } A)$ contains at most one module of $\mathcal{E}_A(\mathcal{X})$.*

Proof. (a) Let Y be an indecomposable module in $\mathcal{E}_A(\mathcal{X}) = \mathcal{E}\mathcal{X}\mathcal{T}_A(\mathcal{X})$ and assume that Y is not a brick. According to (1.5) and (1.6), there exist a quasi-simple regular A-module Z and an infinite chain

$$Z = Z[1] \longrightarrow Z[2] \longrightarrow Z[3] \longrightarrow \cdots \longrightarrow Z[r] \longrightarrow \cdots$$

of irreducible monomorphisms between indecomposable regular modules in mod A such that $Y = Z[r]$, for some $r \geq 1$.

Assume, to the contrary, that $Y = Z[r]$ is not quasi-simple, that is, $r \geq 2$. Then it follows from (2.18) that $\text{Hom}_A(Y, Z) \neq 0$, and consequently $\text{Hom}_A(X, Z) \neq 0$, for some $X \in \mathcal{X}$. Applying now (4.4), we conclude that the modules Z and $\tau^{-1}Z[r-1]$ belong to $\mathcal{E}_A(\mathcal{X})$. We know from (4.3) that all modules in $\mathcal{E}_A(\mathcal{X})$ are not stones, and hence Z and $\tau^{-1}Z[r-1]$ are not stones. In particular, it follows from (2.16) that the modules $Z[2], \ldots, Z[r] = Y$ are not bricks.

If $r \geq 3$, repeating the preceding arguments for $Y' = \tau^{-1}Z[r-1]$, we conclude that the modules

$$\tau^{-1}Z \quad \text{and} \quad \tau^{-2}Z[r-2] = \tau^{-1}((\tau^{-1}Z)[r-2])$$

belong to $\mathcal{E}_A(\mathcal{X})$. Applying the above arguments again (if $r \geq 4$), we finally show that all modules $Z, \tau^{-1}Z, \ldots, \tau^{-(r-1)}Z$ belong to $\mathcal{E}_A(\mathcal{X})$, and consequently the cone $\mathcal{C}(Y) = \mathcal{C}([Z[r])$ is contained in $\mathcal{E}_A(\mathcal{X})$. Because $r \geq 2$, this contradicts (4.3)(b). Therefore, each indecomposable non-brick Y in $\mathcal{E}_A(\mathcal{X})$ is a quasi-simple regular A-module.

(b) Let $\mathcal{C}$ be a component of $\Gamma(\text{mod } A)$ containing a module Y from $\mathcal{E}_A(\mathcal{X})$. By (1.6), $\mathcal{C}$ is of the form $\mathbb{Z}\mathbb{A}_\infty$ presented in (1.6). We have two cases to consider.

Assume first that Y is not a brick. Then it follows from the part (a) that Y is quasi-simple. Note that C does not contain any brick, because otherwise, in view of (1.6), C contains a brick $X[r]$, where $r \geq 1$ and X is a quasi-simple module. By (2.16), X is a brick and therefore Y is a brick, because C is of the form $\mathbb{Z}A_\infty$ and $Y = \tau^s X$, for some $s \in \mathbb{Z}$. Recall that τ carries regular bricks to bricks. This is a contradiction which shows that C does not contain any brick.

Hence we easily conclude that any common module of $\mathcal{E}_A(\mathcal{X})$ and C has to be quasi-simple. Then (b) follows from (4.3)(b).

Finally, assume that Y is a brick. In view of (1.6), $Y = Z[r]$, for some $r \geq 1$ and a quasi-simple regular A-module Z. We know from (4.3)(a) that Y is not a stone. Then, by (2.16), the modules $Z[1], \ldots, Z[r-1]$ (if $r \geq 2$) are stones but the modules $Z[i]$, with $i > r$, are not bricks. Hence the common modules of $\mathcal{E}_A(\mathcal{X})$ and C can be only contained in the τ-orbit of Y. Our claim follows from (4.3)(b). □

As an application of our considerations we obtain the following already announced fact.

4.6. Theorem. *Let $A = KQ$ be a connected wild hereditary K-algebra and let Λ be an arbitrary finite dimensional local K-algebra. Then there exists a full, faithful, exact and K-linear functor*

$$F : \operatorname{mod} \Lambda \longrightarrow \operatorname{add} \mathcal{R}(A)$$

such that the A-module $M = F(\Lambda)$ is quasi-simple and $\operatorname{End}_A(M) \cong \Lambda$.

Proof. Assume that $\operatorname{rad} \Lambda = 0$, that is, $\Lambda \cong K$. It follows from (1.9) and (2.19) that $\operatorname{mod} A$ admits many quasi-simple regular bricks. Take a quasi-simple brick X in $\operatorname{mod} A$ and consider the full subcategory $\operatorname{add} X$ of $\operatorname{mod} A$ formed by all A-modules which are isomorphic to finite direct sums of the module X. Because $\operatorname{End}_A X \cong K \cong \Lambda$, it follows from (VI.3.1) that the exact K-linear functor

$$\operatorname{Hom}_A(X, -) : \operatorname{add} X \longrightarrow \operatorname{mod} \operatorname{End}_A X \cong \operatorname{mod} \Lambda$$

is an equivalence of categories. It is obvious that the quasi-inverse $F : \operatorname{mod} \Lambda \longrightarrow \operatorname{add} X$ of $\operatorname{Hom}_A(X, -)$ is a full, faithful, exact and K-linear functor such that $X \cong F(\Lambda)$, and we are done.

Assume now that $\operatorname{rad} \Lambda \neq 0$, and let $S = \Lambda/\operatorname{rad} \Lambda$. It follows from (4.2) that there exists a full, faithful, exact, K-linear functor

$$F : \operatorname{mod} \Lambda \hookrightarrow \operatorname{add} \mathcal{R}(A),$$

and therefore $\Lambda \cong \mathrm{End}_A M$, where $M = F(\Lambda)$. Observe that the A-module $X = F(S)$ is a brick, but X is not a stone, because F induces a K-linear monomorphism

$$0 \neq \mathrm{Ext}_\Lambda^1(S, S) \longrightarrow \mathrm{Ext}_A^1(F(S), F(S)).$$

We set $\mathcal{X} = \{X\}$. It follows from (X.2.1) and (4.3)(a) that $\mathcal{E}_A(\mathcal{X})$ is an exact abelian subcategory of add $\mathcal{R}(A) \subset \mathrm{mod}\, A$ which is closed under extensions. Because Λ is a local algebra and S is a unique simple Λ-module, up to isomorphism, then F has a factorisation $F : \mathrm{mod}\,\Lambda \longrightarrow \mathcal{E}_A(\mathcal{X})$ through $\mathcal{E}_A(\mathcal{X}) \subseteq \mathrm{add}\,\mathcal{R}(A)$. By our assumption, the Λ-module Λ is not a brick and, consequently, $F(\Lambda)$ is not a brick in $\mathrm{mod}\, A$. Because $F(\Lambda)$ belongs to $\mathcal{E}_A(\mathcal{X})$ we conclude from (4.5) that $M = F(\Lambda)$ is a quasi-simple regular A-module and obviously $\mathrm{End}_A(M) \cong \mathrm{End}_\Lambda(\Lambda) \cong \Lambda$. This finishes the proof. $\square$

4.7. Corollary. *Let $A = KQ$, where Q is a finite, acyclic and connected quiver. The algebra A is strictly representation-wild if and only if Q is a wild quiver.*

Proof. First assume that Q is a wild quiver. It follows from (4.1) that the path K-algebra $A = KQ$ is strictly representation-wild.

Conversely, assume, to the contrary, that Q is not a wild quiver and the algebra $A = KQ$ is strictly representation-wild. It follows that Q is either Dynkin or Euclidean and, for the finite dimensional local K-algebra $\Lambda = K[t_1, t_2]/(t_1, t_2)^2$, there exists a full, faithful, exact and K-linear functor $F : \mathrm{mod}\,\Lambda \longrightarrow \mathrm{mod}\, A$. Therefore the A-module $F(\Lambda)$ is indecomposable and F induces a K-algebra isomorphism $\mathrm{End}_A(F(\Lambda)) \cong \Lambda$. This is a contradiction, because A is hereditary algebra of Dynkin or Euclidean type and, according to (VII.5.14), (X.2.7), and (XI.3.5), the endomorphism algebra of any finite dimensional indecomposable A-module is isomorphic to a truncated polynomial algebra $K[t]/(t^r)$, for some $r \geq 1$. $\square$

XVIII.5. Tilted algebras of wild type

The main objective of this section is to describe basic properties of the module categories $\mathrm{mod}\, B$ of concealed algebras B of wild type and exhibit some other classes of tilted algebras of wild type.

Throughout this section, we let $A = KQ$ be the path K-algebra of a connected, acyclic, wild quiver Q with n vertices, that is, $n = |Q_0|$. Then, for any (multiplicity-free) tilting A-module T, the associated tilted algebra

$$B = \mathrm{End}_A(T)$$

is called a **tilted K-algebra of wild type** Q. Recall that by the tilting theorem (VI.3.8) the functor $\mathrm{Hom}_A(T, -) : \mathrm{mod}\, A \longrightarrow \mathrm{mod}\, B$ induces an equivalence from the category

$$\mathcal{T}(T) = \{M \in \operatorname{mod} A; \ \operatorname{Ext}_A^1(T, M) = 0\}$$

of torsion modules in $\operatorname{mod} A$ to the category

$$\mathcal{Y}(T) = \{N \in \operatorname{mod} B; \ \operatorname{Tor}_1^B(N, T) = 0\}$$

of torsion-free modules in $\operatorname{mod} B$, and the functor

$$\operatorname{Ext}_A^1(T, -) : \operatorname{mod} A \longrightarrow \operatorname{mod} B$$

induces an equivalence from the category

$$\mathcal{F}(T) = \{M \in \operatorname{mod} A; \ \operatorname{Hom}_A(T, M) = 0\}$$

of torsion-free modules in $\operatorname{mod} A$ to the category

$$\mathcal{X}(T) = \{N \in \operatorname{mod} B; \ N \otimes_B T = 0\}$$

of torsion modules in $\operatorname{mod} B$. Moreover, the torsion theory $(\mathcal{Y}(T), \mathcal{X}(T))$ in $\operatorname{mod} B$ is splitting, and so every indecomposable B-module belongs either to $\mathcal{Y}(T)$ or $\mathcal{X}(T)$. Further, B is a basic connected algebra of global dimension at most 2, with the acyclic connected quiver Q_B, and there is an isomorphism $f : K_0(A) \longrightarrow K_0(B)$ of the Grothendieck groups of A and B such that

$$f(\mathbf{dim}\, M) = \mathbf{dim}\, \operatorname{Hom}_A(T, M) - \mathbf{dim}\, \operatorname{Ext}_A^1(T, M),$$

for any module M in $\operatorname{mod} A$. In particular, if T is postprojective, then the functor

$$\operatorname{Hom}_A(T, -) : \operatorname{add} \mathcal{R}(A) \xrightarrow{\ \simeq\ } \operatorname{add} \mathcal{R}(B)$$

is an equivalence of the categories of regular modules over A and over the concealed algebra $B = \operatorname{End}_A(T)$ such that $\mathbf{dim}\, \operatorname{Hom}_A(T, M) = f(\mathbf{dim}\, M)$, for any M in $\operatorname{add} \mathcal{R}(A)$. We then obtain the following important information on the module category $\operatorname{mod} B$ of an arbitrary concealed algebra B of wild type.

5.1. Theorem. *Let $A = KQ$ be a connected wild hereditary K-algebra. Let T be a postprojective tilting A-module and $B = \operatorname{End}_A(T)$. Then the following hold.*

(a) *The Auslander–Reiten quiver $\Gamma(\operatorname{mod} B)$ of B consists of a postprojective component $\mathcal{P}(B)$ containing all indecomposable projective B-modules, a preinjective component $\mathcal{Q}(B)$ containing all indecomposable injective B-modules, and $\operatorname{card}(K)$ many regular components of type $\mathbb{Z}\mathbb{A}_\infty$.*

(b) *If $\mathcal{C}$ is a component of $\Gamma(\operatorname{mod} B)$ then all but finitely many modules in $\mathcal{C}$ are sincere.*

(c) *If $\mathcal{C}$ is a component of $\Gamma(\operatorname{mod} B)$ and M, N is a pair of non-isomorphic indecomposable modules in $\mathcal{C}$ then $\mathbf{dim}\, M \neq \mathbf{dim}\, N$.*

Proof. (a) In view of the K-linear equivalence of categories

$$\operatorname{Hom}_A(T, -) : \operatorname{add} \mathcal{R}(A) \xrightarrow{\;\simeq\;} \operatorname{add} \mathcal{R}(B),$$

the statement (a) is a consequence of (1.6), (1.8) and (VIII.4.5).

(b) Observe that if T' is an indecomposable direct summand of T then $T' = \tau_A^{-s} P$, for some $s \geq 0$ and some indecomposable projective A-module P. Therefore,

$$\operatorname{Hom}_A(T', M) \cong \operatorname{Hom}_A(P, \tau_A^s M),$$

for any module M in $\mathcal{R}(A)$, and (2.5) implies that all but finitely many modules in any regular component of $\Gamma(\operatorname{mod} B)$ are sincere. Moreover, by (IX.5.7) and its dual, all but finitely many modules in $\mathcal{P}(B)$ and in $\mathcal{Q}(B)$ are also sincere. Hence (b) follows.

(c) For the components $\mathcal{P}(B)$ and $\mathcal{Q}(B)$, the statement (c) follows from the fact that the directing modules are uniquely determined by their composition factors (see (IX.3.1)). For the regular components in $\Gamma(\operatorname{mod} B)$, the statement (c) is a direct consequence of (1.7) and the equality

$$f(\operatorname{\mathbf{dim}} M) = \operatorname{\mathbf{dim}} \operatorname{Hom}_A(T, M),$$

for any M from $\mathcal{R}(A)$, where $f : K_0(A) \longrightarrow K_0(B)$ is the group isomorphism defined above. $\qquad\square$

5.2. Theorem. *Let $A = KQ$ be a connected wild hereditary K-algebra. Let T be a postprojective tilting A-module, $B = \operatorname{End}_A(T)$, and M, N indecomposable regular B-modules. Then the following hold.*

(a) *There exists a natural number m_0 such that $\operatorname{Hom}_B(M, \tau^{-m} N) = 0$ and $\operatorname{Hom}_B(M, \tau_B^m N) \neq 0$, for all $m \geq m_0$.*

(b) *If $\operatorname{Hom}_B(M, \tau_B^{-r} M) = 0$, for some $r \geq 1$, then $\operatorname{Hom}_B(M, \tau_B^{-s} M) = 0$, for all $s \geq r$.*

(c) *If $\operatorname{Ext}_B^1(M, M) \neq 0$, then $\operatorname{Hom}_B(M, \tau_B^r M) \neq 0$, for all $r \geq 1$.*

(d) *The B-module M is a quasi-simple brick if and only if $\operatorname{Hom}_B(M, \tau_B^{-1} M) = 0$.*

(e) *There exist paths*

$$M \to \ldots \to N \quad \text{and} \quad N \to \ldots \to M$$

in $\operatorname{mod} B$.

(f) *M and N are non-directing B-modules.*

Proof. In view of the K-linear equivalence of categories

$$\operatorname{Hom}_A(T, -) : \operatorname{add} \mathcal{R}(A) \xrightarrow{\;\simeq\;} \operatorname{add} \mathcal{R}(B)$$

the statements (a)–(d) are direct consequences of (2.6)–(2.9) and (2.11). $\quad\square$

5.3. Proposition. *Let $A = KQ$ be a connected wild hereditary K-algebra. Let T be a tilting A-module and*

$$B = \operatorname{End}_A(T).$$

Then every stone in $\operatorname{mod} B$ *is a brick.*

Proof. By (VIII.3.3), every stone in $\operatorname{mod} A$ is a brick. Then the same holds in $\operatorname{mod} B$, because $(\mathcal{Y}(T), \mathcal{X}(T))$ is a splitting torsion theory in $\operatorname{mod} B$ and the functors

$$\operatorname{Hom}_A(T, -) : \mathcal{T}(T) \longrightarrow \mathcal{Y}(T) \text{ and } \operatorname{Ext}^1_A(T, -) : \mathcal{F}(T) \longrightarrow \mathcal{X}(T)$$

are equivalences of categories inducing isomorphisms between the corresponding extension vector spaces and homomorphism vector spaces. $\square$

The following theorem describes properties of regular bricks and stones over concealed algebras of wild type. Note that the rank of the Grothendieck group $K_0(B)$ of any tilted algebra B of type Q is equal to $n = |Q_0|$, the rank of $K_0(A)$.

5.4. Theorem. *Let $A = KQ$ be a connected wild hereditary K-algebra. Let T be a postprojective tilting A-module, $B = \operatorname{End}_A(T)$, M a quasi-simple regular B-module, and $r \geq 1$ a natural number. Then, in the notation of (1.5), we have:*
  (a) *$M[r]$ is a stone if and only if $M[r + 1]$ is a brick;*
  (b) *if $M[r]$ is a brick then $r \leq n - 1$; and*
  (c) *if $M[r]$ is a stone then $r \leq n - 2$.*

Proof. In view of the equivalence $\operatorname{Hom}_A(T, -) : \operatorname{add} \mathcal{R}(A) \xrightarrow{\sim} \operatorname{add} \mathcal{R}(B)$ the statements (a)–(c) follow directly from (2.14) and (2.15). $\square$

The following theorem shows the strict wildness of the category of regular modules, hence all modules, over concealed algebras of a wild type Q.

5.5. Theorem. *Let $A = KQ$ be a connected wild hereditary K-algebra. Let T be a postprojective tilting A-module,*

$$B = \operatorname{End}_A(T),$$

and Λ an arbitrary finite dimensional K-algebra. Then there exists a full, faithful, exact and K-linear functor

$$F : \operatorname{mod} \Lambda \longrightarrow \operatorname{add} \mathcal{R}(B).$$

If, in addition, Λ is a local algebra, then $F(\Lambda)$ is a quasi-simple regular module and there is a K-algebra isomorphism $\operatorname{End}_B(F(\Lambda)) \cong \Lambda$.

Proof. Note that the equivalence $\operatorname{Hom}_A(T, -) : \operatorname{add} \mathcal{R}(A) \xrightarrow{\sim} \operatorname{add} \mathcal{R}(B)$

is a full, faithful, exact and K-linear functor preserving the quasi-simplicity. Then our claim is a direct consequence of (4.1), (4.2) and (4.6).   □

5.6. Example. Let $A = KQ$ be the path algebra of the wild quiver

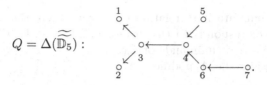

$$Q = \Delta(\widetilde{\mathbb{D}}_5):$$

The standard calculation technique shows that the left hand part of the postprojective component $\mathcal{P}(A)$ of $\Gamma(\mathrm{mod}\,A)$ looks as follows

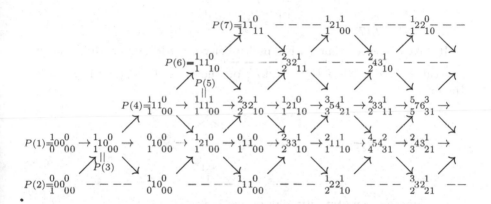

and the indecomposable modules are represented by their dimension vectors.

<u>Case 1°</u> Consider the A-module

$$T_A = T_1 \oplus T_2 \oplus T_3 \oplus T_4 \oplus T_5 \oplus T_6 \oplus T_7,$$

where

$$T_1 = \tau^{-1}P(1),\; T_2 = \tau^{-2}P(2),\; T_3 = \tau^{-1}P(4),\; T_4 = \tau^{-3}P(1),$$
$$T_5 = P(5),\; T_6 = \tau^{-1}P(6),\; \text{and}\; T_7 = \tau^{-1}P(7).$$

It is easy to check that T_A is a multiplicity-free tilting A-module and the concealed algebra $B = \mathrm{End}\,T_A$ of wild type $Q = \Delta(\widetilde{\mathbb{D}}_5)$ is given by the quiver

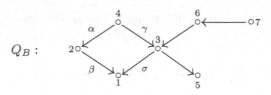

bound by the commutativity relation $\alpha\beta = \gamma\sigma$, where the ordering of the vertices of Q_B corresponds to the ordering $T_1, T_2, T_3, T_4, T_5, T_6, T_7$ of the indecomposable direct summands of T_A.

Case 2° Consider the A-module

$$T'_A = T'_1 \oplus T'_2 \oplus T'_3 \oplus T'_4 \oplus T'_5 \oplus T'_6 \oplus T'_7,$$

where

$$\begin{array}{ll} T'_1 = P(5), & T'_2 = \tau^{-1}P(6), \\ T'_3 = \tau^{-2}P(3), & T'_4 = \tau^{-2}P(4), \\ T'_5 = \tau^{-2}P(1), & T'_6 = \tau^{-3}P(2), \text{ and} \\ T'_7 = P(7). & \end{array}$$

It is easy to check that T'_A is a multiplicity-free tilting A-module and the concealed tilted algebra $B' = \operatorname{End} T'_A$ of wild type $Q = \Delta(\widetilde{\widetilde{\mathbb{D}}}_5)$ is given by the quiver

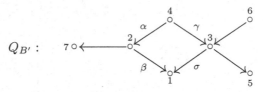

bound by the commutativity relation $\alpha\beta = \gamma\sigma$.

It follows from (5.5) that any concealed K-algebra B of wild type Q is strictly representation-wild, and therefore a classification of all indecomposable B-modules leads to corresponding classifications of the indecomposable modules over all finite dimensional K-algebras, which is an impossible task. On the other hand, the following examples show that there are tilted algebras of wild type Q for which all indecomposable modules can be described explicitly.

5.7. Example. Let B be given by the quiver

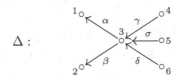

bound by the zero relations $\gamma\alpha = 0$, $\gamma\beta = 0$, $\sigma\alpha = 0$, $\sigma\beta = 0$, $\delta\alpha = 0$ and $\delta\beta = 0$. Then $\Gamma(\operatorname{mod} B)$ is of the form

$$
\begin{array}{ccccccccc}
P(1) & & I(2) & & P(4) & & M/P(4) & & S(4) \\
\searrow & \nearrow & & \searrow & \nearrow & & \searrow & \nearrow & \\
& P(3) & & S(3) \to P(5) \to M \to M/P(5) \to I(3) \to S(5) \\
\nearrow & & \searrow & \nearrow & & \searrow & \nearrow & & \searrow \\
P(2) & & I(1) & & P(6) & & M/P(6) & & S(6)
\end{array}
$$

where M is the indecomposable module with $\operatorname{\mathbf{dim}} M = \begin{smallmatrix} & 0 & 1 \\ & 2 & 1 \\ & 0 & 1 \end{smallmatrix}$. Note that the modules

$$I(1),\, I(2),\, S(3),\, P(4),\, P(5), \text{ and } P(6)$$

form a faithful section Σ such that $\operatorname{Hom}_A(U, \tau_B V) = 0$, for all modules $U, V \in \Sigma$. Then, by (VIII.5.6), B is a representation-finite tilted algebra of the wild type $\Delta = \Sigma^{\operatorname{op}} = Q_B$, and obviously $\mathcal{C} = \Gamma(\operatorname{mod} B)$ is the unique connecting component of $\Gamma(\operatorname{mod} B)$. Moreover, because this component contains projective modules and injective modules, then it follows from (VIII.4.1) that, if $B = \operatorname{End}_{K\Delta}(T)$, for a tilting $K\Delta$-module T, then T contains at least one postprojective and at least one preinjective direct summand.

We also notice that the Euler quadratic form $q_B : \mathbb{Z}^6 \longrightarrow \mathbb{Z}$ of the algebra B is indefinite, because q_B is $\mathbb{Z}$-congruent to the Euler quadratic form $q_{K\Delta} = q_\Delta : \mathbb{Z}^6 \longrightarrow \mathbb{Z}$ of $K\Delta$, see (VI.4.7) and (VII.4.5). On the other hand, q_B is weakly positive, because B is representation-directed and (IX.3.3) applies.

5.8. Example. Let B be given by the quiver

$$\Delta : 1 \circ \underset{\beta}{\overset{\alpha}{\rightleftarrows}} \overset{2}{\circ} \underset{\sigma}{\overset{\gamma}{\rightleftarrows}} \circ\, 3,$$

bound by the zero relations $\gamma\alpha = 0$, $\gamma\beta = 0$, $\sigma\alpha = 0$, and $\sigma\beta = 0$. Then $\Gamma(\operatorname{mod} B)$ has a component $\mathcal{C}$ of the form

$$
\begin{array}{c}
P(3) \;---\tau^{-1}P(3)--- \quad \cdots \\
\nearrow \quad \searrow \quad \nearrow \quad \searrow \quad \nearrow \\
---\; \tau^2 S(2) \;---\; \tau S(2) \;---\; S(2) \;---\tau^{-1}S(2)---\tau^{-2}S(2) \;--- \\
\nearrow \quad \searrow \quad \nearrow \quad \searrow \quad \nearrow \\
\cdots \quad ---\; \tau I(1) \;---\; I(1)
\end{array}
$$

and obviously the modules $I(1)$, $S(2)$, $P(3)$ form a faithful section Σ in $\mathcal{C}$ such that $\operatorname{Hom}_B(U, \tau V) = 0$, for all modules $U, V \in \Sigma$. Hence, B is a representation-infinite tilted algebra of wild type

$$\Delta = \Sigma^{\operatorname{op}} = Q_B$$

and, according to (VIII.5.6), $\mathcal{C}$ is the connecting component $\mathcal{C}_T$ in $\Gamma(\operatorname{mod} B)$ determined by a tilting module T over $K\Delta$ such that $B = \operatorname{End}_{K\Delta}(T)$.

Moreover, if H' and H'' denote the Kronecker algebras given respectively by the vertices 1 and 2, and 2 and 3 then, according to (VIII.3.5), the quiver $\Gamma(\operatorname{mod} B)$ is of the form

$$\Gamma(\operatorname{mod} B) = \mathcal{P}(H') \cup \mathcal{R}(H') \cup \mathcal{C}_T \cup \mathcal{R}(H'') \cup \mathcal{Q}(H''),$$

and $\mathcal{C} = \mathcal{C}_T$ is a glueing of $\mathcal{Q}(H')$ with $\mathcal{P}(H'')$ using the simple module $S(2)$. In particular, B is a representation-infinite tilted algebra of wild type Δ for which all indecomposable modules can be described explicitly. Finally, we note also that the tilting $K\Delta$-module T has both a postprojective and a preinjective direct summand, because the connecting component $\mathcal{C} = \mathcal{C}_T$ contains a projective module and an injective module (see (VIII.4.1)).

It follows from (VI.4.7) that $K_0(B) = \mathbb{Z}^3$ and the Euler quadratic form $q_B : \mathbb{Z}^3 \longrightarrow \mathbb{Z}$ of the algebra B is indefinite, because q_B is $\mathbb{Z}$-congruent to the indefinite Euler quadratic form

$$q_{K\Delta} = q_\Delta : \mathbb{Z}^3 \longrightarrow \mathbb{Z}$$

of $K\Delta$, see (VII.4.5).

Now we show that $q_B : \mathbb{Z}^3 \longrightarrow \mathbb{Z}$ is weakly non-negative, that is, $q_B(\mathbf{x}) \geq 0$, for any non-negative vector $\mathbf{x} \in \mathbb{Z}^3$. To see this we note that the Cartan matrix $\mathbf{C}_B$ of B and its inverse $\mathbf{C}_B^{-1}$ are of the forms

$$\mathbf{C}_B = \begin{bmatrix} 1 & 2 & 0 \\ 0 & 1 & 2 \\ 0 & 0 & 1 \end{bmatrix} \text{ and } \mathbf{C}_B^{-1} = \begin{bmatrix} 1 & -2 & 4 \\ 0 & 1 & -2 \\ 0 & 0 & 1 \end{bmatrix}.$$

Then the Euler quadratic form $q_B : \mathbb{Z}^3 \longrightarrow \mathbb{Z}$ of the algebra B is defined by the formula

$$\begin{aligned} q_B(\mathbf{x}) = \mathbf{x}^t \cdot (\mathbf{C}_B^{-1})^t \cdot \mathbf{x} &= x_1^2 + x_2^2 + x_3^2 - 2x_1x_2 - 2x_2x_3 + 4x_1x_3 \\ &= (x_1 - x_2 + x_3)^2 + 2x_1x_3, \end{aligned}$$

for $\mathbf{x} = [x_1 \; x_2 \; x_3]^t \in \mathbb{Z}^3 = K_0(B)$. It follows easily that $q(\mathbf{x}) \geq 0$, for each vector $\mathbf{x} \in \mathbb{Z}^3$, with $x_1 \geq 0$, $x_2 \geq 0$, and $x_3 \geq 0$, that is, $q_B : \mathbb{Z}^3 \longrightarrow \mathbb{Z}$ is weakly non-negative. Because $q_B(\mathbf{y}) = -1$, for $\mathbf{y} = [1 \; 1 \; -1]^t \in \mathbb{Z}^3$, then q_B is indefinite.

Our next aim is to describe the regular components of tilted algebras of wild type Q given by tilting modules without preinjective (respectively, postprojective) direct summands. We need the following technical lemma.

5.9. Lemma. *Let $A = KQ$ be a connected representation-infinite hereditary algebra, T be a tilting A-module without preinjective direct summands, $\mathcal{T} = \mathcal{T}(T)$ be the associated torsion class and $B = \operatorname{End}_A(T)$.*

(a) *If $Z \in \mathcal{T}$ is an indecomposable module then there exists a short exact sequence*

$$0 \longrightarrow \tau_T Z \longrightarrow \tau_A Z \longrightarrow F \longrightarrow 0,$$

where $\tau_{\mathcal{T}}Z$ is the largest submodule of $\tau_A Z$ from $\mathcal{T}$ and $F \in \mathrm{add}(\tau_A T)$.

(b) *If $Z \in \mathcal{T}$ is an indecomposable module with $\tau_{\mathcal{T}}^m Z \neq 0$, for some positive integer m, then, for each $1 \leq i \leq m$, there exists a short exact sequence*

$$0 \longrightarrow \tau_A^{-i}\tau_{\mathcal{T}}^i Z \longrightarrow \tau_A^{-i+1}\tau_{\mathcal{T}}^{i-1}Z \longrightarrow \tau_A^{-i}F^{(i)} \longrightarrow 0,$$

with $F^{(i)} \in \mathrm{add}(\tau_A T)$.

Proof. (a) It follows from (VI.1.5) that there exists a canonical exact sequence $0 \longrightarrow \tau_{\mathcal{T}}Z \longrightarrow \tau_A Z \longrightarrow F \longrightarrow 0$, with $F \in \mathcal{F} = \mathcal{F}(T)$. Because $D(A) \in \mathcal{T} = \mathcal{T}(T)$, F has no injective direct summands, and applying the functor τ_A^{-1} we obtain a short exact sequence

$$0 \longrightarrow \tau_A^{-1}\tau_{\mathcal{T}}Z \longrightarrow Z \longrightarrow \tau_A^{-1}F \longrightarrow 0.$$

Observe that $Z \in \mathcal{T}$ implies $\tau_A^{-1}F \in \mathcal{T}$. Therefore $\tau_A^{-1}F \in \mathcal{T}$, $\tau_A(\tau_A^{-1}F) = F \in \mathcal{F}$, and invoking (VI.1.11) we conclude that $\tau_A^{-1}F$ is an Ext-projective module in $\mathcal{T}$. Applying now (VI.2.5) we obtain $\tau_A^{-1}F \in \mathrm{add}(T)$, and hence $F \in \mathrm{add}(\tau_A T)$.

(b) Assume that Z is an indecomposable module in $\mathcal{T}$ with $\tau_{\mathcal{T}}^m Z \neq 0$, for some $m \geq 1$. For $i = 1$, we may take $F^{(1)} = F$ and the above sequence ending at $\tau_A^{-1}F$ satisfies the required conditions. Assume that there exists a short exact sequence

$$0 \longrightarrow \tau_A^{-i}\tau_{\mathcal{T}}^i Z \longrightarrow \tau_A^{-i+1}\tau_{\mathcal{T}}^{i-1}Z \longrightarrow \tau_A^{-i}F^{(i)} \longrightarrow 0,$$

with $F^{(i)} \in \mathrm{add}(\tau_A T)$, and $i < m$. Applying now (a) to the module $\tau_{\mathcal{T}}^i Z \in \mathcal{T}$ we obtain a short exact sequence

$$0 \longrightarrow \tau_{\mathcal{T}}^{i+1}Z \longrightarrow \tau_A \tau_{\mathcal{T}}^i Z \longrightarrow F^{(i+1)} \longrightarrow 0,$$

with $F^{(i+1)} \in \mathrm{add}(\tau_A T)$. Because $F^{(i+1)} \in \mathcal{F} = \mathcal{F}(T)$ and all preinjective A-modules belong to $\mathcal{T} = \mathcal{T}(T)$, applying the functor $\tau_A^{-(i+1)}$ to the above exact sequence, we obtain the required exact sequence

$$0 \longrightarrow \tau_A^{-(i+1)}\tau_{\mathcal{T}}^{i+1}Z \longrightarrow \tau_A^{-i}\tau_{\mathcal{T}}^i Z \longrightarrow \tau_A^{-(i+1)}F^{(i+1)} \longrightarrow 0.$$

Therefore our claim follows by induction on $i \in \{1, \ldots, m\}$. □

Given a component $\mathcal{C}$ of $\Gamma(\mathrm{mod}\, A)$ and a module X in $\mathcal{C}$, we denote

- by $\mathcal{C}(\longrightarrow X)$ the full translation subquiver of $\mathcal{C}$ formed by all predecessors of X in $\mathcal{C}$, and
- by $\mathcal{C}(X \longrightarrow)$ the full translation subquiver of $\mathcal{C}$ formed by all successors of X in $\mathcal{C}$.

5.10. Proposition. *Let $A = KQ$ be a connected wild hereditary K-algebra. Let T be a tilting A-module without preinjective direct summands and $B = \operatorname{End}_A(T)$.*

   (a) *For each regular component $\mathcal{C}$ in $\Gamma(\operatorname{mod} A)$, there exists a quasi-simple module X in $\mathcal{C}$ such that $\mathcal{C}(\longrightarrow X)$ is contained in $\mathcal{T}(T)$, and the image of $\mathcal{C}(\longrightarrow X)$ via the functor*

$$\operatorname{Hom}_A(T, -) : \operatorname{mod} A \longrightarrow \operatorname{mod} B$$

   is the full translation subquiver $\mathcal{D}(\longrightarrow \operatorname{Hom}_A(T, X))$ of $\Gamma(\operatorname{mod} B)$, where $\mathcal{D} = \operatorname{Hom}_A(T, \mathcal{C})$.

   (b) *For each regular component $\mathcal{D}$ of $\Gamma(\operatorname{mod} B)$ completely contained in $\mathcal{Y}(T)$, there exists a module $Y = \operatorname{Hom}_A(T, X)$ in $\mathcal{D}$ such that X is a quasi-simple module of a regular component $\mathcal{C}$ of $\Gamma(\operatorname{mod} A)$ and $\mathcal{D}(\longrightarrow Y)$ is the image of $\mathcal{C}(\longrightarrow X)$ via the functor*

$$\operatorname{Hom}_A(T, -) : \operatorname{mod} A \longrightarrow \operatorname{mod} B.$$

   In particular, $\mathcal{D}$ is of type $\mathbb{Z}\mathbb{A}_\infty$.

   (c) *The quiver $\Gamma(\operatorname{mod} B)$ contains a $\operatorname{card}(K)$-family of regular components of type $\mathbb{Z}\mathbb{A}_\infty$, that is, a family of cardinality $\operatorname{card}(K)$.*

Proof. If T is postprojective, then

$$\operatorname{Hom}_A(T, -) : \operatorname{add} \mathcal{R}(A) \xrightarrow{\;\simeq\;} \operatorname{add} \mathcal{R}(B)$$

is an equivalence of categories, and the statements (a)–(c) follow from the corresponding results proved for $A = KQ$ in previous sections.

Then, we may assume that T is not postprojective. Suppose that $T = T^{pp} \oplus T^{rg}$, where T^{pp} is postprojective and $T^{rg} \neq 0$ is regular.

(a) Let $\mathcal{C}$ be a regular component of $\Gamma(\operatorname{mod} A)$. First note that, in view of (VIII.2.3) and (VIII.2.5), we have $\operatorname{Ext}^1_A(T^{pp}, M) \cong D\operatorname{Hom}_A(M, \tau_A T^{pp}) = 0$, for any regular A-module M. By applying (2.6) to the regular module T^{rg} we conclude that $\operatorname{Hom}_A(\tau_A^m X, \tau_A T^{rg}) = 0$, for some quasi-simple module X in $\mathcal{C}$ and all $m \geq 0$.

Then $\operatorname{Ext}^1_A(T, \tau_A^m X) \cong D\operatorname{Hom}_A(\tau_A^m X, \tau_A T) = 0$, for all $m \geq 0$. Observe also that if $\operatorname{Hom}_A(Z, \tau_A T) \neq 0$, for some Z in $\mathcal{C}(\longrightarrow X)$, then $\operatorname{Hom}_A(\tau_A^r X, \tau_A T) \neq 0$, for some $r \geq 0$. Therefore

$$\operatorname{Ext}^1_A(T, Z) = D\operatorname{Hom}_A(Z, \tau_A T) = 0,$$

for any module Z in $\mathcal{C}(\longrightarrow X)$, and so $\mathcal{C}(\longrightarrow X)$ is contained in $\mathcal{T}(T)$. Hence the image of $\mathcal{C}(\longrightarrow X)$ via $\operatorname{Hom}_A(T, -)$ is a full translation subquiver of $\Gamma(\operatorname{mod} B)$ which is closed under predecessors, and so is equal to $\mathcal{D}(\longrightarrow \operatorname{Hom}_A(T, X))$, where $\mathcal{D} = \operatorname{Hom}_A(T, \mathcal{C})$, see (VI.5.2).

(b) Let $\mathcal{D}$ be a regular component of $\Gamma(\text{mod } B)$ which is contained in $\mathcal{Y}(T)$. Let M be a module in $\mathcal{D}$. Then there exists an indecomposable regular A-module $Z \in \mathcal{T}(T)$ such that $M = \text{Hom}_A(T, Z)$.

Let $\mathcal{C}$ be the regular component of $\Gamma(\text{mod } A)$ containing Z. Because $\mathcal{D}$ is a regular component and $\mathcal{Y}(T)$ is closed under predecessors, for all $i \geq 0$, we have $0 \neq \tau_B^i M = \text{Hom}_A(T, \tau_T^i Z)$, and hence $\tau_T^i Z \neq 0$. By applying (5.9)(b), we construct an infinite chain of monomorphisms

$$\cdots \longrightarrow \tau_A^{-i-1}\tau_T^{i+1} Z \longrightarrow \tau_A^{-i}\tau_T^i Z \longrightarrow \cdots \longrightarrow \tau_A^{-2}\tau_T^2 Z \longrightarrow \tau_A^{-1}\tau_T Z \longrightarrow Z.$$

It follows that there exists an $m \geq 0$ such that $\tau_A^{-m-r}\tau_T^{m+r} Z \cong \tau_A^{-m}\tau_T^m Z$, for all $r \geq 0$. For $N = \tau_T^m Z$, we then have $\tau_A^{-m-r}\tau_T^r N \cong \tau_A^{-m} N$, and hence $\tau_T^r N \cong \tau_A^r N$, for all $r \geq 0$. In particular, the module N is not postprojective.

Because T has no preinjective direct summands, the whole preinjective component $\mathcal{Q}(A)$ of $\Gamma(\text{mod } A)$ is mapped by the functor $\text{Hom}_A(T, -)$ to the connecting component $\mathcal{C}_T$ of $\Gamma(\text{mod } B)$ determined by the tilting module T. Moreover, $\mathcal{C}_T$ is not a regular component contained in $\mathcal{Y}(T)$. Therefore, N is a regular A-module.

Finally, because the modules $\tau_A^r N \cong \tau_T^r N$, with $r \geq 0$, belong to $\mathcal{T} = \mathcal{T}(T)$ we conclude that $\mathcal{C}(-\!\!\rightarrow\!N)$ is entirely contained in $\mathcal{T}$. Take now a quasi-simple module X in $\mathcal{C}(-\!\!\rightarrow\!N)$, and set

$$Y = \text{Hom}_B(T, X).$$

Then X is a quasi-simple module of the regular component $\mathcal{C}$, Y belongs to $\mathcal{D}$ and $\mathcal{D}(-\!\!\rightarrow\!Y)$ is the image of $\mathcal{C}(-\!\!\rightarrow\!X)$ via the functor $\text{Hom}_A(T, -)$. In particular, $\mathcal{D}$ is type $\mathbb{Z}\mathbb{A}_\infty$.

(c) In view of (a) and (b), the statement (c) is a consequence of (1.8) and (5.1). This completes the proof. $\square$

We have also the following fact that is dual to (5.10).

5.11. Proposition. *Let $A = KQ$ be a connected wild hereditary K-algebra, T a tilting A-module without postprojective direct summands, and $B = \text{End}_A(T)$.*

(a) *For each regular component $\mathcal{C}$ in $\Gamma(\text{mod } A)$, there exists a quasi-simple module X in $\mathcal{C}$ such that $\mathcal{C}(X\!\!\longrightarrow\!)$ is contained in $\mathcal{F}(T)$, and the image of $\mathcal{C}(X\!\!\longrightarrow\!)$ via the functor*

$$\text{Ext}_A^1(T, -) : \text{mod } A \longrightarrow \text{mod } B$$

is the full translation subquiver $\mathcal{D}(\text{Ext}_A^1(T, X)\!\!\longrightarrow\!)$ in $\Gamma(\text{mod } B)$, where $\mathcal{D} = \text{Ext}_A^1(T, \mathcal{C})$.

(b) *For each regular component $\mathcal{D}$ of $\Gamma(\mathrm{mod}\, B)$ completely contained in $\mathcal{X}(T)$, there exists a module $Y = \mathrm{Ext}^1_A(T, X)$ in $\mathcal{D}$ such that X is a quasi-simple module of a regular component $\mathcal{C}$ of $\Gamma(\mathrm{mod}\, A)$ and $\mathcal{D}(Y \longrightarrow)$ is the image of $\mathcal{C}(X \longrightarrow)$ via the functor*

$$\mathrm{Ext}^1_A(T, -) : \mathrm{mod}\, A \longrightarrow \mathrm{mod}\, B.$$

In particular, $\mathcal{D}$ is of type $\mathbb{Z}\mathbb{A}_\infty$.

(c) *The quiver $\Gamma(\mathrm{mod}\, B)$ contains a $\mathrm{card}(K)$-family of regular components of type $\mathbb{Z}\mathbb{A}_\infty$.*

Proof. Modify the arguments given in the proof of (5.10).   □

5.12. Corollary. *Let $A = KQ$ be a connected wild hereditary K-algebra, T a regular tilting A-module, and $B = \mathrm{End}_A(T)$.*

(a) *For each regular component $\mathcal{C}$ of $\Gamma(\mathrm{mod}\, A)$, there exists a quasi-simple module X in $\mathcal{C}$ and a natural number m such that $\mathcal{C}(\longrightarrow X)$ is contained in $\mathcal{T}(T)$ and $\mathcal{C}(\tau^{-m} X \longrightarrow)$ is contained in $\mathcal{F}(T)$. The image of $\mathcal{C}(\longrightarrow X)$ via $\mathrm{Hom}_A(T, -)$ is a full translation subquiver of $\Gamma(\mathrm{mod}\, B)$, and the image of $\mathcal{C}(\tau^{-m} X \longrightarrow)$ via $\mathrm{Ext}^1_A(T, -)$ is a full translation subquiver of $\Gamma(\mathrm{mod}\, B)$.*

(b) *Every regular component $\mathcal{D}$ of $\Gamma(\mathrm{mod}\, B)$ different from the connecting component $\mathcal{C}_T$ is of type $\mathbb{Z}\mathbb{A}_\infty$ and is contained either in $\mathcal{Y}(T)$ or in $\mathcal{X}(T)$. In particular, there exists a regular component $\mathcal{C}$ of $\Gamma(\mathrm{mod}\, A)$ and a quasi-simple module X in $\mathcal{C}$ such that either $X \in \mathcal{T}(T)$ and $\mathcal{D}(\longrightarrow \mathrm{Hom}_A(T, X))$ is a full translation subquiver of $\mathcal{D}$ or $X \in \mathcal{F}(T)$ and $\mathcal{D}(\mathrm{Ext}^1_A(T, X) \longrightarrow)$ is a full translation subquiver of $\mathcal{D}$.*

(c) *The quiver $\Gamma(\mathrm{mod}\, A)$ contains two different $\mathrm{card}(K)$-families of regular components of type $\mathbb{Z}\mathbb{A}_\infty$.*

Proof. The statement (a) follows from (5.10) and (5.11). To prove (b), observe that every regular component of $\Gamma(\mathrm{mod}\, B)$ different from $\mathcal{C}_T$ lies entirely in $\mathcal{Y}(T)$ or in $\mathcal{X}(T)$. Then, in view of (a) and (b), the statement (c) is a consequence of (5.10) and (5.11).   □

We also note that if a tilting A-module T contains a non-zero postprojective (respectively, preinjective) direct summand then, for each regular component $\mathcal{C}$ of $\Gamma(\mathrm{mod}\, A)$, the set $\mathcal{F}(T) \cap \mathcal{C}$ (respectively, $\mathcal{T}(T) \cap \mathcal{C}$) has at most finitely many modules. Indeed, it is a direct consequence of the fact (2.4) that all but finitely many modules in $\mathcal{C}$ are sincere.

5.13. Example. Let A be a connected wild hereditary algebra and assume that X is a quasi-simple regular stone in $\operatorname{mod} A$.

It follows from (3.10), (3.11), and (3.15) that there exists a multiplicity-free tilting A-module $T = X \oplus Y$ such that Y is a projective generator of $X^{\perp}$, and $\operatorname{Hom}_A(Y, -)$ establishes an equivalence

$$X^{\perp} \xrightarrow{\ \simeq\ } \operatorname{mod} C,$$

where

$$C = \operatorname{End}_A(Y)$$

is a connected wild hereditary algebra. The Grothendieck group $K_0(C)$ of C is of rank $n - 1$, where n is the rank of $K_0(A)$. Moreover, the module $R = [2]X$ is a quasi-simple regular brick in $\operatorname{mod} C$.

Because $\operatorname{Hom}_A(Y, \tau_A X) \cong D\operatorname{Ext}_A^1(X, Y) = 0$, then the irreducible epimorphism $R \longrightarrow X$ induces the isomorphism $\operatorname{Hom}_A(Y, R) \cong \operatorname{Hom}_A(Y, X)$.

Consider the tilted algebra

$$B = \operatorname{End}_A(T).$$

Because $\operatorname{Hom}_A(X, Y) = 0$, then B is the one-point extension $C[R]$ of C by the quasi-simple regular C-module R, that is,

$$B \cong C[R].$$

Denote by ω the extension vertex of $C[R]$. Because $\operatorname{Hom}_C(R, M) = 0$, for any postprojective C-module M, then (XV.1.7) implies that the postprojective component $\mathcal{P}(C)$ of $\Gamma(\operatorname{mod} C)$ is a postprojective component of $\Gamma(\operatorname{mod} B)$ containing all indecomposable projective B-modules, except the module $P(\omega)$.

Now we describe the shape of the component of $\Gamma(\operatorname{mod} B)$ containing the projective module $P(\omega)$.

Because R is a quasi-simple regular brick in $\operatorname{mod} C$, we know by (2.10) that

$$\operatorname{Hom}_C(R, \tau_C^{-m} R) = 0, \quad \text{for all } m \geq 1.$$

Consequently, the full translation subquiver $\mathcal{D}$ of $\Gamma(\operatorname{mod} C)$ formed by all successors of the module $\tau_C^{-1} R$ in $\Gamma(\operatorname{mod} C)$ remains a full translation subquiver (of type $(-\mathbb{N})\mathbb{A}_\infty$) of $\Gamma(\operatorname{mod} B)$, and clearly is closed under successors in $\Gamma(\operatorname{mod} B)$. Applying the notation of (1.5) and the formula (XV.1.6) on the lifting of almost split sequences for the one-point extension $B \cong C[R]$, we deduce that there are almost split sequences of the forms

$$0 \longrightarrow R[i] \longrightarrow R[i+1] \oplus \overline{R[i]} \longrightarrow \overline{R[i+1]} \longrightarrow 0,$$

$$0 \longrightarrow \overline{R[i+1]} \longrightarrow \tau_C^{-1} R[i] \oplus \overline{R[i+2]} \longrightarrow \tau_C^{-1} R[i+1]] \longrightarrow 0,$$

$$0 \longrightarrow \tau_B R[i+1] \longrightarrow R[i] \oplus \tau_B R[i+2] \longrightarrow R[i+1] \longrightarrow 0,$$

in $\text{mod } B$, for all $i \geq 1$, and almost split sequences

$$0 \longrightarrow \overline{R[1]} \longrightarrow \overline{R[2]} \longrightarrow \tau_C^{-1} R[1] \longrightarrow 0,$$
$$0 \longrightarrow \tau_B R[1] \longrightarrow \tau_B R[2] \longrightarrow R[1] \longrightarrow 0,$$

where $R = R[1]$, and $P(\omega) = \overline{R[1]}$. Because B is a tilted algebra, then the only component of $\Gamma(\text{mod } B)$ containing both a projective module and an injective module is the connecting one. Therefore, the component of $\Gamma(\text{mod } B)$ containing $P(\omega)$ is of the form

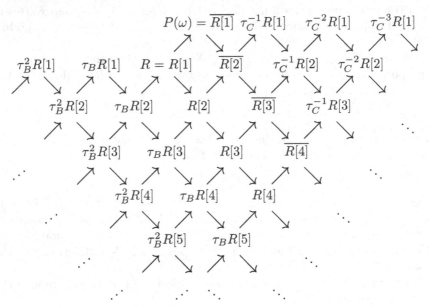

and consists of the τ_B-orbits of the modules $\overline{R[i]}$, $i \geq 1$. From (2.6) we also know that if $\mathcal{C}$ is a regular component on $\Gamma(\text{mod } C)$ then there exists a quasi-simple module Z in $\mathcal{C}$ such that $\text{Hom}_{\mathcal{C}}(R, \tau_C^{-m} Z) = 0$, for all $m \geq 0$, and consequently, by (XV.1.7), the full translation subquiver of $\mathcal{C}$ formed by all successors of Z in $\mathcal{C}$ is a full translation subquiver of $\Gamma(\text{mod } B)$ closed under successors in $\Gamma(\text{mod } B)$. Therefore, $\Gamma(\text{mod } B)$ admits a $\text{card}(K)$-family of regular components of type $\mathbb{Z}\mathbb{A}_\infty$.

5.14. Example. Let $n \geq 1$ and let $H = K\Delta$, where Δ is the wild quiver

$$0 \Longleftarrow 1 \longleftarrow 2 \longleftarrow \cdots \longleftarrow n \Longleftarrow \omega.$$

For $i, j \in \{1, \ldots, n\}$ with $i \leq j$, denote by $[i, j]$ the indecomposable H-module

$$[i,j]: \quad 0 \Longleftarrow 0 \leftarrow \ldots \leftarrow 0 \leftarrow K \leftarrow \ldots \leftarrow K \leftarrow 0 \leftarrow \ldots \leftarrow 0 \Longleftarrow 0$$

whose support is the subquiver $i \leftarrow (i+1) \leftarrow \ldots \leftarrow j$ of Δ.

Let Σ be the full subquiver of Δ given by the vertices $0, \ldots, n$ and Ω the full Dynkin subquiver of Δ formed by the vertices $1, \ldots, n$. We set

$$D = K\Sigma \quad \text{and} \quad \Lambda = K\Omega.$$

Then the Auslander–Reiten quiver $\Gamma(\operatorname{mod}\Lambda)$ of the algebra Λ is finite of the form

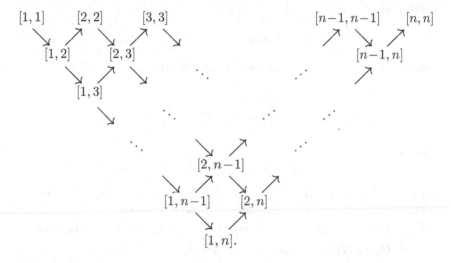

Calculating the end of the preinjective component $\mathcal{Q}(D)$ of D, we conclude that $\Gamma(\operatorname{mod}\Lambda)$ is a full translation subquiver of $\mathcal{Q}(D)$ which is closed under successors.

Observe now that $H \cong D[I_D(0) \oplus I_D(0)]$ that is, the algebra H is a one-point extension of D by the direct sum of two copies of the injective module $I_D(0)$. Because

$$I_D(0)/\operatorname{soc} I_D(0) \cong [1,n] \oplus [1,n] \quad \text{and} \quad \operatorname{Hom}_\Lambda([1,n],[i,j]) = 0,$$

for each $[i,j] \notin \{[1,n], [2,n], \ldots, [n,n]\}$, then from the formula (XV.1.6) on the lifting of almost split sequences for one-point extensions, we conclude that $\Gamma(\operatorname{mod}\Lambda)$ is a full translation subquiver of $\Gamma(\operatorname{mod}H)$.

Moreover, a direct analysis of the beginning of the postprojective component $\mathcal{P}(H)$ and the end of the preinjective component $\mathcal{Q}(H)$ shows that the simple Λ-modules $S(1) = [1,1], \ldots, S(n) = [n,n]$ are neither in $\mathcal{P}(H)$ nor in $\mathcal{Q}(H)$. Therefore, $\Gamma(\operatorname{mod}\Lambda)$ is the cone $\mathcal{C}([1,n])$ of a regular component of $\Gamma(\operatorname{mod}H)$. Let

$$M = [1,n] \oplus [2,n] \oplus \ldots \oplus [n-1,n] \oplus [n,n] \quad \text{and} \quad N = \tau_H^{-1}M.$$

Because the Λ-module M is injective, then $\operatorname{Ext}_H^1(N, N) \cong \operatorname{Ext}_\Lambda^1(M, M) = 0$, and therefore N is a partial tilting H-module. It then follows from (3.10) that there exists a basic hereditary algebra C such that $N^\perp \cong \operatorname{mod} C$, and the Grothendieck group $K_0(C)$ of C is of rank $2 = (n+2) - n$. Moreover,

- $\operatorname{Ext}_H^1(N, P(0) \oplus P(\omega)) \cong D\operatorname{Hom}_H(P(0) \oplus P(\omega), \tau_H N) = 0$, and
- $\operatorname{Hom}_H(N, P(0) \oplus P(\omega)) = 0$,

because $\tau_H N \cong M$ is a Λ-module, N is a regular H-module, and $P(0) \oplus P(\omega)$ lies in $\mathcal{P}(H)$. Hence the H-module $P(0) \oplus P(\omega)$ is a minimal projective generator in $N^\perp$, and there is a K-algebra isomorphism

$$C \cong \operatorname{End}_H(P(0) \oplus P(\omega)).$$

It follows that C is the path algebra of the wild quiver

$$0 \circ \qquad\qquad \circ \omega.$$

Further,

$$T = P(0) \oplus P(\omega) \oplus N$$

is a tilting H-module without preinjective direct summands. The torsion-free class $\mathcal{F}(T)$ in $\operatorname{mod} H$ determined by T is of the form

$$\mathcal{F}(T) = \{X \in \operatorname{mod} H \mid \operatorname{Hom}_A(T, H) = 0\} = \operatorname{Cogen}(\tau_A T) = \operatorname{Cogen}(M)$$
$$= \operatorname{Cogen}(D(\Lambda)) = \operatorname{mod} \Lambda = \operatorname{add} \mathcal{C}([1, n]).$$

It is clear that the torsion class

$$\mathcal{T}(T) = \{X \in \operatorname{mod} H \mid \operatorname{Ext}_H^1(T, X) = 0\}$$

contains the category $N^\perp \cong \operatorname{mod} C$.

Consider the tilted algebra $B = \operatorname{End}_H(T)$. We describe the shape of all connected components in $\Gamma(\operatorname{mod} B)$.

First we prove that the connecting component $\mathcal{C}_T$ of $\Gamma(\operatorname{mod} B)$ determined by T is a preinjective component containing all indecomposable injective B-modules, and all but finitely many modules in $\mathcal{C}_T$ are sincere B-module. Because T has no preinjective direct summands, we know from (VIII.4.1) that $\mathcal{C}_T$ does not contain projective modules.

Moreover, by (VIII.3.5), the modules

$$\operatorname{Hom}_H(T, I_H(i)), \quad \text{with} \quad i \in \{0, 1, \ldots, n, \omega\},$$

form a section $\Sigma \cong \Delta^{\operatorname{op}}$ of $\mathcal{C}_T$ such that the set of all predecessors of modules from Σ in $\mathcal{C}_T$ is exactly the set $\mathcal{Y}(T) \cap \mathcal{C}_T$ of all torsion-free modules in $\mathcal{C}_T$.

Further, the fact that T has no preinjective direct summands implies that the whole preinjective component $\mathcal{Q}(H)$ is contained in $\mathcal{T}(T)$. Therefore, we infer that $\mathcal{Y}(T) \cap \mathcal{C}_T$ is the image of $\mathcal{Q}(H)$ via $\operatorname{Hom}_H(T, -)$.

Finally, because the category $\mathcal{F}(T) = \operatorname{mod} \Lambda$ has only finitely many isomorphism classes of indecomposable objects, the functor $\operatorname{Ext}_H^1(T, -)$ restricts to an equivalence of categories

$$\operatorname{Ext}_H^1(T, -) : \mathcal{F}(T) \xrightarrow{\;\cong\;} \mathcal{X}(T),$$

and every indecomposable B-module belongs either to $\mathcal{Y}(T)$ or $\mathcal{X}(T)$, we conclude from (VI.5.2) that $\mathcal{C}_T$ is a preinjective component consisting of the indecomposable modules from $\mathcal{Y}(T) \cap \mathcal{C}_T$ and $\mathcal{X}(T)$. Moreover, all indecomposable injective B-modules belong to $\mathcal{C}_T$, because the section Σ of $\mathcal{C}_T$ consists of $n + 2$ modules. Let

$$T_0 = P(0), T_1 = \tau_H^{-1}[1, n], T_2 = \tau_H^{-1}[2, n-1], \dots, T_n = \tau_H^{-1}[n, n], T_\omega = P(\omega),$$

and we set

$$T = T_0 \oplus T_1 \oplus \dots \oplus T_n \oplus T_\omega.$$

Then the modules

$$\operatorname{Hom}_A(T, T_0), \operatorname{Hom}_A(T, T_1), \dots, \operatorname{Hom}_A(T, T_n), \operatorname{Hom}_A(T, T_\omega)$$

form a complete set of indecomposable projective B-modules. Hence, to show that all but finitely many modules in $\mathcal{C}_T$ are sincere (equivalently, in

$$\mathcal{Y}(T) \cap \mathcal{C}_T = \mathcal{C}_T \setminus \mathcal{X}(T)$$

are sincere) it is sufficient to prove that $\operatorname{Hom}_H(T_i, X) \neq 0$, for each $i \in \{0, 1, \dots, n, \omega\}$ and all but finitely many modules X in $\mathcal{Q}(H)$.

Because all but finitely many modules in $\mathcal{Q}(H)$ are sincere H-modules, in view of (VIII.5.6), we infer that

- $\operatorname{Hom}_H(T_0, X) \cong \operatorname{Hom}_H(P(0), X) \neq 0$, and
- $\operatorname{Hom}_H(T_\omega, X) \cong \operatorname{Hom}_H(P(\omega), X) \neq 0$, for all but finitely many modules X from $\mathcal{Q}(H)$.

Fix now $i \in \{1, \dots, n\}$. It follows from (2.3) that all but finitely many modules in the τ_A-orbit $\{\tau_A^r T_i, \; r \in \mathbb{Z}\}$ are sincere, and consequently $\operatorname{Hom}_A(\tau_H^r T_i, I(j)) \neq 0$, for all $j \in \{0, 1, \dots, n, \omega\}$ and all but finitely many $r \in \mathbb{Z}$. Because there is an isomorphism

$$\operatorname{Hom}_A(T_i, \tau_H^s I(j)) \cong \operatorname{Hom}_H(\tau_H^{-s} T_i, I(j)),$$

for any $s \geq 0$, and any module in $\mathcal{Q}(H)$ is of the form $\tau_H^s I(j)$, for some $s \geq 0$ and $j \in \{0, 1, \dots, n, \omega\}$, then $\operatorname{Hom}_H(T_i, X) \neq 0$, for all but finitely many modules X in $\mathcal{Q}(H)$. Therefore, all but finitely many modules in $\mathcal{C}_T$ are sincere B-modules.

Observe also that, in the notation of (1.5), we have

$$T_n = [1]T_n, \quad T_{n-1} = [2]T_n, \quad \ldots, \quad T_2 = [n-1]T_n, \quad T_1 = [n]T_n.$$

Our next objective is to show that the module $R = [n+1]T_n$ is a quasi-simple regular brick in the category

$$(T_1 \oplus \ldots \oplus T_n)^\perp = N^\perp = \operatorname{mod} C.$$

Because the modules $N = \tau_H^{-1}M$ and M belong to the standard cones $\tau_H^{-1}\mathcal{C}([1,n])$ and $\mathcal{C}([1,n])$, respectively, then $T_i \in (T_{i+1} \oplus \ldots \oplus T_n)^\perp$, for any $i \in \{1, \ldots, n-1\}$. Hence, by applying (3.15), we deduce that

- the module T_{n-1} is a quasi-simple regular brick in $T_n^\perp$,
- the module T_{n-2} is a quasi-simple regular brick in $(T_{n-1} \oplus T_n)^\perp$,
- the module T_j is a quasi-simple regular brick in $(T_{j+1} \oplus \ldots \oplus T_n)^\perp$, for each $j \in \{2, \ldots, n-3\}$,
- the module T_1 is a quasi-simple regular brick in $(T_2 \oplus \ldots \oplus T_n)^\perp$, and finally
- R is a quasi-simple regular brick in $(T_1 \oplus \ldots \oplus T_n)^\perp = N^\perp = \operatorname{mod} C$.

Observe also that we have canonical irreducible epimorphisms (see (1.5))

$$R \longrightarrow T_1 \longrightarrow T_2 \longrightarrow \cdots \longrightarrow T_{n-1} \longrightarrow T_n$$

which induce the sequence of isomorphisms of vector spaces

$$\operatorname{Hom}_H(P(0) \oplus P(\omega), R) \xrightarrow{\simeq} \operatorname{Hom}_H(P(0) \oplus P(\omega), T_1)$$
$$\xrightarrow{\simeq} \cdots \xrightarrow{\simeq} \operatorname{Hom}_H(P(0) \oplus P(\omega), T_n).$$

Therefore, we get a sequence of isomorphisms

$$\operatorname{Hom}_H(T, R) \cong \operatorname{rad} \operatorname{Hom}_H(T, T_1),$$
$$\operatorname{Hom}_H(T, T_1) \cong \operatorname{rad} \operatorname{Hom}_H(T, T_2),$$

$$\vdots \qquad\qquad \vdots \qquad\quad \vdots$$

$$\operatorname{Hom}_H(T, T_{n-1}) \cong \operatorname{rad} \operatorname{Hom}_H(T, T_n),$$

and

$$\operatorname{Hom}_H(T, R) = \operatorname{Hom}_H(P(0) \oplus P(\omega), R) \cong R_C$$

is the largest C-module among the projective B-modules $\operatorname{Hom}_H(T, T_i)$, with $1 \le i \le n$.

It follows that the algebra $B = \mathrm{End}_H(T)$ is obtained from the one point extension $C[R]$ by identifying the extension vertex of $C[R]$ with the vertex 1 of the equioriented quiver

$$1 \longleftarrow 2 \longleftarrow \cdots \longleftarrow n{-}1 \longleftarrow n,$$

and without extra relations. Because R is a quasi-simple brick in $\mathrm{mod}\,C$ then, by applying (2.10), we get $\mathrm{Hom}_C(R, \tau_C^{-m} R) = 0$, for all integers $m \geq 1$, and, consequently, $\mathrm{Hom}_C(R, X) = 0$, for any successor X of $\tau_C^{-1} R$ in $\Gamma(\mathrm{mod}\,C)$.

Thus, by (XV.1.7), the full translation subquiver of $\Gamma(\mathrm{mod}\,C)$ formed by all successors of $\tau_C^{-1} R$ remains a full translation subquiver of $\Gamma(\mathrm{mod}\,B)$. Moreover, we know that all injective B-modules lie in the connecting component $\mathcal{C}_T$.

Then, similarly as in Example 5.13, we deduce that $\Gamma(\mathrm{mod}\,B)$ contains a component of the form

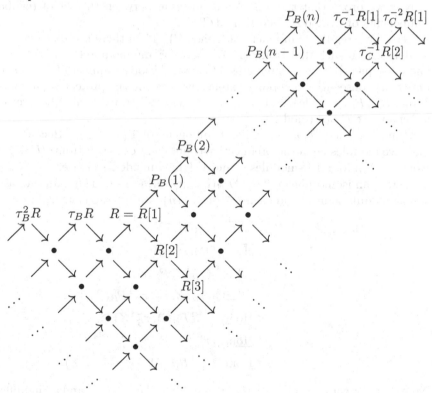

By (XV.1.7), the postprojective component $\mathcal{P}(C)$ of $\Gamma(\mathrm{mod}\,C)$ remains a complete component in $\Gamma(\mathrm{mod}\,B)$, because $\mathrm{Hom}_C(R, Y) = 0$, for any postprojective C-module Y.

Therefore, all indecomposable projective B-modules lie either in the post-projective component $\mathcal{P}(C)$ or in the component described above. We know also that all indecomposable injective B-modules lie in the preinjective connecting component $\mathcal{C}_T$. Because $\mathcal{X}(T)$ is finite we also conclude that every regular component $\mathcal{D}$ of $\Gamma(\operatorname{mod} B)$ is entirely contained in $\mathcal{Y}(T)$.

Then, by (5.10), for each regular component $\mathcal{D}$ of $\Gamma(\operatorname{mod} B)$, there exists a regular component $\mathcal{C}$ of $\Gamma(\operatorname{mod} H)$ and a quasi-simple module X in $\mathcal{C}$ such that $Y = \operatorname{Hom}_H(T, X)$ belongs to $\mathcal{D}$ and $(\longrightarrow Y)$ is the image of $(\longrightarrow X)$ via the functor $\operatorname{Hom}_H(T, -)$. In particular, $\mathcal{D}$ is of the type $\mathbb{Z}\mathbb{A}_\infty$.

Conversely, for every regular component $\mathcal{C}$ of $\Gamma(\operatorname{mod} H)$, there exists a quasi-simple module X in $\mathcal{C}$ such that the image of $\mathcal{C}(\longrightarrow X)$ via the functor $\operatorname{Hom}_H(T, -)$ is a full translation subquiver of a component $\mathcal{D}$ of $\Gamma(\operatorname{mod} B)$ which is closed under predecessors. Therefore, there is a bijection between the components of $\Gamma(\operatorname{mod} H)$ and $\Gamma(\operatorname{mod} B)$.

Now we prove that there is also a bijection between the set of regular components of $\Gamma(\operatorname{mod} H)$ and $\Gamma(\operatorname{mod} C)$.

Let Γ be a regular component of $\Gamma(\operatorname{mod} C)$. Then there is a quasi-simple module Z in Γ such that $\operatorname{Hom}_H(R, W) = 0$, for any module W in $(Z \longrightarrow)$, and consequently $\Gamma(Z \longrightarrow)$ remains a full translation subquiver of $\Gamma(\operatorname{mod} B)$ under the consecutive one-point extensions creating the projective modules $P_B(1), \dots, P_B(n)$. Clearly, $\Gamma(Z \longrightarrow)$ is then a full translation subquiver of a component $\mathcal{D}$ of $\Gamma(\operatorname{mod} B)$.

Conversely, let $\mathcal{D}$ be a regular component of $\Gamma(\operatorname{mod} B)$. Because all injective modules lie in the preinjective component $\mathcal{C}_T$, then $\operatorname{Hom}_B(U, V) = \overline{\operatorname{Hom}}_B(U, V)$, for all B-modules U and V lying outside $\mathcal{C}_T$. For each $1 \leq i \leq n$, there is an isomorphism $\tau_B^{i-n-1} P_B(i) \cong \tau_C^{-1} R[n+1-i]$. Then, for a quasi-simple module Z in $\mathcal{D}$ and each $i \in \{1, \dots, n\}$, we get the isomorphisms

$$
\begin{aligned}
\operatorname{Hom}_B(P_B(i), Z) &= \overline{\operatorname{Hom}}_B(P_B(i), Z) \\
&= \overline{\operatorname{Hom}}_B(P_B(i), \tau_B(\tau_B^{-1} Z)) \\
&\cong D\operatorname{Ext}_B^1(\tau_B^{-1} Z, P_B(i)) \\
&= D\operatorname{Ext}_B^1(\tau_B^{-1} Z, \tau_B(\tau_B^{-1} P_B(i))) \\
&\cong \underline{\operatorname{Hom}}_B(\tau_B^{-1} P_B(i), \tau_B^{-1} Z) \cong \dots \dots \\
&\cong \underline{\operatorname{Hom}}_B(\tau_B^{i-n-1} P_B(i), \tau_B^{i-n-1} Z) \\
&\cong \underline{\operatorname{Hom}}_B(\tau_C^{-1} R[n+1 -i], \tau_B^{i-n-1} Z).
\end{aligned}
$$

Note that, for each $i \in \{1, \dots, n\}$, the module $\tau_C^{-1} R[n+1-i]$ and all modules lying in $\mathcal{D}$ are images of regular H-modules via the functor $\operatorname{Hom}_H(T, -)$. Hence, we conclude that there is a quasi-simple module Y in $\mathcal{D}$ such that $\operatorname{Hom}_B(P_B(i), V) = 0$, for any module V in $\mathcal{D}(Y \longrightarrow)$ and any $i \in \{1, \dots, n\}$.

Therefore, $\mathcal{D}(Y \longrightarrow)$ is a full translation subquiver of $\mathcal{D}$ consisting entirely of regular C-modules. This shows our claim, and finishes the example.

5.15. Remarks. *Assume that A is a hereditary algebra, T a tilting A-module, and $B = \operatorname{End} T_A$ the associated tilted algebra.*

(i) *The connecting component $\mathcal{C}_T$ of $\Gamma(\operatorname{mod} B)$ determined by T is regular (without projective and injective modules) if and only if T is a regular module, see* (VIII.4.2).

(ii) *If T_A is a regular tilting A-module, then*
- *the postprojective component $\mathcal{P}(A)$ is contained in the torsion-free part $\mathcal{F}(T)$ of $\Gamma(\operatorname{mod} A)$,*
- *the preinjective component $\mathcal{Q}(A)$ is contained in the torsion part $\mathcal{T}(T)$ of $\Gamma(\operatorname{mod} A)$, and*
- *the connecting component $\mathcal{C}_T$ of $\Gamma(\operatorname{mod} B)$ determined by T is the glueing of the torsion-free part*

$$\mathcal{Y}(T) \cap \mathcal{C}_T = \operatorname{Hom}_A(T, \mathcal{Q}(A))$$

of $\mathcal{C}_T$ with the torsion part

$$\mathcal{X}(T) \cap \mathcal{C}_T = \operatorname{Ext}_A^1(T, \mathcal{P}(A))$$

of $\mathcal{C}_T$ via the connecting almost split sequences

$$0 \to \operatorname{Hom}_A(T, I) \to \operatorname{Hom}_A(T, I/\operatorname{soc} I) \oplus \operatorname{Ext}_A^1(T, \operatorname{rad} P) \to \operatorname{Ext}_A^1(T, P) \to 0,$$

where $I = I(a)$, $P = P(a)$, and a runs over all vertices of Q_A, see (VI.5.2).

Now we prove a theorem on the existence of regular tilting modules over hereditary algebras, due to Baer [35] and [36].

5.16. Proposition. *Let A be a wild hereditary algebra and X a regular stone in $\operatorname{mod} A$. Then there exists a regular tilting A-module $T = X \oplus Y$.*

Proof. We know that X is a brick (VIII.3.3) and all but finitely many modules in the τ_A-orbit of X are sincere (2.3), and clearly all of them are stones. Moreover, if $T = X \oplus Y$ is a regular tilting A-module then $\tau_A^m T$ is a regular tilting A-module, for any $m \in \mathbb{Z}$.

Therefore, without loss of generality, we may assume that the modules $\tau_A^m X$, $m \geq 1$ are sincere. From the Bongartz lemma (VI.2.4) we know that there exists an exact sequence $0 \longrightarrow A \longrightarrow E \longrightarrow X^d \longrightarrow 0$, where $d = \dim_K \operatorname{Ext}_A^1(T, A)$ and $X \oplus E$ is a tilting A-module. Moreover, by applying $\operatorname{Hom}_A(X, -)$ to the sequence we derive an exact sequence

$$0 \to \operatorname{Hom}_A(X, A) \to \operatorname{Hom}_A(X, E) \to \operatorname{Hom}_A(X, X^d) \xrightarrow{\partial} \operatorname{Ext}_A^1(X, A) \to 0.$$

Because X is a brick then $\dim_K \mathrm{Hom}_A(X, X^d) = d = \dim_K \mathrm{Ext}_A^1(X, A)$, and hence ∂ is an isomorphism. Further, $\mathrm{Hom}_A(X, A) = 0$, because X is a regular module. Consequently, we get $\mathrm{Hom}_A(X, E) = 0$, and hence $E \in X^\perp$.

Let Y be the direct sum of a complete set of pairwise non-isomorphic indecomposable direct summands of E. Then $T = X \oplus Y$ is a multiplicity-free tilting A-module and $Y \in X^\perp$. Because $\mathrm{Hom}_A(Y', X) \neq 0$, for any indecomposable direct summand Y' of Y, we conclude that T has no preinjective direct summands.

We claim that T has no indecomposable postprojective modules. It is enough to show that $X^\perp$ has no postprojective direct summands. Let Z be an indecomposable postprojective A-module. Then $Z = \tau^{-r} P$, for some indecomposable projective A-module P and some $r \geq 0$. By our assumption, the module $\tau_A^{r+1} X$ is sincere. It follows that $\mathrm{Hom}_A(Z, \tau_A X) = \mathrm{Hom}_A(P, \tau_A^{r+1} X) \neq 0$, and hence $\mathrm{Ext}_A^1(X, Z) \neq 0$. This shows that $Z \notin X^\perp$ and, hence, $T = X \oplus Y$ is a regular tilting A-module.   □

As a consequence of our main results of this chapter we get the following important fact proved by Ringel [527].

5.17. Theorem. *Let Q be a finite connected acyclic quiver and $A = KQ$. There exists a regular tilting module in $\mathrm{mod}\, A$ if and only if Q is neither a Dynkin quiver nor a Euclidean quiver, and Q has at least three vertices.*

Proof. The sufficiency follows from (2.19) and (5.16). To prove the necessity, assume that there exists a regular tilting module in $\mathrm{mod}\, A$. Then A is representation-infinite and Q is not a Dynkin quiver. Moreover, it follows from (XVII.3.5) and (XVII.3.6) that Q is not a Euclidean quiver.

It remains to show that Q has at least three vertices. Assume, to the contrary, that Q is a wild quiver with two vertices, that is, Q has the form

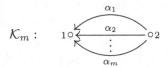

with $m \geq 3$ arrows. Then, according to (2.16), every stone in $\mathrm{mod}\, A$ is either postprojective or preinjective. Consequently, the category $\mathrm{mod}\, A$ does not contain regular tilting modules, and we get a contradiction. This finishes the proof.   □

The following example illustrates the procedure of finding regular tilting modules described in the proof of (5.16).

5.18. Example. Let $A = KQ$ be the path algebra of the quiver

$$Q: \quad 1 \circ \underset{\beta}{\overset{\alpha}{\rightleftarrows}} \overset{2}{\circ} \overset{\gamma}{\longleftarrow} \circ 3.$$

Let X be the indecomposable A-module of the form

$$K \underset{(0,1)}{\overset{(1,0)}{\rightleftarrows}} K^2 \longleftarrow 0.$$

It was shown in (1.10) that X is a quasi-simple regular A-module, and the component of $\Gamma(\mathrm{mod}\,A)$ containing X was described there. In particular, it follows that $\dim \tau X = [3, 4, 4]^t$ and the module $\tau_A^m X$ is sincere, for any $m \geq 1$.

Note that X is a stone, because it is an injective module over the Kronecker algebra H given by the full subquiver of Q consisting of the vertices 1 and 2. Therefore, the procedure described in the proof of (5.16) applies to X.

Because $A = P(1) \oplus P(2) \oplus P(3)$, where

$$\dim P(1) = [1, 0, 0]^t, \ \dim P(2) = [2, 1, 0]^t \text{ and } \dim P(3) = [2, 1, 1]^t,$$

then the Auslander–Reiten formulae (IV.2.13) yields

$$\dim_K \mathrm{Ext}_A^1(X, P(1)) = \dim_K D\mathrm{Hom}_A(P(1), \tau X) = 3,$$
$$\dim_K \mathrm{Ext}_A^1(X, P(2)) = \dim_K D\mathrm{Hom}_A(P(2), \tau X) = 4,$$
$$\dim_K \mathrm{Ext}_A^1(X, P(3)) = \dim_K D\mathrm{Hom}_A(P(3), \tau X) = 4.$$

The universal exact sequence $0 \longrightarrow A \longrightarrow E \longrightarrow X^d \longrightarrow 0$, described in the Bongartz lemma (VI.2.4), is the direct sum of the following three universal exact sequences

$$0 \longrightarrow P(1) \longrightarrow V_1 \longrightarrow X^3 \longrightarrow 0,$$
$$0 \longrightarrow P(2) \longrightarrow V_2 \longrightarrow X^4 \longrightarrow 0,$$
$$0 \longrightarrow P(3) \longrightarrow V_3 \longrightarrow X^4 \longrightarrow 0,$$

and therefore $d = 3 + 4 + 4 = 11$. Moreover, we have

- $\dim V_1 = 3 \cdot \dim X + \dim P(1) = [4, 6, 0]^t,$
- $\dim V_2 = 4 \cdot \dim X + \dim P(2) = [6, 9, 0]^t,$
- $\dim V_3 = 4 \cdot \dim X + \dim P(3) = [6, 9, 1]^t.$

By (5.16), the A-modules V_1, V_2, and V_3 are regular and belong to $X^\perp$. In (1.10), we have exhibited the dimension vectors of the modules in the postprojective component $\mathcal{P}(A)$ of A, and in the preinjective component $\mathcal{Q}(A)$ of A. Hence, we easily deduce that $V_1 \cong Z^2$ and $V_2 \cong Z^3$, where Z is an indecomposable regular A-module with $\dim Z = [2, 3, 0]^t$. It is easy to see that the module Z is of the form

$$K^2 \xleftarrow{\begin{bmatrix} 1 & 0 & 0 \\ 0 & 1 & 0 \\ 0 & \frac{1}{2} & 0 \\ 0 & 0 & 1 \end{bmatrix}} K^3 \longleftarrow 0,$$

and there is an almost split sequence $0 \longrightarrow Z \longrightarrow X^2 \longrightarrow S(2) \longrightarrow 0$ in mod H. We describe now the module V_3. Note that

$$q_A(x_1, x_2, x_3) = q_Q(x_1, x_2, x_3) = x_1^2 + x_2^2 + x_3^2 - 2x_1 x_2 - x_2 x_3.$$

In view of (III.3.13) and (VII.4.1), we get $q_A(\mathbf{dim}\, V_3) = 1$. Because $\mathrm{Ext}^1_A(V_3, V_3) = 0$, we get $\mathrm{End}_A(V_3) \cong K$. It follows that the module V_3 is indecomposable. Further, the restriction of V_3 to H is isomorphic to $V_2 \cong Z^3$. Then a simple analysis shows that V_3, viewed as a representation of Q, is of the form

$$K^6 \underset{\Psi}{\overset{\Phi}{\longleftarrow\!\!\!\longleftarrow}} K^9 \xleftarrow{\Omega} K$$

where Φ, Ψ and Ω are given in the canonical bases by the matrices

$$\Phi = \begin{bmatrix} 1 & 0 & 0 & 0 & 0 & 0 & 0 & 0 & 0 \\ 0 & 1 & 0 & 0 & 0 & 0 & 0 & 0 & 0 \\ 0 & 0 & 0 & 1 & 0 & 0 & 0 & 0 & 0 \\ 0 & 0 & 0 & 0 & 1 & 0 & 0 & 0 & 0 \\ 0 & 0 & 0 & 0 & 0 & 0 & 1 & 0 & 0 \\ 0 & 0 & 0 & 0 & 0 & 0 & 0 & 1 & 0 \end{bmatrix}, \quad \Psi = \begin{bmatrix} 0 & 1 & 0 & 0 & 0 & 0 & 0 & 0 & 0 \\ 0 & 0 & 1 & 0 & 0 & 0 & 0 & 0 & 0 \\ 0 & 0 & 0 & 0 & 1 & 0 & 0 & 0 & 0 \\ 0 & 0 & 0 & 0 & 0 & 1 & 0 & 0 & 0 \\ 0 & 0 & 0 & 0 & 0 & 0 & 0 & 1 & 0 \\ 0 & 0 & 0 & 0 & 0 & 0 & 0 & 0 & 1 \end{bmatrix}, \quad \Omega = \begin{bmatrix} 1 \\ 0 \\ 0 \\ 1 \\ 1 \\ 1 \\ 0 \\ 0 \\ 1 \end{bmatrix}.$$

Then it follows from (5.16) that $T = X \oplus Z \oplus V_3$ is a regular tilting A-module.

Let $T_1 = Z$, $T_2 = V_3$ and $T_3 = X$. Then $\mathrm{End}_A(T_i) \cong K$, for $i \in \{1, 2, 3\}$, and $\mathrm{Hom}_A(T_i, T_j) = 0$, for all $i, j \in \{1, 2, 3\}$ such that $i > j$. Moreover, a simple analysis shows that

$$\dim_K \mathrm{Hom}_A(T_1, T_2) = 3, \quad \dim_K \mathrm{Hom}_A(T_2, T_3) = 4, \quad \text{and} \quad \dim_K \mathrm{Hom}_A(T_1, T_3) = 2,$$

and the tilted algebra

$$B = \mathrm{End}_A(T) = \mathrm{End}_A(T_1 \oplus T_2 \oplus T_3)$$

is given by the quiver

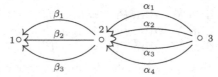

bound by the relations

- $\alpha_1 \beta_2 = \alpha_1 \beta_3 = \alpha_4 \beta_1 = \alpha_4 \beta_2 = 0,$
- $-\alpha_2 \beta_1 = \alpha_2 \beta_2 = -\alpha_3 \beta_2 = -\alpha_3 \beta_1 = \alpha_4 \beta_3,$ and
- $\alpha_1 \beta_1 = -\alpha_2 \beta_3 = \alpha_3 \beta_2 = -\alpha_3 \beta_3.$

We also note that the regular connecting component $\mathcal{C}_T$ of $\Gamma(\operatorname{mod} B)$ determined by T has a section

$$\operatorname{Hom}_A(T, I_A(3))$$
$$\nearrow$$
$$\operatorname{Hom}_A(T, I_A(2))$$
$$\nearrow\!\!\!\nearrow$$
$$\operatorname{Hom}_A(T, I_A(1))$$

of type Q^{op} with

- $\dim \operatorname{Hom}_A(T, I_A(1)) = [2, 6, 1]^t$,
- $\dim \operatorname{Hom}_A(T, I_A(2)) = [3, 9, 2]^t$,
- $\dim \operatorname{Hom}_A(T, I_A(3)) = [0, 1, 0]^t$.

Denote by C the path algebra of the full subquiver of Q_B given by the vertices 1 and 2, and by D the path algebra of the full subquiver of Q_B given by the vertices 2 and 3. Then

- $B = C[R]$, where $R = \operatorname{rad} P_B(3)$, and
- $B = [S]C$, where $S = I_B(1)/\operatorname{soc} I_B(1)$.

Note that $\dim P_B(3) = [2, 4, 1]^t$ and $\dim I_B(1) = [1, 3, 2]^t$. Invoking now the descriptions of the postprojective components and preinjective components of enlarged Kronecker algebras given in (1.9), we conclude that R is a regular C-module and S is a regular D-module.

Together with (XV.1.7) and its dual, this shows that the postprojective component $\mathcal{P}(C)$ of $\Gamma(\operatorname{mod} C)$ is also the unique postprojective component in $\Gamma(\operatorname{mod} B)$ and the preinjective component $\mathcal{Q}(D)$ of $\Gamma(\operatorname{mod} D)$ is also the unique preinjective component of $\Gamma(\operatorname{mod} B)$.

Moreover, we note that, by (5.12), each regular component of $\Gamma(\operatorname{mod} B)$ that is different from $\mathcal{C}_T$ is of type $\mathbb{Z}\mathbb{A}_\infty$.

Finally, the Auslander–Reiten quiver $\Gamma(\operatorname{mod} B)$ of B contains a component containing the projective module $P_B(3)$ and a component containing the injective module $I_B(1)$. This finishes the example.

We end this section with an example showing that the Auslander–Reiten quiver $\Gamma(\operatorname{mod} B)$ of a tilted algebra B of wild type Q may have both stable tubes and components of type $\mathbb{Z}\mathbb{A}_\infty$.

5.19. Example. Let B be the algebra given by the quiver

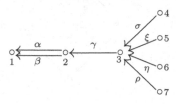

bound by the four zero relations $\sigma\gamma = 0$, $\xi\gamma = 0$, $\eta\gamma = 0$, and $\rho\gamma = 0$. Consider

- the full subquiver

$$\Delta : \quad 1 \circ \underset{\beta}{\overset{\alpha}{\rightleftarrows}} \overset{2}{\circ} \overset{\gamma}{\longleftarrow} \circ 3,$$

of Q, and
- the full subquiver Ω of Q given by the vertices 3, 4, 5, 6, and 7,

and consider the three path algebras

$$A = KQ, \ H = K\Delta, \text{ and } \Lambda = K\Omega.$$

Obviously, there are fully faithful exact embeddings

$$\mathrm{mod}\,H \hookrightarrow \mathrm{mod}\,B \hookleftarrow \mathrm{mod}\,\Lambda$$

induced by the algebra surjections $H \longleftarrow B \longrightarrow \Lambda$. The strong zero relations defining the algebra B force that every indecomposable B-module belongs to the subcategory $\mathrm{mod}\,H$ of $\mathrm{mod}\,B$, or to the subcategory $\mathrm{mod}\,\Lambda$ of $\mathrm{mod}\,B$. The simple B-module $S(3) = e_3 B$ at the vertex 3 is a unique indecomposable B-module that belongs to the intersection $(\mathrm{mod}\,H) \cap (\mathrm{mod}\,\Lambda)$.

Then the Auslander–Reiten quiver $\Gamma(\mathrm{mod}\,B)$ of B has a disjoint union decomposition

$$\Gamma(\mathrm{mod}\,B) = \mathcal{P}(H) \cup \mathcal{R}(H) \cup \mathcal{C} \cup \mathcal{R}(\Lambda) \cup \mathcal{Q}(\Lambda),$$

where

- $\mathcal{P}(H)$ is the postprojective component of $\Gamma(\mathrm{mod}\,H)$,
- $\mathcal{R}(H)$ is the regular part of $\Gamma(\mathrm{mod}\,H)$ consisting of components of type $\mathbb{Z}\mathbb{A}_\infty$,
- $\mathcal{R}(\Lambda)$ is a $\mathbb{P}_1(K)$-family $\boldsymbol{\mathcal{T}}^\Lambda = \{\mathcal{T}_\lambda^\Lambda\}_{\lambda \in \mathbb{P}_1(K)}$ of pairwise orthogonal standard stable tubes of the tubular type $r^\Lambda = \widehat{r}^\Lambda = (2,2,2)$,
- $\mathcal{Q}(\Lambda)$ is the preinjective component of $\Gamma(\mathrm{mod}\,\Lambda)$, and
- $\mathcal{C}$ is an acyclic component of $\Gamma(\mathrm{mod}\,B)$ obtained from the preinjective component $\mathcal{Q}(H)$ of $\Gamma(\mathrm{mod}\,H)$ and the postprojective component $\mathcal{P}(\Lambda)$ of $\Gamma(\mathrm{mod}\,\Lambda)$ by glueing at the simple B-module $S(3)$ as follows

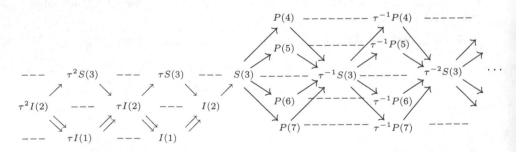

see also (1.10) and (XIII.3.6) for a classification of indecomposable modules in this component.

Then $\mathcal{C}$ admits a unique section $\Sigma \cong Q^{\mathrm{op}}$ of the form

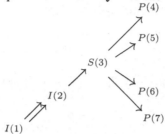

with the property $\mathrm{Hom}_A(U, \tau_A V) = 0$, for each pair of modules U and V lying on Σ. Note also that the section Σ is faithful, because the module $I(1) \oplus I(2) \oplus S(3) = D(H)$ is a faithful H-module and the module $S(3) \oplus P(4) \oplus P(5) \oplus P(6) \oplus P(7) = \Lambda_\Lambda$ is a faithful Λ-module.

Hence, by applying (VIII.5.6), we conclude that B is a tilted algebra of the wild type $Q \cong \Sigma^{\mathrm{op}}$, because there are isomorphisms of algebras $B \cong \mathrm{End}\, T_A^*$ and $\mathrm{End}\, T_B \cong A = KQ$, where $T_A^* = D(_AT)$ and T is the direct sum of all modules lying on the section Σ.

XVIII.6. Exercises

1. Let A be the path algebra KQ of a connected acyclic quiver Q with n points. Let $q_Q : \mathbb{Z}^n \longrightarrow \mathbb{Z}$ be the quadratic form of Q. Show that the algebra A is representation-wild if and only if there exists a non-zero vector $\mathbf{y} \in \mathbb{Z}^n$ such that $q_Q(\mathbf{y}) < 0$.

2. Let $A = KQ$ be the path algebra of the quiver

$$Q:$$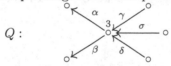

Find a tilting A-module T_A such that the tilted algebra $B = \mathrm{End}\, T_A$ is given by the same quiver Q and bound by the six zero relations $\gamma\alpha = 0$, $\gamma\beta = 0$, $\sigma\alpha = 0$, $\sigma\beta = 0$, $\delta\alpha = 0$ and $\delta\beta = 0$, see (5.7).

3. Let $A = KQ$ be the path algebra of the quiver

$$Q: \quad 1 \circ \underset{\beta}{\overset{\alpha}{\rightleftarrows}} \overset{2}{\circ} \underset{\sigma}{\overset{\gamma}{\rightleftarrows}} \circ 3.$$

Find a tilting A-module T_A such that the tilted algebra $B = \mathrm{End}\, T_A$ is given by the same quiver Q and bound by the four zero relations $\gamma\alpha = 0$, $\gamma\beta = 0$, $\sigma\alpha = 0$, $\sigma\beta = 0$, see (5.8).

4. Prove that the algebra C given by the quiver

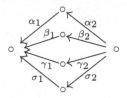

bound by the relation $\alpha_2\alpha_1 + \beta_2\beta_1 + \gamma_2\gamma_1 + \sigma_2\sigma_1 = 0$ is a concealed algebra of the wild type Δ, where

$$\Delta:$$

Show that every basic concealed algebra of type Δ is isomorphic either to C or to the path algebra KQ of a quiver such that the underlying graphs $\overline{Q}$ and $\overline{\Delta}$ coincide.

5. Prove that the algebra B given by the quiver

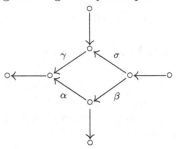

bound by the commutativity relation $\sigma\gamma = \beta\alpha$ is a concealed algebra of the wild type

$$\Delta(\widetilde{\widetilde{\mathbb{E}}}_6):$$

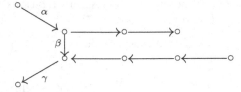

6. Prove that the algebra B given by the quiver

bound by the zero relation $\alpha\beta\gamma = 0$ is a concealed algebra of the wild type

$\Delta(\widetilde{\widetilde{\mathbb{E}}}_7)$:

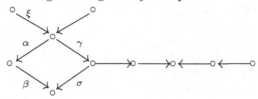

7. Prove that the algebra B given by the quiver

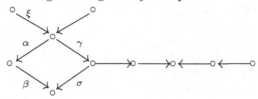

bound by the relations $\alpha\beta = \gamma\sigma$ and $\xi\alpha = 0$ is a concealed algebra of the wild type

$\Delta(\widetilde{\widetilde{\mathbb{E}}}_8)$:

8. Prove that the algebra B given by the quiver

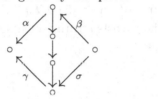

and bound by the commutativity relation $\beta\alpha = \sigma\gamma$ is a concealed algebra of the wild type

$\Delta(\widetilde{\widetilde{\mathbb{A}}}_{1,4})$:

9. Prove that the algebra given by the quiver

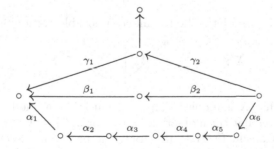

bound by the relation $\alpha_6\alpha_5\alpha_4\alpha_3\alpha_2\alpha_1 + \beta_2\beta_1 + \gamma_2\gamma_1 = 0$ is a concealed

algebra of the wild type

$$\Delta(\widetilde{\widetilde{\mathbb{D}}}_8):$$

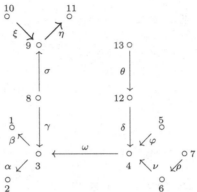

10. Let B be the algebra given by the quiver

bound by the relations $\gamma\beta = 0$, $\gamma\alpha = 0$, $\xi\eta = 0$, and $\delta\omega = 0$.

(i) Prove that there exists a tilting module T_A over the path algebra A of the wild quiver

$$\Delta(\widetilde{\widetilde{\mathbb{D}}}_{11}):$$

(ii) Describe the shape of the component $\mathcal{C}$ of $\Gamma(\mathrm{mod}\,B)$ containing the simple B-module $S(3)$ at the vertex 3.

  Hint: Consult (5.14) and Examples (XV.4.8) and (XVII.2.5).

11. Let $A = K\Delta$ be the path algebra of the wild quiver

$$\Delta:$$

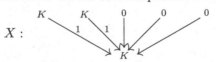

of type $\widetilde{\widetilde{\mathbb{D}}}_4$ and let X be the indecomposable A-module of the form

$$X:$$

(a) Prove that X is a quasi-simple regular stone in $\mathrm{mod}\,A$.
(b) Find a regular tilting module of the form $T_A = X \oplus Y$.
  Hint: Consult (XIII.3) and apply the procedure described in the proof of (5.16).

Chapter XIX

Tame and wild representation type of algebras

Throughout, we let Q be a connected and acyclic quiver Q with n vertices, that is, $n = |Q_0|$. We assume that K is an algebraically closed field and we denote by KQ the (hereditary) path K-algebra of Q.

We have seen in Chapters VII and XIII that, for any hereditary algebra KQ such that the underlying graph $\overline{Q}$ of Q is Dynkin or Euclidean, there exists an explicit description of the isomorphism classes of the indecomposable A-modules and a description of the components of the Auslander–Reiten quiver $\Gamma(\operatorname{mod} KQ)$ of KQ. On the other hand, we have seen in Chapter XVIII that, for any connected hereditary algebra KQ such that Q is acyclic and the underlying graph $\overline{Q}$ of Q contains a Euclidean graph as proper subgraph, then the module category $\operatorname{mod} KQ$ is very complicated and if we are able to classify the indecomposable modules in $\operatorname{mod} KQ$ then, for any finite dimensional algebra B, we are also able to classify the indecomposable modules in $\operatorname{mod} B$. Such a behaviour of the module category $\operatorname{mod} KQ$ is called wild.

In the present chapter, following Drozd [201] and [202], we split all finite dimensional K-algebras A over an algebraically closed field K in two classes: those having tame representation type and those having wild representation type. Precise definitions of tameness and wildness are given below.

Intuitively, the tameness of A means that there is a classification of the isomorphism classes of the indecomposable modules in $\operatorname{mod} A$ in the sense that, for each integer $d \geq 1$, the indecomposable modules in $\operatorname{mod} A$ of dimension d form at most finitely many one-parameter families. The wildness of A means that the category $\operatorname{mod} A$ has a wild behaviour mentioned above. Moreover, any algebra of finite representation type is of tame representation type.

We prove that concealed algebras of Euclidean type are of tame representation type and, in particular, that any hereditary algebra KQ such that the underlying graph $\overline{Q}$ of Q is Dynkin or Euclidean, is of tame representation type. The converse implication is proved by applying the tame-wild dichotomy of Drozd, see Theorem (3.4). It follows that if Q is a finite connected and acyclic quiver such that the path algebra $A = KQ$ is of tame representation type then A is not wild and hence we conclude that the underlying graph $\overline{Q}$ of Q is Dynkin or Euclidean. We also present a characterisation of the group algebras KG of finite groups G that are of tame representation type.

Although we are dealing in this book with finite dimensional K-algebras, in the context of the wild representation type and of the tame representation type we use also infinite dimensional K-algebras and, in particular, the infinite dimensional free associative K-algebra $K\langle t_1, \ldots, t_n \rangle$ of polynomials of the non-commuting indeterminates $t_1, \ldots, t_n$, where $n \geq 1$.

Throughout, given an arbitrary (not necessarily finite dimensional) K-algebra Λ, we denote by $\operatorname{mod} \Lambda$ the category of finitely generated right Λ-modules, and by $\operatorname{fin} \Lambda$ the full exact subcategory of $\operatorname{mod} \Lambda$ whose objects are the finite dimensional modules.

XIX.1. Wild representation type

Throughout this section, we let A be a finite dimensional K-algebra.

1.1. Definition. Let Λ and Λ' be arbitrary (not necessarily finite dimensional) K-algebras and assume that $\mathcal{A} \subseteq \operatorname{Mod} \Lambda$ and $\mathcal{A}' \subseteq \operatorname{Mod} \Lambda'$ are additive full exact subcategories of $\operatorname{Mod} \Lambda$ and $\operatorname{Mod} \Lambda'$, respectively, that are closed under direct summands. Let $T : \mathcal{A} \longrightarrow \mathcal{A}'$ be a K-linear functor.

(a) The functor T **respects the isomorphism classes** if, for each pair of modules X and Y in $\mathcal{A}$, the existence of an isomorphism $T(X) \cong T(Y)$ in $\mathcal{A}'$ implies the existence of an isomorphism $X \cong Y$ in $\mathcal{A}$.

(b) The functor T is defined to be a **representation embedding** if T is exact, respects the isomorphism classes, and carries indecomposable modules to indecomposable ones.

(c) A functor $T : \mathcal{A} \longrightarrow \mathcal{A}'$ is defined to be a **representation equivalence** if T is full, dense, and **respects the isomorphisms**, that is, a homomorphism $f : X \longrightarrow Y$ in $\mathcal{A}$ is an isomorphism if and only if the induced homomorphism $F(f) : F(X) \longrightarrow F(Y)$ in $\mathcal{A}'$ is an isomorphism.

It is easy to prove that a representation equivalence respects the isomorphism classes and carries indecomposable modules to indecomposable ones. The following simple observation is also of importance.

1.2. Lemma. *Let Λ, Λ' be arbitrary (not necessarily finite dimensional) K-algebras and assume that $\mathcal{A}$ is an additive full exact subcategory of $\operatorname{Mod}\Lambda$ that is closed under direct summands. Let $T : \mathcal{A} \longrightarrow \operatorname{Mod}\Lambda'$ be a K-linear functor. If T is full, faithful, and exact then T is a representation embedding.*

Proof. Assume that the functor T is full, faithful and exact. It follows that, for each module X in $\mathcal{A}$, the K-linear map $\operatorname{End} X \longrightarrow \operatorname{End} T(X)$ defined by the formula $\varphi \mapsto T(\varphi)$ is a K-algebra isomorphism. Hence, if X is an indecomposable module, then the algebra $\operatorname{End} X$ has no non-trivial idempotents. Therefore, also the algebra $\operatorname{End} T(X)$ has no non-trivial idempotents and, hence, the module $T(X)$ is indecomposable. This shows that the functor T carries indecomposable modules to indecomposable ones.

It remains to prove that T respects the isomorphism classes. For this purpose, assume that X, Y are modules in $\mathcal{A}$ and $f : T(X) \longrightarrow T(Y)$ is an isomorphism of Λ'-modules. Because the functor T is full then there exist $\varphi \in \operatorname{Hom}_\Lambda(X, Y)$ and $\psi \in \operatorname{Hom}_\Lambda(Y, X)$ such that $f = T(\varphi)$ and $f^{-1} = T(\psi)$. It follows that
$$T(\varphi\psi) = T(\varphi)T(\psi) = ff^{-1} = 1_{T(Y)} = T(1_Y),$$
$$T(\psi\varphi) = T(\psi)T(\varphi) = f^{-1}f = 1_{T(X)} = T(1_X).$$
Because T is faithful then the equalities yield $\varphi\psi = 1_Y$ and $\psi\varphi = 1_X$. Consequently, φ is an isomorphism. This finishes the proof. $\qquad\square$

Note that if A and B are finite dimensional K-algebras such that there exists a representation embedding functor $T : \operatorname{mod} B \longrightarrow \operatorname{mod} A$ then T induces an injection from the set of the isomorphism classes of the indecomposable modules in $\operatorname{mod} B$ to the set of the isomorphism classes of the indecomposable modules in $\operatorname{mod} A$, and T reduces the classification of the indecomposable modules in $\operatorname{mod} B$ to the classification of some indecomposable modules in $\operatorname{mod} A$. $\qquad\square$

1.3. Definition. Let A be a finite dimensional K-algebra.

(a) A is of **wild representation type** (or **representation-wild**; shortly **wild**) if, for each finite dimensional K-algebra B, there exists a representation embedding functor $T : \operatorname{mod} B \longrightarrow \operatorname{mod} A$. In this case we also say that the category $\operatorname{mod} A$ is of **wild representation type**.

(b) If, for each finite dimensional K-algebra B, there exists a fully faithful exact K-linear functor $T : \operatorname{mod} B \longrightarrow \operatorname{mod} A$ then the algebra A is called of **strictly wild representation type** (or **strictly representation-wild**; shortly **strictly wild**) and the category $\operatorname{mod} A$ is called of **strictly wild representation type**; shortly **strictly wild**.

It is clear that any algebra of strictly wild representation type is of wild representation type. It follows from the definition that the classification of the indecomposable modules in mod A, where A is a representation-wild algebra, is very complicated and, for each finite dimensional algebra B, it contains the classification of the indecomposable modules in mod B.

The next lemma shows that the preceding definitions are right-left symmetric.

1.4. Lemma. *Let A be a finite dimensional K-algebra. If A is of wild representation type (or of strictly wild representation type) then the algebra A^{op} opposite to A is also of wild representation type (or of strictly wild representation type, respectively).*

Proof. Assume that A is of wild representation type. Then, for each finite dimensional K-algebra B, there exists a representation embedding functor $T : \operatorname{mod} B^{op} \longrightarrow \operatorname{mod} A$. It follows that the composite functor

$$\operatorname{mod} B \xrightarrow[\cong]{D} \operatorname{mod} B^{op} \xrightarrow{T} \operatorname{mod} A \xrightarrow[\cong]{D} \operatorname{mod} A^{op}$$

is a representation embedding, where D is the standard duality. This shows that the algebra A^{op} opposite to A is also of wild representation type. The remaining statement follows in a similar way. $\square$

The results below show that, for any finite dimensional representation-wild algebra A, the classification of the indecomposable modules in mod A is much more complicated than can be expected directly from the definition, because we show that, for each algebra quotient Λ (not necessarily finite dimensional) of the free associative algebra $K\langle t_1, \dots, t_n \rangle$, with $n \geq 1$, the classification of the indecomposable modules in mod A contains the classification of the indecomposable modules in the category fin Λ of finite dimensional right Λ-modules. In particular it contains the classification of the indecomposable modules in fin $K[t_1, \dots, t_n]$ and in fin $K\langle t_1, \dots, t_n \rangle$ for any $n \geq 1$.

The following lemma due to S. Eilenberg [210] and C.E. Watts [672] shows that the representation embedding functors between module categories that commute with arbitrary direct sums can be viewed as a tensoring by a bimodule. We recall that an additive functor $T : \operatorname{Mod} A \longrightarrow \operatorname{Mod} B$ commutes with arbitrary direct sums if, for each family $\{M_j\}_{j \in J}$ of right A-modules M_j, with the direct summand embeddings $u_j : M_j \longrightarrow \bigoplus_{j \in J} M_j$, the sections $T(u_j) : T(M_j) \longrightarrow T(\bigoplus_{j \in J} M_j)$, with $j \in J$, induce the isomorphism of B-modules

$$\left(T(u_j) \right)_{j \in J} : \bigoplus_{j \in J} T(M_j) \xrightarrow{\simeq} T\left(\bigoplus_{j \in J} M_j \right).$$

The following lemma shows that the class of strictly wild algebras coincides with the class of **fully wild** algebras in the sense of [591], see also [6], [327], [577], and [593].

1.5. Lemma. *Assume that* R *and* Λ *are arbitrary (not necessarily finite dimensional) K-algebras.*

(a) *If* $T : \mathrm{Mod}\,\Lambda \longrightarrow \mathrm{Mod}\,R$ *is a right exact additive functor that commutes with arbitrary direct sums then there exists a functorial isomorphism* $- \otimes_\Lambda M_R \cong T$, *where* $_\Lambda M_R = T(\Lambda)$ *is viewed as a Λ-R-bimodule in a natural way.*

(b) *If* Λ *is a right noetherian K-algebra and* $T : \mathrm{mod}\,\Lambda \longrightarrow \mathrm{mod}\,R$ *is a right exact additive functor then there exists a functorial isomorphism* $- \otimes_\Lambda M_R \cong T$, *where* $_\Lambda M_R = T(\Lambda)$ *is viewed as a Λ-R-bimodule in a natural way.*

(c) *If* T *is a functor as in* (a) *or in* (b), *and the left Λ-module* $_\Lambda M = T(\Lambda)$ *is free then* T *is a faithful functor.*

Proof. (a) First we note that the right R-module $T(\Lambda)$ has a natural left Λ-module structure defined by the formula $\lambda \cdot y = T(\widehat{\lambda})(y)$, for any $\lambda \in \Lambda$ and any $y \in T(\Lambda)$, where $\widehat{\lambda} : \Lambda_\Lambda \longrightarrow \Lambda_\Lambda$ is the homomorphism of right Λ-modules defined by the formula $\widehat{\lambda}(x) = \lambda \cdot x$, for any $x \in \Lambda$. It is easy to check that we have defined a Λ-R-bimodule structure on the vector space $_\Lambda M_R = T(\Lambda)$.

Now, for each module X in $\mathrm{Mod}\,\Lambda$, we define a homomorphism

$$\Phi_X : X \otimes_\Lambda M_R \longrightarrow T(X)$$

of right R-modules as follows. Given $u \in X$, we denote by $\widehat{u} : \Lambda \longrightarrow X$ the unique Λ-homomorphism such that $\widehat{u}(1) = u$. For any $m \in T(\Lambda)$, we set $\Phi_X(u \otimes m) = T(\widehat{u})(m)$. It is clear that Φ_X is functorial at X. Let

$$F_1 \xrightarrow{h_1} F_0 \xrightarrow{h_0} X \longrightarrow 0$$

be an exact sequence in $\mathrm{mod}\,\Lambda$, where F_0 and F_1 are free Λ-modules. We derive the commutative diagram

$$
\begin{array}{ccccccc}
F_1 \otimes_\Lambda T(\Lambda) & \xrightarrow{h_1 \otimes \mathrm{id}} & F_0 \otimes_\Lambda T(\Lambda) & \xrightarrow{h_0 \otimes \mathrm{id}} & X \otimes_\Lambda T(\Lambda) & \longrightarrow & 0 \\
\Big\downarrow{\Phi_{F_1}} & & \Big\downarrow{\Phi_{F_0}} & & \Big\downarrow{\Phi_X} & & \\
T(F_1) & \xrightarrow{T(h_1)} & T(F_0) & \xrightarrow{T(h_0)} & T(X) & \longrightarrow & 0
\end{array}
$$

and, by our hypothesis, the rows are exact. Because T commutes with arbitrary direct sums then the functorial homomorphisms Φ_{F_1} and Φ_{F_0} are isomorphisms. Hence Φ_X is an isomorphism too. It is easy to see that the family $\Phi = \{\Phi_X\}_X$ defines an isomorphism of functors $- \otimes_\Lambda M_R \cong T$ and (a) follows.

(b) The arguments given in the proof of (a) apply. Note only that, given a module X in mod Λ, the free modules F_0 and F_1 in a free presentation of X can be chosen to be finitely generated. Moreover, because T is assumed to be additive then T commutes with finite direct sums. Hence we conclude, as in (a), that the family $\Phi = \{\Phi_X\}_X$ defines a functorial isomorphism $- \otimes_\Lambda M_R \cong T$ and (b) follows.

(c) By (a) (or (b), respectively), there is a functorial isomorphism $- \otimes_\Lambda M_R \cong T$. Because the left Λ-module $T(\Lambda)$ is free then the right R-module $T(X) \cong X \otimes_\Lambda T(\Lambda)$ is a direct sum of $\mathrm{rank}_\Lambda T(\Lambda)$ copies of X, for any non-zero Λ-module X. Hence (c) easily follows and the proof is complete. □

1.6. Corollary. *Let A be a finite dimensional K-algebra.*

(a) *The algebra A is representation-wild if and only if, for any finite dimensional K-algebra B, there exists a B-A-bimodule ${}_BM_A$ such that the left B-module ${}_BM$ is finitely generated projective and the induced functor $- \otimes_B M_A :$ mod $B \longrightarrow$ mod A respects the isomorphism classes and carries indecomposable modules to indecomposable ones.*

(b) *The algebra A is strictly representation-wild if and only if, for any finite dimensional K-algebra B, there exists a B-A-bimodule ${}_BM_A$ such that the left B-module ${}_BM$ is finitely generated projective and the induced functor $- \otimes_B M_A :$ mod $B \longrightarrow$ mod A full, faithful and exact.*

Proof. (a) Because the sufficiency is obvious, we prove the necessity. Assume that, for any finite dimensional K-algebra B, there exists a representation embedding K-linear functor

$$T : \mathrm{mod}\, B \longrightarrow \mathrm{mod}\, A.$$

By (1.5), there exists a finite dimensional B-A-bimodule ${}_BM_A$ and an isomorphism of functors $- \otimes_B M_A \cong T$. Because T is exact, then the functor $- \otimes_B M_A$ is exact and, hence, the left B-module ${}_BM$ is flat in the sense that $\mathrm{Tor}_1^B(-, {}_BM) = 0$. Because B is a finite dimensional K-algebra then B is a perfect ring and, according to the well-known result of H. Bass (see [2, 28.4] and [102, Exercise I.15]) the left B-module ${}_BM$ is finitely generated and projective. This finishes the proof of (a). The proof of (b) is similar and we leave it to the reader. □

The following theorem shows that the category fin $K\langle t_1, t_2 \rangle$ of finite dimensional right modules over the infinite dimensional algebra $K\langle t_1, t_2 \rangle$ is strictly representation-wild. To formulate it, we recall that an arbitrary (not necessarily finite dimensional) K-algebra Λ is **finitely generated as an algebra** if there exist an integer $m \geq 1$ and a K-algebra surjection

$$\varphi : K\langle t_1, \ldots, t_m \rangle \longrightarrow A$$

or, equivalently, if there exist elements $a_1, \ldots, a_n \in A$ (called K-algebra generators of A) such that any element of A is a K-linear combination of finite products of the elements of the form $a_1^{s_1}, a_2^{s_2}, \ldots, a_m^{s_m}$, where $s_1 \geq 0, s_2 \geq 0, \ldots, s_m \geq 0$ are integers.

A correspondence between the K-algebra surjections φ and the sets of K-algebra generators $a_1, \ldots, a_n$ of A is given by the formula

$$\varphi(t_1) = a_1, \ldots, \varphi(t_n) = a_n.$$

1.7. Theorem. *Let Λ be an arbitrary K-algebra which is finitely generated as an algebra over K. Then there exists a Λ-$K\langle t_1, t_2 \rangle$-bimodule $_\Lambda M_{K\langle t_1, t_2 \rangle}$ such that*

(a) *the left Λ-module $_\Lambda M$ is finitely generated and free,*
(b) *the induced functor*

$$(-) \otimes_\Lambda M : \mathrm{Mod}\,\Lambda \longrightarrow \mathrm{Mod}\,K\langle t_1, t_2 \rangle$$

is full, faithful, exact and restricts to the full and faithful representation embedding

$$(-) \otimes_\Lambda M : \mathrm{fin}\,\Lambda \longrightarrow \mathrm{fin}\,K\langle t_1, t_2 \rangle.$$

Proof. Let $\mathbb{L}^{(2)}$ be the two loop quiver

$$\mathbb{L}^{(2)} : \qquad \alpha_1 \;\; {\circ} \;\; \alpha_2.$$

We know from (II.1.3) that the path algebra $K\mathbb{L}^{(2)}$ is isomorphic to the free associative algebra $K\langle t_1, t_2 \rangle$ of polynomials in two non-commuting indeterminates t_1 and t_2. The isomorphism $K\mathbb{L}^{(2)} \cong K\langle t_1, t_2 \rangle$ is defined by setting $\varepsilon_1 \mapsto 1, \alpha_1 \mapsto t_1, \alpha_2 \mapsto t_2$.

We recall that the category $\mathrm{Rep}_K(\mathbb{L}^{(2)})$ of all K-linear representations of the quiver $\mathbb{L}^{(2)}$ consists of the triples

$$\mathbb{X} = (X; \varphi_{\alpha_1}, \varphi_{\alpha_2} : X \longrightarrow X),$$

where X is a K-vector space and $\varphi_{\alpha_1}, \varphi_{\alpha_2}$ are K-linear endomorphisms. Let

$$F : \mathrm{Mod}\,K\langle t_1, t_2 \rangle \longrightarrow \mathrm{Rep}_K(\mathbb{L}^{(2)})$$

be the functor that associates to each module X in $\mathrm{mod}\,K\langle t_1, t_2 \rangle$ the representation $F(X) = (X; \varphi_{\alpha_1}, \varphi_{\alpha_2} : X \longrightarrow X)$ of $\mathbb{L}^{(2)}$, where $\varphi_{\alpha_1}(x) = x \cdot t_1$

and $\varphi_{\alpha_2}(x) = x \cdot t_2$, for any $x \in X$. Moreover, F associates to each homomorphism $f : X \longrightarrow Y$ of $K\langle t_1, t_2 \rangle$-modules the underlying K-linear map $F(f) = f : X \longrightarrow Y$.

It follows from (III.1.6) that F is a K-linear equivalence of categories and restricts to the equivalence

$$F' : \operatorname{fin} K\langle t_1, t_2 \rangle \xrightarrow{\;\simeq\;} \operatorname{rep}_K(\mathbb{L}^{(2)}).$$

Throughout, we identify $\operatorname{Mod} K\langle t_1, t_2 \rangle$ with the category $\operatorname{Rep}_K(\mathbb{L}^{(2)})$ along the functor F.

To prove the theorem, we assume that Λ is a finitely generated K-algebra and we define a K-linear fully faithful embedding functor

$$G : \operatorname{Mod} \Lambda \longrightarrow \operatorname{Rep}_K(\mathbb{L}^{(2)}) \cong \operatorname{Mod} K\langle t_1, t_2 \rangle$$

by applying the arguments used in the proof of (XVIII.4.1) as follows. We fix a set $\{a_1, \dots, a_n\}$ of K-algebra generators of Λ. Given a module X in $\operatorname{Mod} \Lambda$, we put

$$G(X) = (X^{n+2}; \quad \varphi_{\alpha_1}^X, \varphi_{\alpha_2}^X : X^{n+2} \longrightarrow X^{n+2}),$$

where X^{n+2} is the direct sum of $n + 2$ copies of X and the K-linear endomorphisms $\varphi_{\alpha_1}^X, \varphi_{\alpha_2}^X$ of X^{n+2} are defined by the matrices

$$A_1 = \begin{bmatrix} 0 & 1 & 0 & \cdots & 0 & 0 \\ 0 & 0 & 1 & \cdots & 0 & 0 \\ \vdots & \vdots & & \ddots & \vdots & \vdots \\ 0 & 0 & 0 & \cdots & 1 & 0 \\ 0 & 0 & 0 & \cdots & 0 & 1 \\ 0 & 0 & 0 & \cdots & 0 & 0 \end{bmatrix} \quad \text{and} \quad A_2 = \begin{bmatrix} 0 & 0 & \cdots & 0 & 0 & 0 \\ 1 & 0 & \cdots & 0 & 0 & 0 \\ \widehat{a}_1 & 1 & \cdots & 0 & 0 & 0 \\ 0 & \widehat{a}_2 & \cdots & 0 & 0 & 0 \\ \vdots & \vdots & \ddots & \vdots & \vdots & \vdots \\ 0 & 0 & \cdots & \widehat{a}_n & 1 & 0 \end{bmatrix},$$

where $\widehat{a}_1, \dots, \widehat{a}_n : X \longrightarrow X$ are K-linear endomorphisms defined by the formula $\widehat{a}_j(x) = x \cdot a_j$, for $x \in X$ and $j = 1, \dots, n$. This means that the K-linear endomorphisms $\varphi_{\alpha_1}^X, \varphi_{\alpha_2}^X : X^{n+2} \longrightarrow X^{n+2}$ are defined by the formulae

$$\varphi_{\alpha_1}^X(\mathbf{x}) = (x_2, \dots, x_{n+1}, x_{n+2}, 0) = (A_1 \cdot \mathbf{x}^t)^t,$$

$$\varphi_{\alpha_2}^X(\mathbf{x}) = (0, x_1, x_1 \cdot a_1 + x_2, x_2 \cdot a_2 + x_3, \dots, x_n \cdot a_n + x_{n+1}) = (A_2 \cdot \mathbf{x}^t)^t,$$

where $\mathbf{x} = (x_1, \dots, x_{n+2}) \in X^{n+2}$ and $x_1, \dots, x_{n+2} \in X$.

Given a homomorphism $f : X \longrightarrow Y$ of Λ-modules, we define the K-linear map $G(f) : X^{n+2} \longrightarrow Y^{n+2}$ by the formula

$$G(f)(x_1, \dots, x_{n+2}) = (f(x_1), \dots, f(x_{n+2})),$$

for $x_1, \dots, x_{n+2} \in X$.

It is easy to check, using the formulae above, that $G(f)$ is a morphism of quiver representations, that is, the following equalities hold

$$G(f) \circ \varphi_{\alpha_1}^X = \varphi_{\alpha_1}^Y \circ G(f) \quad \text{and} \quad G(f) \circ \varphi_{\alpha_2}^X = \varphi_{\alpha_2}^Y \circ G(f).$$

It follows immediately from the definition that the functor G is exact and faithful. To show that the functor G is full, we take a homomorphism $h : G(X) \longrightarrow G(Y)$ of right $K\langle t_1, t_2 \rangle$-modules. Then h can be viewed as a K-linear map $X^{n+2} \longrightarrow Y^{n+2}$ such that

$$h \circ \varphi_{\alpha_1}^X = \varphi_{\alpha_1}^Y \circ h \quad \text{and} \quad h \circ \varphi_{\alpha_2}^X = \varphi_{\alpha_2}^Y \circ h.$$

Assume that h is given by an $(n+2) \times (n+2)$-matrix

$$\widetilde{h} = \begin{bmatrix} h_{11} & \cdots & h_{1\,n+2} \\ \vdots & \ddots & \vdots \\ h_{n+2\,1} & \cdots & h_{n+2\,n+2} \end{bmatrix},$$

where $h_{ij} : X \longrightarrow Y$ is a K-linear map. Then $h(\mathbf{x}) = (\widetilde{h} \cdot \mathbf{x}^t)^t$, for any $\mathbf{x} \in X^{n+2}$, and the equalities $h \circ \varphi_{\alpha_1}^X = \varphi_{\alpha_1}^Y \circ h$ and $h \circ \varphi_{\alpha_2}^X = \varphi_{\alpha_2}^Y \circ h$ are equivalent to the matrix equalities

$$A_1 \cdot \widetilde{h} = \widetilde{h} \cdot A_1 \quad \text{and} \quad A_2 \cdot \widetilde{h} = \widetilde{h} \cdot A_2.$$

Because

$$\widetilde{h} \cdot A_1 = \begin{bmatrix} 0 & h_{11} & \cdots & h_{1\,n+1} \\ \vdots & \vdots & \ddots & \vdots \\ 0 & h_{n+2\,1} & \cdots & h_{n+2\,n+1} \end{bmatrix} \quad \text{and} \quad A_1 \cdot \widetilde{h} = \begin{bmatrix} h_{21} & h_{22} & \cdots & h_{2\,n+2} \\ \vdots & \vdots & \ddots & \vdots \\ h_{n+2\,1} & h_{n+2\,2} & \cdots & h_{n+2\,n+2} \\ 0 & 0 & \cdots & 0 \end{bmatrix}$$

then the equality $\widetilde{h} \cdot A_1 = A_1 \cdot \widetilde{h}$ yields
- $h_{ij} = 0$, for all $i > j$,
- $h_{11} = \cdots = h_{n+2\,n+2}$, $h_{12} = h_{23} = \cdots = h_{n+1\,n+2}$, and
- $h_{1j} = h_{2\,j+1} = \cdots = h_{n-j+3\,n+2}$, for any $j \leq n+1$.

It follows that the matrix $\widetilde{h}$ is of the form

$$\widetilde{h} = \begin{bmatrix} h_{11} & h_{12} & h_{13} & \cdots & h_{1\,n+1} & h_{1\,n+2} \\ 0 & h_{11} & h_{12} & \ddots & h_{1\,n} & h_{1\,n+1} \\ \vdots & \vdots & \ddots & \ddots & \ddots & \vdots \\ 0 & 0 & 0 & \ddots & h_{12} & h_{13} \\ 0 & 0 & 0 & \cdots & h_{11} & h_{12} \\ 0 & 0 & 0 & \cdots & 0 & h_{11} \end{bmatrix}.$$

On the other hand, we have

$$A_2 \cdot \widetilde{h} = \begin{bmatrix} 0 & 0 & \cdots & 0 \\ h_{11} & h_{12} & \cdots & h_{1\,n+2} \\ \widehat{a}_1 h_{11}+h_{21} & \widehat{a}_1 h_{12}+h_{22} & \cdots & \widehat{a}_1 h_{1\,n+2}+h_{2\,n+2} \\ \vdots & \vdots & \ddots & \vdots \\ \widehat{a}_n h_{n1}+h_{n+1\,1} & \widehat{a}_n h_{n2}+h_{n+1\,2} & \cdots & \widehat{a}_n h_{n\,n+2}+h_{n+1\,n+2} \end{bmatrix}, \quad \text{and}$$

$$\widetilde{h} \cdot A_2 = \begin{bmatrix} h_{12}+h_{13}\widehat{a}_1 & h_{13}+h_{14}\widehat{a}_2 & \cdots & h_{1\,n+1}+h_{1\,n+2}\widehat{a}_n & h_{1\,n+2} & 0 \\ h_{22}+h_{23}\widehat{a}_1 & h_{23}+h_{24}\widehat{a}_2 & \cdots & h_{2\,n+1}+h_{2\,n+2}\widehat{a}_n & h_{2\,n+2} & 0 \\ \vdots & \vdots & \ddots & \vdots & \vdots & \\ h_{n+2\,2}+h_{n+2\,3}\widehat{a}_1 & h_{n+2\,3}+h_{n+2\,4}\widehat{a}_2 & \cdots & h_{n+2\,n+1}+h_{n+2\,n+2}\widehat{a}_n & h_{n+2\,n+2} & 0 \end{bmatrix}.$$

Then the equality $\widetilde{h} \cdot A_2 = A_2 \cdot \widetilde{h}$ yields

- $h_{12} = h_{13} = \cdots = h_{1\,n+2} = 0$, and
- $\widehat{a}_j h_{11} = h_{11}\widehat{a}_j$, for $j = 1, \ldots, n$.

It follows that $h_{11}(x \cdot \lambda) = (h_{11}(x)) \cdot \lambda$, for all $x \in X$ and $\lambda \in \Lambda$, because the elements $a_1, \ldots, a_n$ generate the algebra Λ. Consequently, h_{11} is a Λ-homomorphism and $\widetilde{h}$ has the diagonal form

$$\widetilde{h} = \begin{bmatrix} h_{11} & 0 & \cdots & 0 \\ 0 & h_{11} & \cdots & 0 \\ \vdots & \vdots & \ddots & \vdots \\ 0 & 0 & \cdots & h_{11} \end{bmatrix}.$$

This shows that $h = G(h_{11})$ and that the functor G is full, as we required. Because the functor G is faithful and it is easy to check that G commutes with arbitrary direct sums then the composition

$$G' : \operatorname{Mod}\Lambda \longrightarrow \operatorname{Mod} K\langle t_1, t_2 \rangle$$

of the functor $G : \operatorname{Mod}\Lambda \longrightarrow \operatorname{Rep}_K(\mathbb{L}^{(2)})$ with the equivalence $\operatorname{Rep}_K(\mathbb{L}^{(2)}) \longrightarrow \operatorname{Mod} K\langle t_1, t_2 \rangle$ is full, faithful, and exact. Hence, according to (1.2), the restriction

$$G'' : \operatorname{fin}\Lambda \longrightarrow \operatorname{fin} K\langle t_1, t_2 \rangle$$

of G' to $\operatorname{fin}\Lambda$ is a fully faithful representation embedding functor. It follows from (1.5) that there is a functorial isomorphism

$$G' \cong - \otimes_\Lambda M,$$

where

$$_\Lambda M_{K\langle t_1, t_2 \rangle} = G(\Lambda) = (\Lambda^{n+2}; \quad \varphi_{\alpha_1}^\Lambda, \varphi_{\alpha_2}^\Lambda : \Lambda^{n+2} \longrightarrow \Lambda^{n+2})$$

is viewed as a Λ-$K\langle t_1, t_2 \rangle$-bimodule in a natural way. Because the left Λ-module $_\Lambda M$ is finitely generated free of rank $n + 2$, then the theorem is proved. □

1.8. Corollary. *The category* fin $K\langle t_1, t_2 \rangle$ *of finite dimensional right* $K\langle t_1, t_2 \rangle$-*modules is strictly representation-wild.*

Proof. Apply (1.7) and the definition of strictly wild representation type.   □

1.9. Corollary. *Let* $B = \begin{bmatrix} K & 0 \\ K^3 & K \end{bmatrix}$ *be the enlarged Kronecker K-algebra of dimension 5.*

(a) *There exists a* $K\langle t_1, t_2 \rangle$-$B$-*bimodule* $_{K\langle t_1, t_2 \rangle} N_B$ *such that*
  (i) *the left* $K\langle t_1, t_2 \rangle$-*module* $_{K\langle t_1, t_2 \rangle} N$ *is finitely generated free,*
  (ii) *the induced functor*

$$- \otimes_{K\langle t_1, t_2 \rangle} N_B : \operatorname{Mod} K\langle t_1, t_2 \rangle \longrightarrow \operatorname{Mod} B$$

  is full, faithful, exact and restricts to the full and faithful representation embedding

$$- \otimes_{K\langle t_1, t_2 \rangle} N_B : \operatorname{fin} K\langle t_1, t_2 \rangle \longrightarrow \operatorname{mod} B.$$

(b) *The algebra B is strictly representation-wild.*

Proof. (a) First we note that there is a K-algebra isomorphism $B \cong K\mathcal{K}_3$, where $\mathcal{K}_3$ is the enlarged Kronecker quiver

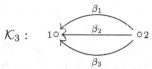

$$\mathcal{K}_3: \quad 1 \circ \underset{\beta_3}{\overset{\beta_1}{\rightleftarrows}} \circ 2$$

It follows from (III.1.6) that there is a K-linear equivalence of categories $\operatorname{Rep}_K(\mathcal{K}_3) \cong \operatorname{Mod} B$, that restricts to the equivalence $\operatorname{rep}_K(\mathcal{K}_3) \simeq \operatorname{mod} B$. On the other hand, we have observed in the proof of (1.7) that there exists a K-linear equivalence of categories $\operatorname{Mod} K\langle t_1, t_2 \rangle \longrightarrow \operatorname{Rep}_K(\mathbb{L}^{(2)})$, that restricts to the equivalence of categories $\operatorname{fin} K\langle t_1, t_2 \rangle \longrightarrow \operatorname{rep}_K(\mathbb{L}^{(2)})$, where $\mathbb{L}^{(2)}$ is the two loop quiver

$$\mathbb{L}^{(2)}: \quad \alpha_1 \,\circlearrowleft \circ \circlearrowright\, \alpha_2.$$

Then, to prove (a), we construct a K-linear functor

$$R : \operatorname{Rep}_K(\mathbb{L}^{(2)}) \longrightarrow \operatorname{Rep}_K(\mathcal{K}_3)$$

as follows. Let

$$\mathbb{X} = (X; \varphi_{\alpha_1}, \varphi_{\alpha_2} : X \longrightarrow X)$$

be an object in $\operatorname{Rep}_K(\mathbb{L}^{(2)})$, where X is a K-vector space and $\varphi_{\alpha_1}, \varphi_{\alpha_2}$ are K-linear endomorphisms. We set

$$R(\mathbb{X}) = (X_1, X_2; \; \varphi_{\beta_1}, \varphi_{\beta_2}, \varphi_{\beta_3}),$$

where $X_1 = X$, $X_2 = X$, $\varphi_{\beta_1} = \varphi_{\alpha_1}$, $\varphi_{\beta_2} = \varphi_{\alpha_2}$, and $\varphi_{\beta_3} = 1_X$.

If $f : \mathbb{X} \longrightarrow \mathbb{Y}$ is a morphism in $\mathrm{Rep}_K(\mathbb{L}^{(2)})$, we set $R(f) = (f_1, f_2)$, where $f_1 = f$ and $f_2 = f$.

It is clear that R is an exact K-linear functor. A simple calculation shows that R is full, faithful and commutes with arbitrary direct sums.

We denote by R' the composite functor

$$\mathrm{Mod}\, K\langle t_1, t_2\rangle \xrightarrow{\;\;\cong\;\;} \mathrm{Rep}_K(\mathbb{L}^{(2)}) \xrightarrow{\;\;R\;\;} \mathrm{Rep}_K(\mathcal{K}_3) \xrightarrow{\;\;\cong\;\;} \mathrm{Mod}\, B.$$

It follows that R' is a full, faithful, K-linear functor and it commutes with arbitrary direct sums. Hence, by (1.5), there exists a functorial isomorphism $- \otimes_{K\langle t_1, t_2\rangle} N_B \cong R'$, where

$$_{K\langle t_1, t_2\rangle}N = R'(K\langle t_1, t_2\rangle) = K\langle t_1, t_2\rangle \oplus K\langle t_1, t_2\rangle$$

is viewed as a $K\langle t_1, t_2\rangle$-B-bimodule, with the right B-module structure given by the formula

$$(f, g) \cdot \begin{bmatrix} x' & 0 \\ x_1\beta_1 + x_2\beta_2 + x_3\beta_3 & x'' \end{bmatrix} = \big(fx' + g(x_1t_1 + x_2t_2 + x_3), \; gx''\big),$$

for $f, g \in K\langle t_1, t_2\rangle$ and $x', x'', x_1, x_2, x_3 \in K$. This means that the right B-module N_B, viewed as a representation of $\mathcal{K}_3$, has the form

$$N_B = (K\langle t_1, t_2\rangle, K\langle t_1, t_2\rangle, \varphi_{\beta_1}, \varphi_{\beta_2}, \varphi_{\beta_3}),$$

where $\varphi_{\beta_1}, \varphi_{\beta_2} : K\langle t_1, t_2\rangle \longrightarrow K\langle t_1, t_2\rangle$ are the K-linear maps defined by formulae $\varphi_{\beta_1}(f) = f \cdot t_1$, $\varphi_{\beta_2}(f) = f \cdot t_2$, and $\varphi_{\beta_3} : K\langle t_1, t_2\rangle \longrightarrow K\langle t_1, t_2\rangle$ is the identity map. The left $K\langle t_1, t_2\rangle$-module structure on N_A is the obvious one. This finishes the proof of (a).

(b) By (ii) of the statement (a), there is a full, faithful, exact and K-linear functor $\mathrm{fin}\, K\langle t_1, t_2\rangle \longrightarrow \mathrm{mod}\, B$. On the other hand, it follows from (1.8) that the category $\mathrm{fin}\, K\langle t_1, t_2\rangle$ is of strictly wild representation type, that is, for any finite dimensional K-algebra A, there is a full, faithful, exact and K-linear functor $\mathrm{mod}\, A \longrightarrow \mathrm{fin}\, K\langle t_1, t_2\rangle$. Consequently, for any finite dimensional K-algebra A, there is a full, faithful, exact and K-linear functor $\mathrm{mod}\, A \longrightarrow \mathrm{mod}\, B$. This shows that the algebra B is strictly representation-wild. $\qquad\square$

Following [577], we prove the following useful result.

1.10. Proposition. *Let A be a finite dimensional K-algebra.*

(a) *There exists a representation embedding functor*
$$\operatorname{fin} K\langle t_1, t_2\rangle \longrightarrow \operatorname{mod} A$$
if and only if there exists a $K\langle t_1, t_2\rangle$-A-bimodule ${}_{K\langle t_1,t_2\rangle}M_A$ such that the left $K\langle t_1, t_2\rangle$-module ${}_{K\langle t_1,t_2\rangle}M$ is finitely generated free and the functor
$$- \otimes_{K\langle t_1,t_2\rangle} M_A : \operatorname{fin} K\langle t_1, t_2\rangle \longrightarrow \operatorname{mod} A$$
respects isomorphism classes and carries indecomposable modules to indecomposable ones.

(b) *There exists a full, faithful, and exact functor*
$$\operatorname{fin} K\langle t_1, t_2\rangle \longrightarrow \operatorname{mod} A,$$
that is K-linear, if and only if there exists a $K\langle t_1, t_2\rangle$-A-bimodule ${}_{K\langle t_1,t_2\rangle}M_A$ such that the left $K\langle t_1, t_2\rangle$-module ${}_{K\langle t_1,t_2\rangle}M$ is finitely generated free and the functor
$$- \otimes_{K\langle t_1,t_2\rangle} M_A : \operatorname{fin} K\langle t_1, t_2\rangle \longrightarrow \operatorname{mod} A$$

is full.

Proof. (a) The sufficiency is obvious, because we know from (1.5) that the functor $- \otimes_{K\langle t_1,t_2\rangle} M_A : \operatorname{fin} K\langle t_1, t_2\rangle \longrightarrow \operatorname{mod} A$ is exact and faithful if the left $K\langle t_1, t_2\rangle$-module M is free.

To prove the necessity, we assume that there exists a representation embedding functor $\operatorname{fin} K\langle t_1, t_2\rangle \longrightarrow \operatorname{mod} A$. Let

$$B = \begin{bmatrix} K & 0 \\ K^3 & K \end{bmatrix}$$

be the enlarged Kronecker K-algebra of dimension 5. It follows from (1.2) and (1.7) that there exists a representation embedding functor $\operatorname{mod} B \longrightarrow \operatorname{fin} K\langle t_1, t_2\rangle$, and, hence, there exists a representation embedding functor $T : \operatorname{mod} B \longrightarrow \operatorname{mod} A$.

By (1.6), there exists an B-A-bimodule ${}_BL_A$ and an isomorphism of functors $- \otimes_B L_A \cong T$ such that the left B-module ${}_BL$ is finitely generated projective and the functor $- \otimes_B L_A : \operatorname{mod} B \longrightarrow \operatorname{mod} A$ is a representation embedding. Further, by (1.9), there exists a $K\langle t_1, t_2\rangle$-B-bimodule ${}_{K\langle t_1,t_2\rangle}N_B$ such that the left $K\langle t_1, t_2\rangle$-module ${}_{K\langle t_1,t_2\rangle}N$ is finitely generated free and the induced functor

$$- \otimes_{K\langle t_1,t_2\rangle} N_B : \operatorname{fin} K\langle t_1, t_2\rangle \longrightarrow \operatorname{mod} B$$

is full. Consider the $K\langle t_1, t_2\rangle$-A-bimodule ${}_{K\langle t_1,t_2\rangle}M_A = {}_{K\langle t_1,t_2\rangle}N \otimes {}_BL_A$.

Because the left B-module $_BL$ is finitely generated projective and the left $K\langle t_1, t_2 \rangle$-module $_{K\langle t_1,t_2\rangle}N$ is finitely generated free then the left $K\langle t_1, t_2 \rangle$-module $_{K\langle t_1,t_2\rangle}M$ is finitely generated projective. It follows that the $K\langle t_1, t_2 \rangle$-module $_{K\langle t_1,t_2\rangle}M$ is a finitely generated free module, because, by a well-known result of P. M. Cohn, every submodule of a free left $K\langle t_1, t_2 \rangle$-module is a free module (see [146]). This finishes the proof of (a), because the functor $- \otimes_{K\langle t_1,t_2\rangle} M_A : \mathrm{fin}\, K\langle t_1, t_2 \rangle \longrightarrow \mathrm{mod}\, A$ is a composition of two representation embedding functors and therefore respects isomorphism classes and carries indecomposable modules to indecomposable ones. The proof of (b) is similar. $\square$

Now we present a characterisation of representation-wild algebras.

1.11. Theorem. *Let A be a finite dimensional K-algebra. The following conditions are equivalent.*
(a) *The algebra A is representation-wild.*
(b) *There exists a representation embedding functor*
$$\mathrm{mod}\, \begin{bmatrix} K & 0 \\ K^3 & K \end{bmatrix} \longrightarrow \mathrm{mod}\, A.$$
(c) *There exists a representation embedding functor*
$$\mathrm{fin}\, K\langle t_1, t_2 \rangle \longrightarrow \mathrm{mod}\, A.$$
(d) *There exists a $K\langle t_1, t_2 \rangle$-$A$-bimodule $_{K\langle t_1,t_2\rangle}M_A$ such that the left $K\langle t_1, t_2 \rangle$-module $_{K\langle t_1,t_2\rangle}M$ is finitely generated free and the functor*
$$- \otimes_{K\langle t_1,t_2\rangle} M_A : \mathrm{fin}\, K\langle t_1, t_2 \rangle \longrightarrow \mathrm{mod}\, A$$
respects the isomorphism classes and carries indecomposable modules to indecomposable ones.
(e) *For every finitely generated K-algebra Λ there exists a representation embedding functor $\mathrm{fin}\, \Lambda \longrightarrow \mathrm{mod}\, A$.*
(f) *There exists a representation embedding functor*
$$\mathrm{fin}\, K[t_1, t_2] \longrightarrow \mathrm{mod}\, A.$$
(g) *There exists a representation embedding functor*
$$H : \mathrm{mod}\, R \longrightarrow \mathrm{mod}\, A,$$
where $C' = K[t_1, t_2, t_3]/(t_1, t_2, t_3)^2$ is a commutative local K-algebra of dimension four.

Proof. It follows from the definition that (a) implies (b). By (1.9) and (1.7), (b) implies (c) and (c) implies (e), respectively. The equivalence of the statements (c) and (d) is a consequence of (1.10). Because the implications (e)$\Rightarrow$(a), (e)$\Rightarrow$(f), (e)$\Rightarrow$(g) are obvious, then the statements (a)–(e) are equivalent and it remains to prove the implications (g)$\Rightarrow$(c) and (f)$\Rightarrow$(c).

(g)$\Rightarrow$(c) In Remark 1.13 and Example 1.17, we construct a faithful representation embedding functor

$$F : \operatorname{fin} K\langle t_1, t_2 \rangle \longrightarrow \operatorname{mod} C',$$

see also [104] and [202]. Hence, if $H : \operatorname{mod} C' \longrightarrow \operatorname{mod} A$ is a representation embedding functor, then the composite functor

$$H \circ F : \operatorname{fin} K\langle t_1, t_2 \rangle \longrightarrow \operatorname{mod} A$$

is a representation embedding functor and (c) follows.

(f)$\Rightarrow$(c) Assume that there exists a representation embedding functor

$$T' : \operatorname{fin} K[t_1, t_2] \longrightarrow \operatorname{mod} A.$$

We construct a representation embedding

$$T'' : \operatorname{fin} K\langle t_1, t_2 \rangle \longrightarrow \operatorname{fin} K[t_1, t_2]. \tag{1.11a}$$

Hence $T'' \circ T' : \operatorname{fin} K\langle t_1, t_2 \rangle \longrightarrow \operatorname{fin} A$ is a representation embedding and (c) follows.

To construct the functor T'', we note that there is a K-algebra isomorphism $K[t_1, t_2] \cong K\langle t_1, t_2 \rangle / (t_1 t_2 - t_2 t_1)$ and the algebra surjection $K\langle t_1, t_2 \rangle \longrightarrow K[t_1, t_2]$ induces the embedding $\operatorname{fin} K[t_1, t_2] \hookrightarrow \operatorname{fin} K\langle t_1, t_2 \rangle$. Moreover, we make the identifications

$$\operatorname{fin} K\langle t_1, t_2 \rangle = \operatorname{rep}_K(\mathbb{L}^{(2)}) \text{ and } \operatorname{fin} K[t_1, t_2] = \operatorname{rep}_K(\mathbb{L}^{(2)}, \mathcal{I}) \subseteq \operatorname{rep}_K(\mathbb{L}^{(2)}),$$

where $\mathbb{L}^{(2)}$ is the two loop quiver

$$\mathbb{L}^{(2)} : \qquad \alpha \; \overset{\circ}{\underset{\curvearrowleft}{\bigcirc}} \; \beta$$

and $\mathcal{I}$ is the two-sided ideal of $K\mathbb{L}^{(2)} \cong K\langle t_1, t_2 \rangle$ generated by the commutativity relation $\alpha\beta - \beta\alpha$.

We construct a K-linear representation embedding endofunctor

$$R : \operatorname{Rep}_K(\mathbb{L}^{(2)}) \longrightarrow \operatorname{Rep}_K(\mathbb{L}^{(2)}) \tag{1.11b}$$

as follows. Let

$$\mathbb{X} = (X; \; \varphi_\alpha, \varphi_\beta : X \longrightarrow X)$$

be an object in $\operatorname{Rep}_K(\mathbb{L}^{(2)})$, where X is a K-vector space and $\varphi_\alpha, \varphi_\beta$ are K-linear endomorphisms. We set

$$R(\mathbb{X}) = (X^4; \; \widehat{\varphi}_\alpha, \widehat{\varphi}_\beta : X^4 \longrightarrow X^4),$$

where $\widehat{\varphi}_\alpha, \widehat{\varphi}_\beta : X^4 \longrightarrow X^4$ are K-linear endomorphisms of the vector space X^4 defined by the matrices

$$\Phi_\alpha = \begin{bmatrix} 0 & 0 & 0 & 0 \\ 0 & 0 & 0 & 0 \\ 1_X & 0 & 0 & 0 \\ 0 & \varphi_\alpha & \varphi_\beta & 0 \end{bmatrix} \quad \text{and} \quad \Phi_\beta = \begin{bmatrix} 0 & 0 & 0 & 0 \\ 1_X & 0 & 0 & 0 \\ 0 & 0 & 0 & 0 \\ 0 & 1_X & \varphi_\alpha & 0 \end{bmatrix}.$$

If $f : X \longrightarrow Y$ is a morphism in $\mathrm{Rep}_K(\mathbb{L}^{(2)})$, we set $R(f) = (f, f, f, f)$.

It is clear that R is an exact additive faithful K-linear functor. A simple calculation shows (as in the proof of (1.7)) that R is a representation embedding (but is not full), see [147, p. 479]. Because the equality $\Phi_\alpha \cdot \Phi_\beta = \Phi_\beta \cdot \Phi_\alpha$ yields the equality $\widehat{\varphi}_\alpha \circ \widehat{\varphi}_\beta = \widehat{\varphi}_\beta \circ \widehat{\varphi}_\alpha$, then the representation $R(X)$ belongs to the subcategory $\mathrm{Mod}\, K[t_1, t_2] = \mathrm{Rep}_K(\mathbb{L}^{(2)}, \mathcal{I}) \subseteq \mathrm{Rep}_K(\mathbb{L}^{(2)})$ of $\mathrm{Mod}\, K\langle t_1, t_2 \rangle = \mathrm{Rep}_K(\mathbb{L}^{(2)})$. It follows that R has a factorisation through a representation embedding $\mathrm{Rep}_K(\mathbb{L}^{(2)}) \longrightarrow \mathrm{Rep}_K(\mathbb{L}^{(2)}, \mathcal{I}) = \mathrm{Mod}\, K[t_1, t_2]$ that restricts to a representation embedding

$$T'' : \mathrm{fin}\, K\langle t_1, t_2 \rangle = \mathrm{rep}_K(\mathbb{L}^{(2)}) \longrightarrow \mathrm{rep}_K(\mathbb{L}^{(2)}, \mathcal{I}) = \mathrm{fin}\, K[t_1, t_2].$$

This finishes the proof of the implication (f)⇒(c) and of the theorem. □

We also have a strictly representation-wild version of (1.11).

1.12. Theorem. *Let A be a finite dimensional K-algebra. The following conditions are equivalent.*
  (a) *The algebra A is strictly representation-wild.*
  (b) *There exists a full faithful and exact functor*
$$\mathrm{mod}\, \begin{bmatrix} K & 0 \\ K^3 & K \end{bmatrix} \longrightarrow \mathrm{mod}\, A.$$
  (c) *There exists a full faithful and exact functor*
$$\mathrm{fin}\, K\langle t_1, t_2 \rangle \longrightarrow \mathrm{mod}\, A.$$
  (d) *There exists a $K\langle t_1, t_2 \rangle$-$A$-bimodule $_{K\langle t_1, t_2 \rangle}M_A$ such that the left $K\langle t_1, t_2 \rangle$-module $_{K\langle t_1, t_2 \rangle}M$ is finitely generated free and the functor*
$$- \otimes_{K\langle t_1, t_2 \rangle}M_A : \mathrm{fin}\, K\langle t_1, t_2 \rangle \longrightarrow \mathrm{mod}\, A$$
  is full.
  (e) *For every finitely generated K-algebra Λ there exists a a full faithful and exact functor $\mathrm{fin}\, \Lambda \longrightarrow \mathrm{mod}\, A$.*

Proof. In view of (1.2) and (1.5), the proof of (1.11) modifies almost verbatim. □

1.13. Remark. The reader might observe that, by (XVIII.4.1), a path algebra $A = KQ$ of an acyclic quiver Q is representation-wild if and only if there exists a full, faithful, exact, K-linear functor

$$H : \mathrm{mod}\, C \longrightarrow \mathrm{mod}\, A,$$

where $C = K[t_1, t_2]/(t_1, t_2)^2$ is a commutative local K-algebra of dimension three. This equivalence is somewhat surprising, because the local K-algebra C has rather simple structure of the module category and $\mathrm{mod}\, C$ is close to the category of Kronecker modules studied in detail in Section XI.4, compare with (1.14) and the proof of the implication (a)$\Rightarrow$(d) in (1.14). It follows that the wildness of $A = KQ$ is a consequence of the fact that the functor H is full, faithful, exact, and the structure of $\mathrm{mod}\, C$ described later is a special one. To see this, we make the following observations.

- The radical $J = \mathrm{rad}\, C$ of the algebra C is the unique maximal ideal of C and is isomorphic to K^2, as a K-vector space,
- $J^2 = 0$, the quotient algebra C/J of C is isomorphic to the field K, and the C/J-module structure on J coincides with the K-vector space structure on K^2, under the algebra isomorphism $C/J \cong K$.
- The matrix algebra $C_J = \begin{bmatrix} C/J & 0 \\ J & C/J \end{bmatrix}$ is hereditary and there is an isomorphism

$$C_J \cong \begin{bmatrix} K & 0 \\ K^2 & K \end{bmatrix}$$

  of C_J with the Kronecker algebra $\begin{bmatrix} K & 0 \\ K^2 & K \end{bmatrix}$.
- There is a K-linear equivalence of categories $\mathrm{mod}\, C_J \cong \mathrm{rep}_K(\mathcal{K}_2)$, where

$$\mathcal{K}_2: \quad 1 \circ \underset{\beta}{\overset{\alpha}{\Longleftarrow}} \circ\, 2$$

  is the Kronecker quiver.
- There exists a full K-linear reduction functor

$$\mathbb{F}_J : \mathrm{mod}\, C \longrightarrow \mathrm{mod}\, C_J \cong \mathrm{rep}_K(\mathcal{K}_2)$$

  that establishes a bijection between the isomorphism classes of the indecomposable modules in $\mathrm{mod}\, C$ and the isomorphism classes of the indecomposable Kronecker modules in $\mathrm{rep}_K(\mathcal{K}_2) \cong \mathrm{mod}\, C_J$ that are not isomorphic to the simple injective Kronecker module $I(2) = (0 \Longleftarrow K)$, and hence
- the reduction functor $\mathbb{F}_J : \mathrm{mod}\, C \longrightarrow \mathrm{mod}\, C_J \cong \mathrm{rep}_K(\mathcal{K}_2)$ establishes a representation equivalence between $\mathrm{mod}\, C$ and the image subcategory $\mathrm{Im}\,\mathbb{F}_J$ of $\mathrm{mod}\, C_J$.

The reduction functor $\mathbb{F}_J$ is defined by attaching to any module X in $\mathrm{mod}\, C$ the triple

$$\mathbb{F}_J(X) = (X', X'', \varphi),$$

where $X' = X/XJ$, $X'' = XJ$ are viewed as right C/J-modules and

$$\varphi : X' \otimes_{C/J} J_{C/J} \longrightarrow X''_{C/J}$$

is a homomorphism of C/J-modules defined by the formula $\varphi(\overline{x} \otimes b) = x \cdot b$, for the coset $\overline{x} = x + XJ$ and $b \in J$. The details can be found in [234], see also [34, Section X.2].

It follows from the properties of the reduction functor $\mathbb{F}_J$ that the classification of the indecomposable modules in $\mathrm{mod}\, C$ is known and is very close to that for Kronecker modules established in Section XI.4. The Auslander–Reiten quiver $\Gamma(\operatorname{mod} C)$ of C is obtained from the Auslander–Reiten quiver

$$\Gamma(\operatorname{mod} C_J) = \mathcal{P}(C_J) \cup \{\mathcal{T}_\lambda^{C_J}\}_{\lambda \in \mathbb{P}_1(K)} \cup \mathcal{Q}(C_J)$$

of $C_J \cong \left[\begin{smallmatrix} K & 0 \\ K^2 & K \end{smallmatrix}\right]$ by keeping unchanged the $\mathbb{P}_1(K)$-family

$$\boldsymbol{\mathcal{T}}^{C_J} = \{\mathcal{T}_\lambda^{C_J}\}_{\lambda \in \mathbb{P}_1(K)}$$

of standard stable tubes and by making the identification of the simple projective Kronecker module $P(1) = (K \rightleftarrows 0)$, that is the source vertex of the postprojective component $\mathcal{P}(C_J)$ of C_J, with the simple injective Kronecker module $I(2) = (0 \rightleftarrows K)$, that is the sink vertex of the preinjective component $\mathcal{Q}(C_J)$ of C_J.

Note that the functor $\mathbb{F}_J$ admits a partial section

$$\mathbb{F}_2^\bullet : \operatorname{rep}_K(\mathcal{K}_2) \longrightarrow \operatorname{mod} C$$

defined by attaching to any $\mathbb{V} = (V_1 \underset{h_2}{\overset{h_1}{\rightleftarrows}} V_2)$ in $\operatorname{rep}_K(\mathcal{K}_2)$ the C-module $\mathbb{F}_2^\bullet(\mathbb{V}) = V_1 \oplus V_2$ with the action of the cosets $\bar{t}_1, \bar{t}_2 \in C$ on $V_1 \oplus V_2$ given by the K-linear endomorphisms $\widetilde{h}_1 = \left[\begin{smallmatrix} 0 & 0 \\ h_1 & 0 \end{smallmatrix}\right]$ and $\widetilde{h}_2 = \left[\begin{smallmatrix} 0 & 0 \\ h_2 & 0 \end{smallmatrix}\right]$ of $V_1 \oplus V_2$. The linear maps $\widetilde{h}_1, \widetilde{h}_2 : V_1 \oplus V_2 \longrightarrow V_1 \oplus V_2$ are defined by the formulae $(v_1, v_2) \mapsto (h_1(v_2), 0)$ and $(v_1, v_2) \mapsto (h_2(v_2), 0)$, for $(v_1, v_2) \in V_1 \oplus V_2$. It is easy to see that $\mathbb{F}_2^\bullet$ restricts to the exact representation embedding of the full subcategory of $\operatorname{rep}_K(\mathcal{K}_2)$ whose objects are representations $\mathbb{V}$ having no summand isomorphic to the simple injective representation $I(2)$, see [581, 1.8].

Similarly, for the commutative local algebra $C' = K[t_1, t_2, t_3]/(t_1, t_2, t_3)^2$ of $(1.11)(g)$, we can construct a K-linear functor

$$\mathbb{F}_3^\bullet : \operatorname{rep}_K(\mathcal{K}_3) \longrightarrow \operatorname{mod} C'$$

such that $\mathbb{F}_3^\bullet$ restricts to the exact representation embedding of the full subcategory of $\operatorname{rep}_K(\mathcal{K}_3)$ whose objects are representations $\mathbb{V}$ having no summand isomorphic to the simple injective representation $I(2)$, where $\mathcal{K}_3$ is the enlarged Kronecker quiver $1 \circ \Rrightarrow \circ 2$, see also (1.17). Because $I(2)$ does not lie in the image of the representation embedding functor

$$R : \operatorname{fin} K\langle t_1, t_2 \rangle \longrightarrow \operatorname{rep}_K(\mathcal{K}_3)$$

constructed in the proof of (1.9) then the composite functor

$$\mathbb{F}_3^\bullet \circ R : \operatorname{fin} K\langle t_1, t_2 \rangle \longrightarrow \operatorname{mod} C'$$

is a representation embedding and, in view of (1.8), the algebra $C' = K[t_1, t_2, t_3]/(t_1, t_2, t_3)^2$ is representation-wild.

We recall from (1.2) that any strictly representation-wild algebra is representation-wild. The following characterisation of wild concealed algebras shows that the inverse implication holds for concealed algebras of acyclic quivers.

1.14. Theorem. *Let B be a concealed algebra of type Q, where Q is a finite, connected, and acyclic quiver that is not a Dynkin quiver. The following conditions are equivalent.*

   (a) *The algebra B is representation-wild.*
   (b) *The algebra B is strictly representation-wild.*
   (c) *The Euler quadratic form $q_B : K_0(B) \longrightarrow \mathbb{Z}$ of B is indefinite.*
   (d) *Q is a wild quiver.*

Proof. Assume that
$$B = \operatorname{End} T_A$$
is a tilted algebra of type Q, where Q is a finite, connected, and acyclic quiver that is not of Dynkin type, and T_A is a multiplicity-free postprojective tilting module over the path algebra $A = KQ$ of Q. Let $n = |Q_0|$ be the number of vertices of Q.

It follows from (VI.4.5) and (VI.4.7) that there are abelian group isomorphisms
$$K_0(B) \cong \mathbb{Z}^n \cong K_0(A)$$
and the Euler quadratic form $q_B : K_0(B) \longrightarrow \mathbb{Z}$ of B is $\mathbb{Z}$-congruent with the Euler quadratic form $q_A : K_0(A) \longrightarrow \mathbb{Z}$ of A. Moreover, by (VII.4.1), the quadratic form q_A coincides with the quadratic form $q_Q : \mathbb{Z}^n \longrightarrow \mathbb{Z}$ of Q. Then the equivalence of (c) and (d) follows from (VII.4.5).

The implication (b)$\Rightarrow$(a) follows from (1.2).

(d)$\Rightarrow$(b) Assume that Q is a wild quiver, that is, Q is not a Dynkin quiver nor a Euclidean quiver. Hence, by (XVIII.5.5), for an arbitrary algebra Λ of finite K-dimension, there exists a fully faithful exact functor $F : \operatorname{mod} \Lambda \longrightarrow \operatorname{mod} B$. Then the algebra B is strictly representation-wild and (b) follows.

(a)$\Rightarrow$(d) Assume, to the contrary, that the algebra B is representation-wild and Q is not a wild quiver. Because Q is not a Dynkin quiver, then Q is a Euclidean quiver and B is a concealed algebra of the Euclidean type Q. It follows from the structure theorem (XII.3.4) that every component of the Auslander–Reiten quiver $\Gamma(\operatorname{mod} B)$ of B is generalised standard, that is, $\operatorname{rad}_B^\infty(X, X) = 0$, for every indecomposable B-module X in $\operatorname{mod} B$, where $\operatorname{rad}_B^\infty$ is the infinite radical of the category $\operatorname{mod} B$.

Because the algebra B is representation-wild then, by (1.11), there exists a $K\langle t_1, t_2\rangle$-B-bimodule $_{K\langle t_1,t_2\rangle}M_B$ such that the left $K\langle t_1, t_2\rangle$-module $_{K\langle t_1,t_2\rangle}M$ is finitely generated free and the functor
$$G = - \otimes_{K\langle t_1,t_2\rangle}M_B : \operatorname{fin} K\langle t_1, t_2\rangle \longrightarrow \operatorname{mod} B$$
respects the isomorphism classes and carries indecomposable modules to indecomposable ones. Then G is exact and faithful, by (1.5).

Let
$$\Lambda = K[t_1, t_2]/(t_1, t_2)^2.$$

We recall that Λ is a commutative local K-algebra of dimension three and, by (1.7), there exists a fully faithful embedding $H : \operatorname{mod}\Lambda \longrightarrow \operatorname{fin} K\langle t_1, t_2\rangle$. Hence, the functor

$$F = G \circ H : \operatorname{mod}\Lambda \longrightarrow \operatorname{mod} B$$

is a faithful exact functor that respects the isomorphism classes and carries indecomposable Λ-modules to indecomposable B-modules. Clearly, Λ is isomorphic to the algebra given by the two loop quiver

bound by the relations $\alpha\beta = 0$, $\beta\alpha = 0$, $\alpha^2 = 0$, and $\beta^2 = 0$. It follows from (X.4.8) that the Auslander–Reiten quiver $\Gamma(\operatorname{mod}\Lambda)$ of Λ admits a homogeneous tube $\mathcal{T}_0^\Lambda$ with the mouth module E of the form

$$\begin{bmatrix} 0 & 0 \\ 1 & 0 \end{bmatrix} \bigcirc K^2 \bigcirc \begin{bmatrix} 0 & 0 \\ 1 & 0 \end{bmatrix}.$$

Denote by $S = \Lambda/\operatorname{rad}\Lambda$ the unique simple Λ-module, up to isomorphism. The module S does not belong to the tube $\mathcal{T}_0^\Lambda$, because $\dim_K S = 1$ and $\dim_K E[m] = 2m$, for any Λ-module $E[m]$ lying on the unique ray of $\mathcal{T}_0^\Lambda$ starting from $E = E[1]$. Note that $S \cong E/\operatorname{rad} E$ and $S \cong \operatorname{soc} E$. Let $v : E \longrightarrow S$ be a canonical surjection and $w : S \longrightarrow E$ a canonical embedding. Because S does not lie on the tube $\mathcal{T}_0^\Lambda$ then, according to (IV.5.1), for each $t \geq 1$, there exists a path of irreducible morphisms
$$E = U_0 \xrightarrow{f_1} U_1 \xrightarrow{f_2} U_2 \longrightarrow \ldots \longrightarrow U_{t-1} \longrightarrow U_t$$
between indecomposable modules in $\mathcal{T}_0^\Lambda$ and a non-isomorphism $g_t : U_t \to S$ such that $g_t \cdot f_t \cdot \ldots \cdot f_2 \cdot f_1 \neq 0$. By applying the functor F, we get the induced path
$$F(E) = F(U_0) \xrightarrow{F(f_1)} F(U_1) \xrightarrow{F(f_2)} F(U_2) \longrightarrow \ldots \longrightarrow F(U_{t-1}) \xrightarrow{F(f_t)} F(U_t)$$
of non-isomorphisms between indecomposable B-modules and a non-isomorphism $F(g_t) : F(U_t) \longrightarrow F(S)$ such that
$$F(g_t) \cdot F(f_t) \cdot \ldots \cdot F(f_2) \cdot F(f_1) = F(g_t \cdot f_t \cdot \ldots \cdot f_2 \cdot f_1) \neq 0,$$
because F is faithful, respects isomorphism classes, and carries indecomposable Λ-modules to indecomposable B-modules. Moreover, for each $t \geq 1$, we have the induced non-zero homomorphism $F(w) : F(S) \longrightarrow F(E)$ such that
$$\varphi_t = F(w) \cdot F(g_t) \cdot F(f_t) \cdot \ldots \cdot F(f_2) \cdot F(f_1) = F(w \cdot g_t \cdot f_t \cdot \ldots \cdot f_2 \cdot f_1) \neq 0.$$

It follows that the right B-module $X = F(E)$ is indecomposable and $\varphi_t \in \operatorname{rad}_B^t(X, X)$, for each $t \geq 1$. Hence, we get the contradiction

$$0 = \operatorname{rad}_B^\infty(X, X) = \operatorname{rad}_B^m(X, X) \neq 0,$$

for some $m \geq 0$, see (X.1.5). This finishes the proof of the implication (a)$\Rightarrow$(d) and completes the proof of the theorem. $\qquad\square$

1.15. Corollary. *Let Q be a finite connected acyclic quiver with $n = |Q_0|$ points and let $q_Q : \mathbb{Z}^n \longrightarrow \mathbb{Z}$ be the quadratic form of Q. The following four conditions are equivalent.*

(a) *The path algebra KQ is representation-wild.*

(b) *The path algebra KQ is strictly representation-wild.*

(c) *There exists a positive vector $v \in \mathbb{Z}^n$ such that $q_Q(v) < 0$.*

(d) *Q is a wild quiver.*

Proof. The equivalence of (a), (b), and (d) follows from (1.14).

(c)$\Rightarrow$(d) If (c) holds then q_Q is not positive semidefinite, and it follows from (VII.4.5) that the underlying graph $\overline{Q}$ of Q is neither a Dynkin diagram, nor $\overline{Q}$ is Euclidean. Thus Q is a wild quiver and (d) follows.

(d)$\Rightarrow$(c) Assume that Q is a wild quiver. By (XVIII.4.1), for any finite dimensional K-algebra Λ, there exists a full, faithful, exact, K-linear functor $F : \operatorname{mod} \Lambda \longrightarrow \operatorname{mod} KQ$ such that $q_Q(\dim F(M)) < 0$, for all non-zero modules M in $\operatorname{mod} \Lambda$. Hence, there exists a positive vector $v \in \mathbb{Z}^n$ such that $q_Q(v) < 0$ and (c) follows. There is also a simple direct proof of the implication (d)$\Rightarrow$(c). $\qquad\square$

In relation to the study of representation-wild algebras, except of the notion of strictly wildness, there are various concepts that are close to wildness. Here we only mention the following:

(i)   controlled wildness (see [271] and [542]),

(ii) wildness mod p (see [5]),

(iii) Corner type Endo-Wildness (see [544] and [590]), and

(iv) endo-wildness (see [590] and [6]).

The reader is referred to [542], [544], [591], and [593] for a discussion of these concepts.

The wildness for coalgebras is introduced in [588]–[589], and is discussed in [316], [317], [592]–[599].

Following [590], an algebra A is defined to be **endo-wild** if any finite dimensional K-algebra C is of the form

$$C \cong \operatorname{End}_A(M),$$

where M is a module in $\operatorname{mod} A$. In other words, A is endo-wild if any finite dimensional K-algebra C can be realised as the endomorphism algebra of some finite dimensional A-module.

Now we show that any strictly representation-wild algebra is endo-wild, and that local algebras and commutative algebras are not strictly representation-wild.

1.16. Proposition. *Assume that A is a strictly representation-wild K-algebra.*

   (i) *For each integer $d \geq 1$, there exists an indecomposable finite dimensional A-module N such that $\dim_K N \geq d$ and $\operatorname{End} N \cong K$.*
   (ii) *The algebra A is endo-wild.*
   (iii) *The algebra A is not local and is not commutative.*

Proof. (i) Given an integer $d \geq 1$, we define a $K\langle t_1, t_2 \rangle$-module U_d of dimension d to be the vector space K^d equipped with the right $K\langle t_1, t_2 \rangle$-module structure defined by the action of t_1 and t_2 on U_d by the formulae $\mathbf{x} \cdot t_1 = \mathbf{x} \cdot J_d(0)$ and $\mathbf{x} \cdot t_2 = \mathbf{x} \cdot J_d(0)^t$, where $\mathbf{x} = [x_1 \ \ldots \ x_d] \in K^d$ and

$$
J_d(0) = \begin{bmatrix} 0 & \cdots & 0 & 0 \\ 1 & \ddots & & \vdots \\ \vdots & \ddots & \ddots & \vdots \\ 0 & \cdots & 1 & 0 \end{bmatrix} \in \mathbb{M}_d(\mathbb{Z})
$$

is the $d \times d$ Jordan block with the eigenvalue 0 and $J_d(0)^t$ is the transpose of $J_d(0)$.

It is easy to see that the endomorphism algebra $\operatorname{End} U_d$ of the $K\langle t_1, t_2 \rangle$-module U_d is isomorphic to K. Because the algebra A is assumed to be strictly representation-wild then, according to (1.11), there exists a $K\langle t_1, t_2 \rangle$-$A$-bimodule $_{K\langle t_1,t_2 \rangle}M_A$ such that the left $K\langle t_1, t_2 \rangle$-module $_{K\langle t_1,t_2 \rangle}M$ is finitely generated free and the exact K-linear functor

$$
T = - \otimes_{K\langle t_1,t_2 \rangle} M_A : \operatorname{fin} K\langle t_1, t_2 \rangle \longrightarrow \operatorname{mod} A
$$

is full and faithful. It follows that there is an isomorphism

$$
K \cong \operatorname{End}_{K\langle t_1,t_2 \rangle} U_d \cong \operatorname{End}_A T(U_d)
$$

of K-algebras and $\dim_K T(U_d) = \dim_K U_d \otimes_{K\langle t_1,t_2 \rangle} M_A = d \cdot r_M \geq d$, where r_M is the rank of the free left $K\langle t_1, t_2 \rangle$-module M.

   (ii) Let C be an arbitrary algebra of finite K-dimension. Because the algebra A is assumed to be strictly representation-wild then, according to (1.11), there exists a fully faithful exact K-linear functor $T : \operatorname{mod} C \longrightarrow \operatorname{mod} A$. The right A-module $M = T(C_C)$ is of finite K-dimension and the K-algebra homomorphism $C \cong \operatorname{End} C_C \longrightarrow \operatorname{End}_A M$, given by assigning to each endomorphism $h : C_C \longrightarrow C_C$ of C the endomorphism $T(h) : M \longrightarrow M$ of right A-modules, is an isomorphism of algebras. Hence, $C \cong \operatorname{End}_A M$ and (ii) follows.

(iii) Because A is assumed to be strictly representation-wild then, according to (i), there exists a non-simple indecomposable module X in mod A such that $\mathrm{End}_A X \cong K$. Then, to prove (iii), it is sufficient to show that if A is a commutative algebra or a local algebra and X is an A-module in mod A such that $\mathrm{End}_A X \cong K$ then $\dim_K X = 1$ and X is simple.

Assume that A is commutative and let X be an A-module in mod A such that $\mathrm{End}_A X \cong K$. Because the elements of A act on X as A-endomorphisms then there is a K-algebra surjection

$$\varphi : A \longrightarrow \mathrm{End}_A X \cong K$$

such that $L = \mathrm{Ker}\,\varphi$ annihilates X. It follows that $A/L \cong K$ and there is a K-algebra isomorphism $\mathrm{End}_A X \cong \mathrm{End}_{A/L} X \cong K$ and, hence, $\dim_K X = 1$ and X is a simple module.

Finally, assume that A is a local algebra and let X be a module in mod A such that $\mathrm{End}_A X \cong K$. We prove by induction on the K-dimension of A that $X \cong A/\mathrm{rad}\,A$, where rad A is the Jacobson radical of A. Because A is local and K is an algebraically closed field then rad A is the unique maximal ideal of A, every simple A-module is isomorphic to $A/\mathrm{rad}\,A$ and $\dim_K(A/\mathrm{rad}\,A) = 1$.

If $\dim_K A = 1$ then there is nothing to prove. Assume that $\dim_K A = n \geq 2$ and that the claim is proved for all algebras of K-dimension smaller than n.

Let $Z(A)$ be the centre of A and let $m \geq 1$ be the integer such that $\mathrm{rad}^m A \neq 0$ and $\mathrm{rad}^{m+1} A = 0$. We claim that $\mathrm{rad}^m A \subseteq Z(A)$. By the assumption that the field K is algebraically closed, each element $a \in A$ has the form

$$a = \lambda \cdot 1_A + a_1,$$

where $\lambda \in K$ and $a_1 \in \mathrm{rad}\,A$. Because, for each element $s \in \mathrm{rad}^m A$, we have $s \cdot a_1 = 0$ and $a_1 \cdot s = 0$, and then $s \cdot \lambda = \lambda \cdot s$. This shows that $\mathrm{rad}^m A \subseteq Z(A)$. Moreover, the ideal $\mathrm{rad}^m A$ annihilates the A-module X, because the elements of $Z(A)$ act on X as A-endomorphisms and the algebra End $X \cong K$ has no non-zero nilpotent elements. Consequently, X is a module over the quotient algebra $\overline{A} = A/\mathrm{rad}^m A$. In view of the assumption End $X \cong K$, the induction hypothesis implies that X is simple when viewed as an $\overline{A}$-module, that is, $X \cong \overline{A}/\mathrm{rad}\,\overline{A}$. Because rad $\overline{A} \cong \mathrm{rad}\,A/\mathrm{rad}^m A$, then

$$X \cong \overline{A}/\mathrm{rad}\,\overline{A} \cong A/\mathrm{rad}\,A.$$

This finishes the proof.   □

Now we present an example of a representation-wild algebra that is not strictly representation-wild.

1.17. Example. Consider the K-algebra

$$C' = K[t_1, t_2, t_3]/(t_1, t_2, t_3)^2$$

of dimension four. Because the algebra C' is commutative and local then C' is not strictly representation-wild, by (1.16).

Now we show that C' is representation-wild by constructing a faithful representation embedding

$$F : \operatorname{fin} K\langle t_1, t_2 \rangle \longrightarrow \operatorname{mod} C'.$$

Given a right $K\langle t_1, t_2 \rangle$-module X, we define the C'-module $F(X)$ to be the K-vector space $X \oplus X$ equipped with the right C'-module structure given by the formulae

$$x \cdot \overline{t_1} = [x_1 \, x_2] \cdot \begin{bmatrix} 0 & 0 \\ t_1 & 0 \end{bmatrix}, \quad x \cdot \overline{t_2} = [x_1 \, x_2] \cdot \begin{bmatrix} 0 & 0 \\ t_2 & 0 \end{bmatrix}, \quad x \cdot \overline{t_3} = [x_1 \, x_2] \cdot \begin{bmatrix} 0 & 0 \\ 1_X & 0 \end{bmatrix},$$

where $x = [x_1 \, x_2] \in X \oplus X$, $\overline{t_1}$, $\overline{t_2}$, $\overline{t_3}$ are cosets in R of the elements $t_1, t_2, t_3 \in K[t_1, t_2, t_3]$, respectively. Given a homomorphism $f : X \longrightarrow Y$ of $K\langle t_1, t_2 \rangle$-modules, we set

$$F(f) = \begin{bmatrix} f & 0 \\ 0 & f \end{bmatrix} : X \oplus X \longrightarrow Y \oplus Y.$$

It is obvious that F is an exact, faithful and K-linear functor. A routine matrix calculation shows that F carries indecomposable modules to indecomposable ones, and $F(X) \cong F(X)$ implies $X \cong Y$, for any pair of indecomposable $K\langle t_1, t_2 \rangle$-modules X and Y, see the proof of (1.7). This means that F is a representation embedding functor and, by (1.11), the algebra R is representation-wild.

To see that F carries indecomposable modules to indecomposable ones we show, by applying the matrix calculation used in the proof of (1.7), that, given an indecomposable module X in $\operatorname{fin} K\langle t_1, t_2 \rangle$, the algebra $\operatorname{End} F(X)$ has no non-trivial idempotents.

XIX.2. Indecomposable modules over the polynomial algebra $K[t]$

In the next section various equivalent forms of the definition of a representation-tame K-algebra are given. Here we collect elementary facts on the category $\operatorname{fin} K[t]$ of finite dimensional modules over the polynomial algebra $K[t]$ in one indeterminate t with coefficients in the field K we need later. Here we follow [575, Section 14.3].

One of the main results of this section asserts that

- for every indecomposable module M in fin $K[t]$, there exist almost split sequences in fin $K[t]$:

$$0 \longrightarrow M \longrightarrow N \longrightarrow M \longrightarrow 0,$$

- the Auslander–Reiten quiver of fin $K[t]$ has the form

$$\Gamma(\text{fin } K[t]) = \boldsymbol{\mathcal{T}}^{K[t]},$$

where $\boldsymbol{\mathcal{T}}^{K[t]} = \{\mathcal{T}_\lambda^{K[t]}\}_{\lambda \in K}$ is a K-family of standard stable rank one tubes.

As usual, we assume that K is an algebraically closed field, and we denote by

$$J_m(\lambda) = \begin{bmatrix} \lambda & \cdots & \cdots & 0 & 0 \\ 1 & \ddots & & \vdots & \vdots \\ \vdots & \ddots & \ddots & \vdots & \vdots \\ 0 & & \ddots & \lambda & 0 \\ 0 & \cdots & \cdots & 1 & \lambda \end{bmatrix} \in \mathbb{M}_m(K)$$

the canonical Jordan $n \times n$ matrix with the eigenvalue $\lambda \in K$.

Following the results of Chapter II, the modules over $K[t]$ can be viewed as K-linear representations of the one loop quiver

$$\mathbb{L}^{(1)}: \qquad 1 \!\circ \!\!\!\bigcirc\!\!\! \beta,$$

under the equivalence of categories fin $K[t] \cong \text{rep}_K(\mathbb{L}^{(1)})$.

For any $\lambda \in K$ and $m \geq 1$, we consider the $K[t]$-module

$$K_\lambda^m = K[t]/(t-\lambda)^m$$

of dimension m and note that K_λ^m can be viewed as the K-linear representation

$$K_\lambda^m = (K^m, \varphi_\beta : K^m \to K^m)$$

of the loop $\mathbb{L}^{(1)}$, where $K^m \cong K_\lambda^m$ and φ_β is the K-linear map corresponding, via the isomorphism $K^m \cong K_\lambda^m$, to the K-linear map $\cdot t : K_\lambda^m \longrightarrow K_\lambda^m$ given by the multiplication by t. It is clear that the matrix of the map φ_β in the basis of K_λ^m, given by the cosets $\bar{1}, \overline{t-\lambda}, \ldots, \overline{(t-\lambda)}^{m-1}$ of the elements $1, t - \lambda, \ldots, (t-\lambda)^{m-1}$, is just the canonical Jordan form $J_m(\lambda)$. Note that the module K_λ^m is simple if and only if $m = 1$.

For any $m \geq 1$ and $\lambda \in K$, consider the exact sequences in fin $K[t]$

$$0 \longrightarrow K_\lambda^m \xrightarrow{\ u_m\ } K_\lambda^{m+1} \longrightarrow K_\lambda^1 \longrightarrow 0,$$

$$0 \longrightarrow K_\lambda^1 \longrightarrow K_\lambda^{m+1} \xrightarrow{\ \pi_m\ } K_\lambda^m \longrightarrow 0,$$

where π_m is the natural epimorphism and $u_m(\bar{1}) = \overline{t-\lambda}$. We often denote π_m and u_m simply by π and u, respectively. Note that $\text{Ker } \pi_m = \overline{(t-\lambda)}^{m-1} \cong K_\lambda^1$.

2.1. Lemma. *Given $j \geq 1$ and $\lambda \in K$, we set*

$$(\overline{t - \lambda})^j = K[t](t-\lambda)^j / K[t](t-\lambda)^m.$$

(a) *The chain*

$$(0) \subset (\overline{t - \lambda})^{m-1} \subset (\overline{t - \lambda})^{m-2} \subset \cdots \subset (\overline{t - \lambda}) \subset K_\lambda^m$$

of $K[t]$-submodules of K_λ^m is a unique composition series of K_λ^m and there is an isomorphism $(\overline{t - \lambda})^j / (\overline{t - \lambda})^{j+1} \cong K_\lambda^1$ of $K[t]$-modules, for $j = 0, 1, \ldots, m$.

(b) *There exists a commutative diagram*

$$
\begin{array}{ccccccc}
(\overline{t - \lambda})^{m-1} & \subset & (\overline{t - \lambda})^{m-2} & \subset & \cdots & \subset & (\overline{t - \lambda}) & \subset & K_\lambda^m \\
\cong \uparrow \sigma_1 & & \cong \uparrow \sigma_2 & & \cdots & & \cong \uparrow \sigma_{m-1} & & 1 \uparrow \\
K_\lambda^1 & \xrightarrow{u_1} & K_\lambda^2 & \xrightarrow{u_2} & \cdots & \longrightarrow & K_\lambda^{m-1} & \xrightarrow{u_{m-1}} & K_\lambda^m,
\end{array}
$$

where $\sigma_1, \ldots, \sigma_{m-1}$ are isomorphisms of $K[t]$-modules and $u_j : K_\lambda^j \longrightarrow K_\lambda^{j+1}$ is the canonical embedding, for $j \in \{1, \ldots, m-1\}$.

Proof. (a) Let $L \subseteq K_\lambda^m$ be a $K[t]$-submodule of K_λ^m. Because $K[t]$ is a principal ideal domain then $L = (\overline{g})$, for some $g \in K[t]$, where $\overline{g}$ is the coset in the algebra $K[t]/(t - \lambda)^m$ represented by g. If $t - \lambda$ does not divide g then $1 = gh + r(t - \lambda)^m$, for some $h, r \in K[t]$. Hence $\overline{g}$ is invertible in the algebra $K[t]/(t - \lambda)^m$ and $L = K_\lambda^m$. If $t - \lambda$ divides g then g has the form $g = h(t - \lambda)^s$, where $s \geq 1$ and $h \in K[t]$ is relatively prime to $t - \lambda$. Hence we get $L = (\overline{g}) = (\overline{t - \lambda})^s$, and (a) follows.

(b) We define σ_j by setting $\sigma_j(\overline{1}) = \overline{[t - \lambda]}^{m-j+1}$. Then σ_j is surjective and the dimension argument shows that it is bijective. The commutativity of the diagram in (b) is obvious. $\qquad \square$

2.2. Lemma. *Let $m \geq 1$, $\lambda \in K$, and $K_\lambda^m = K[t]/(t - \lambda)^m$.*

(a) *There is an isomorphism $\mathrm{End}_{K[t]}(K_\lambda^m) \cong K[t]/(t - \lambda)^m$ of algebras.*

(b) *The algebra $K[t]/(t - \lambda)^m$ is local and the module K_λ^m is indecomposable.*

(c) *$\mathrm{Hom}_{K[t]}(K_\lambda^m, K_\mu^n) = 0$, for all $m, n \geq 1$ and $\lambda \neq \mu$ in K.*

(d) *For any $m, n \geq 1$ and $\lambda \in K$ there are direct sum decompositions*
  - *$\mathrm{Hom}_{K[t]}(K_\lambda^m, K_\lambda^n) = Ku^{n-m} \oplus Ku^{n-m+1}\pi \oplus \cdots \oplus Ku^{n-1}\pi^{m-1}$, if $m \leq n$,*
  - *$\mathrm{Hom}_{K[t]}(K_\lambda^m, K_\lambda^n) = Ku^{n-1}\pi^{m-1} \oplus Ku^{n-2}\pi^{m-2} \oplus \cdots \oplus K\pi^{m-n}$, if $m \geq n$,*
  - *$\mathrm{End}_{K[t]}K_\lambda^m = K \cdot 1_{K_\lambda^m} \oplus K(u\pi) \oplus \cdots \oplus K(u\pi)^{m-1}$, and the equality $u\pi = \pi u$ holds.*

Proof. (a) Fix $m \geq 1$ and $\lambda \in K$. Because $K_\lambda^m \cdot (t - \lambda)^m = 0$ then the $K[t]$-module K_λ^m is a module over the K-algebra $K[t]/(t - \lambda)^m$ and there is an isomorphism $K_\lambda^m \cong K[t]/(t - \lambda)^m$ of $K[t]/(t - \lambda)^m$-modules. Hence there exist K-algebra isomorphisms

$$\mathrm{End}_{K[t]} K_\lambda^m \cong \mathrm{Hom}_{K[t]}(K_\lambda^m, K_\lambda^m) \cong K[t]/(t - \lambda)^m.$$

(b) Note that the mth power of the ideal $J_m = (\overline{t - \lambda})$ of $K[t]/(t - \lambda)^m$ generated by $\overline{t - \lambda}$ is zero. Then (I.1.4) yields $J_m \subseteq \mathrm{rad}\, K[t]/(t - \lambda)^m$. Because the quotient algebra $[K[t]/(t-\lambda)^m]/J_m$ of $K[t]/(t-\lambda)^m$ is isomorphic with K then J_m is a maximal ideal and therefore $J_m = \mathrm{rad}\, K[t]/(t - \lambda)^m$. It follows from (I.4.6) that $K[t]/(t - \lambda)^m$ is a local algebra and, by (I.4.8), the $K[t]$-module K_λ^m is indecomposable.

To prove (c) and (d), we note that $u\pi = \pi u$ and every homomorphism $f : K_\lambda^m \longrightarrow K_\mu^n$ of $K[t]$-modules is uniquely determined by the element $f(\overline{1}) \in K_\mu^n$.

If $\lambda \neq \mu$, then $(t - \lambda)^m$ and $(t - \mu)^n$ are relatively prime in $K[t]$ and there exist $h, r \in K[t]$ such that $1 = h \cdot (t - \lambda)^m + r \cdot (t - \mu)^n$. It follows that

$$f(\overline{1}) = f(\overline{1}) \cdot 1 = f(\overline{1}) \cdot h \cdot (t - \lambda)^m = f(\overline{1} \cdot h \cdot (t - \lambda)^m) = f(0) = 0.$$

Hence $f = 0$ and (c) follows. Now assume that $m \leq n$. Because $\mathrm{Im}\, f$ is a $K[t]$-submodule of K_λ^n of length at most m, then (2.1)(a) yields $\mathrm{Im}\, f \subseteq (\overline{t - \lambda})^{n-m} = (\overline{t - \lambda})^{n-m} K_\lambda^n$. Hence

$$f(\overline{1}) = a_{n-m}\overline{[t - \lambda]}^{n-m} \oplus a_{n-m+1}\overline{[t - \lambda]}^{n-m+1} \oplus \cdots \oplus a_{n-1}\overline{[t - \lambda]}^{n-1},$$

where $a_{n-m}, \ldots, a_n \in K$ are uniquely determined by f. Because

$$u^{n-m+j}\pi^j(\overline{1}) = (\overline{t - \lambda})^{n-m+j},$$

for $j = 0, 1, \ldots n - 1$, then

$$f(\overline{1}) = (a_{n-m} \cdot 1_{K_\lambda^n} \oplus a_{n-m+1}u^{n-m+1}\pi \oplus \cdots \oplus a_{n-1}u^{n-1}\pi^{m-1})(\overline{1})$$

and, hence,

$$f = a_{n-m} \cdot 1_{K_\lambda^n} \oplus a_{n-m+1}u^{n-m+1}\pi \oplus \cdots \oplus a_{n-1}u^{n-1}\pi^{m-1}.$$

This shows the direct sum decomposition in (d), for $m \leq n$. The direct sum decomposition, for $m \geq n$ follows in a similar way. The remaining statement of (d) follows immediately from the previous ones. $\square$

Now we establish the existence of almost split sequences in the category $\mathrm{fin}\, K[t]$.

2.3. Proposition. *Let $\lambda \in K$ and let $\mathrm{fin}_\lambda K[t]$ be the full subcategory of $\mathrm{fin}\, K[t]$ consisting of modules M such that $M \cdot (t - \lambda)^s = 0$, for some $s \geq 1$.*

(a) *The exact sequences $\mathrm{fin}\, K[t]$*

$$0 \longrightarrow K_\lambda^1 \overset{u_1}{\longrightarrow} K_\lambda^2 \overset{\pi_1}{\longrightarrow} K_\lambda^1 \longrightarrow 0,$$

$$0\longrightarrow K_\lambda^m \xrightarrow{\left[\begin{smallmatrix}\pi_{m-1}\\ u_m\end{smallmatrix}\right]} K_\lambda^{m-1}\oplus K_\lambda^{m+1}\xrightarrow{[u_{m-1}\ -\pi_m]} K_\lambda^m\longrightarrow 0$$

are almost split sequences in $\mathrm{fin}_\lambda K[t]$ and in $\mathrm{fin}\,K[t]$.

(b) There exists an irreducible morphism $K_\lambda^m \longrightarrow K_\mu^n$ in $\mathrm{fin}\,K[t]$ if and only if $\lambda = \mu$ and $n = m-1$ or $m = n+1$.

(c) The Auslander–Reiten quiver $\Gamma(\mathrm{fin}_\mu K[t])$ of $\mathrm{fin}_\mu K[t]$ has the form $\mathcal{T}_\mu = \mathbb{Z}\mathbb{A}_\infty/(\tau)$ of a stable tube of rank one.

Proof. (a) Because of the equality $u\pi = \pi u$, it is easy to check that the homomorphisms u_1 and $\left[\begin{smallmatrix}\pi_{m-1}\\ u_m\end{smallmatrix}\right]$ are injective, the homomorphisms π_1 and $[u_{m-1}\ -\pi_m]$ are surjective,

$$\mathrm{Im}\,u_1 \subseteq \mathrm{Ker}\,\pi_1 \quad\text{and}\quad \mathrm{Im}\,\left[\begin{smallmatrix}\pi_{m-1}\\ u_m\end{smallmatrix}\right] \subseteq \mathrm{Ker}\,[u_{m-1}\ -\pi_m],$$

and the following equalities hold

- $\dim_K \mathrm{Im}\,u_1 = \dim_K \mathrm{Ker}\,\pi_1 = 1$, and
- $\dim_K \mathrm{Im}\,\left[\begin{smallmatrix}\pi_{m-1}\\ u_m\end{smallmatrix}\right] = m = 2m - m = \dim_K \mathrm{Ker}\,[u_{m-1}\ -\pi_m]$, for $m \geq 1$.

It follows that the sequences in (a) are exact. By (2.2), the module K_λ^m is indecomposable, for any $m \geq 1$ and $\lambda \in K$. Then, by (IV.1.13), to show that the sequences are almost split it is sufficient to prove that they are right almost split, or equivalently, that any non-invertible homomorphism $f : K_\mu^n \longrightarrow K_\lambda^m$ of $K[t]$-modules has a factorisation through

$$[u_{m-1}\ -\pi_m] : K_\lambda^{m-1}\oplus K_\lambda^{m+1}\longrightarrow K_\lambda^m,$$

in the notation of (2.3). But this follows immediately from (2.2)(c) and (2.2)(d).

The statement (b) follows from (a) and (IV.1.10). Because (c) follows from (b) the proposition is proved. □

We now summarise the preceding results as follows.

2.4. Theorem. *In the notation introduced above, the category* $\mathrm{fin}\,K[t]$ *of finite dimensional* $K[t]$*-modules has the following properties.*

(a) *The natural embeddings* $\mathrm{fin}_\lambda K[t] \hookrightarrow \mathrm{fin}\,K[t]$, *with* $\lambda \in K$, *induce the categorical coproduct decomposition*

$$\mathrm{fin}\,K[t] \cong \coprod_{\lambda \in K} \mathrm{fin}_\lambda K[t].$$

(b) *For each* $\lambda \in K$, *the full subcategory* $\mathrm{ind}\,(\mathrm{fin}_\lambda K[t])$ *of* $\mathrm{fin}\,K[t]$ *given by the indecomposable modules* K_λ^m, *with* $m \geq 1$, *in* $\mathrm{fin}\,K[t]$ *has the form*

$$K_\lambda^1 \underset{\pi_1}{\overset{u_1}{\rightleftarrows}} K_\lambda^2 \underset{\pi_2}{\overset{u_2}{\rightleftarrows}} \cdots \underset{\pi_{m-1}}{\overset{u_{m-1}}{\rightleftarrows}} K_\lambda^m \underset{\pi_m}{\overset{u_m}{\rightleftarrows}} K_\lambda^{m+1} \rightleftarrows \cdots$$

that is, every homomorphism in the category is a K*-linear combination of compositions of the homomorphisms* $u_1, u_2, u_3, \ldots$ *and*

$\pi_1, \pi_2, \pi_3, \ldots$. *Moreover, every $K[t]$-module K_λ^m is uniserial (that is, has a unique composition series) and every $K[t]$-submodule of K_λ^m is of the form $\mathrm{Im}\,(u_{m-1} \cdots u_j)$, where $j \geq 1$.*

(c) *The categories $\mathrm{fin}\,K[t]$ and $\mathrm{fin}_\lambda K[t]$ have almost split sequences.*

(d) *For each $\lambda \in K$, the Auslander–Reiten quiver $\Gamma(\mathrm{fin}_\mu K[t])$ of $\mathrm{fin}_\mu K[t]$ is a standard stable tube $\mathcal{T}_\lambda^{K[t]} \cong \mathbb{Z}\mathbb{A}_\infty/(\tau)$ of rank one.*

(e) *The Auslander–Reiten quiver $\Gamma(\mathrm{fin}\,K[t])$ of $\mathrm{fin}\,K[t]$ has the form*

$$\Gamma(\mathrm{fin}\,K[t]) = \bigcup_{\mu \in K} \Gamma(\mathrm{fin}_\mu K[t]) = \bigcup_{\mu \in K} \mathcal{T}_\lambda^{K[t]},$$

that is, the quiver $\Gamma(\mathrm{fin}\,K[t])$ is a disjoint union of a K-family $\mathcal{T}^{K[t]} = \{\mathcal{T}_\lambda^{K[t]}\}_{\lambda \in K}$ of standard stable tubes $\mathcal{T}_\lambda^{K[t]}$ of rank one.

Proof. Apply (2.1), (2.2), and (2.3). □

2.5. Corollary. *Let $h \neq 0$ be a polynomial $h = \lambda(t - \lambda_1)^{s_1} \cdots (t - \lambda_m)^{s_m}$, with $s_1, \ldots, s_m \geq 1$, $\lambda, \lambda_1, \ldots, \lambda_m \in K$, $\lambda \neq 0$, and let $K[t]_h$ be the localisation of $K[t]$ with respect to the multiplicative system $\{h^j\}_{j \in \mathbb{N}}$.*

(a) *There exists a categorical coproduct decomposition*

$$\mathrm{fin}(K[t]_h) = \coprod_{\mu \notin \{\lambda_1, \ldots, \lambda_m\}} \mathrm{fin}_\mu K[t].$$

(b) *The Auslander–Reiten quiver $\Gamma(\mathrm{fin}(K[t]_h)$ is a disjoint union of the standard stable tubes $\mathcal{T}_\mu^{K[t]}$, with $\mu \notin \{\lambda_1, \ldots, \lambda_m\}$, of the family $\mathcal{T}^{K[t]}$.*

Proof. Apply (2.4), the arguments used in the proof of (2.3), and the fact that the algebra $K[t]_h$ consists of all fractions of the form $\frac{f}{g} \in K(t)$, where g has the form $g = \rho(t - \lambda_1)^{u_1} \cdots (t - \lambda_m)^{u_m}$, with $u_1, \ldots, u_m \geq 0$ and $\rho \in K \setminus \{0\}$. □

XIX.3. Tame representation type

Before we give a definition of tame representation type of a finite dimensional K-algebra A, we introduce some concepts and we present some relevant results. We assume that K is an algebraically closed field.

Let R be an arbitrary K-algebra (not necessarily finite dimensional). We denote by $\mathrm{fin}\,R$ the full subcategory of $\mathrm{Mod}\,R$ whose objects are the modules of finite K-dimension, and by $\mathrm{ind}\,R$ the full subcategory of $\mathrm{fin}\,R$ whose objects are the indecomposable modules. Finally, given an integer $d \geq 1$, we denote by $\mathrm{ind}_d R$ the full subcategory of $\mathrm{ind}\,R$ whose objects are the indecomposable modules M with $\dim_K M = d$.

3.1. Definition. Let A be a finite dimensional K-algebra, $m \geq 1$ an integer, $R_1, \ldots, R_m$ a finite set of K-algebras, and $\mathcal{A}$ a full subcategory of $\mathrm{mod}\, A$. Suppose that, for each $j \in \{1, \ldots, m\}$, $\mathcal{B}_j$ is a full subcategory of $\mathrm{fin}\, R_j$ and we have a functor

$$\widehat{N}^{(j)} = (-) \otimes_{R_j} N_A^{(j)} : \mathcal{B}_j \longrightarrow \mathrm{Mod}\, A,$$

where $_{R_j}N_A^{(j)}$ is an R_j-A-bimodule.

(a) The category $\mathcal{A} \subseteq \mathrm{mod}\, A$ is **almost parametrised** by the family of functors $\widehat{N}^{(1)}, \ldots, \widehat{N}^{(m)}$ if all but a finite number of the isomorphism classes of modules in $\mathcal{A}$ are isomorphic to modules of the form $\widehat{N}^{(j)}(X)$, where $j \in \{1, \ldots, m\}$ and X is a module in $\mathcal{B}_j$.

(b) The category $\mathcal{A} \subseteq \mathrm{mod}\, A$ is **parametrised** by the family of functors $\widehat{N}^{(1)}, \ldots, \widehat{N}^{(m)}$ if all modules in $\mathcal{A}$ are isomorphic to modules of the form $\widehat{N}^{(j)}(X)$, where $j \in \{1, \ldots, m\}$ and X is a module in $\mathcal{B}_j$.

The following proposition provides equivalent conditions that are basic for the definition of tame representation type.

3.2. Proposition. *Let A be a finite dimensional K-algebra and $d \geq 1$ an integer. The following conditions are equivalent.*

(a) *The category $\mathrm{ind}_d A$ is almost parametrised by a family of functors*

$$\widehat{N}^{(1)}, \ldots, \widehat{N}^{(m_d)} : \mathrm{ind}_1 K[t] \longrightarrow \mathrm{mod}\, A,$$

where

$$\widehat{N}^{(j)} = (-) \otimes_{K[t]} N_A^{(j)}, \text{ for } j \in \{1, \ldots, m_d\},$$

and $_{K[t]}N_A^{(1)}, \ldots, {}_{K[t]}N_A^{(m_d)}$ are $K[t]$-A-bimodules that are finitely generated and free left $K[t]$-modules.

(b) *The category $\mathrm{ind}_d A$ is parametrised by a family of functors*

$$\widehat{N}^{(1)}, \ldots, \widehat{N}^{(m'_d)} : \mathrm{ind}_1 K[t] \longrightarrow \mathrm{mod}\, A,$$

where

$$\widehat{N}^{(j)} = (-) \otimes_{K[t]} N_A^{(j)}, \text{ for } j \in \{1, \ldots, m'_d\},$$

and $_{K[t]}N_A^{(1)}, \ldots, {}_{K[t]}N_A^{(m'_d)}$ are $K[t]$-A-bimodules that are finitely generated and free left $K[t]$-modules.

(c) *The category $\mathrm{ind}_d A$ is parametrised by a finite family of functors*

$$\widehat{N}^{(j)} = (-) \otimes_{R_j} N_A^{(j)} : \mathrm{ind}\, R_j \longrightarrow \mathrm{mod}\, A, \text{ with } j \in \{1, \ldots, n_d\},$$

where $R_j = K[t]_{h_j}$ is a localisation of $K[t]$ and $_{R_j}N_A^{(j)}$ are R_j-A-bimodules that are finitely generated left R_j-modules, for all $j \in \{1, \ldots, n_d\}$.

(d) *The category* $\mathrm{ind}_d A$ *is parametrised by a finite family of functors*

$$\widehat{N}^{(j)} = (-) \otimes_{R_j} N_A^{(j)} : \mathrm{ind}\, R_j \longrightarrow \mathrm{mod}\ A, \text{ with } j \in \{1, \dots, n_d'\},$$

where $R_j = K[t]_{h_j}$ *is a localisation of* $K[t]$ *and* $_{R_j}N_A^{(j)}$ *are* R_j-A-*bimodules that are finitely generated and free left* R_j-*modules, for* $j \in \{1, \dots, n_d'\}$.

(e) *The category* $\mathrm{ind}_d A$ *is parametrised by a finite family of functors*

$$\widehat{N}^{(j)} = (-) \otimes_{R_j} N_A^{(j)} : \mathrm{ind}_1 R_j \longrightarrow \mathrm{mod}\ A, \text{ with } j \in \{1, \dots, u_d\},$$

where $R_j = K[t]_{h_j}$ *is a localisation of* $K[t]$ *and* $_{R_j}N_A^{(j)}$ *are* R_j-A-*bimodules that are finitely generated and free left* R_j-*modules, for* $j \in \{1, \dots, u_d\}$.

(f) *The category* $\mathrm{ind}_d A$ *is parametrised by a finite family of functors*

$$\widehat{N}^{(j)} = (-) \otimes_{R_j} N_A^{(j)} : \mathrm{ind}_1 R_j \longrightarrow \mathrm{mod}\ A, \text{ with } j \in \{1, \dots, v_d\},$$

where, for each $j \in \{1, \dots, v_d\}$, *we have*

- $R_j = K$ *or* $R_j = K[t]_{h_j}$ *is a localisation of* $K[t]$,
- $_{R_j}N_A^{(j)}$ *is an* R_j-A-*bimodule, that is a finitely generated and left free* R_j-*module, and*
- *the functor* $\widehat{N}^{(j)} = (-) \otimes_{R_j} N_A^{(j)} : \mathrm{ind}_1 R_j \longrightarrow \mathrm{mod}\ A$ *is a representation embedding.*

Proof. The proof is given in [186] and in [575, Section 14.4], by applying elementary algebraic geometry arguments. $\square$

The following important concept of tame representation type of a finite dimensional K-algebra is due to Drozd [201] and [202].

3.3. Definition. Let A be a finite dimensional K-algebra. Then A is defined to be of **tame representation type** (or **representation-tame**, or shortly **tame**) if, for any integer $d \geq 1$, the category $\mathrm{ind}_d A$ is almost parametrised by a finite family of functors

$$\widehat{N}^{(1)}, \dots, \widehat{N}^{(m_d)} : \mathrm{ind}_1 K[t] \longrightarrow \mathrm{mod}\ A,$$

where $\widehat{N}^{(j)} = (-) \otimes_{K[t]} N_A^{(j)}$, for $j \in \{1, \dots, m_d\}$, and $_{K[t]}N_A^{(1)}, \dots, _{K[t]}N_A^{(m_d)}$ are $K[t]$-A-bimodules that are finitely generated and free left $K[t]$-modules.

The following tame-wild dichotomy result due to Drozd [202] is fundamental for the representation theory of finite dimensional algebras.

3.4. Theorem. *Every finite dimensional algebra* A *over an algebraically closed field* K *is representation-tame or representation-wild, and these two types of algebras are mutually exclusive.*

The proof of the theorem is rather long. It involves representations of a bimodule $\mathcal{B}$ over a category with a coalgebra structure (bocs) and some reduction algorithms for bocses $\mathcal{B}$. The reader is referred to [147] (see also [151]) for a proof invoking the concept of a free triangular bocs $\mathcal{B}$ and a reduction of the problem from the module categories mod A to the categories $\mathrm{rep}(\mathcal{B}, K)$ of K-linear representations of bocses $\mathcal{B}$. For more elementary proofs the reader is referred to [239] and [565, Section 3].

The tame-wild dichotomy for K-coalgebras is discussed in [588], [589], [592], [596], [597], and an fc-version of it is established in [599].

Now we present (without proof) another fundamental result for the representation theory of finite dimensional algebras due to Crawley-Boevey [147].

3.5. Theorem. *If A is a representation-tame algebra then, for each integer $d \geq 1$, all but a finite number of indecomposable modules X in mod A, with $\dim_K X = d$, lie on homogeneous tubes of the Auslander–Reiten quiver $\Gamma(\mathrm{mod}\, A)$ of A.*

It follows from the Definition (3.3) that A is a representation-tame algebra if, for each integer $d \geq 1$, the indecomposable modules in $\mathrm{ind}_d A$ occur in a finite number of discrete (finite) families and in a finite number of one-parameter families.

The counting of the number of one-parameter families of indecomposable A-modules of a fixed dimension $d \geq 1$ leads to the following hierarchy of representation-tame algebras introduced by Skowroński in [606] and [610], see also [446]–[448], [682]–[685].

3.6. Definition. Let A be a representation-tame K-algebra.

(a) Given an integer $d \geq 1$, we denote by $\mu_A(d)$ the minimal number $m_d \geq 0$ of functors

$$\widehat{N}^{(1)}, \ldots, \widehat{N}^{(m_d)} : \mathrm{ind}_1 K[t] \longrightarrow \mathrm{mod}\ A$$

of the form $\widehat{N}^{(j)} = (-) \otimes_{K[t]} N_A^{(j)}$, for $j \in \{1, \ldots, m_d\}$, that almost parametrise the category $\mathrm{ind}_d A$, where $_{K[t]}N_A^{(1)}, \ldots, {}_{K[t]}N_A^{(m_d)}$ are $K[t]$-A-bimodules that are finitely generated and free, when viewed as left $K[t]$-modules.

(b) The tame algebra A is defined to be of **finite growth** if there exists an integer $m \geq 0$ such that $\mu_A(d) \leq m$, for each $d \geq 1$.

(c) The tame algebra R is defined to be of **linear growth** if there exists an integer $m \geq 0$ such that $\mu_A(d) \leq m \cdot d$, for all $d \geq 1$.

(d) The tame algebra R is defined to be of **polynomial growth** if there exists an integer $m \geq 1$ such that $\mu_A(d) \leq d^m$, for all $d \geq 1$.

It follows from Definition (3.3) that every representation-finite algebra A is representation-tame and $\mu_A(d) = 0$, for each $d \geq 1$.

Now we show that the Kronecker algebra is representation-tame, and even one-parametric.

3.7. Example. Let

$$A = KQ \cong \begin{bmatrix} K & 0 \\ K^2 & K \end{bmatrix}$$

be the path algebra of the Kronecker quiver

$$Q: \quad 1 \circ \underset{\beta}{\overset{\alpha}{\longleftarrow\!\!\!=\!\!\!=}} \circ \, 2 \, .$$

Consider the $K[t]$-A-bimodule M of the form

$$M: \quad K[t] \overset{1}{\underset{t\cdot}{\longleftarrow\!\!\!=\!\!\!=}} K[t]$$

Note that M, viewed as a left $K[t]$-module, is free of rank 2. Take an indecomposable module X in $\operatorname{ind} K[t]$. Then $X \cong K_\lambda^m = K[t]/(t - \lambda)^m$, for some $m \geq 1$ and $\lambda \in K$. A direct checking shows that the induced functor

$$\widehat{M} = (-) \otimes_{K[t]} M_A : \operatorname{ind} K[t] \longrightarrow \operatorname{mod} A$$

carries the module X to the indecomposable A-module $\widehat{M}(X)$ of the form

$$E_\lambda[m]: \quad (K^m \underset{\begin{bmatrix} \lambda & 1 & 0 & \cdots & 0 \\ 0 & \lambda & 1 & \cdots & 0 \\ \vdots & & & \ddots & \vdots \\ 0 & 0 & 0 & & 1 \\ 0 & 0 & 0 & \cdots & \lambda \end{bmatrix}}{\overset{\begin{bmatrix} 1 & 0 & \cdots & 0 \\ 0 & 1 & \cdots & 0 \\ \vdots & & \ddots & \vdots \\ 0 & 0 & \cdots & 1 \end{bmatrix}}{\longleftarrow\!\!\!=\!\!\!=\!\!\!\longrightarrow}} K^m)$$

in the notation of Section XI.4. Also, it is shown there that $\operatorname{ind}_d A$ admits infinitely many isomorphism classes of modules if and only if $d = 2m$, for some $m \geq 1$. Hence, it follows from (XI.4.5) and (XI.4.6) that, for any fixed $m \geq 1$, the functor $\widehat{M}$ defines an equivalence between the category $\operatorname{ind}_m K[t]$ and the category of regular A-modules of regular length m lying in the standard stable tubes $\mathcal{T}_\lambda^A$, with $\lambda \in K = \mathbb{P}_1(K) \setminus \{\infty\}$. It follows that, for each integer $d \geq 1$, the category $\operatorname{ind}_d A$ is almost parametrised by the functor $\widehat{M}$, and consequently, the Kronecker algebra $A = \begin{bmatrix} K & 0 \\ K^2 & K \end{bmatrix}$ is representation-tame of linear growth, and even one-parametric, see also the proof of (3.13) presented later.

One of the remarkable results on representation-finite algebras is the following theorem that establishes the validity of the second Brauer-Thrall conjecture, see Section IV.5.

3.8. Theorem. *An algebra A is representation-finite if and only if* $\mu_A(d) = 0$, *for each* $d \geq 1$.

A proof of the theorem was announced by Nazarova and Roiter in 1973, see [444] and [445]. For algebras over a field of characteristic different from 2, the first complete proof of the second Brauer-Thrall conjecture has been given only in 1985 by Bautista [41]. This was extended by Bongatz [91] to any algebraically closed field of arbitrary characteristic. Moreover, some modifications of the proof have been published by Fichbacher [229] and Bretcher-Todorov [113]. All these proofs rely on the following result of Bautista, Gabriel, Roiter, and Salmeron [43].

3.9. Theorem. *Every representation-finite algebra admits a multiplicative basis.*

By a multiplicative basis of an algebra A we mean a K-vector space basis of A such that the multiplication of any two basis vectors is either zero, or is a vector of the given basis.

As a consequence of (3.9) we get that, for each $d \geq 1$, there is only a finite number of the isomorphism classes of representation-finite K-algebras A such that $\dim_K A = d$.

The following important class of representation-tame algebras was introduced by Ringel [525].

3.10. Definition. Let A be a finite dimensional K-algebra.

(a) A is defined to be **representation-domestic**, (or shortly **domestic**) if there exists a finite family of functors

$$\widehat{N}^{(1)}, \dots, \widehat{N}^{(n)} : \operatorname{ind} K[t] \longrightarrow \operatorname{mod} A,$$

such that $_{K[t]}N_A^{(1)}, \dots, {}_{K[t]}N_A^{(n)}$ are $K[t]$-A-bimodules that are finitely generated and free left $K[t]$-modules and, for any integer $d \geq 1$, the category $\operatorname{ind}_d A$ is almost parametrised by the functors $\widehat{N}^{(1)}, \dots, \widehat{N}^{(n)}$.

(b) A representation-domestic algebra A is defined to be n-**parametric**, if $n \geq 0$ is the minimal number of functors $\widehat{N}^{(1)}, \dots, \widehat{N}^{(n)}$ with the properties listed in (a).

The following simple lemma shows that every representation-domestic algebra is representation-tame and of finite growth.

3.11. Lemma. *If A is an n-parametric algebra then A is of finite growth and $\mu_A(d) \leq n$, for all $d \geq 1$.*

Proof. Let $_{K[t]}M_A$ be a $K[t]$-A-bimodule such that M viewed as a left $K[t]$-module is finitely generated free, say of $K[t]$-rank $r \geq 1$. Let

$$\widehat{M} = (-) \otimes {}_{K[t]}M_A : \operatorname{ind} K[t] \longrightarrow \operatorname{mod} A.$$

It is easy to see that, for any module $K_\lambda^m = K[t]/(t-\lambda)^m$ in $\operatorname{ind} K[t]$, with $m \geq 1$ and $\lambda \in K$, we have

$$\dim_K \widehat{M}(K_\lambda^m) = \dim_K(K_\lambda^m \otimes {}_{K[t]}M_A) = m \cdot r.$$

Consider the polynomial K-algebras $K[t_2]$ and $K[t, t_2]$. Given $m \geq 1$, we view the vector space

$$M^\bullet = K[t, t_2]/(t - t_2)^m \otimes {}_{K[t]}M_A$$

as a $K[t_2]$-A-bimodule. It is clear that $M^\bullet$ viewed as a left $K[t_2]$-module is finitely generated free of rank $m \cdot r$ and the induced functor

$$\widehat{M^\bullet} = (-) \otimes {}_{K[t_2]}M_A^\bullet : \operatorname{ind}_1 K[t_2] \longrightarrow \operatorname{mod} A,$$

carries any simple $K[t_2]$-module $S_\lambda = K[t_2]/(t_2 - \lambda)$, with $\lambda \in K$, to the A-module

$$\widehat{M^\bullet}(S_\lambda) \cong \widehat{M}(K_\lambda^m)$$

of dimension $\dim_K \widehat{M}(K_\lambda^m) = m \cdot r$.

Now assume that A is a representation-domestic algebra and

$$\widehat{N}^{(1)}, \ldots, \widehat{N}^{(n)} : \operatorname{ind} K[t] \longrightarrow \operatorname{mod} A,$$

is a finite family of functors such that ${}_{K[t]}N_A^{(1)}, \ldots, {}_{K[t]}N_A^{(n)}$ are $K[t]$-A-bimodules that are finitely generated and free left $K[t]$-modules and, for any integer $d \geq 1$, the category $\operatorname{ind}_d A$ is almost parametrised by the functors $\widehat{N}^{(1)}, \ldots, \widehat{N}^{(n)}$. By applying the first part of the proof to $M = N^{(j)}$, for $j \in \{1, \ldots, n\}$, we construct a new family of functors

$$\widehat{N^\bullet}^{(1)}, \ldots, \widehat{N^\bullet}^{(n)} : \operatorname{ind}_1 K[t_2] \longrightarrow \operatorname{mod} A,$$

that parametrise the category $\operatorname{ind}_d A$, for any $d \geq 1$. It follows that $\mu_A(d) \leq n$, for any $d \geq 1$. $\square$

It was conjectured in [610] that the converse implication to that in (3.11) also holds, that is, any algebra of finite growth is representation-domestic. This was proved by Crawley-Bocvey in [152], and together with (3.11) yields the following fact.

3.12. Theorem. *A K-algebra A is representation-domestic if and only if A is of finite growth.*

The question, raised in [610], whether or not all algebras of polynomial growth are of linear growth, remains still an open problem.

We may visualise the hierarchy of classes of representation-tame algebras defined in (3.6) as follows

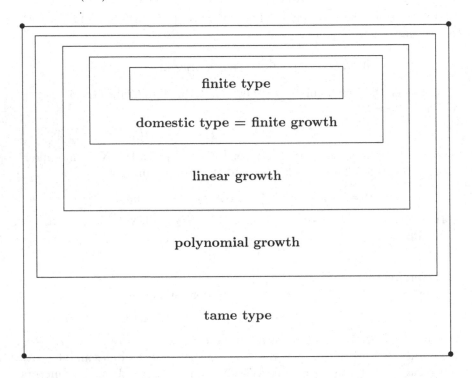

In the final part of this section we exhibit some classes of representation-infinite tame algebras that illustrate the hierarchy visualised in the figure.

Now we show that concealed algebras of Euclidean type are one-parametric and, hence, representation-tame.

3.13. Proposition. *Let B be a concealed algebra of Euclidean type. Then there exists a $K[t]$-B-bimodule $_{K[t]}N_B$ such that*

(i) *N viewed as a left $K[t]$-module is finitely generated free,*

(ii) *the image of the functor $\widehat{M} = (-) \otimes _{K[t]}N_B : \operatorname{ind} K[t] \longrightarrow \operatorname{mod} B$, is contained in the subcategory $\mathcal{R}(B)$ of $\operatorname{mod} B$ of regular modules,*

(iii) *for each integer $d \geq 1$, the category $\operatorname{ind}_d B$ is almost parametrised by the functor $\widehat{N}$.*

In particular, the algebra B is one-parametric and, hence, representation-tame.

Proof. Assume that B is a concealed algebra of Euclidean type. It follows from (XII.3.1) that B is a tilted algebra of the form

$$B \cong \operatorname{End} T_A,$$

where T_A is a multiplicity-free postprojective tilting module over the path algebra $A = K\Delta$ of a canonically oriented Euclidean quiver

$$\Delta \in \{\Delta(\widetilde{\mathbb{A}}_{p,q}), 1 \leq p \leq q, \ \Delta(\widetilde{\mathbb{D}}_m), m \geq 4, \ \Delta(\widetilde{\mathbb{E}}_6), \Delta(\widetilde{\mathbb{E}}_7), \Delta(\widetilde{\mathbb{E}}_8)\}.$$

First, for the hereditary algebra $A = K\Delta$, we construct a $K[t]$-A-bimodule $_{K[t]}M_A$ such that

(i') M viewed as a left $K[t]$-module is finitely generated free,

(ii') the image of the induced functor

$$\widehat{M} = (-) \otimes _{K[t]} M_A : \operatorname{ind} K[t] \longrightarrow \operatorname{mod} A,$$

is contained in the subcategory $\mathcal{R}(A)$ of $\operatorname{mod} A$ of regular modules, and

(iii') for each integer $d \geq 1$, the category $\operatorname{ind}_d A$ is almost parametrised by the functor $\widehat{M}$.

We do it by a case by case inspection of the five types of canonically oriented Euclidean quivers. Assume that $A = K\Delta$.

Case 1°. If $\Delta = \Delta(\widetilde{\mathbb{A}}_{p,q})$, where $1 \leq p \leq q$, then, for $_{K[t]}M_{K\Delta}$, we take the $K[t]$-$K\Delta$-bimodule given by the diagram, see (XII.2.4),

Case 2°. If $\Delta = \Delta(\widetilde{\mathbb{D}}_m)$, where $m \geq 4$, then, for $_{K[t]}M_{K\Delta}$, we take the $K[t]$-$K\Delta$-bimodule given by the diagram, see (XIII.2.6)(d),

Case 3°. If $\Delta = \Delta(\widetilde{\mathbb{E}}_6)$ then, for $_{K[t]}M_{K\Delta}$, we take the $K[t]$-$K\Delta$-bimodule given by the diagram, see (XIII.2.12)(d),

$$K[t]$$
$$\Big\downarrow {\scriptsize\begin{bmatrix}1\\0\end{bmatrix}}$$
$$K[t]^2$$
$$\Big\downarrow {\scriptsize\begin{bmatrix}t\cdot 1\\1\;1\\1\;0\end{bmatrix}}$$

$$K[t] \xrightarrow{\;{\scriptsize\begin{bmatrix}1\\0\end{bmatrix}}\;} K[t]^2 \xrightarrow{\;{\scriptsize\begin{bmatrix}1\;0\\0\;1\\0\;0\end{bmatrix}}\;} K[t]^3 \xleftarrow{\;{\scriptsize\begin{bmatrix}0\;0\\1\;0\\0\;1\end{bmatrix}}\;} K[t]^2 \xleftarrow{\;{\scriptsize\begin{bmatrix}0\\1\end{bmatrix}}\;} K[t]$$

<u>Case 4°</u>.   If $\Delta = \Delta(\widetilde{\mathbb{E}}_7)$ then, for $_{K[t]}M_{K\Delta}$, we take the $K[t]$-$K\Delta$-bimodule given by the diagram, see (XIII.2.16)(d),

$$K[t]^2$$
$$\Big\downarrow {\scriptsize\begin{bmatrix}1\;t\cdot\\1\;0\\1\;1\\0\;1\end{bmatrix}}$$

$$K[t] \xrightarrow{\;{\scriptsize\begin{bmatrix}1\\0\end{bmatrix}}\;} K[t]^2 \xrightarrow{\;{\scriptsize\begin{bmatrix}1\;0\\0\;1\\0\;0\end{bmatrix}}\;} K[t]^3 \xrightarrow{\;{\scriptsize\begin{bmatrix}1\;0\;0\\0\;1\;0\\0\;0\;1\\0\;0\;0\end{bmatrix}}\;} K[t]^4 \xleftarrow{\;{\scriptsize\begin{bmatrix}0\;0\;0\\1\;0\;0\\0\;1\;0\\0\;0\;1\end{bmatrix}}\;} K[t]^3 \xleftarrow{\;{\scriptsize\begin{bmatrix}0\;0\\1\;0\\0\;1\end{bmatrix}}\;} K[t]^2 \xleftarrow{\;{\scriptsize\begin{bmatrix}0\\1\end{bmatrix}}\;} K[t]$$

<u>Case 5°</u>.   If $\Delta = \Delta(\widetilde{\mathbb{E}}_8)$ then, for $_{K[t]}M_{K\Delta}$, we take the $K[t]$-$K\Delta$-bimodule given by the diagram, see (XIII.2.20)(d),

$$K[t]^3$$
$$\Big\downarrow {\scriptsize\begin{bmatrix}t\cdot\;1\;0\\0\;0\;1\\1\;1\;0\\1\;0\;1\\1\;1\;0\\0\;1\;0\end{bmatrix}}$$

$$K[t]^2 \xrightarrow{\;{\scriptsize\begin{bmatrix}0\;0\\0\;0\\1\;0\\0\;1\end{bmatrix}}\;} K[t]^4 \xrightarrow{\;{\scriptsize\begin{bmatrix}0\;0\;0\;0\\0\;0\;0\;0\\1\;0\;0\;0\\0\;1\;0\;0\\0\;0\;1\;0\\0\;0\;0\;1\end{bmatrix}}\;} K[t]^6 \xleftarrow{\;{\scriptsize\begin{bmatrix}1\;0\;0\;0\;0\\0\;1\;0\;0\;0\\0\;0\;1\;0\;0\\0\;0\;0\;1\;0\\0\;0\;0\;0\;1\\0\;0\;0\;0\;0\end{bmatrix}}\;} K[t]^5 \xleftarrow{\;{\scriptsize\begin{bmatrix}1\;0\;0\;0\\0\;1\;0\;0\\0\;0\;1\;0\\0\;0\;0\;1\\0\;0\;0\;0\end{bmatrix}}\;} K[t]^4 \xleftarrow{\;{\scriptsize\begin{bmatrix}1\;0\;0\\0\;1\;0\\0\;0\;1\\0\;0\;0\end{bmatrix}}\;} K[t]^3 \xleftarrow{\;{\scriptsize\begin{bmatrix}1\;0\\0\;1\\0\;0\end{bmatrix}}\;} K[t]^2 \xleftarrow{\;{\scriptsize\begin{bmatrix}1\\0\end{bmatrix}}\;} K[t]$$

It is clear that M, viewed as a left $K[t]$-module is finitely generated free, and hence the condition (i') is satisfied. Further, it follows from (XIII.2.4), (XIII.2.6), (XIII.2.12), (XIII.2.16), and (XIII.2.20) that, for any indecomposable $K[t]$-module $K_\lambda^n = K[t]/(t-\lambda)^n$, with $n \geq 1$ and $\lambda \in K$, the image of K_λ^n under the functor $\widehat{M} : \operatorname{fin} K[t] \longrightarrow \operatorname{mod} A$ is an indecomposable regular A-module, because there is an isomorphism

$$\widehat{M}(K_\lambda^n) \cong K[t]/(t-\lambda)^n \otimes_{K[t]} M_A$$

of A-modules and the A-module $K[t]/(t-\lambda)^n \otimes_{K[t]} M_A$, viewed as a representation of the quiver Δ, is obtained from the diagram defining the bimodule M by interchanging $K[t]$ with the vector space K^n. It follows that the A-defect $\partial_A(\mathbf{dim}\,\widehat{M}(K_\lambda^n))$ of the dimension vector $\mathbf{dim}\,\widehat{M}(K_\lambda^n)$ is zero and, by (XI.2.3), the A-module $M(K_\lambda^n)$ is regular. It is easy to see that, in the notation of Section XIII.2,

(1) the regular A-module $M(K_\lambda^n)$ belongs to the stable tube $\mathcal{T}_\lambda^A$ of rank one, with the mouth module $F^{(\lambda)}$, for $\lambda \in K \setminus \{0,1\}$,

(2) there is an isomorphism $\widehat{M}(K^n_\lambda) \cong F^{(\lambda)}[n]$, if $\lambda \subset K \setminus \{0, 1\}$, and hence

(3) $\widehat{M}(K^n_\lambda)$ is a regular A-module of regular length n in the homogeneous tube $\mathcal{T}^A_\lambda$, if $\lambda \in K \setminus \{0, 1\}$.

The statement (1) implies (ii'). To prove (iii'), we note that, according to (XIII.2.5), (XIII.2.9), (XIII.2.15), (XIII.2.19), and (XIII.2.23), all indecomposable modules in the homogeneous tubes $\mathcal{T}^A_\lambda$, with $\lambda \in K \setminus \{0, 1\}$, are of the form $F^{(\lambda)}[n]$, where $n \geq 1$, and the remaining stable tubes $\mathcal{T}^A_\infty$, $\mathcal{T}^A_0$, and $\mathcal{T}^A_1$ of $\Gamma(\mathrm{mod}\, A)$ are nonhomogeneous and contain only finitely many modules of any fixed dimension $d \geq 1$.

Because the indecomposable postprojective A-modules and the indecomposable preinjective A-modules are uniquely determined by their dimension vectors, by (IX.3.1), then, for each $d \geq 1$, all but a finite number of the isomorphism classes of the modules in $\mathrm{ind}_d A$ are represented by the indecomposable regular modules lying in the homogeneous tubes $\mathcal{T}^A_\lambda$, with $\lambda \in K \setminus \{0, 1\}$. It follows that, for each $d \geq 1$, the category $\mathrm{ind}_d A$ is parametrised by the functor $\widehat{M}$, and (iii') follows.

By our assumption, $B \cong \mathrm{End}\, T_A$, where T_A is a multiplicity-free postprojective tilting module over the path algebra $A = K\Delta$ of a canonically oriented Euclidean quiver Δ. We recall that the functor

$$\mathrm{Hom}_A(T, -) : \mathrm{mod}\, A \longrightarrow \mathrm{mod}\, B$$

restricts to the equivalence of categories

$$G = \mathrm{Hom}_A(T, -) : \mathcal{R}(A) \overset{\simeq}{\longrightarrow} \mathcal{R}(B)$$

of regular modules. The functor G carries short exact sequences in $\mathcal{R}(A)$ to the exact sequences in $\mathrm{mod}\, B$, because the module T is postprojective and (VIII.2.13) implies that $\mathcal{R}(A)$ is a subcategory of

$$\mathcal{T}(T) = \{X_A | \, \mathrm{Ext}^1_A(T, X) = 0\} = \{X_A | \, \mathrm{Hom}_A(X, \tau T) = 0\},$$

that is, $\mathrm{Ext}^1_A(T, X) = 0$, for each regular A-module X_A. It then follows that the composite functor

$$F = \mathrm{Hom}_A(T, -) \circ \widehat{M} : \mathrm{ind}\, K[t] \longrightarrow \mathrm{mod}\, B,$$

is exact and, by (1.5), there is a functorial isomorphism $F \cong (-) \otimes_{K[t]} N_B = \widehat{N}$, where $_{K[t]}N_B$ is a $K[t]$-B-bimodule that is finitely generated free, when viewed as a left $K[t]$-module. Obviously, the image of F is contained in the category

$$\mathcal{R}(B) = \mathrm{add}\, \boldsymbol{\mathcal{T}}^B$$

of regular B-modules and, by (VIII.4.5), $\mathcal{T}_\lambda^B = \operatorname{Hom}_A(T, \mathcal{T}_\lambda^A)$ is a stable tube of $\Gamma(\operatorname{mod} B)$ and the rank of $\mathcal{T}_\lambda^B$ equals the rank of $\mathcal{T}_\lambda^A$, for each $\lambda \in \mathbb{P}_1(K)$. It follows that all indecomposable B-modules lying in the homogeneous tubes $\mathcal{T}_\lambda^B$ of $\Gamma(\operatorname{mod} B)$, with $\lambda \in K \setminus \{0, 1\}$, are contained in the image of the functor $\widehat{N} : \operatorname{ind} K[t] \longrightarrow \operatorname{mod} B$. This shows that, for each $d \geq 1$, the category $\operatorname{ind}_d B$ is almost parametrised by the functor $\widehat{N}$, and the proposition follows. $\square$

3.14. Theorem. *If B is a representation-infinite tilted algebra of Euclidean type then B is one-parametric, and hence representation-tame.*

Proof. Assume that B is a representation-infinite tilted algebra of Euclidean type Q, where Q is a Euclidean quiver. It follows from the classification theorem (XVII.5.1) that B is a domestic tubular extension or a domestic tubular coextension of a concealed algebra C of Euclidean type. Moreover, the canonical surjective homomorphism of algebras $B \longrightarrow C$ induces a fully faithful exact embedding $\operatorname{mod} C \hookrightarrow \operatorname{mod} B$. It follows from (XVII.6.1) that, for each dimension $d \geq 1$, all but a finite number of modules in $\operatorname{ind}_d B$ lie in the subcategory $\operatorname{ind}_d C$ of $\operatorname{ind}_d B$, up to isomorphism. Therefore, it is sufficient to show that the concealed algebra C of Euclidean type is one-parametric. Hence, the theorem follows from (3.13). $\square$

Now we give a characterisation of path algebras KQ of tame representation type.

3.15. Theorem. *Let Q be an acyclic, finite, and connected quiver, and $A = KQ$ the path algebra of Q. The following conditions are equivalent.*

(a) *The algebra A is representation-tame.*

(b) *The algebra A is representation-domestic, and at most one-parametric.*

(c) *The algebra A is representation-finite, or there exists a $K[t]$-A-bimodule $_{K[t]}M_A$ such that*

    (i) *M viewed as a left $K[t]$-module is finitely generated free,*

    (ii) *the image of the functor*

$$\widehat{M} = (-) \otimes {}_{K[t]}M_A : \operatorname{ind} K[t] \longrightarrow \operatorname{mod} A,$$

    is contained in the subcategory $\mathcal{R}(A)$ of $\operatorname{mod} A$ of regular modules,

    (iii) *for each integer $d \geq 1$, the category $\operatorname{ind}_d A$ is almost parametrised by the functor $\widehat{M}$.*

(d) *The Euler quadratic form $q_A : K_0(A) \longrightarrow \mathbb{Z}$ of A is positive semidefinite.*

(e) *The underlying graph $\overline{Q}$ of Q is any of the Dynkin diagrams $\mathbb{A}_m$, $\mathbb{D}_m$,*
    $\mathbb{E}_6$, $\mathbb{E}_7$, $\mathbb{E}_8$ or any of the Euclidean diagrams $\widetilde{\mathbb{A}}_m$, $\widetilde{\mathbb{D}}_m$, $\widetilde{\mathbb{E}}_6$, $\widetilde{\mathbb{E}}_7$, $\widetilde{\mathbb{E}}_8$.
(f) *The algebra A is not endo-wild.*
(g) *The algebra A is not representation-wild.*

Proof. The implications (c)$\Rightarrow$(b)$\Rightarrow$(a) are obvious, the equivalence (d)$\Leftrightarrow$(e) follows from (VII.4.5), and the implication (f)$\Rightarrow$(g) is a consequence of (1.14) and (1.16). We recall from (1.14) that $A = KQ$ is representation-wild if and only if Q is a wild quiver. Hence, the equivalence (g)$\Leftrightarrow$(e) follows.

(e)$\Rightarrow$(c) If the graph $\overline{Q}$ is any of the Dynkin diagrams $\mathbb{A}_m$, $\mathbb{D}_m$, $\mathbb{E}_6$, $\mathbb{E}_7$, $\mathbb{E}_8$ then A is representation-finite, by Gabriel's theorem (VII.5.10), and (c) follows.

Assume that the graph $\overline{Q}$ is any of the Euclidean diagrams $\widetilde{\mathbb{A}}_m$, $\widetilde{\mathbb{D}}_m$, $\widetilde{\mathbb{E}}_6$, $\widetilde{\mathbb{E}}_7$, $\widetilde{\mathbb{E}}_8$. Then (3.13) applies and (c) follows.

(e)$\Rightarrow$(f) It follows from (VII.5.14) that if $\overline{Q}$ is a Dynkin graph then any indecomposable A-module X_A is a brick, that is, $\operatorname{End} X_A \cong K$. Assume that $\overline{Q}$ is a Euclidean diagram. It follows from (XIII.2.1) and (XI.1.4) that any indecomposable A-module X_A in $\operatorname{mod} A$ is either postprojective with $\operatorname{End} X_A \cong K$, or preinjective with $\operatorname{End} X_A \cong K$, or regular. If X_A is regular it lies in a standard stable tube and, by (X.2.7), $\operatorname{End} X_A$ is isomorphic to the algebra $K[t]/(t^m)$, for some $m \geq 1$. It follows that, given an indecomposable A-module X_A in $\operatorname{mod} A$, the local algebra $\operatorname{End} X_A$ is isomorphic to the field K, or to the algebra $K[t]/(t^m)$, for any $m \geq 2$.

Assume, to the contrary, that A is endo-wild. Then the local algebra $C = k[t_1, t_2]/(t_1, t_2)^2$ is isomorphic to the endomorphism algebra $\operatorname{End} X_A$ of some module X_A in $\operatorname{mod} A$. Because $C \cong \operatorname{End} X_A$ is local then the module X_A is indecomposable, and we get a contradiction, because the algebra C is obviously not isomorphic to the algebra $K[t]/(t^m)$, for some $m \geq 1$. Hence the implication (e)$\Rightarrow$(f) follows.

To finish the proof, it remains to show that (a) implies (e). Assume that the algebra A is representation-tame. By the tame-wild dichotomy (3.4), A is representation-wild and (e) follows form (1.14). Instead of applying (3.4), we can conclude from the dimension arguments in [457] (see also [575, pp. 317–321]) that $q_A = q_Q$ is positive semidefinite, if $A = KQ$ is tame. Hence (e) easily follows. $\qquad\square$

3.16. Remarks. (a) It follows from (1.15) and (3.15) that, for the path algebras $A = KQ$ of acyclic and finite quivers Q, the wildness, strictly wildness, and the endo-wildness coincide.

(b) Assume that $A = KQ$ is the path algebra of a Euclidean quiver Q. We know from (3.15) that A is representation-tame and one-parametric.

For the convenience of the reader, we give an outline of the proof involving only reflection functors and we show that the proof is a constructive one.

First we prove the result in case $Q = \Delta$ is any of the canonically oriented Euclidean quivers $\Delta(\widetilde{\mathbb{A}}_{p,q}), 1 \leq p \leq q$, $\Delta(\widetilde{\mathbb{D}}_m), m \geq 4$, $\Delta(\widetilde{\mathbb{E}}_6)$, $\Delta(\widetilde{\mathbb{E}}_7)$, and $\Delta(\widetilde{\mathbb{E}}_8)$ by constructing a bimodule $_{K[t]}M_{K\Delta}$ such that the image of the functor

$$\widehat{M} = (-) \otimes {}_{K[t]}M_{K\Delta} : \operatorname{ind} K[t] \longrightarrow \operatorname{mod} K\Delta,$$

is contained in the category $\mathcal{R}(K\Delta)$ of regular A-modules and, for each dimension $d \geq 1$, the category $\operatorname{ind}_d K\Delta$ of indecomposable $K\Delta$-modules of the dimension d is almost parametrised by the functor $\widehat{M}$. Here we apply the simple construction given in the proof of (3.13).

Now, assume that $A = KQ$ and Q is an arbitrary acyclic Euclidean quiver. It was shown in (VII.5.2) and (VIII.1.8) that Q can be obtained from the canonically oriented quiver Δ, with $\overline{\Delta} = \overline{Q}$, by a finite number of reflections and the categories of modules are connected by a sequence of reflection functors

$$\operatorname{mod} A \underset{S_{a_1}^-}{\overset{S_{a_1}^+}{\rightleftarrows}} \operatorname{mod} A_1 \underset{S_{a_2}^-}{\overset{S_{a_2}^+}{\rightleftarrows}} \cdots\cdots \underset{S_{a_s}^-}{\overset{S_{a_s}^+}{\rightleftarrows}} \operatorname{mod} K\Delta.$$

By applying (VII.5.3) and (VII.5.6), one shows that the functors restrict to equivalence of categories

$$\mathcal{R}(A) \xrightarrow[\simeq]{\widetilde{S}_{a_1}^+} \mathcal{R}(A_1) \xrightarrow[\simeq]{\widetilde{S}_{a_2}^+} \cdots\cdots \xrightarrow[\simeq]{\widetilde{S}_{a_s}^+} \mathcal{R}(K\Delta)$$

between the categories of regular modules such that the composite equivalence $S : \mathcal{R}(A) \xrightarrow{\simeq} \mathcal{R}(K\Delta)$ is an exact functor. It follows that the composite functor

$$\operatorname{ind} K[t] \xrightarrow{\widehat{M}} \mathcal{R}(K\Delta) \xrightarrow{S^{-1}} \mathcal{R}(A) \hookrightarrow \operatorname{mod} A$$

is of the form $\widehat{N} \cong (-) \otimes {}_{K[t]}N_A$, where $_{K[t]}N_A$ is a $K[t]$-A-bimodule such that N viewed as a left $K[t]$-module is finitely generated free. It follows that, for each dimension $d \geq 1$, the category $\operatorname{ind}_d A$ of indecomposable A-modules of the dimension d is almost parametrised by the functor $\widehat{N}$, see the proof of (3.13). In fact, the bimodule N is obtained from M by applying reflection functors.

For the convenience of the reader we state without proof the following general result on the representation type of tilted algebras, proved by Kerner in [343].

3.17. Theorem. *Let Q be an acyclic, finite, and connected quiver, and B a tilted algebra of the type Q. The following conditions are equivalent.*

    (a) *The tilted algebra B is representation-tame.*

    (b) *The tilted algebra B is representation-domestic.*

(c) *The Euler quadratic form* $q_B : K_0(B) \longrightarrow \mathbb{Z}$ *of* B *is weakly non-negative.*

(d) *The category* $\operatorname{mod} B$ *is controlled by the Euler quadratic form* $q_B : K_0(B) \longrightarrow \mathbb{Z}.$

(e) *The tilted algebra* B *is not representation-wild.*

(f) *The tilted algebra* B *is not strictly representation-wild.*

The following example shows that, for any $r \geq 2$, there exists an r-parametric representation-domestic tilted algebra B.

3.18. Example. Let $r \geq 2$ be an integer and let $B^{(r)}$ be the path algebra of the quiver

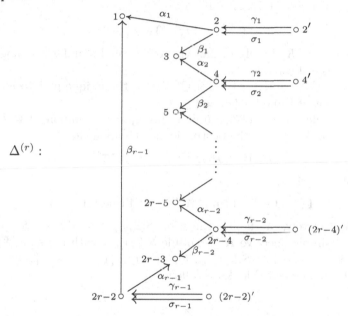

bound by the zero relations $\gamma_j \alpha_j = 0$, $\sigma_j \alpha_j = 0$, $\gamma_j \beta_j = 0$, and $\sigma_j \beta_j = 0$, for $j \in \{1, \ldots, r-1\}$. We note that for $r = 2$, the algebra $B^{(2)}$ is that one considered in (XVIII.5.8).

Denote by $H^{(r)}$ the path algebra of the full subquiver $Q^{(r)}$ of $\Delta^{(r)}$ given by the vertices $1, 2, 3, 4, \ldots, 2r-3, 2r-2$.

It is clear that $H^{(r)}$ is a concealed algebra of the Euclidean type $\Delta(\widetilde{\mathbb{A}}_{r-1,r-1})$, because $r-1$ is the number of counterclockwise oriented arrows, as well as the number of clockwise oriented arrows of $Q^{(r)}$. Further, for each $j \in \{1, \ldots, r-1\}$, denote by $A^{(j)}$ the Kronecker algebra given by the arrows γ_j and σ_j.

A simple checking shows that every indecomposable $B^{(r)}$-module is either an indecomposable $H^{(r)}$-module or is an indecomposable $A^{(j)}$-module, for some $j \in \{1, \ldots, r-1\}$. The Auslander–Reiten quiver $\Gamma(\mathrm{mod}\, B^{(r)})$ is of the form

$$\Gamma(\mathrm{mod}\, B^{(r)}) = \mathcal{P}(H^{(r)}) \cup \boldsymbol{T}^{H^{(r)}} \cup \mathcal{C} \cup \left(\bigcup_{j=1}^{r-1} \boldsymbol{T}^{A^{(j)}}\right) \cup \left(\bigcup_{j=1}^{r-1} \mathcal{Q}(A^{(j)})\right)$$

where

- $\mathcal{P}(H^{(r)})$ is the unique postprojective component of $\Gamma(\mathrm{mod}\, H^{(r)})$,
- $\boldsymbol{T}^{H^{(r)}} = \{\mathcal{T}_\lambda^{H^{(r)}}\}_{\lambda \in \mathbb{P}_1(K)}$ is the $\mathbb{P}_1(K)$-family of pairwise orthogonal standard stable tubes of $\Gamma(\mathrm{mod}\, H^{(r)})$, with the tubes $\mathcal{T}_\infty^{H^{(r)}}$ and $\mathcal{T}_0^{H^{(r)}}$ of rank r, and the homogeneous tubes $\mathcal{T}_\lambda^{H^{(r)}}$, for $\lambda \in K \setminus \{0\}$,
- for each $j \in \{1, \ldots, r-1\}$,

$$\boldsymbol{T}^{A^{(j)}} = \{\mathcal{T}_\lambda^{A^{(j)}}\}_{\lambda \in \mathbb{P}_1(K)}$$

is the $\mathbb{P}_1(K)$-family of pairwise orthogonal standard homogeneous tubes of $\Gamma(\mathrm{mod}\, H^{(j)})$,

- for each $j \in \{1, \ldots, r-1\}$, $\mathcal{Q}(A^{(j)})$ is the unique preinjective component of $\Gamma(\mathrm{mod}\, A^{(j)})$, and
- $\mathcal{C}$ is the glueing of the unique preinjective component $\mathcal{Q}(H^{(r)})$ of $\Gamma(\mathrm{mod}\, H^{(r)})$ with the postprojective components

$$\mathcal{P}(A^{(1)}), \mathcal{P}(A^{(2)}), \ldots, \mathcal{P}(A^{(r-1)})$$

of the quivers

$$\Gamma(\mathrm{mod}\, A^{(1)}), \Gamma(\mathrm{mod}\, A^{(2)}), \ldots, \Gamma(\mathrm{mod}\, A^{(r-1)}),$$

respectively, by the identification $S(2j)_{H^{(r)}} = S(2j) = S(2j)_{A^{(j)}}$ of the simple injective $H^{(r)}$-module $S(2j)_{H^{(r)}}$ with the simple projective $A^{(j)}$-module $S(2j)_{A^{(j)}}$, for any $j \in \{1, \ldots, r-1\}$. For $r = 4$, the component $\mathcal{C}$ looks as follows

Observe that the modules

$$I(1), S(2), P(2'), I(3), S(4), P(4'), \ldots\ldots, I(2r{-}3), S(2r{-}2), P((2r{-}2)')$$

form a faithful section Σ of $\mathcal{C}$ isomorphic to $(\Delta^{(r-1)})^{\mathrm{op}}$. It is easy to check that, for any pair of modules U and V lying on the section Σ, we have $\mathrm{Hom}_{B^{(r)}}(U, \tau_{B^{(r)}} V) = 0$. Hence, by the criterion (VIII.5.6), $B^{(r)} \cong \mathrm{End}\, T^{(r)}$ is a tilted algebra of the wild type $\Delta^{(r)}$ and $\mathcal{C}$ is the connecting component $\mathcal{C}_{T^{(r)}}$, for a tilting module $T^{(r)}$ over the path algebra $K\Delta^{(r)}$. On the other hand, the algebra $B^{(r)}$ is r-parametric, because the hereditary algebras $H^{(r)}, A^{(1)}, \ldots, A^{(r-1)}$ are one-parametric, by (3.12).

In the representation theory of tame algebras an important rôle is played by the class of tubular algebras, introduced by Ringel in [525].

3.19. Definition. Let C be a concealed algebra of Euclidean type and let

$$B = C[E_1, \mathcal{L}^{(1)}, E_2, \mathcal{L}^{(2)}, \ldots, E_s, \mathcal{L}^{(s)}]$$

be a tubular (branch) extension of C in the sense of (XV.4.1). The algebra B is defined to be a **tubular algebra** if the tubular type $r^B = \widehat{r}^B$ is one of the following four types $(3,3,3)$, $(2,4,4)$, $(2,3,6)$, and $(2,2,2,2)$.

Examples of tubular algebras are
- the **canonical tubular algebras** $C(3,3,3)$, $C(2,4,4)$, and $C(2,3,6)$ of tubular types $(3,3,3)$, $(2,4,4)$, and $(2,3,6)$, as defined in (XII.1.2), and
- for each $\lambda \in K \setminus \{0\}$, the canonical tubular algebra $C(2,2,2,2,\lambda)$ of the tubular type $(2,2,2,2)$ given by the quiver

$$\Delta(2,2,2,2):$$
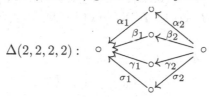

bound by the relations $\alpha_2\alpha_1 + \beta_2\beta_1 + \gamma_2\gamma_1 = 0$ and $\alpha_2\alpha_1 + \beta_2\beta_1 + \lambda\sigma_2\sigma_1 = 0$.

It is easy to check that (p, q, r) is a triple of integers such that $r \geq q \geq p \geq 2$ and $\frac{1}{p} + \frac{1}{q} + \frac{1}{r} = 1$ if and only if it is one of the following triples $(3,3,3)$, $(2,4,4)$, and $(2,3,6)$.

For the convenience of the reader, we state without proof the following theorem of Ringel [525] that collects the basic properties of the tubular algebras.

3.20. Theorem. *Let B be a tubular algebra and let $r^B = \hat{r}^B$ be its tubular type.*

(a) *The algebra B^{op} is a tubular algebra and its tubular type coincides with the tubular type $r^B = \hat{r}^B$ of B.*

(b) *The algebra B is representation-tame, non-domestic, and of linear growth.*

(c) *The Euler quadratic form $q_B : K_0(B) \longrightarrow \mathbb{Z}$ of the algebra B is positive semidefinite of corank 2.*

(d) *The category $\operatorname{mod} B$ is controlled by the Euler quadratic form $q_B : K_0(B) \longrightarrow \mathbb{Z}$.*

(e) *The Auslander–Reiten quiver $\Gamma(\operatorname{mod} B^{(r)})$ is of the form*

$$\Gamma(\operatorname{mod} B) = \mathcal{P}(B) \cup \boldsymbol{T}_0^B \cup \left(\bigcup_{q \in \mathbb{Q}^+} \boldsymbol{T}_q^B \right) \cup \boldsymbol{T}_\infty^B \cup \mathcal{Q}(B),$$

where $\mathbb{Q}^+$ is the set of all positive rational numbers and

- $\mathcal{P}(B)$ *is the unique postprojective component of $\Gamma(\operatorname{mod} B)$,*
- $\mathcal{Q}(B)$ *is the unique preinjective component of $\Gamma(\operatorname{mod} B)$,*
- $\boldsymbol{T}_0^B = \{\mathcal{T}_{0,\lambda}^B\}_{\lambda \in \mathbb{P}_1(K)}$ *is a $\mathbb{P}_1(K)$-family of pairwise orthogonal standard ray tubes of tubular type r^B, containing at least one indecomposable projective B-module,*
- $\boldsymbol{T}_\infty^B = \{\mathcal{T}_{\infty,\lambda}^B\}_{\lambda \in \mathbb{P}_1(K)}$ *is a $\mathbb{P}_1(K)$-family of pairwise orthogonal standard coray tubes of tubular type r^B, containing at least one indecomposable injective B-module,*
- *for each $q \in \mathbb{Q}^+$, $\boldsymbol{T}_q^B = \{\mathcal{T}_{q,\lambda}^B\}_{\lambda \in \mathbb{P}_1(K)}$ is a $\mathbb{P}_1(K)$-family of pairwise orthogonal standard stable tubes of tubular type r^B,*
- *for each $q \in \mathbb{Q}^+ \cup \{0, \infty\}$, the $\mathbb{P}_1(K)$-family $\boldsymbol{T}_q^B = \{\mathcal{T}_{q,\lambda}^B\}_{\lambda \in \mathbb{P}_1(K)}$ separates $\mathcal{P}(B) \cup \left(\bigcup_{p < q} \boldsymbol{T}_p^B \right)$ from $\left(\bigcup_{p > q} \boldsymbol{T}_p^B \cup \mathcal{Q}(B) \right)$,*
- $\operatorname{gl.dim} B = 2$,
- $\operatorname{pd} X \leq 1$, *for any indecomposable module X in*

$$\mathcal{P}(B) \cup \boldsymbol{T}_0^B \cup \left(\bigcup_{q \in \mathbb{Q}^+} \boldsymbol{T}_q^B \right),$$

- $\operatorname{id} Y \leq 1$, *for any indecomposable module Y in*

$$\left(\bigcup_{q \in \mathbb{Q}^+} \boldsymbol{T}_q^B \right) \cup \boldsymbol{T}_\infty^B \cup \mathcal{Q}(B).$$

We note that the class of algebras B^{op} opposite to the tubular algebras B coincides with the class of tubular (branch) coextensions

$$[E_1, \mathcal{L}^{(1)}, E_2, \mathcal{L}^{(2)}, \dots, E_s, \mathcal{L}^{(s)}]C$$

of concealed algebras C of Euclidean type. The tubular type of any such an algebra is one of the types $(3,3,3)$, $(2,4,4)$, $(2,3,6)$, and $(2,2,2,2)$.

3.21. Example. Let $\Lambda_n = K[t_1, t_2]/(t_1t_2, t_1^n, t_2^n)$, with $n \geq 3$, be the algebra studied by Gelfand and Ponomariev in [259]. For each $n \geq 3$, the algebra Λ_n is representation-tame, and is not of polynomial growth, see [187], [603], and [606].

We recall that, by Higman's theorem (V.5.6), if G is a finite group and K is an algebraically closed field of characteristic $p > 0$ dividing the order of the group G then the group algebra KG is representation-finite if and only if the Sylow p-subgroups of G are cyclic.

The following theorem, proved by Bondarenko and Drozd in [86], classifies the representation-infinite group algebras KG of tame representation type.

3.22. Theorem. *Let G be a finite group and let K be an algebraically closed field of characteristic $p > 0$ dividing the order of G. The group K-algebra KG of G is representation-infinite of tame representation type if and only if $p = 2$ and any Sylow 2-subgroup of G is of one of the groups*
  (i) *The dihedral group*

$$\mathbf{D}_m = \langle g, h \mid g^2 = h^{2^m} = 1, hgh = g \rangle, \quad m \geq 1,$$

  (ii) *The semidihedral group*

$$\mathbf{S}_m = \langle g, h \mid g^2 = h^{2^m} = 1, hg = gh^{2^{m-1}-1} \rangle, \ m \geq 3,$$

  (iii) *The quaternion group*

$$\mathbf{Q}_m = \langle g, h \mid g^2 = h^{2^{m-1}}, g^4 = 1, hgh = g \rangle, \quad m \geq 2.$$

As a consequence of Theorem (3.22) one gets the following characterisation of group algebras of polynomial growth, see Skowroński [606] for a proof.

3.23. Corollary. *Let G be a finite group and let K be an algebraically closed field of characteristic $p > 0$ dividing the order of G. Let KG be the group algebra of G. The following three condition are equivalent.*
  (a) *The algebra KG is representation-infinite tame of polynomial growth.*
  (b) *The algebra KG is representation-infinite and representation-domestic.*
  (c) *$p = 2$ and any Sylow 2-subgroup of G is the Klein group $D_2 \cong \mathbb{Z}_2 \times \mathbb{Z}_2$.*

We end this section with relevant information on the classification of the indecomposable representations of $\mathbf{D}_m, \mathbf{S}_m, \mathbf{Q}_m$, and modules over related group algebras.
  (i) The reader is referred to Ringel [511] for a classification of the indecomposable finite dimensional $K\mathbf{D}_m$-modules, in case (3.22)(i) with $p = 2$, see also Bondarenko [84].

(ii) For a classification of the indecomposable finite dimensional $K\mathbf{S}_m$-modules, in the case (3.22)(ii) with $p = 2$, we refer to Crawley-Boevey [148]–[150], see also Erdmann [211], and Brenner [103].

(iii) The classification of the indecomposable finite dimensional $K\mathbf{Q}_m$-modules in the case (3.22)(iii) remains an open problem.

(iv) For the classification of related representation-tame algebras of the dihedral, semidihedral, and quaternion type, the reader is referred to Erdmann [211], Erdmann–Skowroński [219], [220], and Holm [295].

(v) Representation-finite group algebras AG of finite groups G with coefficients in an arbitrary algebra A are completely described by Meltzer–Skowroński [438], see also Dowbor–Simson [183] for a special case when A is a local algebra. Representation-infinite group algebras AG of tame representation type are described by Skowroński [606] and Leszczyński–Skowroński [410].

(vi) Finite dimensional cocommutative Hopf algebras H of finite representation type are completely described by Farnsteiner [222] and Farnsteiner–Voigt [227], and the representation-infinite cocommutative Hopf algebras of tame representation type in odd characteristic are completely described by Farnsteiner–Skowroński [224], [225], [226], and Farnsteiner–Voigt [228].

(vii) For a complete description of finite dimensional representation-tame Hecke algebras of classical type the reader is referred to Ariki [4] and Erdmann–Nakano [218].

XIX.4. Exercises

1. Assume that Λ and Λ' are arbitrary (not necessarily finite dimensional) K-algebras and let $\mathcal{A}$ be an additive full exact subcategory of mod Λ that is closed under direct summands. Prove that any representation equivalence $T : \mathcal{A} \longrightarrow \text{mod } \Lambda'$ respects the isomorphism classes and carries indecomposable modules to indecomposable ones.

2. Let A be a finite dimensional K-algebra and $e \in A$ an idempotent. Consider the K-algebra $B = eAe \cong \text{End } eA$ and let
$$T_e, L_e : \text{mod } B \longrightarrow \text{mod } A$$
be the K-linear covariant functors defined by the formulae $T_e(-) = -\otimes_B eA$ and $L_e(-) = \text{Hom}_B(Ae, -)$.

(a) Show that the functors T_e and L_e are full, faithful, carry indecomposable modules to indecomposable ones and respect the isomorphism classes.

(b) Give examples of algebras A and idempotents e such that the functor T_e is not a representation embedding (L_e is not a representation embedding).

3. (a) Show that the path K-algebra of the wild quiver

$$\Delta : \quad \begin{array}{c} 2 \\ \downarrow \\ 1 \to 6 \leftarrow 3 \\ \nearrow \quad \nwarrow \\ 5 \qquad 4 \end{array}$$

is isomorphic to the algebra $A = \begin{bmatrix} K & 0 & 0 & 0 & 0 & K \\ & K & 0 & 0 & 0 & K \\ & & K & 0 & 0 & K \\ & \mathbf{O} & & K & 0 & K \\ & & & & K & K \\ & & & & & K \end{bmatrix}$.

(b) Show that there is a functorial isomorphism $\mathrm{mod}\, A \cong \mathrm{rep}_K(\Delta)$.

(c) Let $\mathbb{L}^{(2)}$ be the two loops quiver

$$\mathbb{L}^{(2)} : \qquad \alpha_1 \, \bigcirc\!\bigcirc \, \alpha_2.$$

Consider the functor $G : \mathrm{Rep}_K(\mathbb{L}^{(2)}) \to \mathrm{Rep}_K(\Delta)$ defined by attaching to any representation $\mathbb{X} = (X, \varphi_{\alpha_1}, \varphi_{\alpha_2})$ of $\mathbb{L}^{(2)}$ the representation of Δ given by the diagram

$$G(\mathbb{X}) : \qquad X \xrightarrow{\begin{bmatrix} 1 \\ \varphi_{\alpha_1} \end{bmatrix}} \begin{array}{c} X \\ \downarrow d \\ X \oplus X \end{array} \xleftarrow{\begin{bmatrix} 1 \\ \varphi_{\alpha_2} \end{bmatrix}} X$$
$$\qquad\qquad u_1 \nearrow \qquad \nwarrow u_2 \\ \qquad\qquad X \qquad\qquad X$$

where $d(x) = (x, x)$, $u_1(x) = (x, 0)$, $u_2(x) = (0, x)$, for $x \in X$. Given a morphism $f : \mathbb{X} \to \mathbb{Y}$ in $\mathrm{Rep}_K(\mathbb{L}^{(2)})$, we put $G(f) = (f, f, f, f, f, f \oplus f)$.

   (i) Show that the functor G is exact, full, faithful and commutes with arbitrary direct sums.
   (ii) Let $G' : \mathrm{rep}_K(\mathbb{L}^{(2)}) \longrightarrow \mathrm{rep}_K(\Delta)$ be the restriction of G to $\mathrm{rep}_K(\mathbb{L}^{(2)})$. Prove that the functor $\mathrm{fin}\, K\langle t_1, t_2\rangle \cong \mathrm{rep}_K(\mathbb{L}^{(2)}) \xrightarrow{G'} \mathrm{rep}_K(\Delta) \cong \mathrm{mod}\, A$ is a representation embedding and there is an isomorphism of functors $G' \cong - \otimes_{K\langle t_1, t_2\rangle} M_A$, where $_{K\langle t_1, t_2\rangle} M_A$ is a $K\langle t_1, t_2\rangle$-A-bimodule such that the left $K\langle t_1, t_2\rangle$-module $_{K\langle t_1, t_2\rangle} M$ is finitely generated free of rank 7.
   (iii) Conclude that the algebra A is strictly representation-wild.

4. Let $A = KQ$ be the path algebra of the four subspace quiver

$$Q : \qquad \overset{\circ}{\underset{\alpha}{\searrow}} \ \overset{\circ}{\underset{\beta}{\searrow}} \ \overset{\circ}{\underset{\gamma}{\swarrow}} \ \overset{\circ}{\underset{\delta}{\swarrow}} \\ \qquad\qquad\qquad \circ$$

and let M be a $K[t]$-A-bimodule

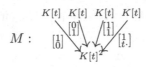

$$M :$$

(i) Prove that the algebra A is one-parametric, by showing that the bi-module M satisfies the conditions (i), (ii), and (iii) listed in Proposition (3.13).

   Hint: Apply (XIII.3) and consult the proof of (3.13).

(ii) By applying the reflection functor technique explained in Remark (3.16)(b), find a $K[t]$-KQ'-bimodule M' satisfying the conditions (i), (ii), and (iii) listed in (3.13), for any quiver Q' obtained from the four subspace quiver Q by at most four element sequence of reflections.

5. Assume that B is a tubular algebra and use the notation of (3.20). By applying (3.20)(e), prove that

(i) gl.dim $B = 2$,

(ii) pd $X \leq 1$, for any indecomposable module X in
$$\mathcal{P}(B) \cup \boldsymbol{T}_0^B \cup \left(\bigcup_{q \in \mathbb{Q}^+} \boldsymbol{T}_q^B \right),$$

(iii) id $Y \leq 1$, for any indecomposable module Y in
$$\left(\bigcup_{q \in \mathbb{Q}^+} \boldsymbol{T}_q^B \right) \cup \boldsymbol{T}_\infty^B \cup \mathcal{Q}(B).$$

6. Let C be any of the canonical tubular algebras $C(3,3,3)$, $C(2,4,4)$, $C(2,3,6)$, or $C(2,2,2,2,\lambda)$, see (3.19) and Section XII.1.

(a) Determine the Euler quadratic form $q_C : K_0(C) \longrightarrow \mathbb{Z}$ of C.

(b) Prove that $q_C : K_0(C) \longrightarrow \mathbb{Z}$ is semidefinite of corank 2.

(c) Prove that the radical rad q_C of q_C is generated by two positive vectors of the Grothendieck group $K_0(C)$ of C.

7. Let B be a tubular algebra and C the canonical tubular algebra with $r^C = r^B$.

(a) Prove that $\Gamma(\operatorname{mod} C)$ admits a $\mathbb{P}_1(K)$-family $\boldsymbol{T}_1^C = \{\mathcal{T}_{1,\lambda}^C\}_{\lambda \in \mathbb{P}_1(K)}$ of pairwise orthogonal standard stable tubes of tubular type r^C such that the mouth modules of each of the tubes $\mathcal{T}_{1,\lambda}^C$ have dimension vectors with coordinates in the set $\{0, 1\}$.

   Hint: Consult Section XII.2.

(b) Prove that there exists a multiplicity-free tilting C-module T_C in the category add $\left(\mathcal{P}(C) \cup \boldsymbol{T}_0^B \cup \bigcup_{q \in \mathbb{Q}^+} \boldsymbol{T}_q^B\right)$ such that $B \cong \operatorname{End} T_C$.

   Hint: Consult the proof of (X.4.2).

Chapter XX

Perspectives

The aim of this chapter is to present (without proofs) some old and new results of the representation theory of finite dimensional algebras, related to the material presented in the book. This, together with the long list of complementary references included in the bibliography, gives good perspectives for further study and interesting research directions.

In Section 1, results on the shape of the connected components of the Auslander–Reiten quiver $\Gamma(\operatorname{mod} A)$ of an algebra A are presented. In particular, the possible shapes and the structure of generalised standard components of $\Gamma(\operatorname{mod} A)$ are discussed.

In Section 2, the Tits quadratic form $\widehat{q}_A : K_0(A) \longrightarrow \mathbb{Z}$ of an algebra A, with the ordinary quiver acyclic, is defined. A geometric nature and an importance of the Tits form $\widehat{q}_A$ for the representation theory of representation-tame simply connected algebras is shown. Several characterisations of representation-tame simply connected algebras of polynomial growth are presented.

In Section 3, we briefly present some results on tilted algebras of wild type, that are complementary to the results of Chapter XVIII. We also outline the representation theory of quasitilted algebras, that is, the algebras of the form $\operatorname{End}_{\mathcal{H}}(T)$, where T is a tilting object of a hereditary abelian K-category $\mathcal{H}$.

Section 4 is devoted to the representation theory of algebras of small homological dimensions and their characterisations. In particular, we investigate the classes of double tilted algebras and generalised double tilted algebras.

In Section 5, we discuss the importance of the tilted algebras and quasitilted algebras for the representation theory of self-injective (Frobenius) algebras. We show that the class of basic, connected, self-injective algebras Λ of polynomial growth, with $\dim_K \Lambda \geq 2$, coincides with the class of the socle deformations of the orbit algebras of the repetitive categories of the tilted algebras of Dynkin or Euclidean type, and the tubular algebras. The structure and the main properties of self-injective algebras Λ, having a generalised standard component in $\Gamma(\operatorname{mod} \Lambda)$, are also discussed.

In the final Section 6, we indicate some of the important research directions of the modern representation theory of finite dimensional algebras.

XX.1. Components of the Auslander–Reiten quiver of an algebra

The aim of this section is to present some results on the shape and the structure of components of the Auslander–Reiten quiver $\Gamma(\operatorname{mod} A)$ of an algebra A.

1.1. Definition. Let A be an algebra and $\mathcal{C}$ a component of the quiver $\Gamma(\operatorname{mod} A)$.

    (a) The component $\mathcal{C}$ is defined to be **regular** if $\mathcal{C}$ contains neither a projective module nor an injective module.

    (b) The component $\mathcal{C}$ is defined to be **semiregular** if $\mathcal{C}$ does not contain both a projective and an injective module.

The possible shapes of regular components of an Auslander–Reiten quiver describe the following theorem, proved independently by Liu [414] and Zhang [689].

1.2. Theorem. *Let A be an algebra and $\mathcal{C}$ be a regular component of $\Gamma(\operatorname{mod} A)$.*

    (a) *$\mathcal{C}$ contains an oriented cycle if and only if $\mathcal{C}$ is a stable tube.*

    (b) *$\mathcal{C}$ is acyclic if and only if $\mathcal{C}$ is of the form $\mathbb{Z}\Delta$, for some locally finite acyclic quiver Δ.* $\square$

One should remark that the part (a) of (1.2) implies that a regular component $\mathcal{C}$ of the Auslander–Reiten quiver $\Gamma(\operatorname{mod} A)$ of A is a stable tube if and only if $\mathcal{C}$ contains a τ_A-periodic module (a theorem by Happel–Preiser–Ringel in [281]).

The possible shapes of the semiregular components of an Auslander–Reiten quiver describe the following theorems, proved by Liu in [414] and [415].

1.3. Theorem. *Let A be an algebra and $\mathcal{C}$ be a component of $\Gamma(\operatorname{mod} A)$ without injective modules.*

    (a) *$\mathcal{C}$ contains an oriented cycle if and only if $\mathcal{C}$ is a ray tube.*

    (b) *$\mathcal{C}$ is acyclic if and only if there exists a locally finite acyclic quiver Δ such that $\mathcal{C}$ is isomorphic to a full translation subquiver of $\mathbb{Z}\Delta$ that is closed under τ_A^{-1}-shifts.* $\square$

1.4. Theorem. *Let A be an algebra and $\mathcal{C}$ be a component of $\Gamma(\operatorname{mod} A)$ without projective modules.*

    (a) *$\mathcal{C}$ contains an oriented cycle if and only if $\mathcal{C}$ is a coray tube.*

(b) C *is acyclic if and only if there exists a locally finite acyclic quiver* Δ *such that* C *is isomorphic to a full translation subquiver of* $\mathbb{Z}\Delta$ *which is closed under* τ_A-*shifts.*   □

It follows from Chapters XII and XVII that an arbitrary stable tube, ray tube and coray tube occurs as a component of the Auslander–Reiten quiver $\Gamma(\mathrm{mod}\, B)$ of a representation-infinite tilted algebra B of a Euclidean type $\widetilde{\mathbb{A}}_m$ or $\widetilde{\mathbb{D}}_m$. On the other hand, it is not clear which acyclic quivers $\mathbb{Z}\Delta$ occur as regular components of an Auslander–Reiten quiver $\Gamma(\mathrm{mod}\, A)$.

It is expected that, for any representation-tame algebra A, each acyclic regular component of $\Gamma(\mathrm{mod}\, A)$ is of one of the forms $\mathbb{A}_\infty^\infty$ and $\mathbb{Z}\mathbb{D}_\infty$, where

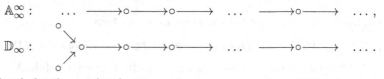

By [211], [219], and [673], this is the case, when $A = KG$ is a representation-tame group algebra of a finite group G. We should also mention the following two facts

- a theorem of Webb [673] asserting that, given an arbitrary finite group G, any acyclic regular component of $\Gamma(\mathrm{mod}\, A)$ is of one of the forms $\mathbb{Z}\mathbb{A}_\infty$, $\mathbb{Z}\mathbb{A}_\infty^\infty$, or $\mathbb{Z}\mathbb{D}_\infty$, and
- a theorem of Erdmann [212] asserting that, if G is a finite group such that the group algebra KG is representation-wild then $\Gamma(\mathrm{mod}\, A)$ contains components of type $\mathbb{Z}\mathbb{A}_\infty$.

The following theorem is proved by Ringel [527], see also (XVIII.5.17).

1.5. Theorem. *Let* Δ *be a connected, finite, acyclic quiver with at least three vertices. The translation quiver* $\mathbb{Z}\Delta$ *occurs as a regular component of an Auslander–Reiten quiver* $\Gamma(\mathrm{mod}\, A)$ *if and only if* Δ *is neither a Dynkin nor a Euclidean quiver.*   □

The following theorem, proved by Crawley–Boevey and Ringel [162], shows that there exist large acyclic regular components of the Auslander–Reiten quivers.

1.6. Theorem. *Let* Δ *be a locally finite, connected, acyclic quiver such that after deleting finitely many vertices and arrows* Δ *becomes a disjoint union of quivers of type* $\mathbb{A}_\infty$. *Then there exists an algebra* A *such that* $\mathbb{Z}\Delta$ *occurs as a regular component of* $\Gamma(\mathrm{mod}\, A)$.   □

In general, not much is known on the structure of components of an Auslander–Reiten quiver containing both a projective module and an injective module. To present a few general results in this direction we need some concepts.

1.7. Definition [170]. Let A be an algebra. A component $\mathcal{C}$ of $\Gamma(\mathrm{mod}\,A)$ is defined to be **coherent** if the following two conditions are satisfied:

(c1) for each projective module P in $\mathcal{C}$, there is an infinite sectional path

$$P = X_1 \longrightarrow X_2 \longrightarrow \cdots \longrightarrow X_m \longrightarrow X_{m+1} \longrightarrow \cdots .$$

(c2) for each injective module I in $\mathcal{C}$, there is an infinite sectional path

$$\cdots \longrightarrow Y_{m+1} \longrightarrow Y_m \longrightarrow \cdots \longrightarrow Y_2 \longrightarrow Y_1 = I.$$

We note that every ray tube and every coray tube is coherent.

1.8. Definition. Let A be an algebra and $\mathcal{C}$ be a component of $\Gamma(\mathrm{mod}\,A)$.

(a) The component $\mathcal{C}$ is defined to be **cyclic** if every module X of $\mathcal{C}$ lies on an oriented cycle of $\mathcal{C}$.
(b) The component $\mathcal{C}$ is defined to be **almost cyclic** if all but finitely many modules of $\mathcal{C}$ lie on oriented cycles of $\mathcal{C}$.
(c) The component $\mathcal{C}$ is said to be **almost periodic** if all but finitely many τ_A-orbits in $\mathcal{C}$ are periodic.

In [425], Malicki and Skowroński introduce the concept of a **generalised multicoil**, extending the concept of a **coil** from [17] and [20]. Roughly speaking a generalised multicoil is a connected translation quiver obtained from a finite family of stable tubes by a sequence of admissible operations. A generalised multicoil is coherent, almost cyclic and contains usually many projective vertices and many injective vertices. Moreover, a generalised multicoil $\mathcal{C}$ is semiregular if and only if $\mathcal{C}$ is a ray tube or a coray tube.

The following theorem, proved in [425], may be viewed as an extension of (1.3)(a) and (1.4)(a).

1.9. Theorem. *Let A be an algebra and $\mathcal{C}$ be a component of $\Gamma(\mathrm{mod}\,A)$.*

(a) *$\mathcal{C}$ is almost cyclic and coherent if and only if $\mathcal{C}$ is a generalised multicoil.*
(b) *$\mathcal{C}$ is cyclic and coherent if and only if $\mathcal{C}$ is a cyclic generalised multicoil.* $\qquad\square$

One can prove that, given an algebra, every almost cyclic and coherent component of an Auslander–Reiten quiver $\Gamma(\mathrm{mod}\,A)$ is almost periodic. The following theorem of Skowroński [615] contains important information on the dimensions of the indecomposable modules lying in almost periodic components.

1.10. Theorem. *Let A be an algebra and C be an almost periodic component of $\Gamma(\operatorname{mod} A)$. For each integer $d \geq 1$, the number of indecomposable modules X in C with $\dim_K X = d$ is finite (or zero).* $\square$

It follows from (XVIII.1.6) and (XVIII.1.7) that the regular components of the Auslander–Reiten quiver $\Gamma(\operatorname{mod} A)$ of any wild hereditary algebra A are of type $\mathbb{Z}\mathbb{A}_\infty$ and, hence, they are not almost periodic, but admit at most a finite number of indecomposable modules of a fixed dimension $d \geq 1$. Surprisingly, Liu and Schultz [421] construct examples of algebras Λ such that $\Gamma(\operatorname{mod} \Lambda)$ contains a component C of type $\mathbb{Z}\mathbb{A}_\infty$ and every τ_A-orbit of C consists of modules of the same dimension, see also [535].

An interesting open problem concerns the structure of generalised standard components of an Auslander–Reiten quiver.

The following theorem, proved by Skowroński in [615], describes the shape of arbitrary generalised standard components.

1.11. Theorem. *Let A be an algebra and C be a generalised standard component of $\Gamma(\operatorname{mod} A)$. Then C is almost periodic.* $\square$

As a consequence, we obtain the following description of the shapes of the regular generalised standard components.

1.12. Corollary. *Let A be an algebra and C be a regular generalised standard component of $\Gamma(\operatorname{mod} A)$. Then C is either a stable tube or is of the form $\mathbb{Z}\Delta$, for a connected, finite, acyclic quiver Δ.* $\square$

The following results, proved in [615], describe the structure of semiregular acyclic generalised standard components.

1.13. Theorem. *Let A be an algebra, C a component of $\Gamma(\operatorname{mod} A)$ and $B = A/\operatorname{Ann}_A C$. The following conditions are equivalent.*

  (a) *C is generalised standard, acyclic, and has no injective modules.*

  (b) *B is isomorphic to a tilted algebra $\operatorname{End} T_H$, where H is a hereditary algebra, T is a tilting H-module without non-zero postprojective direct summands, and C is the connecting component C_T of $\Gamma(\operatorname{mod} B)$ determined by T.* $\square$

1.14. Theorem. *Let A be an algebra, C a component of $\Gamma(\operatorname{mod} A)$, and $B = A/\operatorname{Ann}_A C$. The following conditions are equivalent.*

  (a) *C is generalised standard, acyclic, and has no projective modules.*

  (b) *B is isomorphic to a tilted algebra $\operatorname{End} T_H$, where H is a hereditary algebra, T is a tilting H-module without non-zero preinjective direct summands, and C is the connecting component C_T of $\Gamma(\operatorname{mod} B)$ determined by T.* $\square$

As a consequence, we get the following useful fact.

1.15. Corollary. *Let A be an algebra, C a component of $\Gamma(\operatorname{mod} A)$, and $B = A/\operatorname{Ann}_A C$. Then C is generalised standard regular and acyclic if and only if B is isomorphic to a tilted algebra $\operatorname{End} T_H$, where H is a wild hereditary algebra, T is a regular tilting H-module, and C is the connecting component C_T of $\Gamma(\operatorname{mod} B)$ determined by T.* $\qquad\square$

The following characterisation of acyclic regular generalised standard components is established by Skowroński in [613].

1.16. Theorem. *Let A be an algebra and C a regular component of $\Gamma(\operatorname{mod} A)$. Then C contains a directing module if and only if C is an acyclic generalised standard component.* $\qquad\square$

In this context, we should also mention the following result proved indepedently by Peng and Xiao in [472] and Skowroński [613].

1.17. Theorem. *Given an arbitrary algebra A, the quiver $\Gamma(\operatorname{mod} A)$ contains at most a finite number of τ_A-orbits containing directing A-modules.* $\qquad\square$

A structure of algebras such that all indecomposable projective modules are directing is described by Skowroński and Wenderlich [647].

It follows from (XVIII.5.16) and (XVIII.5.17) that there exist many regular tilting modules over any wild hereditary algebra H, with $K_0(H)$ of rank at least 3, and hence many generalised standard, regular acyclic components of Auslander–Reiten quivers of algebras. However, the following result, proved in [615], shows that we may have only finitely many such components, for a given algebra A.

1.18. Theorem. *For any algebra A, all but finitely many generalised standard components in the Auslander–Reiten quiver $\Gamma(\operatorname{mod} A)$ of A are stable tubes.* $\qquad\square$

The structure of arbitrary generalised standard acyclic components is described completely in [612]. In particular, the following result is proved in [612].

1.19. Theorem. *Let A be an algebra.*

(a) *The Auslander–Reiten quiver $\Gamma(\operatorname{mod} A)$ of A contains at most two faithful generalised standard acyclic components.*

(b) *A is a concealed algebra if and only if the Auslander–Reiten quiver $\Gamma(\operatorname{mod} A)$ of A contains precisely two faithful generalised standard acyclic components.*
$\qquad\square$

It follows from (XII.3.4) that every component of the Auslander–Reiten quiver of a concealed algebra of Euclidean type is generalised standard. Conversely, we have the following result proved in [617].

1.20. Theorem. *Let A be an algebra. The following conditions are equivalent.*

(a) *A is a concealed algebra of Euclidean type.*
(b) *A satisfies the following two conditions*
  (i) *A is representation-infinite and A/I is representation-finite, for every non-zero two-sided ideal I of A.*
  (ii) *Every component of $\Gamma(\operatorname{mod} A)$ is generalised standard.* $\square$

The problem of describing the algebras A for which the Auslander–Reiten quiver $\Gamma(\operatorname{mod} A)$ admits a component $\mathcal{C}$ with a faithful (generalised) standard stable tube seems to be difficult. The class of these algebras includes:

- the canonical algebras [525], [528],
- concealed canonical algebras [394],
- supercanonical algebras [397],
- generalised canonical algebras [625], and
- concealed generalised canonical algebras [427],
  see Section 3, for some results in this direction.

Here, we only note the following consequences of the main result of [625].

1.21. Theorem. *Given $0 \neq g \in \mathbb{N} \cup \{\infty\}$ and a sequence $r_1,\ldots,r_m \geq 2$ of integers, there exists a generalised canonical algebra C such that*

(a) *$\operatorname{gl.dim} C = g$, and*
(b) *the quiver $\Gamma(\operatorname{mod} C)$ admits a K-family*

$$\boldsymbol{\mathcal{T}}^C = \{\mathcal{T}_\lambda^C\}_{\lambda \in K}$$

of pairwise orthogonal standard faithful stable tubes, with m tubes $\mathcal{T}_{\lambda_1}^C,\ldots,\mathcal{T}_{\lambda_m}^C$ of ranks $r_1,\ldots,r_m$, respectively, and the remaining tubes homogeneous. $\square$

1.22. Theorem. *Given an arbitrary basic algebra B and a sequence of integers $r_1,\ldots,r_m \geq 2$, there exists a generalised canonical algebra C such that*

(a) *B is a quotient algebra of C, and*
(b) *the quiver $\Gamma(\operatorname{mod} C)$ admits a K-family*

$$\boldsymbol{\mathcal{T}}^C = \{\mathcal{T}_\lambda^C\}_{\lambda \in K}$$

of pairwise orthogonal standard faithful stable tubes, with m tubes $\mathcal{T}_{\lambda_1}^C,\ldots,\mathcal{T}_{\lambda_m}^C$ of ranks $r_1,\ldots,r_m$, respectively, and the remaining tubes homogeneous. $\square$

XX.2. The Tits quadratic form of an algebra

Throughout this book, a fundamental rôle is played by the Euler quadratic form $q_A : K_0(A) \longrightarrow \mathbb{Z}$, associated to an algebra A of finite global dimension (III.3.11), which is a homological quadratic form. In the representation theory of algebras an important rôle is also played by the Tits quadratic form $\widehat{q}_A : K_0(A) \longrightarrow \mathbb{Z}$, associated to any algebra A with the acyclic quiver Q_A, which is a geometric quadratic form. To introduce it, we briefly explain the related geometric context.

2.1. A geometric context. Let $A = KQ/I$, where $Q = (Q_0, Q_1, s, t)$ is a finite connected quiver and I is an admissible ideal of the path algebra KQ of Q. Let $n = |Q_0|$ and $Q_0 = \{1, \dots, n\}$. Fix a positive vector $\mathbf{d} = (d_1, \dots, d_n) \in K_0(A) = \mathbb{Z}^n$.

(i) Denote by $\mathrm{mod}_A(\mathbf{d})$ the set of all representations

$$V = (V_i, \varphi_\alpha)_{i \in Q_0, \alpha \in Q_1}$$

in the category $\mathrm{rep}_K(Q, I)$ of finite dimensional K-linear representations of the bound quiver (Q, I) with $V_i = K^{d_i}$, for all $i \in Q_0$.

(ii) A representation V in $\mathrm{mod}_A(\mathbf{d})$ is given by $d_{t(\alpha)} \times d_{s(\alpha)}$-matrices $V(\alpha)$ determining the K-linear maps $\varphi_\alpha : K^{d_{s(\alpha)}} \longrightarrow K^{d_{t(\alpha)}}$, in the canonical bases of $K^{d_i}, i \in Q_0$. Moreover, the matrices $V(\alpha), \alpha \in Q_1$, satisfy the relations

$$\sum_{i=1}^{m} \lambda_i V(\alpha_1^{(i)}) \cdot \dots \cdot V(\alpha_{n_i}^{(i)}) = 0,$$

for all relations $\sum_{i=1}^{m} \lambda_i \alpha_1^{(i)} \dots \alpha_{n_i}^{(i)} \in I$, that are, by (II.2.3), K-linear combinations of paths in Q with a common source and a common target.

(iii) View $\mathrm{mod}_A(\mathbf{d})$ as a subset of the affine space

$$\mathbb{A}(\mathbf{d}) = \prod_{\alpha \in Q_1} K^{d_{t(\alpha)} \times d_{s(\alpha)}}$$

defined by vanishing a finite number of polynomials, given by the matrix relations in (ii). Hence, $\mathrm{mod}_A(\mathbf{d})$ a closed subset of the affine space $\mathbb{A}(\mathbf{d})$ in the Zariski topology.

We call $\mathrm{mod}_A(\mathbf{d})$ the **affine variety** of A-modules of dimension vector $\mathbf{d}$.

(iv) Define the action $\cdot : G(\mathbf{d}) \times \mathrm{mod}_A(\mathbf{d}) \longrightarrow \mathrm{mod}_A(\mathbf{d})$ of the affine algebraic group $G(\mathbf{d}) = \prod_{i \in Q_0} \mathrm{GL}_{d_i}(K)$ on the variety $\mathrm{mod}_A(\mathbf{d})$ by the conjugation formula

$$(g \cdot V)(\alpha) = g_{t(\alpha)} \cdot V(\alpha) \cdot g_{s(\alpha)}^{-1},$$

for $g = (g_i) \in G(\mathbf{d}), V \in \mathrm{mod}_A(\mathbf{d}), \alpha \in Q_1$. $\square$

It is clear that two representations M and N in $\text{mod}_A(\mathbf{d})$ are isomorphic if and only if M and N belong to the same $G(\mathbf{d})$-orbit.

An algebra $A = KQ/I$, with the quiver $Q = Q_A$ acyclic, is said to be **triangular**. We note that every triangular algebra is of finite global dimension.

2.2. Definition. Let $A = KQ/I$ be a triangular algebra and $n = |Q_0|$. The **Tits quadratic form** $\widehat{q}_A$ of A is the integral quadratic form $\widehat{q}_A : \mathbb{Z}^n \longrightarrow \mathbb{Z}$ defined by

$$\widehat{q}_A(\mathbf{x}) = \sum_{i \in Q_0} x_i^2 - \sum_{\alpha \in Q_1} x_{s(\alpha)} x_{t(\alpha)} + \sum_{i,j \in Q_0} r_{ij} x_i x_j,$$

for $\mathbf{x} \in \mathbb{Z}^n = K_0(A)$, where r_{ij} is the number of K-linear relations with source i and target j, for a minimal (finite) set $\mathcal{R}$ of K-linear relations generating the ideal I.

In [88], Bongartz proves that $r_{ij} = \dim_K \text{Ext}_A^2(S(i), S(j))$, and hence r_{ij} does not depend on the choice of $\mathcal{R}$ (the triangularity of A is essential here). It follows from (III.2.12) that the number of arrows in Q with source i and target j is equal to $\dim_K \text{Ext}_A^1(S(i), S(j))$. Therefore, given $\mathbf{x} \in \mathbb{Z}^n = K_0(A)$, we have

$$\widehat{q}_A(\mathbf{x}) = \sum_{i \in Q_0} x_i^2 - \sum_{i,j \in Q_0} \dim_K \text{Ext}_A^1(S(i), S(j)) x_i x_j$$

$$+ \sum_{i,j \in Q_0} \dim_K \text{Ext}_A^2(S(i), S(j)) x_i x_j.$$

In particular, for a triangular algebra $A = KQ/I$, we have
- $\widehat{q}_A = q_A$, if gl.dim $A \leq 2$, and
- $\widehat{q}_A = q_Q$, if $A = KQ$ (that is, $I = 0$).

Let $A = KQ/I$ be a triangular algebra. Then, the Krull's Principal Ideal Theorem yields

$$\widehat{q}_A(\mathbf{d}) \geq \dim G(\mathbf{d}) - \dim \text{mod}_A(\mathbf{d}),$$

for any positive vector $\mathbf{d} \in K_0(A)$, and hence $\widehat{q}_A$ is a geometric form.

The following geometric results are proved by Bongartz [88] (in the finite type case) and de la Peña [457] (in the tame case).

2.3. Proposition. *Let $A = KQ/I$ be a triangular algebra and $\mathbf{d} \in K_0(A)$ a positive vector.*
   (a) *If $A = KQ/I$ is representation-finite then $\dim G(\mathbf{d}) > \dim \text{mod}_A(\mathbf{d})$.*
   (b) *If A is representation-tame then $\dim G(\mathbf{d}) \geq \dim \text{mod}_A(\mathbf{d})$.* $\square$

As a direct consequence we obtain the following facts.

2.4. Corollary. *Let $A = KQ/I$ be a triangular algebra.*

(a) *If A is representation-finite then the Tits form $\widehat{q}_A$ is weakly positive.*

(b) *If A is representation-tame then the Tits form $\widehat{q}_A$ is weakly non-negative.* $\square$

The reverse implications are proved for the following classes of algebras of small homological dimensions:

- tilted algebras [343],
- double tilted algebras [490],
- quasitilted algebras [622],
- coil enlargements of concealed algebras [21], and
- generalised multicoil algebras [426].

Unfortunately, these implications are not true for arbitrary triangular algebras $A = KQ/I$, because there are wild triangular algebras (even of global dimension 2) with weakly positive Tits form (see [88]). One has to impose some nondegeneracy conditions on a triangular algebra $A = KQ/I$ to recover its representation type from the weak positivity or weak non-negativity of the Tits form $\widehat{q}_A : K_0(A) \longrightarrow \mathbb{Z}$. A natural and important condition is the simple connectedness of the algebra A.

The general definition of a simply connected algebra is due to Assem and Skowroński [13]. To introduce it, we need the concept of the fundamental group of a bound quiver (Q, I), proposed in [264] and [432].

Let (Q, I) be a connected bound quiver. A relation

$$\varrho = \sum_{j=1}^{m} \lambda_j w_j \in I$$

is said to be minimal if $m \geq 2$ and $\sum_{j \in J}^{m} \lambda_j w_j \notin I$, for any proper subset J of $\{1, \ldots, m\}$. Let $m(I)$ be the set of all minimal relations of the ideal I.

Denote by $\Pi_1(Q, x_0)$ the fundamental group of the quiver Q at a fixed vertex $x_0 \in Q_0$, and by $N(Q, m(I), x_0)$ the normal subgroup of $\Pi_1(Q, x_0)$ generated by all homotopy classes of the form $[wvu^{-1}w^{-1}]$ where w is a walk from x_0 to a vertex x and u, v are paths in Q from x to a vertex y such that there exists an element $\varrho = \sum_{j=1}^{m} \lambda_j w_j \in m(I)$ with $v = w_r$ and $u = w_s$, for some $r, s \in \{1, \ldots, m\}$. Then the group

$$\Pi_1(Q, I) = \Pi_1(Q, x_0)/N(Q, m(I), x_0)$$

is said to be the **fundamental group** of the bound quiver (Q, I).

2.5. Definition. A triangular algebra A is said to be **simply connected** if, for any presentation

$$A \cong KQ/I$$

of A as a bound quiver algebra, the fundamental group $\Pi_1(Q, I)$ is trivial.

It is proved in [610] that a triangular algebra $A = KQ/I$ is simply connected if and only if A does not admit a proper Galois covering.

One can show that every algebra satisfying the separation condition or the coseparation condition (IX.4.1) is simply connected (see [611]).

Let $A = KQ/I$ be an algebra. By a **convex subalgebra** of A we mean an algebra of the form

$$C = K\Delta/J,$$

where Δ is a convex subquiver of Q and $J = I \cap K\Delta$. The following concept of a strong simple connectedness introduced in [611] is essential for the representation theory of triangular algebras.

2.6. Definition. A triangular algebra $A = KQ/I$ is said to be **strongly simply connected** if every convex subalgebra C of A is simply connected.

A class of strongly simply connected algebras contains the **tree algebras**, that is, the algebras whose quiver is a tree. In general, we have the following result proved in [611].

2.7. Theorem. *Let $A = KQ/I$ be a triangular algebra. The following conditions are equivalent.*

  (a) *A is strongly simply connected.*
  (b) *Every convex subalgebra of A satisfies the separation condition.*
  (c) *Every convex subalgebra of A satisfies the coseparation condition.*
  (d) *The first Hochschild cohomology space $H^1(C)$ of any convex subalgebra C of A vanishes.*   □

We recall that there is an isomororphism

$$H^1(C) \cong \mathrm{Der}_K(C, C)/\mathrm{Der}_K^0(C, C)$$

of vector spaces, where $\mathrm{Der}_K(C, C)$ is the space of all K-linear derivations $d : C \to C$ and $\mathrm{Der}_K^0(C, C)$ is the subspace of $\mathrm{Der}_K(C, C)$ consisting of the inner derivations of C.

The following classes of strongly simply connected algebras are of importance, see [89], [293], [456], and Chapter XIV.

2.8. Definition. (a) A **critical algebra** is a concealed algebra C of Euclidean type Δ, if the underlying graph $\overline{\Delta}$ is one of the Euclidean trees $\widetilde{\mathbb{D}}_m$, $m \geq 4$, $\widetilde{\mathbb{E}}_6$, $\widetilde{\mathbb{E}}_7$, and $\widetilde{\mathbb{E}}_8$.

(b) A **hypercritical algebra** is defined to be a concealed algebra of wild type Δ, with the underlying graph $\overline{\Delta}$ of any of the following minimal wild tree forms

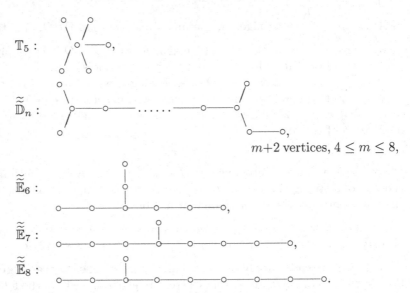

$\mathbb{T}_5$:

$\widetilde{\widetilde{\mathbb{D}}}_n$:

$m+2$ vertices, $4 \leq m \leq 8$,

$\widetilde{\widetilde{\mathbb{E}}}_6$:

$\widetilde{\widetilde{\mathbb{E}}}_7$:

$\widetilde{\widetilde{\mathbb{E}}}_8$:

We recall that a classification of critical algebras is presented in Chapter XIV. A classification of hypercritical algebras $A = KQ/I$ in terms of their bound quivers (Q, I) is given by Unger [665].

Now we present very efficient criteria for the finite representation type of simply connected algebras, proved by Bongartz in [88], [89] and [90].

2.9. Theorem. *Let A be a simply connected algebra. The following statements are equivalent.*

(a) *A is representation-finite.*
(b) *The Tits form $\widehat{q}_A$ of A is weakly positive.*
(c) *A does not admit a convex subalgebra C which is critical.*   □

It is known that every representation-finite simply connected algebra is strongly simply connected. It is shown in [122] that there exist wild simply connected algebras (even of global dimension 2) with weakly non-negative Tits form.

The following criteria for the tame representation type of strongly simply connected algebras are proved recently by Brüstle, de la Peña and Skowroński in [122].

2.10. Theorem. *Let A be a strongly simply connected algebra. The following conditions are equivalent.*

(a) *A is representation-tame.*
(b) *The Tits form $\widehat{q}_A$ of A is weakly non-negative.*
(c) *A does not admit a convex subalgebra C which is hypercritical.*   □

The proof of this theorem relies on the representation theory of strongly simply connected algebras of polynomial growth established by Skowroński in [620], the classification and representation theory of tame minimal non-polynomial growth simply connected algebras (*pg*-**critical algebras**) established by Nörenberg and Skowroński in [451], and the Geiss theorem on degenerations of algebras [248].

In particular, we have the following two theorems characterising the strongly simply connected algebras that are of finite growth or of polynomial growth, established by Skowroński in [620] and completed by Simson-Skowroński in [601], with the parts (d) and (e) of (2.11), and by Wenderlich [674], with the part (f) of (2.11).

2.11. Theorem. *For a strongly simply connected algebra A, the following six statements are equivalent.*

(a) *A is of finite growth (domestic).*

(b) *A does not contain a convex subcategory C which is tubular, pg-critical or hypercritical.*

(c) *The infinite radical $\operatorname{rad}_A^\infty$ of $\operatorname{mod} A$ is nilpotent.*

(d) *The double infinite radical*

$$(\operatorname{rad}_A^\infty)^\infty = \bigcap_{m=1}^\infty (\operatorname{rad}_A^\infty)^m$$

of the category $\operatorname{mod} A$ is zero.

(e) *The square $((\operatorname{rad}_A^\infty)^\infty)^2$ of the double infinite radical $(\operatorname{rad}_A^\infty)^\infty$ of the category $\operatorname{mod} A$ is zero.*

(f) *The Krull-Gabriel dimension $\operatorname{KG}(A)$ of the algebra A is finite.* $\square$

The infinite radical $\operatorname{rad}_A^\infty$ is said to be **nilpotent**, if there is an integer $m \geq 1$ such that $(\operatorname{rad}_A^\infty)^m = 0$.

The reader is referred to Geigle [244] and [245] for basic facts concerning the Krull-Gabriel dimension of an algebra. We refer also to de la Peña [460] for a characterisation of strongly simply connected domestic algebras $A = KQ/I$ by the corank of their Tits forms $\widehat{q}_A : K_0(A) \longrightarrow \mathbb{Z}$.

We also recall from [601] that:

- if A is an arbitrary algebra such that $(\operatorname{rad}_A^\infty)^\infty = 0$ then A is representation-tame, and
- if A is a *pg*-critical algebra then the square of the double infinite radical $(\operatorname{rad}_A^\infty)^\infty$ of $\operatorname{mod} A$ is non-zero.

2.12. Theorem. *For a strongly simply connected algebra A, the following five statements are equivalent.*

(a) *A is of polynomial growth.*

(b) *A is of linear growth.*

(c) *A does not contain a convex subcategory C that is pg-critical or hypercritical.*

(d) *Every component of $\Gamma(\operatorname{mod} A)$ is generalised standard.*

(e) *The infinite radical $\operatorname{rad}_A^\infty$ of $\operatorname{mod} A$ is locally nilpotent.*   □

The infinite radical $\operatorname{rad}_A^\infty$ is called **locally nilpotent** if there exists a positive integer m such that $(\operatorname{rad}_A^\infty(X,X))^m = 0$, for any indecomposable module X in $\operatorname{mod} A$. We note that the statement (d) forces $\operatorname{rad}_A^\infty(X,X) = 0$, for any indecomposable A-module X.

The following theorem proved by de la Peña and Skowroński in [464] provides a geometric and homological characterisation of strongly simply connected algebras of polynomial growth.

2.13. Theorem. *Let A be a strongly simply connected algebra and*

$$q_A, \widehat{q}_A : K_0(A) \longrightarrow \mathbb{Z}$$

be the Euler and the Tits quadratic form of A. The following statements are equivalent.

(a) *A is of polynomial growth.*

(b) *For any indecomposable A-module X and $\mathbf{d} = \dim X$, we have*
$q_A(\mathbf{d}) = \dim G(\mathbf{d}) - \dim_X \operatorname{mod}_A(\mathbf{d}) \geq 0$.

(c) *The Tits quadratic form $\widehat{q}_A : K_0(A) \longrightarrow \mathbb{Z}$ is weakly non-negative and $\operatorname{Ext}_A^2(X,X) = 0$, for any indecomposable A-module X.*

(d) *$\dim_K \operatorname{Ext}_A^1(X,X) \leq \dim_K \operatorname{End}_A(X)$ and $\operatorname{Ext}_A^r(X,X) = 0$, for $r \geq 2$ and any indecomposable A-module X.*   □

In (b), $\dim_X \operatorname{mod}_A(\mathbf{d})$ denotes the local dimension of the module variety $\operatorname{mod}_A(\mathbf{d})$ at the point X, that is the maximum of the dimensions of the irreducible components of $\operatorname{mod}_A(\mathbf{d})$ containing X. Note that the equality $\operatorname{Ext}_A^2(X,X) = 0$ implies that X is a nonsingular point of the variety $\operatorname{mod}_A(\dim X)$.

The following fact on the values of the Euler and Tits forms on the dimension vectors of indecomposable modules is proved by de la Peña and Skowroński in [467].

2.14. Theorem. *Let A be a strongly simply connected algebra of polynomial growth. Then there exists a natural number m such that*

$$0 \leq q_A(\dim X) \leq \widehat{q}_A(\dim X) \leq m,$$

for any indecomposable A-module X.   □

For tame strongly simply connected algebras of non-polynomial growth, the behaviour of the Euler and Tits forms is completely different. Namely, there exists a tame strongly simply connected algebra Λ of global dimension 3 (see [467] for details) which admits a family $\{X_n\}_{n\geq 1}$ of indecomposable A-modules such that

$$\widehat{q}_A(\dim X_n) = 1 + 2n \quad \text{and} \quad q_A(\dim X) = 1 - 3n \ \text{(negative!)}.$$

We show in Section 5 that the positivity of the Euler form q_A of a simply connected algebra A is related to the tameness of some induced selfinjective algebras.

We recall from (XVII.6.1) that the module category mod B of a tilted algebra of Euclidean type is controlled by the Euler form q_B of B, which coincides with the Tits form $\widehat{q}_B$, because gl.dim $B \leq 2$. This is no longer the case for other classes of tame algebras of global dimension 2.

Namely, it is shown in [467] that, for any pair of integers $r \geq 1$ and $s \geq 1$, there exists a one-parametric strongly simply connected algebra A of global dimension 2 (hence $q_A = \widehat{q}_A$) such that, for each $d \in \{1, \ldots, r\}$, there exist pairwise non-isomorphic indecomposable A-modules

$$X_1^{(d)}, \ldots, X_s^{(d)}, \quad \text{with} \quad q_A(\dim X_1^{(d)}) = d, \ \ldots, q_A(\dim X_s^{(d)}) = d.$$

This shows that even one-parametric triangular algebras A may have many discrete indecomposable modules which are not controlled by the Euler (Tits) form of the algebras A.

On the other hand, the following result, proved by Skowroński and Zwara [652], shows that information on the behaviour of discrete indecomposable modules may determine the representation type of any strongly simply connected algebra.

2.15. Theorem. *Let A be a strongly simply connected algebra. The following conditions are equivalent.*

(a) *A is of polynomial growth.*

(b) *A is tame and there exists a natural number $m \geq 1$ such that, for each positive vector $\mathbf{d} \in K_0(A)$, there are at most m pairwise non-isomorphic indecomposable A-modules X with $\dim X = \mathbf{d}$ and $X \not\cong \tau_A X$.*

$\square$

We refer also to Skowroński and Zwara [653] for a geometric characterisation of strongly simply connected algebras in terms of degenerations of indecomposable modules and to Skowroński and Weyman [649] for a characterisation of the path algebras of Dynkin and Euclidean quivers in terms of the associated algebras of semi-invariants.

XX.3. Tilted and quasitilted algebras

In this book we have described the structure of the module category for all tilted algebras of Dynkin and Euclidean type, the concealed algebras of wild type, and discussed the shape of components of the Auslander–Reiten quivers of arbitrary tilted algebras.

In this section we provide complementary information on the structure of the module category of arbitrary tilted algebras. Our second objective is to outline the representation theory of algebras which are homologically very close to the tilted algebras: the quasitilted algebras introduced by Happel, Reiten and Smalø [288].

It follows from the Liu–Skowroński criterion (VIII.5.6) that a basic connected algebra B is a tilted algebra if and only if the Auslander–Reiten quiver $\Gamma(\operatorname{mod} B)$ of B admits a component $\mathcal{C}$ with a faithful section Δ such that $\operatorname{Hom}_B(X, \tau_B Y) = 0$, for all modules X and Y on Δ. Further, in this case, $\mathcal{C}$ is the connecting component $\mathcal{C}_T$ determined by a multiplicity-free tilting module T over a hereditary algebra A such that

$$B = \operatorname{End} T_A.$$

Moreover, $\mathcal{C}_T$ admits the section Σ given by the images $\operatorname{Hom}_A(T, I)$ of the indecomposable injective A-modules I via the tilting functor $\operatorname{Hom}_A(T, -)$, the torsion-free part $\mathcal{Y}(T) \cap \mathcal{C}_T$ of $\mathcal{C}_T$ consists of all predecessors of Σ in $\mathcal{C}_T$ while the torsion part $\mathcal{X}(T) \cap \mathcal{C}_T$ consists of all proper successors of Σ in $\mathcal{C}_T$.

The following theorem proved by Kerner [343] describes the structure of the module category of an arbitrary tilted algebra.

3.1. Theorem. *Let $A = K\Delta$ be a basic connected hereditary algebra, T a multiplicity-free tilting A-module and $B = \operatorname{End} T_A$ the associated tilted algebra. Then the connecting component $\mathcal{C} = \mathcal{C}_T$ of $\Gamma(\operatorname{mod} B)$ determined by T admits a finite (possibly empty) family of pairwise disjoint translation subquivers*

$$\mathcal{D}_1^{(l)}, \ldots, \mathcal{D}_m^{(l)}, \mathcal{D}_1^{(r)}, \ldots, \mathcal{D}_n^{(r)}$$

such that the following statements hold.

(a) *For each $i \in \{1, \ldots, m\}$, there exists an isomorphism*

$$\mathcal{D}_i^{(l)} \cong \mathbb{N}\Delta_i^{(l)},$$

where $\Delta_i^{(l)}$ is a connected subquiver of Δ and $\mathcal{D}_i^{(l)}$ is closed under predecessors in $\mathcal{C}$.

(b) *For each $j \in \{1, \ldots, n\}$, there exists an isomorphism*

$$\mathcal{D}_j^{(r)} \cong (-\mathbb{N})\Delta_j^{(r)},$$

where $\Delta_j^{(r)}$ is a connected subquiver of Δ and $\mathcal{D}_j^{(r)}$ is closed under successors in $\mathcal{C}$.

(c) All but finitely many indecomposable modules of $\mathcal{C}$ lie in

$$\mathcal{D}_1^{(l)} \cup \cdots \cup \mathcal{D}_m^{(l)} \cup \mathcal{D}_1^{(r)} \cup \cdots \cup \mathcal{D}_n^{(r)}.$$

(d) For each $i \in \{1, \dots, m\}$, there exists a tilted algebra

$$B_i^{(l)} = \mathrm{End}_{A_i^{(l)}}(T_i^{(l)}),$$

where $A_i^{(l)}$ is the path algebra $K\Delta_i^{(l)}$ and $T_i^{(l)}$ is a multiplicity-free tilting $A_i^{(l)}$-module without non-zero preinjective direct summands such that

- $B_i^{(l)}$ is a quotient algebra of B and hence there is a fully faithful embedding $\mod B_i^{(l)} \hookrightarrow \mod B$.
- $\mathcal{D}_i^{(l)}$ coincides with the torsion-free part $\mathcal{Y}(T_i^{(l)}) \cap \mathcal{C}_{T_i^{(l)}}$ of the connecting component $\mathcal{C}_{T_i^{(l)}}$ of $\Gamma(\mod B_i^{(l)})$ determined by $T_i^{(l)}$.

(e) For each $j \in \{1, \dots, n\}$, there exists a tilted algebra

$$B_j^{(r)} = \mathrm{End}_{A_j^{(r)}}(T_j^{(r)}),$$

where $A_j^{(r)}$ is the path algebra $K\Delta_j^{(r)}$ and $T_j^{(r)}$ is a multiplicity-free tilting $A_j^{(r)}$-module without non-zero postprojective direct summands such that

- $B_j^{(r)}$ is a quotient algebra of B and hence there is a fully faithful embedding $\mod B_j^{(r)} \hookrightarrow \mod B$.
- $\mathcal{D}_j^{(r)}$ coincides with the torsion part $\mathcal{X}(T_j^{(l)}) \cap \mathcal{C}_{T_j^{(r)}}$ of the connecting component $\mathcal{C}_{T_j^{(r)}}$ of $\Gamma(\mod B_j^{(r)})$ determined by $T_j^{(r)}$.

(f) $\mathcal{Y}(T) = \mathrm{add}\,(\mathcal{Y}(T_1^{(l)}) \cup \cdots \cup \mathcal{Y}(T_m^{(l)}) \cup \mathcal{Y}(T) \cap \mathcal{C}_T).$

(g) $\mathcal{X}(T) = \mathrm{add}\,(\mathcal{X}(T_1^{(r)}) \cup \cdots \cup \mathcal{X}(T_n^{(r)}) \cup \mathcal{X}(T) \cap \mathcal{C}_T).$

(h) The Auslander–Reiten quiver $\Gamma(\mod B)$ of B has the disjoint union form

$$\Gamma(\mod B) = (\bigcup_{i=1}^{m} \mathcal{Y}\Gamma(\mod B_i^{(l)})) \cup \mathcal{C}_T \cup (\bigcup_{j=1}^{n} \mathcal{X}\Gamma(\mod B_j^{(r)})),$$

where

- for each $i \in \{1, \dots, m\}$, $\mathcal{Y}\Gamma(\mod B_i^{(l)})$ is the union of all components of $\Gamma(\mod B_i^{(l)})$ contained entirely in $\mathcal{Y}(T_i^{(l)})$,
- for each $j \in \{1, \dots, n\}$, $\mathcal{X}\Gamma(\mod B_j^{(r)})$ is the union of all components of $\Gamma(\mod B_j^{(r)})$ contained entirely in $\mathcal{X}(T_j^{(r)})$. $\qquad \square$

The following facts follow from (XVII.5.1).

(i) If an algebra $B_i^{(l)}$ is of Euclidean type then the torsion-free part $\mathcal{Y}\Gamma(\text{mod } B_i^{(l)})$ of $\Gamma(\text{mod } B_i^{(l)})$ consists of

    • a unique postprojective component $\mathcal{P}(B_i^{(l)})$ of $\Gamma(\text{mod } B_i^{(l)})$, and

    • a $\mathbb{P}_1(K)$-family $\boldsymbol{T}^{B_i^{(l)}} = \{T_\lambda^{B_i^{(l)}}\}_{\lambda \in \mathbb{P}_1(K)}$ of pairwise orthogonal standard ray tubes.

Moreover, the postprojective component $\mathcal{P}(B_i^{(l)})$ is the unique postprojective component $\mathcal{P}(C_i^{(l)})$ of the quiver $\Gamma(\text{mod } C_i^{(l)})$ of a concealed algebra $C_i^{(l)}$ of Euclidean type, and $C_i^{(l)}$ is a quotient algebra of $B_i^{(l)}$.

(ii) If an algebra $B_j^{(r)}$ is of Euclidean type then the torsion part $\mathcal{X}\Gamma(\text{mod } B_j^{(r)})$ of $\Gamma(\text{mod } B_j^{(r)})$ consists of

    • a unique preinjective component $\mathcal{Q}(B_j^{(r)})$ of $\Gamma(\text{mod } B_j^{(r)})$, and

    • a $\mathbb{P}_1(K)$-family $\boldsymbol{T}^{B_j^{(r)}} = \{T_\lambda^{B_j^{(r)}}\}_{\lambda \in \mathbb{P}_1(K)}$ of pairwise orthogonal standard coray tubes.

Moreover, the preinjective component $\mathcal{Q}(B_j^{(r)})$ is the unique preinjective component $\mathcal{Q}(C_j^{(r)})$ of the quiver $\Gamma(\text{mod } C_j^{(r)})$ of a concealed algebra $C_j^{(r)}$ of Euclidean type, and $C_j^{(r)}$ is a quotient algebra of $B_j^{(r)}$.

The following combination of results proved by Kerner [344], [347], Liu [417], and Strauss [663] completes the description of the shapes of components in the Auslander–Reiten quivers $\Gamma(\text{mod } A)$ of tilted algebras A of wild type, see also (XVIII.5.10) and (XVIII.5.11) for related results.

3.2. Theorem. *Let A be a connected wild hereditary algebra, T a multiplicity-free tilting A-module and $B = \text{End } T_A$ the associated tilted algebra.*

(a) *If T has no non-zero preinjective direct summand then the torsion-free part $\mathcal{Y}\Gamma(\text{mod } B)$ of $\Gamma(\text{mod } B)$ consists of a unique postprojective component $\mathcal{P}(B)$ of $\Gamma(\text{mod } B)$ and a $\text{card}(K)$-family of components obtained from components of type $\mathbb{Z}\mathbb{A}_\infty$ by rectangle (ray) insertions. Moreover, $\mathcal{P}(B)$ is the unique postprojective component $\mathcal{P}(C)$ of the quiver $\Gamma(\text{mod } C)$ of a concealed algebra C of wild type, and C is a quotient algebra of B.*

(b) *If T has no non-zero postprojective direct summand then the torsion-free part $\mathcal{X}\Gamma(\text{mod } B)$ of $\Gamma(\text{mod } B)$ consists of a unique preinjective component $\mathcal{Q}(B)$ of $\Gamma(\text{mod } B)$ and a $\text{card}(K)$-family of components obtained from components of type $\mathbb{Z}\mathbb{A}_\infty$ by rectangle (coray) coinsertions. Moreover, $\mathcal{Q}(B)$ is the unique preinjective component $\mathcal{Q}(C)$*

of the quiver $\Gamma(\operatorname{mod} C)$ *of a concealed algebra* C *of wild type, and* C
is a quotient algebra of B.   □

As a direct consequence of (3.1), (3.2), (VIII.3.6), and (XVII.5.2) we
obtain the following fact.

3.3. Corollary. *If* B *is a tilted algebra then the Auslander–Reiten quiver*
$\Gamma(\operatorname{mod} B)$ *of* B *admits a postprojective component and a preinjective com-*
ponent.   □

For a tilted algebra B, using notation of (3.1), we may call
$$B^{(l)} = B_1^{(l)} \times \cdots \times B_m^{(l)} \quad \text{and} \quad B^{(r)} = B_1^{(r)} \times \cdots \times B_n^{(r)}$$
the **left part** of B and the **right part** of B, respectively. The following
characterisation of representation-tame tilted algebras is complementary to
(XIX.3.17).

3.4. Theorem. *For a tilted algebra* B, *the following conditions are*
equivalent.

(a) B *is representation-tame.*
(b) $B^{(l)}$ *and* $B^{(r)}$ *are representation-tame.*
(c) $B^{(l)}$ *and* $B^{(r)}$ *are products of representation-infinite tilted algebras*
    of Euclidean types or zero.   □

In connection to (XVIII.5.16) and (XVIII.5.17), we present also a result
by Kerner and Skowroński [353] asserting that there are many tilted algebras
with complicated regular connecting components.

3.5. Theorem. *Let* $A = KQ$ *be the path algebra of a connected wild*
quiver Q *with at least three vertices and* $m \geq 1$ *an integer. There ex-*
ist infinitely many pairwise non-isomorphic wild hereditary algebras H *and*
quasi-simple regular H-*modules* M *such that, for each such an algebra* H
and an H-*module* M, *we have*

(i) *the one-point extension* $B = H[M]$ *is a tilted algebra of type* Q *of*
    the form $\operatorname{End}_A T_A$, *where* T_A *is a regular tilting* A-*module, and*
(ii) *the regular connecting component* $\mathcal{C}_T$ *of* $\Gamma(\operatorname{mod} B)$ *determined by*
    T *has the property: for any indecomposable module* X *in* $\mathcal{C}_T$, *each*
    simple B-*module occurs with multiplicity at least* m *as a composition*
    factor of X.   □

We refer to [161], [279], [342], [345], [346], [347], [348], [349], [350], [351],
[354], [663] for further results concerning the module categories of wild
hereditary and wild tilted algebras.

The class of tilted algebras is generalised in [288] to the class of quasitilted
algebras. To define it, we need some category theory concepts.

3.6. Definition. An abelian K-category $\mathcal{H}$ is defined to be **hereditary** if, for each pair of objects X and Y in $\mathcal{H}$, the following two conditions are satisfied:

(i) $\mathrm{Ext}_{\mathcal{H}}^2(X, Y) = 0$, and

(ii) the K-vector spaces $\mathrm{Hom}_{\mathcal{H}}(X, Y)$ and $\mathrm{Ext}_{\mathcal{H}}^1(X, Y)$ are finite dimensional.

3.7. Definition. Let $\mathcal{H}$ be an abelian hereditary K-category. An object T of $\mathcal{H}$ is defined to be a **tilting object** if the following two conditions are satisfied:

(i) $\mathrm{Ext}_{\mathcal{H}}^1(T, T) = 0$, and

(ii) if X is an object of $\mathcal{H}$ such that $\mathrm{Hom}_{\mathcal{H}}(T, X) = 0$ and $\mathrm{Ext}_{\mathcal{H}}^1(T, X) = 0$ then $X = 0$.

3.8. Definition. An algebra B is said to be a **quasitilted algebra** if there exist a hereditary abelian K-category $\mathcal{H}$ and a tilting object T in $\mathcal{H}$ such that $B \cong \mathrm{End}_{\mathcal{H}}(T)$.

The module category $\mathrm{mod}\, A$ of a hereditary algebra A is a hereditary abelian K-category. Moreover, then a module T in $\mathrm{mod}\, A$ is a tilting module in the sense of (VI.2.1) if and only if T is a tilting object in $\mathrm{mod}\, A$ in the sense of (3.7). Hence every tilted algebra is quasitilted.

The quasitilted algebras have several interesting characterisations. To present homological characterisation of these algebras, we need some notation.

3.9. Definition. Let A be an algebra and $\mathrm{ind}\, A$ the full subcategory of $\mathrm{mod}\, A$ consisting of all indecomposable modules.

(a) The category $\mathcal{L}_A$ is defined to be the full subcategory of $\mathrm{ind}\, A$ formed by all modules X such that $\mathrm{pd}_A Y \leq 1$, for every predecessor Y of X in $\mathrm{ind}\, A$.

(b) The category $\mathcal{R}_A$ is defined to be the full subcategory of $\mathrm{ind}\, A$ formed by all modules X such that $\mathrm{id}_A Y \leq 1$, for every successor Y of X in $\mathrm{ind}\, A$.

The following characterisation of quasitilted algebras is due to Happel, Reiten and Smalø [288].

3.10. Theorem. *Let B be an algebra. The following conditions are equivalent.*

(a) *B is a quasitilted algebra.*

(b) *$\mathrm{gl.dim}\, B \leq 2$ and every indecomposable B-module X satisfies $\mathrm{pd}_B X \leq 1$ or $\mathrm{id}_B X \leq 1$.*

(c) $\mathcal{L}_B$ contains all indecomposable projective B-modules.

(d) $\mathcal{R}_B$ contains all indecomposable injective B-modules. $\qquad\square$

The following facts are also proved in [288].

3.11. Proposition. *If B is a quasitilted algebra then*

(a) *B is a triangular algebra,*

(b) *$\operatorname{ind} B = \mathcal{L}_B \cup \mathcal{R}_B$, and*

(c) *if B is representation-finite then B is a tilted algebra.* $\qquad\square$

We have also the following characterisation of tilted algebras in the class of quasitilted algebras established by Skowroński in [626].

3.12. Theorem. *A connected algebra B is a tilted algebra if and only if* gl.dim $B \leq 2$, $\operatorname{ind} B = \mathcal{L}_B \cup \mathcal{R}_B$, *and $\mathcal{L}_B \cap \mathcal{R}_B$ contains a directing module.*
$\qquad\square$

The following general information on the components of the Auslander-Reiten quivers of quasitilted algebras is proved by Coelho and Skowroński in [145].

3.13. Theorem. *Let B be a quasitilted but not tilted algebra. Then every component of $\Gamma(\operatorname{mod} B)$ is semiregular.* $\qquad\square$

A distinguished class of quasitilted algebras is formed by the canonical algebras introduced by Ringel in [525] (see also [528]).

3.14. Definition. Let $m \geq 2$ be an integer, $\mathbf{p} = (p_1, \ldots, p_m)$ an m-tuple of positive integers, and $\underline{\lambda} = (\lambda_1, \ldots, \lambda_m)$ an m-tuple of pairwise different elements of $\mathbb{P}_1(K) = K \cup \{\infty\}$, normalised such that $\lambda_1 = \infty$, $\lambda_2 = 0$, $\lambda_3 = 1$. Consider the quiver

$$\Delta(\mathbf{p}):$$

and define the algebra $C(\mathbf{p}, \underline{\lambda})$ as follows.

(a) For $m = 2$, we set $C(\mathbf{p}, \underline{\lambda}) = K\Delta(\mathbf{p})$.

(b) For $m \geq 3$, we assume that $p_1 \geq 2, \ldots, p_m \geq 2$, and we set

$$C(\mathbf{p}, \underline{\lambda}) = K\Delta(\mathbf{p})/I(\mathbf{p}, \underline{\lambda}),$$

where $I(\mathbf{p}, \underline{\lambda})$ is the ideal of the path algebra $K\Delta(\mathbf{p})$ generated by the elements

$$\alpha_{jp_j} \cdots \alpha_{j2}\alpha_{j1} + \alpha_{1p_1} \cdots \alpha_{12}\alpha_{11} + \lambda_j \alpha_{2p_2} \cdots \alpha_{22}\alpha_{21},$$

with $j \in \{3, \ldots, m\}$.

The algebra $C(\mathbf{p}, \underline{\lambda})$ is said to be the **canonical algebra** of type $(\mathbf{p}, \underline{\lambda})$, $\mathbf{p}$ the **weight sequence** of $C(\mathbf{p}, \underline{\lambda})$, and $\underline{\lambda}$ the **parameter sequence** of $C(\mathbf{p}, \underline{\lambda})$.

It is shown by Ringel in [525] that the Auslander–Reiten quiver $\Gamma(\mathrm{mod}\, C)$ of a canonical algebra $C = C(\mathbf{p}, \underline{\lambda})$ has the disjoint union decomposition

$$\Gamma(\mathrm{mod}\, C) = \boldsymbol{P}^C \cup \boldsymbol{T}^C \cup \boldsymbol{Q}^C$$

where

- $\boldsymbol{P}^C$ is a family of components containing all the indecomposable projective C-modules.
- $\boldsymbol{Q}^C$ is a family of components containing all the indecomposable injective C-modules.
- $\boldsymbol{T}^C = \{\mathcal{T}_\lambda^C\}_{\lambda \in \mathbb{P}_1(K)}$ is a $\mathbb{P}_1(K)$-family of pairwise orthogonal faithful standard stable tubes of tubular type $r^C = (p_1, \ldots, p_m) = \mathbf{p}$.
- $\boldsymbol{T}^C$ separates $\boldsymbol{P}^C$ from $\boldsymbol{Q}^C$.

The following more general class of algebras is introduced by Lenzing and Meltzer in [394].

3.15. Definition. Let $C = C(\mathbf{p}, \underline{\lambda})$ be a canonical algebra. A **concealed canonical algebra** B of type $(\mathbf{p}, \underline{\lambda})$ is defined to be the algebra of the form $B = \mathrm{End}\, T_C$, where T_C is a tilting C-module from the additive category add $\boldsymbol{P}^C$ of $\boldsymbol{P}^C$.

Again the Auslander–Reiten quiver $\Gamma(\mathrm{mod}\, B)$ of a concealed canonical algebra $B = \mathrm{End}\, T_C$ has a disjoint union decomposition

$$\Gamma(\mathrm{mod}\, B) = \boldsymbol{P}^B \cup \boldsymbol{T}^B \cup \boldsymbol{Q}^B,$$

where

- $\boldsymbol{P}^B$ is a family of components containing all the indecomposable projective B-modules.
- $\boldsymbol{Q}^B$ is a family of components containing all the indecomposable injective B-modules.
- $\boldsymbol{T}^B = \mathrm{Hom}_C(T, \boldsymbol{T}^C)$ is a $\mathbb{P}_1(K)$-family $\boldsymbol{T}^B = \{\mathcal{T}_\lambda^B\}_{\lambda \in \mathbb{P}_1(K)}$ of pairwise orthogonal faithful standard stable tubes.
- $\boldsymbol{T}^B$ separates $\boldsymbol{P}^B$ from $\boldsymbol{Q}^B$.

Moreover, we have

- $\mathrm{pd}_B X \leq 1$, for any indecomposable B-module $X \in \boldsymbol{P}^B \cup \boldsymbol{T}^B$.
- $\mathrm{id}_B Y \leq 1$, for any indecomposable B-module $Y \in \boldsymbol{T}^B \cup \boldsymbol{Q}^B$.
- $\mathrm{gl.dim}\, B \leq 2$.

The $\mathbb{P}_1(K)$-family $\boldsymbol{T}^B = \{\mathcal{T}_\lambda^B\}_{\lambda \in \mathbb{P}_1(K)}$ is said to be the **canonical family of stable tubes** of $\Gamma(\mathrm{mod}\, B)$.

Therefore, every concealed canonical algebra is a quasitilted algebra. We refer to [353], [394], [395], [396], [436], [525], [528], for the representation theory of concealed canonical algebras.

The following theorem completely determines the representation type of a concealed canonical algebra B of type $(\mathbf{p}, \underline{\lambda})$ by means of the **genus**

$$g(B) = 1 + \frac{1}{2}\left((m-2)p - \left[\frac{p}{p_1} + \ldots + \frac{p}{p_m} \right] \right)$$

of B, where $\mathbf{p} = (p_1, \ldots, p_m)$ and $p = \ell.c.m.(p_1, \ldots, p_m)$.

3.16. Theorem. *Let B be a concealed canonical algebra of type $(\mathbf{p}, \underline{\lambda})$ with genus $g(B)$. The following equivalences hold.*

(a) *$g(B) < 1$ if and only if B is a concealed algebra of Euclidean type.*
(b) *$g(B) = 1$ if and only if B is a tubular algebra.*
(c) *$g(B) > 1$ if and only if B is wild.* □

In particular, we have the following consequence.

3.17. Corollary. *Let B be a concealed canonical algebra of type $(\mathbf{p}, \underline{\lambda})$ with genus $g(B)$. The algebra B is a tilted algebra if and only if $g(B) < 1$.* □

The following characterisation of concealed canonical algebras is proved by Lenzing and de la Peña [396], see also [619].

3.18. Theorem. *An algebra B is a concealed canonical algebra if and only if the quiver $\Gamma(\mathrm{mod}\, B)$ admits a sincere separating family $\mathbf{T}$ of stable tubes.* □

A family $\mathcal{C}$ of components of an Auslander–Reiten quiver $\Gamma(\mathrm{mod}\, A)$ is said to be **sincere** if any simple A-module occurs as a composition factor of a module in $\mathcal{C}$.

The preceding theorem is deepened in [489], [625] (see also [624]) by showing that a concealed canonical algebra can be recovered from a single stable tube of its Auslander–Reiten quiver satisfying a much weaker assumption than the separating condition. Following [489], a short cycle $X \to Y \to X$ in a module category $\mathrm{mod}\, A$ is said to be an **external short cycle** with respect to a component $\mathcal{C}$ of $\Gamma(\mathrm{mod}\, A)$ if X lies in $\mathcal{C}$ but Y does not lie in $\mathcal{C}$. Then we have the following result.

3.19. Theorem. *An algebra B is a concealed canonical algebra if and only if the quiver $\Gamma(\mathrm{mod}\, B)$ admits a sincere stable tube $\mathbf{T}$ without external short cycles.* □

As a consequence we obtain the following interesting fact.

3.20. Corollary. *Let A be an algebra such that $\Gamma(\operatorname{mod} A)$ admits a sincere stable tube $\mathcal{T}$ without external short cycles. Then A is a concealed canonical algebra and $\mathcal{T}$ is a faithful standard stable tube.* $\qquad\square$

We also note that the existence of a faithful standard stable tube in the Auslander–Reiten quiver $\Gamma(\operatorname{mod} A)$ of an algebra A does not force the algebra A to be concealed canonical, see (1.21) and (1.22).

The following result of Kerner and Skowroński [353] shows that there exist many concealed canonical algebras of a given wild type with complicated separating families of stable tubes.

3.21. Theorem. *Let $C = C(\mathbf{p}, \underline{\lambda})$ be a canonical algebra of wild type and $m \geq 1$ an integer. There exist infinitely many pairwise non-isomorphic wild hereditary algebras H and quasi-simple regular H-modules M such that*

  (i) *the one-point extension $B = H[M]$ is a concealed canonical algebra $\operatorname{End} T_C$ of type $(\mathbf{p}, \underline{\lambda})$, and*

 (ii) *the canonical separating family $\boldsymbol{\mathcal{T}}^B$ of stable tubes in $\Gamma(\operatorname{mod} B)$ has the property: for any indecomposable B-module X in $\boldsymbol{\mathcal{T}}^B$, each simple B-module occurs with multiplicity at least m as a composition factor of X.* $\qquad\square$

We note that every canonical algebra $C = C(\mathbf{p}, \underline{\lambda})$ of wild type is itself a one-point extension $A[R]$, where $A = A(\mathbf{p})$ is the path algebra $KQ(\mathbf{p})$ of the wild subquiver $Q(\mathbf{p})$ of $\Delta(\mathbf{p})$ given by all vertices except the unique source ω, and R is the quasi-simple regular A-module with the dimension vector

$$
\mathbf{dim}\, R = \begin{smallmatrix} 1 & 1 & \cdots & 1 & 1 \\ 2 & 1 & 1 & \cdots & 1 & 1 \\ \vdots & \vdots & & \vdots & \vdots \\ 1 & 1 & \cdots & 1 & 1 \end{smallmatrix}
$$

But in this case, for any indecomposable C-module E lying on the mouth of a stable tube of the canonical family $\boldsymbol{\mathcal{T}}^C$, each simple C-module occurs with multiplicity at most 1 as a composition quotient of E.

Our next objective is to describe the structure of arbitrary quasitilted algebras.

An important homological invariant of a module category $\operatorname{mod} A$ is the **derived category** $D^b(\operatorname{mod} A)$ of bounded complexes of modules in $\operatorname{mod} A$, which is a triangulated category. We refer to the book [276] for basic background on the triangulated categories in the representation theory of algebras.

One of the important questions is to know when two algebras A and B are **derived equivalent**, that is, the derived categories $D^b(\operatorname{mod} A)$ and $D^b(\operatorname{mod} B)$ are equivalent as triangulated categories. One knows that, if A

is an algebra, T is a tilting A-module and $B = \operatorname{End} T_A$, then A and B are derived equivalent [276].

In [495], Rickard proved his celebrated general criterion: two algebras A and B are derived equivalent if and only if B is the endomorphism algebra of a tilting complex over A. It follows from [288] that a quasitilted algebra B is tilted if and only if B is derived equivalent to a hereditary algebra H.

New types of abelian hereditary K-categories $\mathcal{H}$ with tilting objects can be constructed from certain subcategories of the derived categories $D^b(\operatorname{mod} C)$ of canonical algebras $C = C(\mathbf{p}, \underline{\lambda})$, and have the property $D^b(\mathcal{H}) \cong D^b(\operatorname{mod} C)$. These categories $\mathcal{H}$ are called **abelian hereditary categories of canonical type**, and are classified completely by Lenzing and Skowroński in [398]. Then the following concept is natural.

3.22. Definition. An algebra B is said to be a **quasitilted algebra of canonical type** if $B \cong \operatorname{End}_{\mathcal{H}}(T)$ for a tilting object T in a hereditary abelian K-category $\mathcal{H}$ of canonical type.

In particular, every quasitilted algebra B of canonical type is derived equivalent to a canonical algebra. A complete characterisation of the quasitilted algebras of canonical type is established by Lenzing and Skowroński in [398]. To present it, we need some concepts.

3.23. Definition. Let A be a concealed canonical algebra and

$$\boldsymbol{T}^A = \{\mathcal{T}_\lambda^A\}_{\lambda \in \mathbb{P}_1(K)}$$

the canonical $\mathbb{P}_1(K)$-family of standard stable tubes of $\Gamma(\operatorname{mod} A)$.

(a) A branch $\boldsymbol{T}^A$-extension (in the sense of (XV.3.6)(a))
$$B = A[E_1, \mathcal{L}^{(1)}, \ldots, E_s, \mathcal{L}^{(s)}]$$
of A, where $E_1, \ldots, E_s$ are pairwise different mouth modules of $\boldsymbol{T}^A$ and $\mathcal{L}^{(1)}, \ldots, \mathcal{L}^{(s)}$ are branches, is said to be a **branch extension** of A.

(b) A branch $\boldsymbol{T}^A$-coextension (in the sense of (XV.3.6)(b))
$$B = [E_1, \mathcal{L}^{(1)}, \ldots, E_m, \mathcal{L}^{(m)}]A$$
of A, where $E_1, \ldots, E_m$ are pairwise different mouth modules of $\boldsymbol{T}^A$ and $\mathcal{L}^{(1)}, \ldots, \mathcal{L}^{(m)}$ are branches, is said to be a **branch coextension** of A.

(c) A branch extension-coextension B of A of the form
$$B = [E_1, \mathcal{L}^{(1)}, \ldots, E_m, \mathcal{L}^{(m)}]A[\widehat{E}_1, \widehat{\mathcal{L}}^{(1)}, \ldots, \widehat{E}_s, \widehat{\mathcal{L}}^{(s)}],$$
where $E_1, \ldots, E_m, \widehat{E}_1, \ldots, \widehat{E}_s$ are pairwise different mouth modules of $\boldsymbol{T}^A$, $\mathcal{L}^{(1)}, \ldots, \mathcal{L}^{(m)}, \widehat{\mathcal{L}}^{(1)}, \ldots, \widehat{\mathcal{L}}^{(s)}$ are branches, and the tubes of

$\mathcal{T}^A$ containing $E_1, \ldots, E_m$ are disjoint with the tubes of $\mathcal{T}^A$ containing $\widehat{E}_1, \ldots, \widehat{E}_s$, is said to be a **semiregular branch enlargement** of A.

By a **semiregular family of tubes** of an Auslander–Reiten quiver $\Gamma(\operatorname{mod} A)$ we mean a family of components consisting of ray and coray tubes (in the sense of (XV.2.10)).

Recall also that all representation-finite quasitilted algebras are tilted algebras (3.11). Then we have the following theorem proved in [398].

3.24. Theorem. *Let B be a basic connected algebra. The following statements are equivalent.*

(a) *B is representation-infinite and quasitilted of canonical type.*

(b) *B is a semiregular branch enlargement of a concealed canonical algebra A.*

(c) *The quiver $\Gamma(\operatorname{mod} B)$ admits a sincere separating family of semiregular tubes.* $\square$

For a semiregular branch enlargement

$$B = [E_1, \mathcal{L}^{(1)}, \ldots, E_m, \mathcal{L}^{(m)}] A [\widehat{E}_1, \widehat{\mathcal{L}}^{(1)}, \ldots, \widehat{E}_s, \widehat{\mathcal{L}}^{(s)}]$$

of a concealed canonical algebra A, we call the branch coextension

$$B^{(l)} = [E_1, \mathcal{L}^{(1)}, \ldots, E_m, \mathcal{L}^{(m)}] A$$

of A the **left part** of B and the branch extension

$$B^{(r)} = A [\widehat{E}_1, \widehat{\mathcal{L}}^{(1)}, \ldots, \widehat{E}_s, \widehat{\mathcal{L}}^{(s)}]$$

of A the **right part** of B. Observe that $B^{(l)}$ and $B^{(r)}$ are quotient algebras of B and hence we have fully faithful embeddings $\operatorname{mod} B^{(l)} \hookrightarrow \operatorname{mod} B$ and $\operatorname{mod} B^{(r)} \hookrightarrow \operatorname{mod} B$. Moreover, we have the following facts, see [394] and [398].

3.25. Proposition. *An algebra B is a representation-infinite quasitilted algebra of canonical type with $B = B^{(r)}$ if and only if there is a canonical algebra $C = C(\mathbf{p}, \underline{\lambda})$ and a tilting C-module $T \in \operatorname{add}(\mathcal{P}^C \cup \mathcal{T}^C)$ such that $B \cong \operatorname{End} T_C$.* $\square$

We also note that if B is a representation-infinite quasitilted algebra of canonical type then the opposite algebra B^{op} is also a representation-infinite quasitilted algebra of canonical type and $(B^{\mathrm{op}})^{(l)} \cong (B^{(r)})^{\mathrm{op}}$ and $(B^{\mathrm{op}})^{(r)} \cong (B^{(l)})^{\mathrm{op}}$.

The following classification of the representation-tame quasitilted algebras is established by Skowroński in [622].

3.26. Theorem. *Let B be a basic connected quasitilted algebra. The following statements are equivalent.*

(a) *B is representation-tame.*

(b) *B is of linear growth.*

(c) *The Euler form $q_B : K_0(B) \longrightarrow \mathbb{Z}$ of B is weakly non-negative.*

(d) *$\mathrm{mod}\, B$ is controlled by $q_B : K_0(B) \longrightarrow \mathbb{Z}$.*

(e) *$\dim_K \mathrm{Ext}_B^1(X, X) \leq \dim_K \mathrm{End}_B(X)$, for any indecomposable B-module X.*

(f) *Every component of $\Gamma(\mathrm{mod}\, B)$ is generalised standard.*

(g) *The infinite radical rad_B^∞ of $\mathrm{mod}\, B$ is locally nilpotent.*

(h) *B is representation-tame tilted or representation-tame quasitilted of canonical type.*

(i) *Each of the algebras $B^{(l)}$ and $B^{(r)}$ is zero, or is a tubular algebra, or is a product of representation-infinite tilted algebras of Euclidean type.* $\square$

We note that, for any quasitilted algebra B, the Euler quadratic form $q_B : K_0(B) \longrightarrow \mathbb{Z}$ coincides with the Tits form $\widehat{q}_B : K_0(B) \longrightarrow \mathbb{Z}$ of B. Moreover, if B is quasitilted of canonical type then each of the algebras $B^{(l)}$ and $B^{(r)}$ is non-zero and connected.

Finally, Happel proved in [278] (see also [286] for a different proof) that every abelian hereditary K-category $\mathcal{H}$ with a tilting object is either the module category of a hereditary algebra or is of canonical type. This is equivalent to the following classification of arbitrary quasitilted algebras.

3.27. Theorem. *A connected algebra B is quasitilted if and only if B is a tilted algebra or a quasitilted algebra of canonical type.* $\square$

We end this section with the description of the Auslander–Reiten quivers of quasitilted algebras of canonical type, which is a combination of results from [398], [436], [525], as well as (XVII.5.2).

3.28. Theorem. *Let B be a quasitilted algebra of canonical type. The Auslander–Reiten quiver $\Gamma(\mathrm{mod}\, B)$ has a disjoint union decomposition*

$$\Gamma(\mathrm{mod}\, B) = \boldsymbol{\mathcal{P}}^B \cup \boldsymbol{\mathcal{T}}^B \cup \boldsymbol{\mathcal{Q}}^B$$

such that

(a) *$\boldsymbol{\mathcal{T}}^B = \{\mathcal{T}_\lambda^B\}_{\lambda \in \mathbb{P}_1(K)}$ is a sincere semiregular $\mathbb{P}_1(K)$-family of pairwise orthogonal standard tubes, separating $\boldsymbol{\mathcal{P}}^B$ from $\boldsymbol{\mathcal{Q}}^B$.*

(b) *$\boldsymbol{\mathcal{P}}^B = \boldsymbol{\mathcal{P}}^{B^{(l)}}$ is a family of components containing all the indecomposable projective B-modules which are not in $\boldsymbol{\mathcal{T}}^B$.*

(c) *$\boldsymbol{\mathcal{P}}^B$ contains a postprojective component $\mathcal{P}(B)$, and*
   - *$\mathcal{P}(B) = \mathcal{P}(B^{(l)})$ is a unique postprojective component of $\Gamma(\mathrm{mod}\, B)$.*

- B admits a connected concealed quotient algebra $C^{(l)}$ such that $\mathcal{P}(B)$ coincides with the unique postprojective component $\mathcal{P}(C^{(l)})$ of $\Gamma(\operatorname{mod} C^{(l)})$.

(d) $\boldsymbol{Q}^B = \boldsymbol{Q}^{B^{(r)}}$ is a family of components containing all the indecomposable injective B-modules which are not in $\boldsymbol{T}^B$.

(e) $\boldsymbol{Q}^B$ contains a preinjective component $\mathcal{Q}(B)$, and

- $\mathcal{Q}(B) = \mathcal{Q}(B^{(r)})$ is a unique preinjective component of $\Gamma(\operatorname{mod} B)$.
- B admits a connected concealed quotient algebra $C^{(r)}$ such that $\mathcal{Q}(B)$ coincides with the unique preinjective component $\mathcal{Q}(C^{(r)})$ of $\Gamma(\operatorname{mod} C^{(r)})$.

(f) If $B^{(l)}$ is tame then $C^{(l)}$ is tame and every component of $\boldsymbol{P}^B$ different from $\mathcal{P}(B)$ is a standard ray tube.

(g) $\boldsymbol{P}^B = P(B)$ if and only if $B^{(l)}$ is tilted of Euclidean type.

(h) If $B^{(l)}$ is wild then $C^{(l)}$ is wild and every component of $\boldsymbol{P}^B$ different from $\mathcal{P}(B)$ is either of the form $\mathbb{Z}\mathbb{A}_\infty$ or is obtained from a component of type $\mathbb{Z}\mathbb{A}_\infty$ by rectangle (ray) insertions.

(i) If $B^{(r)}$ is tame then $C^{(r)}$ is tame and every component of $\boldsymbol{Q}^B$ different from $\mathcal{Q}(B)$ is a standard coray tube.

(j) $\boldsymbol{Q}^B = Q(B)$ if and only if $B^{(r)}$ is tilted of Euclidean type.

(k) If $B^{(r)}$ is wild then $C^{(r)}$ is wild and every component of $\boldsymbol{Q}^B$ different from $\mathcal{Q}(B)$ is either of the form $\mathbb{Z}\mathbb{A}_\infty$ or is obtained from a component of type $\mathbb{Z}\mathbb{A}_\infty$ by rectangle (coray) coinsertions. $\square$

XX.4. Algebras of small homological dimensions

The aim of this section is to describe some classes of algebras of small homological dimensions which are closely related to the tilted and quasitilted algebras.

We start with a general fact observed in [636].

4.1. Theorem. *Let A be an algebra. Then* $\operatorname{gl.dim} A < \infty$ *if and only if every indecomposable A-module X satisfies* $\operatorname{pd}_A X < \infty$ *or* $\operatorname{id}_A X < 1$. $\square$

It is easy to see that, if every indecomposable A-module X satisfies $\operatorname{pd}_A X \leq 1$ or $\operatorname{id}_A X \leq 1$, then $\operatorname{gl.dim} A \leq 3$, see [288].

Following [139] we have the following concept.

4.2. Definition. Let A be an algebra.

(a) A is said to be an **algebra of small homological dimension** (shortly, **shod**) if every indecomposable A-module X satisfies $\operatorname{pd}_A X \leq 1$ or $\operatorname{id}_A X \leq 1$.

(b) A is said to be a **strict shod** if A is shod and $\operatorname{gl.dim} A = 3$.

Observe that the class of shod algebras consists of the quasitilted algebras and the strict shod algebras.

The following characterisation of shod algebras is obtained by Coelho and Lanzilotta in [139].

4.3. Theorem. *Let A be an algebra. The following conditions are equivalent.*

(a) *A is a shod algebra.*
(b) *ind $A = \mathcal{L}_A \cup \mathcal{R}_A$.*
(c) *There exists a splitting torsion pair $(\mathcal{X}, \mathcal{Y})$ in mod A such that $\mathrm{pd}_A Y \leq 1$, for each module $Y \in \mathcal{Y}$, and $\mathrm{id}_A X \leq 1$, for each module $X \in \mathcal{X}$.* □

As a direct consequence of (3.10) and (4.3) we have the following characterisation of strict shod algebras in the class of shod algebras.

4.4. Theorem. *For a shod algebra A, the following conditions are equivalent.*

(a) *A is a strict shod algebra.*
(b) *$\mathcal{L}_A \setminus \mathcal{R}_A$ contains an indecomposable injective A-module.*
(c) *$\mathcal{R}_A \setminus \mathcal{L}_A$ contains an indecomposable projective A-module.* □

The structure and the representation theory of strict shod algebras is described by Reiten and Skowroński in [490]. To present it, we need the concepts of a double section and a double tilted algebra introduced in [490].

4.5. Definition. Let A be an algebra and $\mathcal{C}$ be a component of $\Gamma(\mathrm{mod}\, A)$.

(a) A full connected subquiver Δ of $\mathcal{C}$ is said to be a **double section** of $\mathcal{C}$ if the following conditions are satisfied:

   (a1) Δ is acyclic.
   (a2) Δ is convex in $\mathcal{C}$.
   (a3) For each τ_A-orbit $\mathcal{O}$ in $\mathcal{C}$, we have $1 \leq |\Delta \cap \mathcal{O}| \leq 2$.
   (a4) If $\mathcal{O}$ is a τ_A-orbit $\mathcal{O}$ in $\mathcal{C}$ and $|\Delta \cap \mathcal{O}| = 2$ then $\Delta \cap \mathcal{O} = \{X, \tau_A X\}$, for some module $X \in \mathcal{C}$, and there exist sectional paths $I \to \cdots \to \tau_A X$ and $X \to \cdots \to P$ in $\mathcal{C}$, with I injective and P projective.

(b) A double section Δ in $\mathcal{C}$ with $|\Delta \cap \mathcal{O}| = 2$, for some τ_A-orbit $\mathcal{O}$ in $\mathcal{C}$, is said to be a **strict double section** of $\mathcal{C}$.

A path $X_0 \to X_1 \to \cdots \to X_m$, with $m \geq 2$, in an Auslander–Reiten quiver $\Gamma(\mathrm{mod}\, A)$ is said to be **almost sectional** if there exists exactly one index $i \in \{2, \ldots, m\}$ such that $X_{i-2} \cong \tau_A X_i$.

4.6. Definition. Let A be an algebra, $\mathcal{C}$ a component of $\Gamma(\operatorname{mod} A)$ and Δ a double section of $\mathcal{C}$. We define the following full subquivers of Δ:

$$\Delta'_l = \{X \in \Delta; \text{there is an almost sectional path } X \to \cdots \to P,$$
$$\text{with } P \text{ projective }\},$$
$$\Delta'_r = \{X \in \Delta; \text{there is an almost sectional path } I \to \cdots \to X,$$
$$\text{with } I \text{ injective }\},$$
$$\Delta_l = (\Delta \setminus \Delta'_r) \cup \tau_A \Delta'_r, \text{ and}$$
$$\Delta_r = (\Delta \setminus \Delta'_l) \cup \tau_A^{-1} \Delta'_l.$$

The subquiver Δ_l is said to be the **left part** of Δ, and the subquiver Δ_r is said to be the **right part** of Δ.

Observe that, if $|\Delta \cap \mathcal{O}| = 1$, for each τ_A-orbit $\mathcal{O}$ in $\mathcal{C}$, then $\Delta = \Delta_l = \Delta_r$ and Δ is a section of $\mathcal{C}$, see (VIII.1.2).

4.7. Definition. A connected algebra B is said to be a **double tilted algebra** if the following conditions are satisfied:

(i) $\Gamma(\operatorname{mod} B)$ admits a component $\mathcal{C}$ with a faithful double section Δ.

(ii) There exists a tilted quotient algebra $B^{(l)}$ of B (not necessarily connected) such that Δ_l is a disjoint union of sections of the connecting components of the connected parts of $B^{(l)}$ and the category of all predecessors of Δ_l in ind B coincides with the category of all predecessors of Δ_l in ind $B^{(l)}$.

(iii) There exists a tilted quotient algebra $B^{(r)}$ of B (not necessarily connected) such that Δ_r is a disjoint union of sections of the connecting components of the connected parts of $B^{(r)}$, and the category of all successors of Δ_r in ind B coincides with the category of all successors of Δ_r in ind $B^{(r)}$.

If moreover the double section is strict, then B is said to be a **strict double tilted algebra**.

For a double tilted algebra B, the algebras $B^{(l)}$ and $B^{(r)}$ are said to be the **left tilted algebra** and the **right tilted algebra** of B, respectively. Observe that a double tilted algebra B is tilted if and only if $B = B^{(l)} = B^{(r)}$. We also note that every double tilted algebra B is triangular, because the quiver Q_B is acyclic.

The following criterion of Reiten and Skowroński [490] extends the criterion of Liu and Skowroński (VIII.5.6) to the double tilted algebras.

4.8. Theorem. *A connected algebra B is a double tilted algebra if and only if the translation quiver $\Gamma(\operatorname{mod} B)$ contains a component $\mathcal{C}$ with a faithful double section Δ such that $\operatorname{Hom}_B(U, \tau_B V) = 0$, for all modules U in Δ_r and V in Δ_l.* $\square$

The second main result of [490] gives the classification of all strict shod algebras.

4.9. Theorem. *For a connected algebra B, the following three conditions are equivalent.*

(a) *B is a strict shod algebra.*

(b) *B is a strict double tilted algebra.*

(c) *$\Gamma(\operatorname{mod} B)$ contains a component $\mathcal{C}$ with a faithful strict double section Δ such that $\operatorname{Hom}_B(U, \tau_B V) = 0$, for all modules $U \in \Delta_r$ and $V \in \Delta_l$.* $\qquad\square$

As a direct consequence of (3.24), (3.27) and (4.9) we obtain the following classification of algebras with small homological dimensions.

4.10. Corollary. *A connected algebra B is a shod algebra if and only if B is either a tilted algebra, a strict double tilted algebra, or a semiregular branch enlargement of a concealed canonical algebra.* $\qquad\square$

Let B be a strict double tilted algebra. It follows from (4.7) that the Auslander–Reiten quiver $\Gamma(\operatorname{mod} B)$ contains a unique component $\mathcal{C}$ with a faithful strict double section Δ. Moreover, every indecomposable B-module is either a predecessor of Δ_l in $\operatorname{ind} B$ or a successor of Δ_r in $\operatorname{ind} B$, and consequently is either an indecomposable $B^{(l)}$-module or an indecomposable $B^{(r)}$-module. Therefore, the category $\operatorname{ind} B$ is a glueing of $\operatorname{ind} B^{(l)}$ and $\operatorname{ind} B^{(r)}$ along the double section Δ of $\mathcal{C}$, and hence $\mathcal{C} = \mathcal{C}_B$ is called the **connecting component** of $\Gamma(\operatorname{mod} B)$. In particular, we obtain the following description of $\Gamma(\operatorname{mod} B)$.

4.11. Corollary. *Let B be a strict shod algebra. Then $\Gamma(\operatorname{mod} B)$ has the disjoint union form*

$$\Gamma(\operatorname{mod} B) = \mathcal{Y}\Gamma(\operatorname{mod} B^{(l)}) \cup \mathcal{C}_B \cup \mathcal{X}\Gamma(\operatorname{mod} B^{(r)}),$$

where

(i) *$\mathcal{Y}\Gamma(\operatorname{mod} B^{(l)})$ is the union of all components of $\Gamma(\operatorname{mod} B^{(l)})$ contained entirely in the torsion-free part $\mathcal{Y}(T^{(l)})$ of $\operatorname{mod} B^{(l)}$, determined by a tilting module $T^{(l)}$ over the path algebra $A^{(l)} = K\Delta_l$ with $B^{(l)} \cong \operatorname{End} T^{(l)}_{A^{(l)}}$.*

(ii) *$\mathcal{X}\Gamma(\operatorname{mod} B^{(r)})$ is the union of all components of $\Gamma(\operatorname{mod} B^{(r)})$ contained entirely in the torsion part $\mathcal{X}(T^{(r)})$ of $\operatorname{mod} B^{(r)}$, determined by a tilting module $T^{(r)}$ over the path algebra $A^{(r)} = K\Delta_r$ with $B^{(r)} \cong \operatorname{End} T^{(r)}_{A^{(r)}}$.* $\qquad\square$

The following characterisation of double tilted algebras in terms of directing modules is given in [626].

4.12. Theorem. *Let B be a connected algebra. The following conditions are equivalent.*

(a) *B is a double tilted algebra.*

(b) *$\operatorname{ind} A = \mathcal{L}_A \cup \mathcal{R}_A$ and $\mathcal{L}_A \cap (\mathcal{R}_A \cup \tau_A \mathcal{R}_A)$ contains a directing module.*

(c) *$\operatorname{ind} A = \mathcal{L}_A \cup \mathcal{R}_A$ and $(\mathcal{L}_A \cup \tau_A^{-1} \mathcal{L}_A) \cap \mathcal{R}_A$ contains a directing module.* $\square$

We conclude the discussion with the following characterisation of representation-tame shod algebras established in [490], extending (3.26).

4.13. Theorem. *Let B be a basic connected shod algebra. The following statements are equivalent.*

(a) *A is representation-tame.*

(b) *A is of linear growth.*

(c) *The Tits form $\widehat{q}_B$ of B is weakly non-negative.*

(d) *$\operatorname{mod} B$ is controlled by $\widehat{q}_B$.*

(e) *$\dim_K \operatorname{Ext}^1_B(X, X) \leq \dim_K \operatorname{End}_B(X)$, for any indecomposable B-module X.*

(f) *Every component of $\Gamma(\operatorname{mod} B)$ is generalised standard.*

(g) *The infinite radical $\operatorname{rad}_B^\infty$ of $\operatorname{mod} B$ is locally nilpotent.*

(h) *The quasitilted algebras $B^{(l)}$ and $B^{(r)}$ are representation-tame.* $\square$

We note also the following consequence of the definition of a double tilted algebra, (3.4) and (XIX.3.17).

4.14. Corollary. *Let B be a double tilted algebra. The following conditions are equivalent.*

(a) *B is representation-tame.*

(b) *B is of finite growth (domestic).*

(c) *The radical $\operatorname{rad}_B^\infty$ is nilpotent.*

(d) *The tilted algebras $B^{(l)}$ and $B^{(r)}$ are representation-tame.* $\square$

The class of double tilted algebras is generalised in [491] to the class of generalised double tilted algebras, containing the class of all representation-finite algebras. This is done on the basis of the concept of a multisection.

4.15. Definition. Let A be an algebra. A full subquiver Σ of $\Gamma(\operatorname{mod} A)$ is said to be **almost acyclic** (or **almost directed**) if all but finitely many modules of Σ do not lie on oriented cycles in $\Gamma(\operatorname{mod} A)$.

4.16. Definition. Let A be an algebra and $\mathcal{C}$ a component of $\Gamma(\operatorname{mod} A)$. A full connected subquiver Δ of $\mathcal{C}$ is said to be a **multisection** of $\mathcal{C}$ if the following conditions are satisfied:

(i) Δ is almost acyclic.

(ii) Δ is convex.

(iii) For each τ_A-orbit $\mathcal{O}$ in $\mathcal{C}$, we have $1 \leq |\Delta \cap \mathcal{O}| < \infty$.

(iv) $|\Delta \cap \mathcal{O}| = 1$, for all but finitely many τ_A-orbits $\mathcal{O}$ in $\mathcal{C}$.

(v) No proper full convex subquiver of Δ satisfies the conditions (i)–(iv).

4.17. Definition. Let A be an algebra, $\mathcal{C}$ a component of $\Gamma(\text{mod } A)$ and Δ a multisection of $\mathcal{C}$. We define the following full subquivers of $\mathcal{C}$:

(i) $\Delta'_l = \{X \in \Delta; \text{there is a nonsectional path } X \to \cdots \to P$
$$\text{with } P \text{ projective } \},$$

(ii) $\Delta'_r = \{X \in \Delta; \text{there is a nonsectional path } I \to \cdots \to X$
$$\text{with } I \text{ injective } \},$$

(iii) $\Delta''_l = \{X \in \Delta'_l; \tau_A^{-1} X \notin \Delta'_l\}, \qquad \Delta''_r = \{X \in \Delta'_r; \tau_A X \notin \Delta'_r\},$

(iv) $\Delta_l = (\Delta \setminus \Delta'_r) \cup \tau_A \Delta''_r, \quad \Delta_c = \Delta'_l \cap \Delta'_r, \quad \Delta_r = (\Delta \setminus \Delta'_l) \cup \tau_A^{-1} \Delta''_l.$

Then Δ_l is said to be the **left part** of Δ, Δ_r the **right part** of Δ and Δ_c the **core** of Δ.

The following two theorems describe basic properties of multisections.

4.18. Theorem. *Let A be an algebra. A component $\mathcal{C}$ of $\Gamma(\text{mod } A)$ is almost acyclic if and only if $\mathcal{C}$ admits a multisection Δ.* $\square$

4.19. Theorem. *Let A be an algebra, $\mathcal{C}$ a component of $\Gamma(\text{mod } A)$ and Δ a multisection of $\mathcal{C}$.*

(a) *Every cycle of $\mathcal{C}$ lies in Δ_c.*

(b) *Δ_c is finite.*

(c) *Every indecomposable module X in $\mathcal{C}$ is in Δ_c, or a predecessor of Δ_l or a successor of Δ_r in $\mathcal{C}$.*

(d) *Δ is faithful if and only if $\mathcal{C}$ is faithful.* $\square$

Moreover, in [491] a numerical invariant $w(\Delta) \in \mathbb{N} \cup \{\infty\}$ of a multisection Δ, called the **width** of Δ, is introduced. Then a multisection Δ with $w(\Delta) = n$, is called an n-**section**. Moreover, for a multisection Δ of a component $\mathcal{C}$ of $\Gamma(\text{mod } A)$, we have

- $w(\Delta) < \infty$ if and only if Δ is acyclic.
- Δ is a 1-section if and only if Δ is a section.
- Δ is a 2-section if and only if Δ is a strict double section.

The following facts also hold.

4.20. Proposition. *Let A be an algebra, $\mathcal{C}$ a component of $\Gamma(\text{mod } A)$ and Δ, Σ are multisections of $\mathcal{C}$. Then $\Delta_c = \Sigma_c$ and $w(\Delta) = w(\Sigma)$.* $\square$

4.21. Definition. Let B be a connected algebra.

(a) B is said to be a **generalised double tilted algebra** if the following conditions are satisfied:

(a1) $\Gamma(\operatorname{mod} B)$ admits a component $\mathcal{C}$ with a faithful multisection Δ.

(a2) There exists a tilted quotient algebra $B^{(l)}$ of B (not necessarily connected) such that Δ_l is a disjoint union of sections of the connecting components of the connected parts of $B^{(l)}$ and the category of all predecessors of Δ_l in $\operatorname{ind} B$ coincides with the category of all predecessors of Δ_l in $\operatorname{ind} B^{(l)}$.

(a3) There exists a tilted quotient algebra $B^{(r)}$ of B (not necessarily connected) such that Δ_r is a disjoint union of sections of the connecting components of the connected parts of $B^{(r)}$, and the category of all successors of Δ_r in $\operatorname{ind} B$ coincides with the category of all successors of Δ_r in $\operatorname{ind} B^{(r)}$.

(b) B is said to be an n-**double tilted algebra** if $\Gamma(\operatorname{mod} B)$ admits a component $\mathcal{C}$ with a faithful n-section Δ and the conditions (a2) and (a3) hold.

The following theorem gives information on the global dimension of generalised double tilted algebras [491].

4.22. Theorem. *If B is an n-double tilted algebra then* $\operatorname{gl.dim} B \leq n{+}1$. □

For each $n \in \mathbb{N} \cup \{\infty\}$, there are many examples of n-double tilted algebras of global dimension $n + 1$. The following theorem is the main result of [491].

4.23. Theorem. *Let B be a connected algebra. The following conditions are equivalent:*

(a) *B is a generalised double tilted algebra.*

(b) *The quiver $\Gamma(\operatorname{mod} B)$ admits a component $\mathcal{C}$ with a faithful multisection Δ such that $\operatorname{Hom}_B(U, \tau_B V) = 0$, for all modules $U \in \Delta_r$ and $V \in \Delta_l$.*

(c) *The quiver $\Gamma(\operatorname{mod} B)$ admits a faithful generalised standard almost cyclic component.* □

Moreover, we have the following consequences of the above theorem.

4.24. Corollary. *Let B be a connected algebra.*

(a) *B is an n-double tilted algebra, for some $n \geq 2$, if and only if $\Gamma(\operatorname{mod} B)$ contains a faithful generalised standard almost cyclic component $\mathcal{C}$ with a nonsectional path from an injective module to a projective module.*

(b) *B is an n-double tilted algebra, for some $n \geq 3$, if and only if $\Gamma(\operatorname{mod} B)$ admits a faithful generalised standard component $\mathcal{C}$ with a multisection Δ such that $\Delta_c \neq \emptyset$.* □

We note also the following consequence of the definition of a generalised double tilted algebra, (3.4), (4.19) and (XIX.3.17).

4.25. Corollary. *Let B be a generalised double tilted algebra. The following conditions are equivalent:*

(a) *B is representation-tame.*
(b) *B is of finite growth (domestic).*
(c) *The radical rad_B^∞ is nilpotent.*
(d) *The tilted algebras $B^{(l)}$ and $B^{(r)}$ are representation-tame.* □

Let A be an algebra and $\mathcal{C}$ a component of $\Gamma(\mathrm{mod}\,A)$. By analogy to (3.9), we denote by $\mathcal{L}_\mathcal{C}$ the set of all modules X in $\mathcal{C}$ such that $\mathrm{pd}_A Y \leq 1$, for any predecessor Y of X in $\mathcal{C}$, and by $\mathcal{R}_\mathcal{C}$ the set of all modules X in $\mathcal{C}$ such that $\mathrm{id}_A Y \leq 1$, for any successor Y of X in $\mathcal{C}$. Observe that $\mathcal{L}_\mathcal{C}$ is closed under predecessors in $\mathcal{C}$ and $\mathcal{R}_\mathcal{C}$ is closed under successors in $\mathcal{C}$. Moreover, if Δ is a multisection of $\mathcal{C}$, then $\Delta_c \subseteq \mathcal{C} \setminus (\mathcal{L}_\mathcal{C} \cup \mathcal{R}_\mathcal{C})$. The following theorem is also proved in [491].

4.26. Theorem. *Let B be a connected algebra, $\mathcal{C}$ a faithful component of the quiver $\Gamma(\mathrm{mod}\,B)$ with a multisection Δ, and $\mathcal{C}$ is not semiregular. Then the component $\mathcal{C}$ is generalised standard if and only if $\mathcal{C} = \mathcal{L}_\mathcal{C} \cup \Delta_c \cup \mathcal{R}_\mathcal{C}$.*
□

We note that, by (4.24), every n-double tilted algebra B, with $n \geq 2$, $\Gamma(\mathrm{mod}\,B)$ admits a unique (connecting) component $\mathcal{C}$ with a faithful multisection and this component is not semiregular.

The following theorem of Skowroński in [628] provides a common homological characterisation of quasitilted and generalised double tilted algebras.

4.27. Theorem. *For any connected algebra B, the following two conditions are equivalent.*

(a) *B is either a generalised double tilted algebra or a quasitilted algebra.*
(b) *The set $\mathrm{ind}\,B \setminus (\mathcal{L}_B \cup \mathcal{R}_B)$ is finite.* □

The class of generalised double tilted algebras has been also investigated by Assem–Coelho [8] and Coelho–Lanzilotta [140], [141].

There is an open problem whether a connected algebra B is generalised double tilted or quasitilted if $\mathrm{pd}_B X \leq 1$ or $\mathrm{id}_B X \leq 1$, for all but finitely many indecomposable B-modules X.

An interesting class of algebras of global dimension at most 3, related with quasitilted algebras, is investigated by Malicki and Skowroński [426]. It follows from (1.9) that a component $\mathcal{C}$ of an Auslander–Reiten quiver $\Gamma(\mathrm{mod}\,A)$ is almost cyclic and coherent if and only if $\mathcal{C}$ is a generalised multicoil, that is, $\mathcal{C}$ can be obtained from a finite family of stable tubes by a sequence of admissible operations.

In [426], the authors introduce the concept of a **generalised multicoil enlargement** of a finite family of concealed canonical algebras, by extending the concept of the coil enlargement of a concealed algebra introduced in [21].

One of the main results of [426] is the following theorem.

4.28. Theorem. *Let B be a connected algebra. The quiver $\Gamma(\mathrm{mod}\,B)$ admits a sincere separating family of almost cyclic coherent components if and only if B is a generalised multicoil enlargement of a product C of concealed canonical algebras.* $\square$

The following theorem describes the structure of the module category of an algebra with a sincere separating family of almost cyclic coherent components.

4.29. Theorem. *Let B be an algebra with a sincere separating family $\mathcal{C}^B$ of almost cyclic coherent components in the Auslander–Reiten quiver $\Gamma(\mathrm{mod}\,B)$ of B. Then the quiver $\Gamma(\mathrm{mod}\,B)$ has the disjoint union decomposition*

$$\Gamma(\mathrm{mod}\,B) = \mathcal{P}^B \cup \mathcal{C}^B \cup \mathcal{Q}^B$$

where $\mathcal{C}^B$ separates $\mathcal{P}^B$ from $\mathcal{Q}^B$, and the families $\mathcal{P}^B$ and $\mathcal{Q}^B$ are described as follows.

  (a) *There is a unique quotient algebra $B^{(l)}$ of B which is a product $B^{(l)} = B_1^{(l)} \times \cdots \times B_m^{(l)}$ of quasitilted algebras $B_1^{(l)}, \ldots, B_m^{(l)}$ of canonical type such that, for each $i \in \{1, \ldots, m\}$, we have*

$$\mathcal{P}^B = \mathcal{P}^{B^{(l)}} = \mathcal{P}^{B_1^{(l)}} \cup \cdots \cup \mathcal{P}^{B_m^{(l)}} \text{ and}$$
$$\Gamma(\mathrm{mod}\,B_i^{(l)}) = \mathcal{P}^{B_i^{(l)}} \cup \mathcal{T}^{B_i^{(l)}} \cup \mathcal{Q}^{B_i^{(l)}},$$

  where $\mathcal{T}^{B_i^{(l)}}$ is a $\mathbb{P}_1(K)$-family of pairwise orthogonal standard coray tubes separating $\mathcal{P}^{B_i^{(l)}}$ from $\mathcal{Q}^{B_i^{(l)}}$.

  (b) *There is a unique quotient algebra $B^{(r)}$ of B which is a product $B^{(r)} = B_1^{(r)} \times \cdots \times B_n^{(r)}$ of quasitilted algebras $B_1^{(r)}, \ldots, B_n^{(r)}$ of canonical type such that, for each $j \in \{1, \ldots, n\}$, we have*

$$\mathcal{Q}^B = \mathcal{Q}^{B^{(r)}} = \mathcal{Q}^{B_1^{(r)}} \cup \cdots \cup \mathcal{Q}^{B_n^{(r)}} \text{ and}$$
$$\Gamma(\mathrm{mod}\,B_j^{(r)}) = \mathcal{P}^{B_j^{(r)}} \cup \mathcal{T}^{B_j^{(r)}} \cup \mathcal{Q}^{B_j^{(r)}},$$

  where $\mathcal{T}^{B_j^{(r)}}$ is a $\mathbb{P}_1(K)$-family of pairwise orthogonal standard ray tubes separating $\mathcal{P}^{B_j^{(r)}}$ from $\mathcal{Q}^{B_j^{(r)}}$.

Moreover, the quiver $\Gamma(\mathrm{mod}\,B)$ contains exactly $m \geq 1$ postprojective components $\mathcal{P}(B_1^{(l)}), \ldots, \mathcal{P}(B_m^{(l)})$ and exactly $n \geq 1$ preinjective components $\mathcal{Q}(B_1^{(r)}), \ldots, \mathcal{Q}(B_n^{(r)})$. $\square$

The algebra $B^{(l)}$ is called the **left quasitilted part** of B and $B^{(r)}$ the **right quasitilted part** of B.

The next theorem describes the homological properties of algebras with sincere separating families of almost cyclic coherent components.

4.30. Theorem. *Assume that B is an algebra such that the Auslander–Reiten translation quiver $\Gamma(\mathrm{mod}\, B)$ of B admits the decomposition (4.29)*

$$\Gamma(\mathrm{mod}\, B) = \boldsymbol{P}^B \cup \boldsymbol{C}^B \cup \boldsymbol{Q}^B,$$

where $\boldsymbol{C}^B$ is a sincere separating family of almost cyclic coherent components.

  (a) *B is a triangular algebra.*
  (b) *$\mathrm{pd}_B X \leq 1$, for any module $X \in \boldsymbol{P}^B$.*
  (c) *$\mathrm{id}_B X \leq 1$, for any module X in $\boldsymbol{Q}^B$.*
  (d) *$\mathrm{pd}_B X \leq 2$ and $\mathrm{id}_B X \leq 2$, for any module $X \in \boldsymbol{C}^B$.*
  (e) *$\mathrm{gl.dim}\, B \leq 3$.* $\qquad\square$

The following theorem follows from the main results of [426] and [428].

4.31. Theorem. *Let B be an algebra with a sincere separating family of almost cyclic coherent components in $\Gamma(\mathrm{mod}\, B)$. The following statements are equivalent.*

  (a) *B is representation-tame.*
  (b) *B is of linear growth.*
  (c) *The Tits quadratic form $\widehat{q}_B : K_0(B) \longrightarrow \mathbb{Z}$ of B is weakly nonnegative.*
  (d) *$\widehat{q}_B(\mathbf{dim}\, X) \geq q_B(\mathbf{dim}\, X) \geq 0$, for any indecomposable B-module X.*
  (e) *$\dim_K \mathrm{Ext}_B^1(X, X) \leq \dim_K \mathrm{End}_B(X)$, and $\mathrm{Ext}_B^2(X, X) = 0$, for any indecomposable B-module X.*
  (f) *Every component of $\Gamma(\mathrm{mod}\, B)$ is generalised standard.*
  (g) *The infinite radical rad_B^∞ of $\mathrm{mod}\, B$ is locally nilpotent.*
  (h) *Each of the algebras $B^{(l)}$ and $B^{(r)}$ is a finite product of tilted algebras of Euclidean types and tubular algebras.* $\qquad\square$

It follows from (3.28) that the Auslander–Reiten quiver of a quasitilted algebra admits a postprojective component and a preinjective component. However, there is an open problem of deciding when the Auslander–Reiten quiver of an algebra A of global dimension 2 admits a postprojective (respectively, preinjective) component. We refer to [134], [353], [354], [356], [386] for some results in this direction.

XX.5. Selfinjective algebras of tilted and quasitilted type

A prominent role in the representation theory of algebras is played by the selfinjective algebras, for which the projective modules are injective. Classical examples of selfinjective algebras are provided by the group algebras of finite groups, the Hecke algebras of classical type, the restricted enveloping algebras of restricted Lie algebras, the finite dimensional Hopf algebras and the Hochschild extensions of algebras. The wide class of selfinjective algebras is formed by the **Frobenius algebras** A for which there exists a nondegenerate K-bilinear form

$$(-,-) : A \times A \longrightarrow K$$

satisfying the associativity condition $(ab, c) = (a, bc)$, for all $a, b, c \in A$. If the bilinear form $(-,-)$ is symmetric the algebra A is said to be **symmetric**. Moreover, every basic selfinjective algebra A is a Frobenius algebra. We refer also to [644] for a general form of non-Frobenius selfinjective algebras.

The main objective of this section is to indicate the importance of the tilted algebras and quasitilted algebras for the representation theory of selfinjective algebras.

The selfinjective algebras are, with the exception of semisimple cases, of infinite global dimension, their quivers are not acyclic, and this complicates the representation theory of these algebras significantly. But frequently, the selfinjective algebras are deformations of selfinjective algebras having triangular Galois coverings, and then the study of such algebras and their module categories may be reduced to that for the corresponding algebras of finite global dimension. We refer to [99], [185], [187], [238] for the covering techniques, see also [177], [178], [179], [180], and [181]. In the theory a crucial role is played by the Galois coverings by the repetitive (locally bounded) categories introduced in [311], see also [600].

5.1. Definition. Let B be a basic connected algebra and $1 = e_1 + \ldots + e_n$ a decomposition of the identity 1 of B into the sum of orthogonal primitive idempotents.

(a) The **repetitive category** $\widehat{B}$ of B is the category with the objects $e_{m,i}$, $(m, i) \in \mathbb{Z} \times \{1, \ldots, n\}$, and the morphism spaces

$$\widehat{B}(e_{m,i}, e_{r,j}) = \begin{cases} e_j B e_i, & \text{if } r = m, \\ D(e_i B e_j), & \text{if } r = m + 1, \\ 0, & \text{otherwise.} \end{cases}$$

(b) A group G of K-linear automorphisms of the category $\widehat{B}$ is said to be **admissible** if G acts freely on the objects of $\widehat{B}$ (that is, $g \cdot e_{m,i} = e_{m,i}$ forces $g = 1$) and has finitely many orbits.

(c) The **orbit category** $\widehat{B}/G$, for an admissible group G of automor-
    phisms of $\widehat{B}$, has a natural structure of a basic connected finite
    dimensional selfinjective K-algebra, called the **orbit algebra** of $\widehat{B}$
    with respect to G. Moreover, there is the induced **Galois covering**
    $F : \widehat{B} \to \widehat{B}/G$, see [238] for details.
(d) The **Nakayama automorphism** $\nu_{\widehat{B}}$ of $\widehat{B}$ is defined by $\nu_{\widehat{B}}(e_{m,i}) = e_{m+1,i}$, for all $(m, i) \in \mathbb{Z} \times \{1, \ldots, n\}$.
(e) The orbit algebra $T(B) = \widehat{B}/(\nu_{\widehat{B}})$ of $\widehat{B}$ with respect to the admis-
    sible infinite cyclic group $(\nu_{\widehat{B}})$ generated by $\nu_{\widehat{B}}$ is the trivial **exten-
    sion algebra** $B \ltimes D(B)$ of B by $D(B) = \mathrm{Hom}_K(B, K)$. This is a
    symmetric algebra with $B \ltimes D(B) = B \oplus D(B)$ as K-vector space
    and the multiplication is given by $(a, \varphi) \cdot (b, \psi) = (ab, a\psi + \varphi b)$, for
    $a, b \in B$ and $\varphi, \psi \in D(B)$.

Let B be a basic connected K-algebra, G an admissible group of automorphisms of $\widehat{B}$ and $A = \widehat{B}/G$ the associated selfinjective algebra. Following
[99] one associates the **push-down functor**

$$F_\lambda : \mathrm{mod}\, \widehat{B} \longrightarrow \mathrm{mod}\, A.$$

If G is torsion-free, then it follows from a theorem of Gabriel [238] that F_λ
induces an injection from the set of G-orbits of isomorphism classes of indecomposable finite dimensional $\widehat{B}$-modules into the set of the isomorphism
classes of indecomposable finite dimensional A-modules, and preserves the
almost split sequences. Moreover, if $\widehat{B}$ is **locally support-finite** (in the
sense of [185]) then, by the density theorem of Dowbor and Skowroński [185]
(or [187]) the functor F_λ is dense.

Combining results of the papers [9], [217], [311], [402], [449], and [609]
one obtain the following.

5.2. Theorem. *Let B be a basic connected quasitilted algebra.*

(a) *The repetitive category $\widehat{B}$ is locally support-finite.*
(b) *Every admissible torsion-free group of automorphisms of $\widehat{B}$ is infi-
    nite cyclic.*
(c) *For an admissible group G of automorphisms of $\widehat{B}$ and $A = \widehat{B}/G$,
    we have*
    - *the push-down functor $F_\lambda : \mathrm{mod}\, \widehat{B} \longrightarrow \mathrm{mod}\, A$ is dense and is
      a Galois covering of categories,*
    - *the Auslander–Reiten quiver $\Gamma(\mathrm{mod}\, A)$ of A coincides with the
      orbit quiver $\Gamma(\mathrm{mod}\, \widehat{B})/G$ of $\Gamma(\mathrm{mod}\, \widehat{B})$.* $\square$

5.3. Definition. Let B be a basic connected algebra and G an admissible infinite cyclic group of automorphisms of $\widehat{B}$.

(a) An orbit algebra $\widehat{B}/G$ with B a quasitilted algebra is said to be a **selfinjective algebra of quasitilted type**.

(b) An orbit algebra $\widehat{B}/G$ with B a tilted algebra is said to be a **selfinjective algebra of tilted type**.

(c) An orbit algebra $\widehat{B}/G$ with B a tilted algebra of Dynkin type is said to be a **selfinjective algebra of Dynkin type**.

(d) An orbit algebra $\widehat{B}/G$ with B a tilted algebra of Euclidean type is said to be a **selfinjective algebra of Euclidean type**.

(e) An orbit algebra $\widehat{B}/G$ with B a tilted algebra of wild type is said to be a **selfinjective algebra of wild tilted type**.

(f) An orbit algebra $\widehat{B}/G$ with B a tubular algebra is said to be a **selfinjective algebra of tubular type**.

(g) An orbit algebra $\widehat{B}/G$ with B a quasitilted algebra of wild canonical type is said to be a **selfinjective algebra of wild canonical type**.

To describe all basic connected selfinjective algebras of polynomial growth, we need an additional concept.

Let A be a non-semisimple connected selfinjective algebra. It is easy to see that the left socle soc $(_A A)$ and the right socle soc (A_A) of A coincide, and we set soc $A = $ soc $(_A A) = $ soc (A_A). A selfinjective algebra Λ is defined to be a **socle deformation** of A if the quotient algebras $A/\text{soc}\,A$ and $\Lambda/\text{soc}\,\Lambda$ are isomorphic. It follows from (IV.3.11) that, for any indecomposable projective (hence injective) A-module P, there is an almost split sequence in mod A of the form

$$0 \longrightarrow \text{rad}\,P \longrightarrow \text{rad}\,P/\text{soc}\,P \oplus P \longrightarrow P/\text{soc}\,P \longrightarrow 0.$$

Therefore, we may recover the Auslander–Reiten quiver $\Gamma(\text{mod}\,A)$ from the quiver $\Gamma(\text{mod}\,A/\text{soc}\,A)$ if we know the positions of the modules $P/\text{soc}\,P$ in $\Gamma(\text{mod}\,A/\text{soc}\,A)$. In particular, the socle deformations of selfinjective algebras A and Λ give very close relationship between their module categories mod A and mod Λ.

The following theorem, due to Riedtmann [499], [500], [501] and Waschbüsch [670], [671] (see also [112], [311]), gives a description of the representation-finite selfinjective algebras.

5.4. Theorem. *Let Λ be a non-simple basic connected selfinjective algebra. Then Λ is representation-finite if and only if Λ is a socle deformation of a selfinjective algebra A of Dynkin type.* $\qquad\square$

We also note that the non-trivial socle deformations of selfinjective algebras of Dynkin type occur only in characteristic 2 and are classified completely by quivers and relations (see [501], [670]). We refer to the survey article [630] for a detailed description of the representation-finite selfinjective algebras, and to [7] for a classification of the derived equivalence classes of these algebras.

The following theorem of Skowroński [631] describes the representation-infinite selfinjective algebras of finite growth (domestic).

5.5. Theorem. *Let Λ be a basic connected selfinjective algebra. Then Λ is representation-infinite domestic (of finite growth) if and only if Λ is a socle deformation of a selfinjective algebra A of Euclidean type.* □

The non-trivial socle deformations of the selfinjective algebras of Euclidean type occur in any characteristic and are classified completely by Bocian and Skowroński in [83] in terms of quivers and relations. The classification of all selfinjective algebras of Euclidean type, initiated in [9], [12], [13], [14], [609], is completed by Bocian and Skowroński in [80], [81], [82], [83], and Lenzing and Skowroński in [399].

The following theorem of Skowroński [632] describes all non-domestic selfinjective algebras of polynomial growth.

5.6. Theorem. *For a basic connected selfinjective algebra Λ, the following conditions are equivalent.*

(a) *Λ is non-domestic of polynomial growth.*

(b) *Λ is non-domestic of linear growth.*

(c) *Λ is a socle deformation of a selfinjective algebra A of tubular type.*
□

The non-trivial socle deformations of selfinjective algebras of tubular type occur only in characteristic 2 and 3 and are classified completely by Białkowski and Skowroński in [64] in terms of quivers and relations. The classification of all selfinjective algebras of tubular type, initiated in [14], [285], [449], [609], is completed recently by Białkowski and Skowroński in [62], [63], [64], and Lenzing and Skowroński in [400] (see also [56], [57]).

The following theorem of Skowroński [633] gives a characterisation of selfinjective algebras of polynomial growth in terms of the infinite radical of the module category.

5.7. Theorem. *Assume that Λ is a selfinjective algebra.*

(a) *Λ is of polynomial growth if and only if the infinite radical rad_B^∞ of $\mathrm{mod}\,B$ is locally nilpotent.*

(b) *Λ is domestic (of finite growth) if and only if the infinite radical rad_B^∞ of $\mathrm{mod}\,B$ is nilpotent.* □

We refer to the survey article [630] for more information on selfinjective algebras of polynomial growth and their module categories. Moreover, we refer to [77], [78], [79], [296] (respectively, to [60], [61], [481]) for some results concerning the derived equivalence classification of selfinjective algebras of domestic type (respectively, non-domestic selfinjective algebras of polynomial growth).

The representation theory of arbitrary representation-tame selfinjective algebras is still only emerging. We refer to [1], [211], [219], [220], and [295] for a classification of representation-tame blocks of group algebras of finite groups and related algebras, and to [496] and [295] for the derived equivalence classification of these algebras B. We present only the following result in this direction, invoking the Euler quadratic form q_B, proved by Skowroński in [634].

5.8. Theorem. *For a simply connected algebra B, the following four conditions are equivalent.*

   (a) *The Euler form $q_B : K_0(B) \longrightarrow \mathbb{Z}$ of B is positive semidefinite.*

   (b) *The trivial extension algebra $\mathrm{T}(B)$ is representation-tame.*

   (c) *The derived category $D^b(\mathrm{mod}\,B)$ is representation-tame.*

   (d) *The algebra B is derived equivalent to a representation-tame simply connected generalised canonical algebra.* $\qquad\square$

We refer to Geiss and Krause [253] for the notion of derived tameness, to Drozd [204] for the derived tame-wild dichotomy of algebras, and to Leszczyński and Skowroński [411] for a classification of all representation-tame generalised canonical algebras.

The module categories of the selfinjective algebras of wild tilted type and wild canonical type, as well as their homological invariants, are described by Erdmann–Kerner–Skowroński [217] and Lenzing–Skowroński [402].

The remaining part of this section is devoted to a discussion of the structure of selfinjective algebras having a generalised standard component in the Auslander–Reiten quiver. We give first a criterion for a selfinjective algebra A to be the orbit algebra $\widehat{B}/G$ of the repetitive algebra $\widehat{B}$ of an algebra B. To present it, we need some concepts.

5.9. Definition. Let B be a basic connected algebra and $\varphi : \widehat{B} \longrightarrow \widehat{B}$ a K-category automorphism of $\widehat{B}$. In the notation of (5.1), the automorphism φ is defined to be

- **positive** if, for each pair $(m, i) \in \mathbb{Z} \times \{1, \ldots, n\}$, we have $\varphi(e_{m,i}) = e_{p,j}$, for some $p \geq m$ and $j \in \{1, \ldots, n\}$.
- **rigid** if, for each pair $(m, i) \in \mathbb{Z} \times \{1, \ldots, n\}$, there exists $j \in \{1, \ldots, n\}$ such that $\varphi(e_{m,i}) = e_{m,j}$.
- **strictly positive** if φ is positive but not rigid.

The following criterion is established by Skowroński and Yamagata in [642], see also [637], [639].

5.10. Theorem. *Let A be a basic connected selfinjective algebra. Then A is isomorphic to an orbit algebra $\widehat{B}/(\varphi\nu_{\widehat{B}})$, for a basic connected algebra B and a positive automorphism $\varphi : \widehat{B} \longrightarrow \widehat{B}$ of $\widehat{B}$, if and only if there is an ideal I of A such that, for some idempotent e of A, the following two conditions are satisfied:*

(i) $\mathrm{Ann}_A (I_A) = eI$,

(ii) *the canonical algebra epimorphism $eAe \rightarrow eAe/eIe$ splits.*

Moreover, in this case, B is isomorphic to the quotient algebra A/I. $\qquad\square$

The criterion (5.10) is the main tool in proving the following result obtained in [637], [639], see also [640].

5.11. Theorem. *Let A be a basic connected selfinjective algebra. The following conditions are equivalent.*

(a) *A is isomorphic to an orbit algebra $\widehat{B}/(\varphi\nu_{\widehat{B}})$, where B is a tilted algebra not of Dynkin type and $\varphi : \widehat{B} \longrightarrow \widehat{B}$ is a positive automorphism of $\widehat{B}$.*

(b) *$\Gamma(\mathrm{mod}\, A)$ admits a generalised standard acyclic full translation subquiver which is closed under predecessors in $\Gamma(\mathrm{mod}\, A)$.*

(c) *$\Gamma(\mathrm{mod}\, A)$ admits a generalised standard acyclic full translation subquiver which is closed under successors in $\Gamma(\mathrm{mod}\, A)$.* $\qquad\square$

As a direct consequence we obtain the following fact.

5.12. Corollary. *Let A be a basic connected selfinjective algebra such that $\Gamma(\mathrm{mod}\, A)$ admits a generalised standard acyclic component. Then A is a selfinjective algebra of tilted type.* $\qquad\square$

Applying the criterion (5.10), Skowroński and Yamagata prove in [645] the following fact.

5.13. Theorem. *For a basic connected algebra A, the following two statements are equivalent.*

(a) *A is representation-infinite and every component of $\Gamma(\mathrm{mod}\, A)$ is generalised standard.*

(b) *The algebra A is isomorphic to an orbit algebra $\widehat{B}/(\varphi\nu_{\widehat{B}})$, where B is a (representation-infinite) tilted algebra of Euclidean type or a tubular algebra, and $\varphi : \widehat{B} \longrightarrow \widehat{B}$ is a strictly positive automorphism of $\widehat{B}$.* $\qquad\square$

We would like to mention that a description of the selfinjective algebras A such that $\Gamma(\mathrm{mod}\, A)$ admits a (generalised) standard stable tube is a difficult problem. It is not true that such an algebra A is selfinjective of quasitilted type. We already indicated in (1.21) and (1.22) that there exist complicated algebras B for which the quiver $\Gamma(\mathrm{mod}\,B)$ admits a faithful (generalised) standard stable tube. The Auslander–Reiten quiver $\Gamma(\mathrm{mod}\,B \ltimes D(B))$ of the trivial extension $B \ltimes D(B)$ of such an algebra B admits also a sincere (generalised) standard stable tube (see [427], [629]). The following interesting result is recently proved by Białkowski, Skowroński and Yamagata in [66].

5.14. Theorem. *If A is a symmetric algebra such that the Auslander–Reiten quiver $\Gamma(\mathrm{mod}\,A)$ admits a (generalised) standard stable tube then the Cartan matrix C_A of A is singular.* $\square$

We end this section with two theorems on the invariance of some classes of selfinjective algebras of tilted and quasitilted type under the stable equivalences.

Let A be a selfinjective algebra. We denote by $\underline{\mathrm{mod}}\,A$ the **stable module category** of A. The objects of $\underline{\mathrm{mod}}\,A$ are the objects of $\mathrm{mod}\,A$ without non-zero projective direct summands, and, for any two objects M and N of $\mathrm{mod}\,A$, the K-space $\underline{\mathrm{Hom}}_A(M,N)$ of morphisms from M to N in $\underline{\mathrm{mod}}\,A$ is the quotient $\mathrm{Hom}_A(M,N)/\mathcal{P}_A(M,N)$, where $\mathcal{P}_A(M,N)$ is the subspace of $\mathrm{Hom}_A(M,N)$ consisting of all homomorphisms that admit a factorisation through a projective A-module.

Two selfinjective algebras A and Λ are defined to be **stably equivalent** if the stable module categories $\underline{\mathrm{mod}}\,A$ and $\underline{\mathrm{mod}}\,\Lambda$ are equivalent.

In [496], Rickard proves that two derived equivalent selfinjective algebras are stably equivalent.

A combination of the results proved by Skowroński and Yamagata in [638], [641], and [643] gives the following theorem.

5.15. Theorem. *The class of selfinjective algebras of tilted type of the form $\widehat{B}/G$, where B is a tilted algebra and G is an infinite cyclic group $(\varphi\nu_{\widehat{B}})$, with a positive automorphism $\varphi : \widehat{B} \longrightarrow \widehat{B}$ of $\widehat{B}$, is invariant under the stable equivalences and under the derived equivalences.* $\square$

We finish this section with the following theorem proved recently by Kerner, Skowroński and Yamagata in [355].

5.16. Theorem. *The class of selfinjective algebras of quasitilted type of the form $\widehat{B}/G$, where B is a quasitilted algebra and G is an infinite cyclic group $(\varphi\nu_{\widehat{B}})$, with a strictly positive automorphism $\varphi : \widehat{B} \longrightarrow \widehat{B}$ of $\widehat{B}$, is invariant under the stable equivalences and under the derived equivalences.* $\square$

XX.6. Related topics and research directions

We list here some of the interesting research directions related to the modern representation theory of finite dimensional algebras. We also list a number of references for each of the topics. Unfortunately our list is far from being complete; we mention only some of the representative references for each of the topics.

- Geometry of module varieties is studied in [40], [67], [68], [71], [72], [73], [92], [96], [156], [157], [158], [163], [254], [326], [373], [457], [538], [560], [564], [661], [662], [695], [697].
- Orbit closures of modules and singularities are studied in [52], [74], [75], [76], [93], [94], [96], [98], [502], [506], [650], [653], [654], [655], [657], [694], [696], [697], [698], [699], [700], [701], [702], [703].
- Semi-invariants of modules and their zero sets are studied in [167], [168], [174], [505], [508], [509], [510], [518], [559], [561], [648], [649].
- Moduli spaces and noncommutative geometry are studied in [159], [175], [252], [360], [439], [440].
- Derived and triangulated categories are studied in [14], [47], [48], [70], [253], [275], [276], [285], [295], [296], [309], [334], [335], [336], [338], [340], [341], [390], [393], [394], [398], [400], [401], [434], [437], [471], [495], [496], [498], [555].
- Ringel-Hall algebras and Lie algebras are studied in [3], [133], [164], [305], [307], [368], [369], [370], [413], [473], [474], [503], [504], [529], [530], [547].
- Quantum groups and Hopf algebras are studied in [3], [115], [133], [214], [267], [485], [486], [529], [530].
- Cluster algebras and cluster tilting theory are studied in [124], [125], [130], [131], [230], [231], [254].
- Hochschild cohomologies of algebras are studied in [126], [127], [128], [221], [243], [256], [266], [277], [337], [429], [435], [476], [611], [661].
- Coverings of algebras and module categories are studied in [18], [33], [99], [177], [178], [179], [180], [181], [185], [187], [225], [226], [237], [263], [264], [325], [432], [573], [574], [576], [639], [642].
- Geometry of algebras and their deformations are studied in [154], [232], [236], [248], [256], [321], [322], [323], [324].
- Representations of Frobenius algebras are studied in [1], [4], [53], [54], [58], [59], [115], [165], [211], [212], [213], [215], [217], [219], [220], [223], [224], [225], [226], [227], [228], [241], [267], [295], [355], [402], [496], [497], [498], [630], [637], [639], [640], [642].
- Periodicity of modules and algebras are studied in [32], [58], [59], [220], [221], [222], [281], [303], [478].

- Applications of integral quadratic forms and their root systems in representation theory presented in [16], [38], [39], [88], [106], [119], [122], [172], [194], [197], [200], [201], [202], [297], [298], [299], [300], [301], [302], [320], [326]–[330], [388], [442], [446], [448], [452], [453], [454], [456], [457], [459]–[461], [462], [466], [467], [468], [514], [525], [575], [579], [584], [586], [589], [596], [597], [634], [682]–[684], [687].

- Representation theory of vector space categories and matrix problems is presented in the books [242] and [575], see also [46], [49], [50], [51], [85], [147]–[151], [198], [200]–[206], [302], [326]–[330], [365], [443]–[448], [462], [516], [517], [525], [552]–[554], [565], [572]–[582], [668], [682]–[686].

- For a combinatorial representation theory we refer to [169], [281], [282], [306], [320], [339], [388], [544], [545], [546], [550].

- For a spectral representation theory we refer to [158], [159], [173], [459], [469], [533].

- Module categories over artin algebras are studied in [34], [114], [142], [143], [244], [245], [280], [396], [420], [426], [492], [562], [563], [612], [613], [614], [615], [616], [617], [619], [628], [635], [640], [647], [656], [658], [660].

- Important classes of infinite dimensional modules over finite dimensional algebras are studied in [24], [110], [152], [153], [155], [203], [205], [206], [308], [374], [375], [378], [379], [381], [392], [482], [483], [488], [513], [531], [536], [540], [542], [543], [587], [627], [681].

- Hereditary abelian categories are studied in [278], [286], [288], [390], [494].

- Basic information on the representation theory of K-coalgebras over a field K can be found in [135], [136], [137], [316], [317], [269], [364], [371], [588], [589], [592]–[599], and [675].

- For a discussion of representation theory of artinian rings the reader is referred to [23], [24], [42], [182], [183], [184], [209], [268], [294], [308], [312], [319], [384], [514], [549], [556], [557], [558], [568], [569], [570], [571], [583], [585], [690], [691] (see also [115], [361], [362], [363], for a discussion of representation theory problems of noetherian rings that are not artinian).

Bibliography

[1] J. L. Alperin, *Local Representation Theory*, Cambridge Studies in Advanced Mathematics 11, Cambridge University Press, 1986.

[2] F. W. Anderson and K. R. Fuller, *Rings and Categories of Modules*, Graduate Texts in Mathematics 13, Springer-Verlag, New York, Heidelberg, Berlin, 1973 (new edition 1991).

[3] S. Ariki, Representations of quantum algebras and combinatorics of Young Tableaux, *University Lecture Series*, Vol. 26, American Math. Soc., 2002, pp. 1–158.

[4] S. Ariki, Hecke algebras of classical type and their representation type, *Proc. London Math. Soc.*, 91(2005), 355–413; Corrigendum 92(2006), 342–344.

[5] D. Arnold, *Abelian Groups and Representations of Finite Partially Ordered Sets*, CMS Books in Math., Springer, New York, 2000.

[6] D. Arnold and D. Simson, Endo-wild representation type and generic representations of finite posets, *Pacific J. Math.*, 219(2005), 101-126.

[7] H. Asashiba, The derived equivalence classification of representation-finite selfinjective algebras, *J. Algebra*, 214(1999), 182–221.

[8] I. Assem and F. U. Coelho, Two-sided gluings of tilted algebras, *J. Algebra*, 269(2003), 456–479.

[9] I. Assem, J. Nehring and A. Skowroński, Domestic trivial extensions of simply connected algebras, *Tsukuba J. Math.*, 13(1989), 31–72.

[10] I. Assem and J. A. de la Peña, The fundamental group of a triangular algebra, *Comm. Algebra,* 24(1996), 187–208.

[11] I. Assem, D. Simson and A. Skowroński, *Elements of the Representation Theory of Associative Algebras, Volume 1: Techniques of Representation Theory*, London Mathematical Society Student Texts 65, Cambridge University Press, 2006.

[12] I. Assem and A. Skowroński, Iterated tilted algebras of type $\widetilde{A}_n$, *Math. Z.*, 195(1987), 269–290.

[13] I. Assem and A. Skowroński, On some classes of simply connected algebras, *Proc. London Math. Soc.*, 56(1988), 417–450.

[14] I. Assem and A. Skowroński, Algebras with cycle-finite derived categories, *Math. Ann.*, 280(1988), 441–463.

[15] I. Assem and A. Skowroński, Algèbres pré-inclinées et catégories dérivées, In: *Séminaire d'Algèbre Paul Dubreil et Marie-Paul Malliavin*, Lecture Notes in Math., No. 1404, Springer-Verlag, Berlin, Heidelberg, New York, 1989, pp. 1–34.

[16] I. Assem and A. Skowroński, Quadratic forms and iterated tilted algebras, *J. Algebra*, 128(1990), 55–85.

[17] I. Assem and A. Skowroński, Multicoil algebras, In: *Representations of Algebras*, Canad. Math. Soc. Conf. Proc., AMS, Vol. 14, 1993, pp. 29–68.

[18] I. Assem and A. Skowroński, On tame repetitive algebras, *Fund. Math,* 142(1993), 59–84.

[19] I. Assem and A. Skowroński, Tilting simply connected algebras, *Comm. Algebra,* 22(1994), 4611–4619.

[20] I. Assem and A. Skowroński, Coil and multicoil algebras, In: *Representation Theory of Algebras and Related Topics*, Canad. Math. Soc. Conf. Proc., AMS, Vol. 19, 1996, pp. 1–24.

[21] I. Assem, A. Skowroński and B. Tomé, Coil enlargements of algebras, *Tsukuba J. Math.,* 19(1995), 453–479.

[22] M. Auslander, Representation dimension of artin algebras, Queen Mary College Mathematical Notes, London, 1971.

[23] M. Auslander, Representation theory of artin algebras II, *Comm. Algebra* 1(1974), 269–310.

[24] M. Auslander, Large modules over artin algebras, In: *Algebra, Topology and Category Theory*, Academic Press, New York, 1976, pp. 3–17.

[25] M. Auslander, Functors and morphisms determined by objects, In: *Proceedings, Conference on Representation Theory, Philadelphia*, 1976, Lecture Notes in Pure Appl. Math., Vol. 37, Marcel-Dekker, 1978, pp. 1–244.

[26] M. Auslander, Applications of morphisms determined by objects, In: *Proceedings, Conference on Representation Theory, Philadelphia*, 1976, Lecture Notes in Pure Appl. Math., Vol. 37, Marcel-Dekker, 1978, pp. 245–327.

[27] M. Auslander and M. Bridger, Stable module theory, *Memoirs Amer. Math. Soc.,* Vol. 94, 1969.

[28] M. Auslander, R. Bautista, M. Platzeck, I. Reiten and S. Smalø, Almost split sequences whose middle term has at most two indecomposable summands, *Canad. J. Math.*, 31(1979), 942–960.

[29] M. Auslander, M. I. Platzeck, and I. Reiten, Coxeter functors without diagrams, *Trans. Amer. Math. Soc.*, 250(1979), 1–46.

[30] M. Auslander and I. Reiten, On the representation type of triangular matrix algebras, *J. London Math. Soc.*, 12(1976), 371–382.

[31] M. Auslander and I. Reiten, Uniserial functors, In: *Representation Theory II*, Lecture Notes in Math. No. 832, Springer-Verlag, Berlin, Heidelberg, New York, 1980, pp. 1–47.

[32] M. Auslander and I. Reiten, DTr-periodic modules and functors, In: *Representation Theory of Algebras*, Canad. Math. Soc. Conf. Proc., AMS, Vol. 18, 1996, pp. 39–50.

[33] M. Auslander, I. Reiten and S. Smalø, Galois actions on rings and finite Galois coverings, *Math. Scand.*, 65(1989), 5–32.

[34] M. Auslander, I. Reiten and S. Smalø, *Representation Theory of Artin Algebras*, Cambridge Studies in Advanced Mathematics 36, Cambridge University Press, 1995.

[35] D. Baer, Wild hereditary algebras and linear methods, *Manuscr. Math.*, 55(1986), 69–82.

[36] D. Baer, A note on wild quiver algebras, *Comm. Algebra*, 17(1989), 751–757.

[37] M. Barot and H. Lenzing, Derived canonical algebras as one-point extensions, In: *Trends in Representation Theory of Finite Dimensional Algebras*, Contemp. Math., 229(1998), 7–15.

[38] M. Barot and J. A. de la Peña, The Dynkin type of a non-negative unit form, *Exposition Math.*, 17(1999), 339–348.

[39] M. Barot and J. A. de la Peña, Algebras whose Euler form is nonnegative, *Colloq. Math.*, 79(1999), 119–131.

[40] M. Barot and J. Schröer, Module varieties over canonical algebras, *J. Algebra*, 246(2001), 175–192.

[41] R. Bautista, On algebras of strongly unbounded representation type, *Comment. Math. Helvetici*, 60(1985), 392–399.

[42] R. Bautista, On some tame and discrete families of modules, In: *Infinite Length Modules*, (Eds: H. Krause and C. M. Ringel), Trends in Mathematics, Birkhäuser Verlag, Basel-Boston-Berlin, 2000, pp. 321–330.

[43] R. Bautista, P. Gabriel, A.V. Roiter and L. Salmerón, Representation-finite algebras and multiplicative basis, *Invent. Math.*, 81(1985), 217–285.

[44] R. Bautista and S. Smalø, Nonexistent cycles, *Comm. Algebra*, 11(1983), 1755–1767.

[45] R. Bautista and R. Zuazua, One-parameter families of modules for tame algebras and bocses, *Algebras and Representation Theory*, 8(2005), 635–677.

[46] V. Bekkert, A characterization of a class of schurian vector space categories of polynomial growth, In: *Representation Theory of Algebras*, Canad. Math. Soc. Conf. Proc., AMS, Vol. 18, 1996, pp. 109–140.

[47] A. Beligiannis, Relative homological algebra and purity in triangulated categories, *J. Algebra*, 227(2000), 268–361.

[48] A. Beligiannis, Auslander-Reiten triangles, Ziegler spectra and Gorenstein rings, *K-Theory*, 32(2004), 1–82.

[49] G. Belitskii, V. M. Bondarenko, R. Lipyanski, V. V. Plahotnik and V. V. Sergeichuk, The problems of classifying the pairs of forms and local algebras with zero cube radical are wild, *Linear Algebra Appl.*, 402(2005), 135–142.

[50] G. Belitskii, R. Lipyanski, and V. V. Sergeichuk, Problems of classifying associative or Lie algebras and triples of symmetric or skew-symmetric matrices are wild, *Linear Algebra Appl.*, 407(2005), 249–262.

[51] G. Belitskii and V. V. Sergeichuk, Complexity of matrix problems, *Linear Algebra Appl.*, 361(2003), 203–222.

[52] J. Bender and K. Bongartz, Minimal singularities in orbit closures of matrix pencils, *J. Algebra Appl.*, 365(2003), 13–24.

[53] D. J. Benson, *Representations and Cohomology I: Basic Representation Theory of Finite Groups and Associative Algebras*, Cambridge Studies in Advanced Mathematics 30, Cambridge University Press, 1991.

[54] D. J. Benson, *Representations and Cohomology II: Cohomology of Groups and Modules*, Cambridge Studies in Advanced Mathematics 31, Cambridge University Press, 1991.

[55] I. N. Bernstein, I. M. Gelfand and V. A. Ponomarev, Coxeter functors and Gabriel's theorem, *Uspiehi Mat. Nauk*, 28(1973), 19–33 (in Russian); English translation in *Russian Math. Surveys*, 28(1973), 17–32.

[56] J. Białkowski, Cartan matrices of selfinjective algebras of tubular type, *Centr. Eur. J. Math.*, 2(2004), 123–142 (electronic).

[57] J. Białkowski, On the trivial extensions of tubular algebras, *Colloq. Math.*, 101(2004), 259–269.

[58] J. Białkowski, K. Erdmann and A. Skowroński, Deformed preprojective algebras of generalized Dynkin type, *Trans. Amer. Math. Soc.*, 359(2007), 2625–2650.

[59] J. Białkowski, K. Erdmann and A. Skowroński, Selfinjective algebras of stable Calabi-Yau dimension two, Preprint, Toruń, 2007.

[60] J. Białkowski, T. Holm and A. Skowroński, Derived equivalences for tame weakly symmetric algebras having only periodic modules, *J. Algebra*, 269(2003), 652–668.

[61] J. Białkowski, T. Holm and A. Skowroński, On nonstandard tame selfinjective algebras having only periodic modules, *Colloq. Math.*, 97(2003), 33–47.

[62] J. Białkowski and A. Skowroński, Selfinjective algebras of tubular type, *Colloq. Math.*, 94(2002), 175–194.

[63] J. Białkowski and A. Skowroński, On tame weakly symmetric algebras having only periodic modules, *Archiv Math.* (*Basel*), 8(2003), 142–154.

[64] J. Białkowski and A. Skowroński, Socle deformations of selfinjective algebras of tubular type, *J. Math. Soc. Japan*, 56(2004), 687–716.

[65] J. Białkowski, and A. Skowroński, Calabi-Yau stable module categories of finite type, *Colloq. Math.* 2007, in press.

[66] J. Białkowski, A. Skowroński and K. Yamagata, Cartan matrices of symmetric algebras having generalized standard stable tubes, *Osaka J. Math.*, 2007, in press.

[67] G. Bobiński, Geometry of decomposable directing modules over tame algebras, *J. Math. Soc. Japan*, 54(2002), 609–620.

[68] G. Bobiński, Geometry of regular modules over canonical algebras, *Trans. Amer. Math. Soc.*, 2006, in press.

[69] G. Bobiński, P. Dräxler and A. Skowroński, Domestic algebras with many nonperiodic Auslander-Reiten components, *Comm. Algebra*, 31(2003), 1881–1926.

[70] G. Bobiński, C. Geiss and A. Skowroński, Classification of discrete derived categories, *Centr. Eur. J. Math.*, 2(2004), 19–49 (electronic).

[71] G. Bobiński and A. Skowroński, Geometry of directing modules over tame algebras, *J. Algebra*, 215(1999), 603–643.

[72] G. Bobiński and A. Skowroński, Geometry of modules over tame quasi-tilted algebras, *Colloq. Math.*, 79(1999), 85–118.

[73] G. Bobiński and A. Skowroński, Geometry of periodic modules over tame concealed and tubular algebras, *Algebras and Representation Theory*, 5(2002), 187–200.

[74] G. Bobiński and G. Zwara, Normality of orbit closures for Dynkin quivers, *Manuscr. Math.*, 105(2001), 103–109.

[75] G. Bobiński and G. Zwara, Schubert varieties and representations of Dynkin quivers, *Colloq. Math.*, 94(2002), 285–309.

[76] G. Bobiński and G. Zwara, Normality of orbit closures for directing modules over tame algebras, *J. Algebra*, 298(2006), 120–133.

[77] R. Bocian, T. Holm and A. Skowroński, Derived equivalence classification of weakly symmetric algebras of Euclidean type, *J. Pure Appl. Algebra,* 191(2004), 43–74.

[78] R. Bocian, T. Holm and A. Skowroński, Derived equivalence classification of nonstandard algebras of domestic type, *Comm. Algebra*, 35(2007), 1–12.

[79] R. Bocian, T. Holm and A. Skowroński, Derived equivalence classification of one-parametric selfinjective algebras, *J. Pure Appl. Algebra*, 207(2006), 491–536.

[80] R. Bocian and A. Skowroński, Symmetric special biserial algebras of Euclidean type, *Colloq. Math.,* 96(2003), 121–148.

[81] R. Bocian and A. Skowroński, Weakly symmetric algebras of Euclidean type, *J. reine angew. Math.*, 580(2005), 157–199.

[82] R. Bocian and A. Skowroński, One-parametric selfinjective algebras, *J. Math. Soc. Japan,* 57(2005), 491–512.

[83] R. Bocian and A. Skowroński, Socle deformations of selfinjective algebras of Euclidean type, *Comm. Algebra*, 34(2006), 4235–4257.

[84] V. M. Bondarenko, Representations of dihedral groups over a field of characteristic 2, *Math. Sbornik*, 96(1975), 63–74 (in Russian).

[85] V. M. Bondarenko, Representations of bundles of semichained sets and their applications, *Algebra i Analiz*, 3(1991), 38–61 (in Russian); English translation: *St. Petersburg Math. J.*, 3(1992), 973–996.

[86] V. M. Bondarenko and Ju. A. Drozd, The representation type of finite groups, In: *Modules and Representations, Zap. Nauchn. Sem. Leningrad. Otdel. Mat. Inst. Steklov. (LOMI)*, 57 (1977), 24–41 (in Russian).

[87] K. Bongartz, Tilted algebras, In: *Representations of Algebras*, Lecture Notes in Math., No. 903, Springer-Verlag, Berlin, Heidelberg, New York, 1981, pp. 26–38.

[88] K. Bongartz, Algebras and quadratic forms, *J. London Math. Soc.*, 28(1983), 461–469.

[89] K. Bongartz, Critical simply connected algebras, *Manuscr. Math.*, 46(1984), 117–136.

[90] K. Bongartz, A criterion for finite representation type, *Math. Ann.*, 269(1984), 1–12.

[91] K. Bongartz, Indecomposable modules are standard, *Comment. Math. Helvetici,* 60(1985), 400–410.

[92] K. Bongartz, A geometric version of the Morita equivalence, *J. Algebra*, 139(1991), 159–179.

[93] K. Bongartz, Minimal singularities for representations of Dynkin quivers, *Comment. Math. Helvetici*, 69(1994), 575–611.

[94] K. Bongartz, Degenerations for representations of tame quivers, *Ann. Sci. École Norm. Sup.*, 28(1995), 647–668.

[95] K. Bongartz, On degenerations and extensions of finite dimensional modules, *Advances Math.*, 121(1996), 245–287.

[96] K. Bongartz, Some geometric aspects of representation theory, In: *Algebras and Modules I*, Canad. Math. Soc. Conf. Proc., AMS, Vol. 23, 1998, pp. 1–27.

[97] K. Bongartz and D. Dudek, Decomposition classes for representations of tame quivers, *J. Algebra*, 240(2001), 268–288.

[98] K. Bongartz and T. Fritzsche, On minimal disjoint degenerations for preprojective representations of quivers, *Math. Comp.*, 72(2003), 2013–2042.

[99] K. Bongartz and P. Gabriel, Covering spaces in representation theory, *Invent. Math.*, 65(1982), 331–378.

[100] K. Bongartz, M. Kettler and C. Riedtmann, On module categories where the hom order and the stable hom order coincide, *J. Algebra*, 299(2006), 219–235.

[101] K. Bongartz and C. M. Ringel, Representation-finite tree algebras, In: *Representations of Algebras*, Lecture Notes in Math., No. 903, Springer-Verlag, Berlin, Heidelberg, New York, 1981, pp. 39–54.

[102] N. Bourbaki, *Algèbres de Lie*, Chapitre IV, Masson, Paris, 1968.

[103] S. Brenner, Modular representations of p-groups, *J. Algebra*, 15(1970), 89–102.

[104] S. Brenner, Decomposition properties of some small diagrams of modules, *Symposia Math. Inst. Naz. Alta Mat.*, 13(1974), 127–141.

[105] S. Brenner, On four subspaces of a vector space, *J. Algebra*, 29(1974), 100–114.

[106] S. Brenner, Quivers with commutativity conditions and some phenomenology of forms, In: *Representations of Algebras*, Lecture Notes in Math. No. 488, Springer-Verlag, Berlin, Heidelberg, New York, 1975, pp. 29–53.

[107] S. Brenner and M. C. R. Butler, Endomorphism rings of vector spaces and torsion free abelian groups, *J. London Math. Soc.*, 40(1965), 183–187.

[108] S. Brenner and M. C. R. Butler, Generalisations of the Bernstein–Gelfand–Ponomarev reflection functors, In: *Representation Theory II*, Lecture Notes in Math. No. 832, Springer-Verlag, Berlin, Heidelberg, New York, 1980, pp. 103–169.

[109] S. Brenner and M. C. R. Butler, Wild subquivers of the Auslander-Reiten quiver of a tame algebra, In: *Trends in Representation Theory of Finite Dimensional Algebras, Contemp. Math.*, 229(1998), pp. 29–48.

[110] S. Brenner and C. M. Ringel, Pathological modules over tame rings, *J. London Math. Soc.*, 14(1976), 207–215.

[111] O. Bretscher and P. Gabriel, The standard form of a representation-finite algebra, *Bull. Soc. Math. France*, 111(1983), 21–40.

[112] O. Bretscher, C. Läser and C. Riedtmann, Selfinjective and simply connected algebras, *Manuscr. Math.*, 36(1981), 253–307.

[113] O. Bretscher and G. Todorov, On a theorem of Nazarova and Roiter, In: *Representation Theory I. Finite Dimensional Algebras*, Lecture Notes in Math. No. 1177, Springer-Verlag, Berlin, Heidelberg, New York, 1986, pp. 50–54.

[114] K. A. Brown, The Artin algebras associated with differential operators, *Math. Z.*, 206(1991), 423–442.

[115] K. A. Brown, Representation theory of Noetherian Hopf algebras satisfying a polynomial identity, In: *Trends in Representation Theory of Finite Dimensional Algebras, Contemp. Math.*, 229(1998), 49–79.

[116] T. Brüstle, On commutative tame algebras, *C. R. Acad. Sci. Paris*, Série I, 319(1994), 1141–1145.

[117] T. Brüstle, On the growth function of tame algebras, *C. R. Acad. Sci. Paris*, Série I, 322(1996), 211–215.

[118] T. Brüstle, Derived-tame tree algebras, *Compositio Math.*, 129(2001), 301–323.

[119] T. Brüstle, On positive roots of pg-critical algebras, *Linear Alg. Appl.* 365(2003), 107–114.

[120] T. Brüstle, Tame tree algebras, *J. reine angew. Math.*, 567(2004), 51–98.

[121] T. Brüstle and Y. Han, Tame two-point algebras without loops, *Comm. Algebra*, 29(2001), 4683–4692.

[122] T. Brüstle, J. A. de la Peña and A. Skowroński, Tame algebras and Tits quadratic forms, Preprint, Toruń 2007.

[123] T. Brüstle and V. V. Sergeichuk, Estimate of the number of one-parameter families of modules over a tame algebra, *Linear Alg. Appl.* 365(2003), 107–114.

[124] A. B. Buan and R. Marsh, Cluster-tilting theory, In: *Trends in Representation Theory of Algebras and Related Topics, Contemp. Math.*, 406(2006), 1–30.

[125] A. B. Buan, R. Marsh, M. Reineke, I. Reiten and G. Todorov, Tilting theory and cluster combinatorics, *Advances Math.*, 204(2006), 572–618.

[126] R.-O. Buchweitz, Finite representation type and periodic Hochschild (co-)homology, In: *Trends in Representation Theory of Finite Dimensional Algebras, Contemp. Math.*, 229(1998), 81–109.

[127] R.-O. Buchweitz and S. Liu, Artin algebras with loops but no outer derivations, *Algebras and Representation Theory*, 5(2002), 149–162.

[128] R.-O. Buchweitz and S. Liu, Hochschild cohomology and representation-finite algebras, *Proc. London Math. Soc.*, 88(2004), 355–380.

[129] M. C. R. Butler and C. M. Ringel, Auslander-Reiten sequences with few middle terms and applications to string algebras, *Comm. Algebra*, 15(1987), 145–179.

[130] P. Caldero and B. Keller, From triangulated categories to cluster algebras, Preprint, Paris, 2005.

[131] P. Caldero and B. Keller, From triangulated categories to cluster algebras II, Preprint, Paris, 2005.

[132] C. Chang and J. Weyman, Representations of quivers with free module of covariants, *J. Pure Appl. Algebra*, 192(2004), 69–94.

[133] X. Chen and J. Xiao, Exceptional sequences in Hall algebras and quantum groups, *Compositio Math.*, 117(1999), 161–187.

[134] C. Chesné, One-point extensions of wild hereditary algebras, *J. Algebra*, 280(2004), 384–393.

[135] W. Chin, A brief introduction to coalgebra representation theory, *Lecture Notes in Pure and Appl. Math.*, Marcel-Dekker, 237(2004), pp. 109–131.

[136] W. Chin, M. Kleiner and D. Quinn, Almost split sequences for comodules, *J. Algebra*, 249(2002), 1–19.

[137] W. Chin, M. Kleiner and D. Quinn, Local theory of almost split sequences for comodules, *Ann. Univ. Ferrara, Sez.VII, Sci. Math.*, 51(2005), 183–196.

[138] F. U. Coelho, D. Happel, and L. Unger, Tilting up algebras of small homological dimensions, *J. Pure Appl. Algebra*, 174(2002), 219–241.

[139] F. U. Coelho and M. Lanzilotta, Algebras with small homological dimensions, *Manuscr. Math.*, 100(1999), 1–11.

[140] F. U. Coelho and M. Lanzilotta, On semiregular components with paths from injective to projective modules, *Comm. Algebra*, 30(2002), 4837–4849.

[141] F. U. Coelho and M. Lanzilotta, Weakly shod algebras, *J. Algebra*, 265(2003), 379–403.

[142] F. U. Coelho, E. N. Marcos, H. A. Merklen and A. Skowroński, Module categories with infinite radical square zero are of finite type, *Comm. Algebra*, 22(1994), 4511–4517.

[143] F. U. Coelho, E. N. Marcos, H. Merklen and A. Skowroński, Module categories with infinite radical cube zero, *J. Algebra*, 183(1996), 1–23.

[144] F. U. Coelho, J. A. de la Peña and B. Tomé, Algebras whose Tits form weakly controls the module category, *J. Algebra*, 191(1997), 89–108.

[145] F. U. Coelho and A. Skowroński, On Auslander-Reiten components for quasi-tilted algebras, *Fund. Math.*, 149(1996), 67–82.

[146] P. M. Cohn, *Skew Fields. Theory of General Division Rings*, Encyclopedia of Mathematics and its Applications, 57, Cambridge University Press, 1995.

[147] W. Crawley-Boevey, On tame algebras and bocses, *Proc. London Math. Soc.,* 56(1988), 451–483.

[148] W. W. Crawley-Boevey, Functorial filtrations and the problem of an idempotent and a square-zero matrix, *J. London Math. Soc.,* 38(1988), 385–402.

[149] W. W. Crawley-Boevey, Functorial filtrations II: clans and the Gelfand problem, *J. London Math. Soc.,* 40(1989), 9–30.

[150] W. W. Crawley-Boevey, Functorial filtrations III: semidihedral algebras, *J. London Math. Soc.,* 40(1989), 31–39.

[151] W. W. Crawley-Boevey, Matrix problems and Drozd's theorem, In: *Topics in Algebra, Part I: Rings and Representations of Algebras,* (Eds: S. Balcerzyk, T. Józefiak, J. Krempa, D. Simson, W. Vogel), Banach Center Publications, Vol. 26, PWN Warszawa, 1990, pp. 199–222.

[152] W. W. Crawley-Boevey, Tame algebras and generic modules, *Proc. London Math. Soc.,* 63(1991), 241–265.

[153] W. Crawley-Boevey, Modules of finite length over their endomorphism rings, In: *Representations of Algebras and Related Topics*, London Math. Soc. Lecture Notes Series 168(1992), pp. 127–184.

[154] W. W. Crawley-Boevey, Tameness of biserial algebras, *Archiv Math. (Basel)*, 65(1995), 399–407.

[155] W. W. Crawley-Boevey, Infinite-dimensional modules in the representation theory of finite-dimensional algebras, In: *Algebras and Modules I*, Canad. Math. Soc. Conf. Proc., AMS, Vol. 23, 1998, pp. 29–54.

[156] W. W. Crawley-Boevey, Geometry of the moment map for representations of quivers, *Compositio Math.*, 126(2001), 257–293.

[157] W. W. Crawley-Boevey, Normality of Marsden-Weinstein reductions for representations of quivers, *Math. Ann.*, 325(2003), 55–79.

[158] W. W. Crawley-Boevey, On matrices in prescribed conjugacy classes with no common invariant subspace and sum zero, *Duke Math. J.*, 118(2003), 339–352.

[159] W. W. Crawley-Boevey and C. Geiss, Horn's problem and semistability for quiver representations, In: *Representations of Algebras*, Vol. II, Proc. Conf. ICRA IX, Beijing 2000, (Eds: D. Happel and Y. B. Zhang), Beijing Normal University Press, 2002, pp. 40–48.

[160] W. W. Crawley-Boevey and M. P. Holland, Noncommutative deformations of Kleinian singularities, *Duke Math. J.*, 92(1998), 605–635.

[161] W. Crawley-Boevey and O. Kerner, A functor between regular modules for wild hereditary algebras, *Math. Ann.*, 298(1994), 481–487.

[162] W. W. Crawley-Boevey and C. M. Ringel, Algebras whose Auslander-Reiten quivers have large regular components, *J. Algebra*, 153(1992), 494–516.

[163] W. W. Crawley-Boevey and J. Schröer, Irreducible components of varieties of modules, *J. reine angew. Math.*, 553(2002), 201–220.

[164] W. W. Crawley-Boevey and M. Van den Bergh, Absolutely indecomposable representations and Kac-Moody Lie algebras. With an appendix by Hiraki Nakajima, *Invent. Math.*, 155(2004), 537–559.

[165] C. W. Curtis and I. Reiner, *Representation Theory of Finite Groups and Associative Algebras*, Wiley (Interscience), New York, 1962.

[166] B. Deng, On a problem of Nazarova and Roiter, *Comment. Math. Helvetici*, 75(2000), 368–409.

[167] H. Derksen and J. Weyman, Semi-invariants of quivers and saturation for Littlewood-Richardson coefficients, *J. Amer. Math. Soc.*, 13(2000), 467–479.

[168] H. Derksen and J. Weyman, Semi-invariants of quivers with relations, *J. Algebra*, 258(2002), 216–227.

[169] H. Derksen and J. Weyman, On the canonical decomposition of quiver representations, *Compositio Math.*, 133(2002), 245–265.

[170] G. D'Este and C. M. Ringel, Coherent tubes, *J. Algebra*, 87(1984), 150–201.

[171] V. Dlab and C. M. Ringel, On algebras of finite representation type, *J. Algebra*, 33(1975), 306–394.

[172] V. Dlab and C. M. Ringel, Indecomposable representations of graphs and algebras, *Memoirs Amer. Math. Soc.*, Vol. 173, 1976.

[173] V. Dlab and C. M. Ringel, Eigenvalues of Coxeter transformations and the Gelfand-Kirilov dimension of the preprojective algebras, *Proc. Amer. Math. Soc.*, 87(1981), 228–232.

[174] M. Domokos and H. Lenzing, Invariant theory of canonical algebras, *J. Algebra*, 228(2000), 738–762.

[175] M. Domokos and H. Lenzing, Moduli spaces for representations of concealed-canonical algebras, *J. Algebra*, 251(2002), 371–394.

[176] P. Donovan and M. R. Freislich, The representation theory of finite graphs and associated algebras, *Carleton Lecture Notes*, 5, Ottawa, 1973.

[177] P. Dowbor, On the category of modules of the second kind for Galois coverings, *Fund. Math.*, 149(1996), 31–54.

[178] P. Dowbor, Stabilizer conjecture for representation-tame Galois coverings of algebras, *J. Algebra*, 239(2001), 112–149.

[179] P. Dowbor, Non-orbicular modules for Galois coverings, *Colloq. Math.*, 89(2001), 241–310.

[180] P. Dowbor, A construction of non-regularly orbicular modules for Galois coverings, *J. Math. Soc. Japan*, 57(2005), 1077–1127.

[181] P. Dowbor, H. Lenzing and A. Skowroński, Galois covering of algebras by locally support-finite categories, In: *Representation Theory I. Finite Dimensional Algebras*, Lecture Notes in Math., No. 1177, Springer-Verlag, Berlin, Heidelberg, New York, 1986, pp. 91–93.

[182] P. Dowbor, C. M. Ringel and D. Simson, Hereditary artinian rings of finite representation type, In: *Representation Theory II*, Lecture Notes in Math. No. 832, Springer-Verlag, Berlin, Heidelberg, New York, 1980, pp. 232–241.

[183] P. Dowbor and D. Simson, Quasi-Artin species and rings of finite representation type, *J. Algebra*, 63(1980), 435–443.

[184] P. Dowbor and D. Simson, A characterization of hereditary rings of finite representation type, *Bull. Amer. Math. Soc.* 2(1980), 300–302.

[185] P. Dowbor and A. Skowroński, On Galois coverings of tame algebras, *Archiv Math. (Basel)*, 44(1985), 522–529.

[186] P. Dowbor and A. Skowroński, On the representation type of locally bounded categories, *Tsukuba J. Math.*, 10(1986), 63–77.

[187] P. Dowbor and A. Skowroński, Galois coverings of representation-infinite algebras, *Comment. Math. Helvetici*, 62(1987), 311–337.

[188] S. R. Doty, K. Erdmann, S. Martin and D. K. Nakano, Representation type of Schur algebras, *Math. Z.*, 232(1999), 137–182.

[189] P. Dräxler, Über einen Zusammenhang zwischen darstellungsendlichen Algebren und geordneten Mengen, *Manuscr. Math.*, 60(1988), 349–377.

[190] P. Dräxler, Completely separating algebras, *J. Algebra*, 165(1994), 550–565.

[191] P. Dräxler, On the density of fiber sum functors, *Math. Z.*, 216(1994), 645–656.

[192] P. Dräxler, Generalized one-point extensions, *Math. Ann.*, 304(1996), 645–667.

[193] P. Dräxler, Cleaving functors and controlled wild algebras, *J. Pure Appl. Algebra*, 169(2002), 33–42.

[194] P. Dräxler, N. Golovachtchuk, S. Ovsienko and J. A. de la Peña, Coordinates of maximal roots of weakly non-negative unit forms, *Colloq. Math.*, 78(1998), 163–193.

[195] P. Dräxler and R. Nörenberg, Thin start modules and representation type, In: *Representations of Algebras*, Canad. Math. Soc. Conf. Proc., AMS, Vol. 14, 1993, pp. 149–163.

[196] P. Dräxler and J. A. de la Peña, On the existence of preprojective components in the Auslander-Reiten quiver of an algebra, *Tsukuba J. Math.*, 20(1996), 457–469.

[197] P. Dräxler and J. A. de la Peña, The homological quadratic form of a biextension algebra, *Archiv Math. (Basel)*, 72(1999), 9–21.

[198] P. Dräxler, I Reiten, S. Smalø and Ø. Solberg, Exact categories and vector space categories, *Trans. Amer. Math. Soc.*, 351(1999), 647–682.

[199] P. Dräxler and A. Skowroński, Biextensions by indecomposable modules of derived regular length 2, *Compositio Math.*, 117(1999), 205–221.

[200] Yu. A. Drozd, Coxeter transformations and representations of partially ordered sets, *Funkc. Anal. i Priložen.*, 8(1974), 34–42 (in Russian).

[201] Yu. A. Drozd, On tame and wild matrix problems, In: *Matrix Problems*, Akad. Nauk Ukr. S.S.R., Inst. Matem, Kiev, 1977, pp. 104-114 (in Russian).

[202] Yu. A. Drozd, Tame and wild matrix problems, In: *Representations and Quadratic Forms*, Akad. Nauk Ukr. S.S.R., Inst. Matem., Kiev 1979, 39–74 (in Russian).

[203] Yu. A. Drozd, Cohen-Macaulay modules and vector bundles, In: *Interactions between Ring Theory and Representations of Algebras*, Lecture Notes in Pure and Appl. Math., Vol. 221, Marcel-Dekker, 2000, pp. 107–130.

[204] Yu. A. Drozd, Derived tame and derived wild algebras, *Algebra and Discr. Math.*, 1(2004), 57–74.

[205] Yu. A. Drozd and G.-M. Greuel, Tame-wild dichotomy for Cohen-Macaulay modules, *Math. Ann.*, 294(1992), 387–394.

[206] Yu. A. Drozd and G.-M. Greuel, Tame and wild projective curves and classification of vector bundles, *J. Algebra*, 246(2001), 1–54.

[207] Yu. A. Drozd and V. V. Kirichenko, *Finite Dimensional Algebras*, Springer-Verlag, Berlin, Heidelberg, New York, 1994.

[208] Yu. A. Drozd, S. A. Ovsienko and B. Ju. Furchin, Categorical constructions in representation theory, In: *Algebraic Structures and their Applications*, University of Kiev, Kiev, UMK VO, 1988, pp. 43–73 (in Russian).

[209] N. V. Dung and D. Simson, The Gabriel-Roiter measure for right pure semisimple rings, *Algebras and Representation Theory*, 2007, to appear.

[210] S. Eilenberg, Abstract description of some basic functors, *J. Indian Math. Soc.*, 24(1960), 231–234.

[211] K. Erdmann, *Blocks of Tame Representation Type and Related Algebras*, Lecture Notes in Math. No. 1428, Springer-Verlag, Berlin-Heidelberg-New York, 1990.

[212] K. Erdmann, On Auslander-Reiten components for group algebras, *J. Pure Appl. Algebra*, 104(1995), 149–160.

[213] K. Erdmann, On tubes for blocks of wild type, *Colloq. Math.*, 82(1999), 261–270.

[214] K. Erdmann, E. L. Green, N. Snashall, Ø. Solberg and R. Taillefer, Representation theory of the Drinfeld doubles of a family of Hopf algebras, *J. Pure Appl. Algebra*, 204(2006), 413–454.

[215] K. Erdmann, M. Holloway, N. Snashall and R. Taillefer, Support varieties for selfinjective algebras, *K-Theory*, 33(2004), 67–87.

[216] K. Erdmann, T. Holm, O. Iyama and J. Schröer, Radical embeddings and representation dimension, *Advances Math.*, 185(2004), 159–177.

[217] K. Erdmann, O. Kerner and A. Skowroński, Self-injective algebras of wild tilted type. *J. Pure Appl. Algebra*, 149(2000), 127–176.

[218] K. Erdmann and D. K. Nakano, Representation type of Hecke algebras of type *A*, *Trans. Amer. Math. Soc.*, 354(2002), 275–285.

[219] K. Erdmann and A. Skowroński, On Auslander-Reiten components of blocks and selfinjective biserial algebras, *Trans. Amer. Math. Soc.*, 330(1992), 165–189.

[220] K. Erdmann and A. Skowroński, The stable Calabi-Yau dimension of tame symmetric algebras, *J. Math. Soc. Japan*, 58(2006), 97–128.

[221] K. Erdmann and N. Snashall, Preprojective algebras of Dynkin type, periodicity and the second Hochschild cohomology, In: *Algebras and Modules II*, Canad. Math. Soc. Conf. Proc., AMS, Vol. 24, 1998, pp. 183–193.

[222] R. Farnsteiner, Periodicity and representation type of modular Lie algebras, *J. reine angew. Math.*, 464(1995), 47–65.

[223] R. Farnsteiner, On the Auslander-Reiten quiver of an infinitesimal group, *Nagoya. J. Math.*, 160(2000), 103–121.

[224] R. Farnsteiner and A. Skowroński, Classification of restricted Lie algebras with tame principal block, *J. reine angew. Math.*, 546(2002), 1–45.

[225] R. Farnsteiner and A. Skowroński, The tame infinitesimal groups of odd characteristic, *Advances Math.*, 205(2006), 229–274.

[226] R. Farnsteiner and A. Skowroński, Galois actions and blocks of tame infinitesimal group schemes, *Trans. Amer. Math. Soc.*, 2007, in press.

[227] R. Farnsteiner and D. Voigt, On cocommutative Hopf algebras of finite representation type, *Advances Math.*, 155(2000), 1–22.

[228] R. Farnsteiner and D. Voigt, On infinitesimal groups of tame representation type, *Math. Z.*, 244(2003), 479–513.

[229] U. Fischbacher, Une nouvelle preuve d'un théoréme de Nazarova et Roiter, *C. R. Acad. Sci. Paris*, Série I, 300(1984), 1–9, 259–263.

[230] S. Fomin and A. Zelevinsky, Cluster algebras I. Foundations, *J. Amer. Math. Soc.*, 15(2002), 497–529.

[231] S. Fomin and A. Zelevinsky, Cluster algebras II. Finite type classification, *Invent. Math.*, 154(2003), 63–121.

[232] H. Fujita, Y. Sakai and D. Simson, Minor degenerations of the full matrix algebra over a field, *J. Math. Soc. Japan,* 2007, in press.

[233] P. Gabriel, Unzerlegbare Darstellungen I, *Manuscr. Math.*, 6(1972), 71–103.

[234] P. Gabriel, Indecomposable representations II, *Symposia Mat. Inst. Naz. Alta Mat.*, 11(1973), 81–104.

[235] P. Gabriel, Représentations indécomposables, In: *Séminaire Bourbaki* (1973–74), Lecture Notes in Math., No. 431, Springer-Verlag, Berlin, Heidelberg, New York, 1975, pp. 143–169.

[236] P. Gabriel, Finite representation type is open, In: *Representations of Algebras,* Lecture Notes in Math., No. 488, Springer-Verlag, Berlin, Heidelberg, New York, 1975, pp. 132–155.

[237] P. Gabriel, Auslander–Reiten sequences and representation-finite algebras, In: *Representation Theory* I, Lecture Notes in Math., No. 831, Springer-Verlag, Berlin, Heidelberg, New York, 1980, pp. 1–71.

[238] P. Gabriel, The universal cover of a representation-finite algebra, In: *Representations of Algebras,* Lecture Notes in Math., No. 903, Springer-Verlag, Berlin, Heidelberg, New York, 1981, pp. 68–105.

[239] P. Gabriel, L. A. Nazarova, A. V. Roiter, V. V. Sergeichuk, D. Vossieck, Tame and wild subspace problems, *Ukrainian J. Math.*, 45(1993), 335–372.

[240] P. Gabriel and J. A. de la Peña, Quotients of representation-finite algebras, *Comm. Algebra*, 15(1987), 279–307.

[241] P. Gabriel and C. Riedtmann, Group representations without groups, *Coment. Math. Helvetici*, 54(1979), 240–287.

[242] P. Gabriel and A. V. Roiter, *Representations of Finite Dimensional Algebras,* Algebra VIII, Encyclopaedia of Math. Sc., Vol. 73, Springer-Verlag, Berlin, Heidelberg, New York, 1992.

[243] S. Gastaminza, J. A. de la Peña, M. I. Platzeck, M. J. Redondo and S. Trepode, Finite dimensional algebras with vanishing Hochschild cohomology, *J. Algebra,* 212(1999), 1–16.

[244] W. Geigle, The Krull-Gabriel dimension of the representation theory of a tame hereditary artin algebra and application to the structure of exact sequences, *Manuscr. Math.*, 54(1985), 83–106.

[245] W. Geigle, Krull dimension of Artin algebras. In: *Representation Theory I, Finite Dimensional Algebras,* Lecture Notes in Math., No. 1177, Springer-Verlag, Berlin, Heidelberg, New York, 1986, pp. 135–155.

[246] W. Geigle and H. Lenzing, A class of weighted projective curves arising in representation theory of finite dimensional algebras. In: *Singularities, Representations of Algebras and Vector Bundles*, Lecture Notes in Math., No. 1273, Springer-Verlag, Berlin, Heidelberg, New York, 1987, pp. 265–297.

[247] W. Geigle and H. Lenzing, Perpendicular categories with applications to representations and sheaves, *J. Algebra*, 144(1991), 273–343.

[248] C. Geiss, On degenerations of tame and wild algebras, *Archiv Math. (Basel)*, 64(1995), 11–16.

[249] C. Geiss, A decomposition theorem for tensor products with bimodules, *Comm. Algebra*, 20(1992), 2991–2998.

[250] C. Geiss, On components of type $\mathbb{Z}[A_\infty^\infty]$ for string algebras, *Comm. Algebra*, 26(1998), 749–758.

[251] C. Geiss, Derived tame algebras and Euler forms (with an appendix by the author and B. Keller), *Math. Z.*, 239(2002), 829–862.

[252] C. Geiss, Introduction to moduli spaces associated to quivers (with an appendix by Lieven Le Bruyn and Markus Reineke), In: *Trends in Representation Theory of Algebras and Related Topics, Contemp. Math.*, 406(2006), 31–50.

[253] C. Geiss and H. Krause, On the notion of derived tameness, *J. Algebra Appl.*, 1(2002), 133–157.

[254] C. Geiss, B. Leclerc and J. Schröer, Semicanonical bases and preprojective algebras, *Ann. Sci. École Norm. Sup.*, 38(2005), 193–253.

[255] C. Geiss, B. Leclerc and J. Schröer, Rigid modules over preprojective algebras, *Invent. Math.*, 165(2006), 589–632.

[256] C. Geiss and J. A. de la Peña, On the deformation theory of finite dimensional algebras, *Manuscr. Math.*, 88(1995), 191–208.

[257] C. Geiss and J. A. de la Peña, Auslander-Reiten components for clans, *Bol. Soc. Mat. Mexicana*, 5(1999), 307–326.

[258] C. Geiss and J. Schröer, Extension-orthogonal components of preprojective algebras, *Trans. Amer. Math. Soc.*, 357(2005), 1953–1962.

[259] I. M. Gelfand and V. A. Ponomarev, Indecomposable representations of the Lorentz group, *Uspechi Mat. Nauk*, 2(1968), 1–60 (in Russian).

[260] I. M. Gelfand and V. A. Ponomarev, Problems of linear algebra and classification of quadruples of subspaces in a finite-dimensional vector space. *Coll. Math. Soc. Bolyai*, Tihany (Hungary), 5(1970), 163–237.

[261] I. M. Gelfand and V. A. Ponomarev, Model algebras and representations of graphs, *Funct. Anal. Appl*, 13(1979), 1–12.

[262] R. Göbel and D. Simson, Rigid families and endomorphism algebras of Kronecker modules, *Israel J. Math.,* 110(1999), 293–315.

[263] E. L. Green, Group-graded algebras and the zero relation problem, In: *Representations of Algebras,* Lecture Notes in Math., No. 903, Springer-Verlag, Berlin, Heidelberg, New York, 1981, pp. 106–115.

[264] E. L. Green, Graphs with relations, coverings and group-graded algebras, *Trans. Amer. Math. Soc.,* 279(1983), 297–310.

[265] E. L. Green and I. Reiner, Integral representations and diagrams, *Michigan Math. J.,* 25(1968), 53–84.

[266] E. L. Green, N. Snashall and Ø. Solberg, The Hochschild cohomology ring of a selfinjective algebra of finite type, *Proc. Amer. Math. Soc.,* 131(2003), 3387–3393.

[267] E. L. Green and Ø. Solberg, Basic Hopf algebras and quantum groups, *Math. Z.,* 229(1998), 45–76.

[268] E. L. Green and B. Zimmermann-Huisgen, Finitistic dimension of Artinian rings with vanishing radical cube, *Math. Z.,* 206(1991), 505–526.

[269] J. A. Green, Locally finite representations, *J. Algebra* 41(1976), 137–171.

[270] S. Gruson, Simple coherent functors, In: *Representations of Algebras,* Lecture Notes in Math. No. 488, Springer-Verlag, Berlin, Heidelberg, New York, 1975, pp. 156–159.

[271] Y. Han, Controlled wild algebras, *Proc. London Math. Soc.,* 83(2001), 279–298.

[272] Y. Han, Strictly wild algebras with radical square zero, *Archiv Math. (Basel),* 76(2001), 95–99.

[273] Y. Han, On wild radical square zero algebras, *Sci. China,* 45(2002), 29–32.

[274] Y. Han, Is tame open?, *J. Algebra,* 284(2005), 801–810.

[275] D. Happel, On the derived category of a finite dimensional algebra, *Comment. Math. Helvetici,* 62(1987), 339–388.

[276] D. Happel, *Triangulated categories in the representation theory of finite dimensional algebras,* London Math. Soc. Lecture Notes Series, Vol. 119, 1988.

[277] D. Happel, Hochschild cohomology of finite-dimensional algebras, In: *Séminaire d'Algèbre Paul Dubreil et Marie-Paul Malliavin,* Lecture Notes in Math., No. 1404, Springer-Verlag, Berlin, Heidelberg, New York, 1989, pp. 108–126.

[278] D. Happel, A characterization of hereditary categories with tilting object, *Invent. Math.,* 144(2001), 381–398.

[279] D. Happel, S. Hartlieb and O. Kerner, On perpendicular categories of stones over quiver algebras, *Comment. Math. Helvetici,* 71(1996) 463–474.

[280] D. Happel and S. Liu, Module categories without short cycles are of finite type, *Proc. Amer. Math. Soc.,* 120(1994), 371–375.

[281] D. Happel, U. Preiser and C. M. Ringel, Vinberg's characterization of Dynkin diagrams using subadditive functions with applications to *DTr*-periodic modules, In: *Representation Theory* II, Lecture Notes in Math., No. 832, Springer-Verlag, Berlin, Heidelberg, New York, 1980, pp. 280–294.

[282] D. Happel, U. Preiser and C. M. Ringel, Binary polyhedral groups and Euclidean diagrams, *Manuscr. Math.,* 31(1980), 317–329.

[283] D. Happel and C. M. Ringel, Construction of tilted algebras, In: *Representations of Algebras,* Lecture Notes in Math., No. 903, Springer-Verlag, Berlin, Heidelberg, New York, 1981, pp. 125–144.

[284] D. Happel and C. M. Ringel, Tilted algebras, *Trans. Amer. Math. Soc.,* 274(1982), 399–443.

[285] D. Happel and C. M. Ringel, The derived category of a tubular algebra, In: *Representation Theory I. Finite Dimensional Algebras,* Lecture Notes in Math., No. 1177, Springer-Verlag, Berlin, Heidelberg, New York, 1986, pp. 156–180.

[286] D. Happel and I. Reiten, Hereditary abelian categories with tilting object over arbitrary basic fields, *J. Algebra,* 256(2002), 414–432.

[287] D. Happel, I. Reiten and S. Smalø, Short cycles and sincere modules, In: *Representations of Algebras,* Canad. Math. Soc. Conf. Proc., AMS, Vol. 14, 1993, pp. 233–236.

[288] D. Happel, I. Reiten and S. Smalø, Tilting in abelian categories and quasitilted algebras, *Memoirs Amer. Math. Soc.,* Vol. 575, 1996.

[289] D. Happel, I. Reiten and S. Smalø, Piecewise hereditary algebras, *Archiv Math. (Basel),* 66(1996), 182–186.

[290] D. Happel and L. Unger, Almost complete tilting modules, *Proc. Amer. Math. Soc.,* 107(1989), 603–610.

[291] D. Happel and L. Unger, On the quiver of tilting modules, *J. Algebra,* 284(2005), 857–868.

[292] D. Happel and L. Unger, On partial order of tilting modules, *Algebras and Representation Theory,* 8(2005), 147–156.

[293] D. Happel and D. Vossieck, Minimal algebras of infinite representation type with preprojective component, *Manuscr. Math.,* 42(1983), 221–243.

[294] I. Herzog, A test for finite representation type, *J. Pure Appl. Algebra*, 95(1994), 151–182.

[295] T. Holm, Derived equivalence classification of algebras of dihedral, semidihedral, and quaternion type, *J. Algebra*, 211(1999), 159–205.

[296] T. Holm and A. Skowroński, Derived equivalence classification of symmetric algebras of domestic type, *J. Math. Soc. Japan*, 58(2006), 1133–1149.

[297] H. J. von Höhne, On weakly positive unit forms, *Comment Math. Helvetici*, 63(1988), 312–336.

[298] H. J. von Höhne, Edge reduction for unit forms, *Archiv Math. (Basel)*, 65(1995), 300–302.

[299] H. J. von Höhne, On weakly non-negative unit forms and tame algebras, *Proc. London Math. Soc.*, 73(1996), 47–67.

[300] H.-J. von Höhne and J. A. de la Peña, Isotropic vectors of non-negative integral quadratic forms, *Europ. J. Combinatorics*, 19(1998), 621–638.

[301] H. J. von Höhne and J. A. de la Peña, Edge reduction for weakly non-negative unit forms, *Archiv Math. (Basel)*, 70(1998), 270–277.

[302] H.-J. von Höhne and D. Simson, Bipartite posets of finite prinjective type, *J. Algebra*, 201(1998), 86–114.

[303] M. Hoshino, DTr-invariant modules, *Tsukuba J. Math.*, 7(1983), 205–214.

[304] M. Hoshino, Modules with self-extensions and Nakayama's conjecture, *Archiv Math. (Basel)*, 43(1984), 493–500.

[305] J. Hua and J. Xiao, On Ringel-Hall algebras of tame hereditary algebras, *Algebras and Representation Theory*, 5(2002), 527–550.

[306] A. Hubery, Quiver representations respecting a quiver automorphism: a generalization of a theorem of Kac, *J. London Math. Soc.*, 69(2004), 79–96.

[307] A. Hubery, From triangulated categories to Lie algebras: a theorem of Peng and Xiao, In: *Trends in Representation Theory of Algebras and Related Topics, Contemp. Math.*, 406(2006), 51–66.

[308] B. Huisgen-Zimmermann, Purity, algebraic compactness, direct sum decompositions, and representation type In: *Infinite Length Modules*, (Eds: H. Krause and C. M. Ringel), Trends in Mathematics, Birkhäuser Verlag, Basel-Boston-Berlin, 2000, pp. 331–367.

[309] B. Huisgen-Zimmermann and M. Saorin, Geometry of chain complexes and outer automorphisms under derived equivalence, *Trans. Amer. Math. Soc.*, 353(2001), 4757–4777.

[310] B. Huisgen-Zimmermann and S. Smalø, The homology of string algebras, *J. reine angew. Math.*, 580(2005), 1–37.

[311] D. Hughes and J. Waschbüsch, Trivial extensions of tilted algebras, *Proc. London Math. Soc.*, 46(1983), 347–364.

[312] O. Iyama, Finiteness of representation dimension, *Proc. Amer. Math. Soc.*, 131(2003), 1011–1014.

[313] O. Iyama, Representation dimension and Solomon zeta function, In: *Representations of Finite Dimensional Algebras and Related Topics in Lie Theory and Geometry*, Fields Inst. Comm., 40(2004), 45–64.

[314] O. Iyama, The relationship between homological properties and representation theoretic realization of artin algebras, *Trans. Amer. Math. Soc.*, 357(2004), 709–734.

[315] K. Igusa, M. I. Platzeck, G. Todorov and D. Zacharia, Auslander algebras of finite representation type, *Comm. Algebra*, 15(1987), 377–424.

[316] P. Jara, L. Merino and G. Navarro, On path coalgebras of quivers with relations, *Colloq. Math.*, 102(2005), 49–65.

[317] P. Jara, L. Merino and G. Navarro, Localization in tame and wild coalgebras, *J. Pure Appl. Algebra*, doi:10.1016/j.jpaa. 2007.01009.

[318] C. U. Jensen and H. Lenzing, *Model Theoretic Algebra With Particular Emphasis on Fields, Rings, Modules*, Algebra, Logic and Applications, Vol. 2, Gordon & Breach Science Publishers, New York, 1989.

[319] S. Jøndrup and D. Simson, Indecomposable modules over semiperfect rings, *J. Algebra*, 73(1981), 23–29.

[320] V. G. Kac, Infinite root systems, representations of graphs and invariant theory, *Invent. Math.*, 56(1980), 57–92; II, *J. Algebra*, 78(1982), 141–162.

[321] S. Kasjan, On the problem of axiomatization of tame representation type, *Fund. Math.*, 171(2002), 53–67.

[322] S. Kasjan, Representation-directed algebras form an open scheme, *Colloq. Math.*, 93(2002), 237–250.

[323] S. Kasjan, Representation-finite triangular algebras form an open scheme, *Centr. Europ. J. Math.*, 1(2003), 97–107 (electronic).

[324] S. Kasjan, Tame strongly simply connected algebras form an open scheme, *J. Pure Appl. Algebra*, 208(2007), 435–443.

[325] S. Kasjan and J. A. de la Peña, Galois coverings and the problem of axiomatization of the representation type, *Extracta Math.*, 20(2005), 137–150.

[326] S. Kasjan and D. Simson, Varieties of poset representations and minimal posets of wild prinjective type, In: *Representations of Algebras*, Canad. Math. Soc. Conf. Proc., AMS, Vol. 14, 1993, pp. 245–284.

[327] S. Kasjan and D. Simson, Fully wild prinjective type of posets and their quadratic forms, *J. Algebra,* 172(1995), 506–529.

[328] S. Kasjan and D. Simson, Tame prinjective type and Tits form of two-peak posets I, *J. Pure Appl. Algebra,* 106(1996), 307–330.

[329] S. Kasjan and D. Simson, Tame prinjective type and Tits form of two-peak posets II, *J. Algebra,* 187(1997), 71–96.

[330] S. Kasjan and D. Simson, A subbimodule reduction, a peak reduction functor and tame prinjective type, *Bull. Polish Acad. Sci., Ser. Math.,* 45(1997), 89–107.

[331] S. Kawata, On Auslander-Reiten components and simple modules for finite group algebras, *Osaka J. Math.,* 34(1997), 681–688.

[332] S. Kawata, G. O. Michler and K. Uno, On simple modules in the Auslander-Reiten components of finite groups, *Math. Z.,* 234(2000), 375–398.

[333] S. Kawata, G. O. Michler and K. Uno, On Auslander-Reiten components and simple modules for finite groups of Lie type, *Osaka J. Math.,* 38(2001), 21–26.

[334] B. Keller, Deriving DG algebras, *Ann. Sci. École Norm. Sup.,* 27(1994), 63–102.

[335] B. Keller, Tilting theory and differential graded algebras, In: *Finite Dimensional Algebras and Related Topics*, NATO ASI Series, Series C: Mathematical and Physical Sciences, Kluwer Acad. Publ., Dordrecht, Vol. 424, 1994, pp. 183–190.

[336] B. Keller, Derived categories and their uses, *in Handbook of Algebra*, (Ed.: M. Hazewinkel), Vol. 1, North-Holland Elsevier, Amsterdam, 1996, pp. 671–701.

[337] B. Keller, Hochschild cohomology and derived Picard groups, *J. Pure Appl. Algebra*, 190(2004), 177–196.

[338] B. Keller, On triangulated orbit categories, *Doc. Math.,* 10(2005), 551–581 (electronic).

[339] B. Keller, A-infinity algebras, modules and functor categories, In: *Trends in Representation Theory of Algebras and Related Topics*, *Contemp. Math.,* 406(2006), 67–93.

[340] B. Keller and D. Vossieck, Sous les catégories dérivées, *C. R. Acad. Sci. Paris,* Série I, 305(1987), 225–228.

[341] B. Keller and D. Vossieck, Aisles in derived categories, *Bull Soc. Math. Belg.,* 40(1988), 239–253.

[342] O. Kerner, Preprojective components of wild tilted algebras, *Manuscr. Math.*, 61(1988) 429–445.

[343] O. Kerner, Tilting wild algebras, *J. London Math. Soc.*, 39(1989), 29–47.

[344] O. Kerner, Stable components of wild tilted algebras, *J. Algebra*, 142(1991), 37–57.

[345] O. Kerner, Exceptional components of wild hereditary algebras, *J. Algebra*, 152(1992), 184–206.

[346] O. Kerner, Elementary stones, *Comm. Algebra*, 22(1994), 1797–1806.

[347] O. Kerner, Wild tilted algebras revisited, *Colloq. Math.*, 73(1997), 67–81.

[348] O. Kerner, Factorizations for morphisms for wild hereditary algebras, In: *Representations of Algebras,* (Sao Paulo, 1999), Lecture Notes in Pure and Appl. Math. (Dekker, New York), 224(2002), pp. 121–127.

[349] O. Kerner, Endomorphism rings of regular modules over wild hereditary algebras, *Colloq. Math.*, 87(2003), 207–220.

[350] O. Kerner and F. Lukas, Regular stones of wild hereditary algebras, *J. Pure Appl. Algebra,* 93(1994), 15–31.

[351] O. Kerner and F. Lukas, Elementary modules, *Math. Z.,* 223(1996), 421–434.

[352] O. Kerner and A. Skowroński, On module categories with nilpotent infinite radical, *Compositio Math.,* 77(1991), 313–333.

[353] O. Kerner and A. Skowroński, Quasitilted one-point extensions of wild hereditary algebras, *J. Algebra,* 244(2001), 785–827.

[354] O. Kerner and A. Skowroński, On the structure of modules over wild hereditary algebras, *Manuscr. Math.,* 108(2002), 369–383.

[355] O. Kerner, A. Skowroński and K. Yamagata, Invariance of selfinjective algebras of quasitilted type under stable equivalences, *Manuscr. Math.,* 119(2006), 359–381.

[356] O. Kerner, A. Skowroński and K. Yamagata, One-point extensions of wild concealed algebras, *J. Algebra,* 301(2006), 627–641.

[357] O. Kerner, A. Skowroński, K. Yamagata and D. Zacharia, Finiteness of the strong global dimension of radical square zero algebras, *Centr. Eur. J. Math.,* 2(2004), 103–111 (electronic).

[358] S. M. Khorochkin, On the category of Harish-Chandra modules of the group SU(1,n), *Funkc. Analiz i Priložen.,* 14(1980), 85–86 (in Russian).

[359] S. M. Khorochkin, Indecomposable representations of Lorentz groups, *Funkc. Anal. i Priložen.,* 15(1981), 50–60 (in Russian).

[360] A. D. King, Moduli of representations of finite-dimensional algebras, *Quart. J. Math. Oxford,* 45(1994), 515–530.

[361] L. Klingler and L. S. Levy, Representation type of commutative noetherian local rings I: local wildness, *Pacific J. Math.,* 200(2001), 345–386.

[362] L. Klingler and L. S. Levy, Representation type of commutative noetherian local rings II: local tameness, *Pacific J. Math.,* 200(2001), 387–483.

[363] L. Klingler and L. S. Levy, Representation type of commutative noetherian local rings III: global wildness and tameness, *Memoirs Amer. Math. Soc.,* 832(2005), pp. 1–170.

[364] M. Kleiner and I. Reiten, Abelian categories, almost split sequences and comodules, *Trans. Amer. Math. Soc.,* 357(2005), 3201–3214.

[365] B. Klemp and D. Simson, Schurian sp-representation-finite right peak PI-rings and their indecomposable socle projective modules, *J. Algebra* 131(1990), 390–468.

[366] J. Kosakowska, Degenerations in a class of matrix varieties and prinjective modules, *J. Algebra,* 263(2003), 262–277.

[367] J. Kosakowska, Generic extensions of prinjective modules, *Algebras and Representation Theory,* 9(2006), 557–568.

[368] J. Kosakowska, The existence of Hall polynomials for posets of finite prinjective type, *J. Algebra,* 308(2007), 654–665.

[369] J. Kosakowska, Prinjective Ringel-Hall algebras for posets of finite prinjective type, 2006, in press.

[370] J. Kosakowska, A specialization of prinjective Ringel-Hall algebra and the associated Lie algebra, 2006, in press.

[371] J. Kosakowska and D. Simson, Hereditary coalgebras and representations of species, *J. Algebra,* 293(2005), 457–505.

[372] H. Kraft, Geometric methods in representation theory, In: *Representations of Algebras,* Lecture Notes in Math., No. 944, Springer-Verlag, Berlin, Heidelberg, New York, 1982, pp. 180–258.

[373] H. Kraft and C. Riedtmann, Geometry of representation of quivers, In: *Representations of Algebras,* London Math. Soc. Lecture Notes Series, 116(1986), pp. 109–145.

[374] H. Krause, Stable equivalence preserves representation type, *Comment. Math. Helvetici,* 72(1997), 266–284.

[375] H. Krause, Generic modules over Artin algebras, *Proc. London Math. Soc.,* 76(1998), 601–606.

[376] H. Krause, Representation type and stable equivalence of Morita type for finite dimensional algebras, *Math. Z.,* 229(1998), 276–306.

[377] H. Krause, Finitistic dimension and Ziegler spectrum, *Proc. Amer. Math. Soc.*, 126(1998), 983–987.

[378] H. Krause, Finite versus infinite dimensional representations - a new definition of tameness, In: *Infinite Length Modules*, (Eds: H. Krause and C. M. Ringel), Trends in Mathematics, Birkhäuser Verlag, Basel-Boston-Berlin, 2000, pp. 393–403.

[379] H. Krause, The spectrum of a module category, *Memoirs Amer. Math. Soc.*, Vol. 707, 2001.

[380] H. Krause and D. Kussin, Rouqier's theorem on representation dimension, In: *Trends in Representation Theory of Algebras and Related Topics*, *Contemp. Math.*, 406(2006), 95–103.

[381] H. Krause and G. Zwara, Stable equivalence and generic modules, *Bull. London Math. Soc.*, 32(2000), 615–618.

[382] D. Kussin, Non-isomorphic derived-equivalent tubular curves and their associated tubular algebras, *J. Algebra*, 236(2000), 436–450.

[383] D. Kussin, On the K-theory of tubular algebras, *Colloq. Math.*, 86(2000), 137–152.

[384] D. Kussin, One-parameter families for finite dimensional algebras, Lecture Notes, Paderborn, 2006, pp. 1–151.

[385] M. Kwiecień and A. Skowroński, On wings of the Auslander-Reiten quivers of selfinjective algebras, *Colloq. Math.*, 103(2005), 265–285.

[386] S. Lache, Piecewise hereditary one-point extensions, *J. Algebra*, 226(2000), 53–70.

[387] H. Lenzing, Wild canonical algebras and automorphic forms, In: *Finite Dimensional Algebras and Related Topics*, NATO ASI Series, Series C: Mathematical and Physical Sciences, Kluwer Acad. Publ., Dordrecht, Vol. 424, 1994, pp. 191–212.

[388] H. Lenzing, A K-theoretic study of canonical algebras, In: *Representation Theory of Algebras*, Canad. Math. Soc. Conf. Proc., AMS, Vol. 18, 1996, pp. 433–454.

[389] H. Lenzing, Generic modules over tubular algebras, In: *Advances in Algebra and Model Theory*, (Eds: M. Droste and R. Göbel), Algebra, Logic and Applications Series, Vol. 9, Gordon & Breach Science Publishers, Australia, 1997, pp. 375–385.

[390] H. Lenzing, Hereditary noetherian categories with a tilting object, *Proc. Amer. Math. Soc.*, 125(1997), 1893–1901.

[391] H. Lenzing, Representations of finite dimensional algebras and singularity theory, In: *Trends in Ring Theory*, Canad. Math. Soc. Conf. Proc., AMS, Vol. 22, 1998, pp. 71–97.

[392] H. Lenzing, Invariance of tameness under stable equivalence: Krause's theorem, In: *Infinite Length Modules,* (Eds: H. Krause and C. M. Ringel), Trends in Mathematics, Birkhäuser Verlag, Basel-Boston-Berlin, 2000, pp. 405–418.

[393] H. Lenzing and H. Meltzer, Sheaves on weighted projective line of genus one and representations of a tubular algebra. In: *Representations of Algebras,* Canad. Math. Soc. Conf. Proc., AMS, Vol. 14, 1993, pp. 313-337.

[394] H. Lenzing and H. Meltzer, Tilting sheaves and concealed-canonical algebras. In: *Representation Theory of Algebras,* Canad. Math. Soc. Conf. Proc., AMS, Vol. 18, 1996, pp. 455–473.

[395] H. Lenzing and J. A. de la Peña, Wild canonical algebras, *Math. Z.,* 224(1997), 403–425.

[396] H. Lenzing and J. A. de la Peña, Concealed-canonical algebras and separating tubular families, *Proc. London Math. Soc.,* 78(1999), 513–540.

[397] H. Lenzing and J. A. de la Peña, Supercanonical algebras, *J. Algebra,* 282(2004), 298–348.

[398] H. Lenzing and A. Skowroński, Quasi-tilted algebras of canonical type, *Colloq. Math.,* 71(1996), 161–181.

[399] H. Lenzing and A. Skowroński, On selfinjective algebras of Euclidean type, *Colloq. Math.,* 79(1999), 71–76.

[400] H. Lenzing and A. Skowroński, Roots of Nakayama and Auslander-Reiten translations, *Colloq. Math.,* 86(2000), 209–230.

[401] H. Lenzing and A. Skowroński, Derived equivalence as iterated tilting, *Archiv Math. (Basel),* 76(2001), 20–24.

[402] H. Lenzing and A. Skowroński, Selfinjective algebras of wild canonical type, *Colloq. Math.,* 96(2003), 245–275.

[403] Z. Leszczyński, On the representation type of tensor product algebras, *Fund. Math.,* 144(1994), 143–151.

[404] Z. Leszczyński, The completely separating incidence algebras of tame representation type, *Colloq. Math.,* 94(2002), 243–262.

[405] Z. Leszczyński, Representation-tame incidence algebras of finite posets, *Colloq. Math.,* 96(2003), 293–306.

[406] Z. Leszczyński, Representation-tame locally hereditary algebras, *Colloq. Math.,* 99(2004), 175–187.

[407] Z. Leszczyński and D. Simson, On triangular matrix rings of finite representation type, *J. London Math. Soc.,* 20(1979), 396–402.

[408] Z. Leszczyński and A. Skowroński, Auslander algebras of tame representation type, In: *Representation Theory of Algebras*, Canad. Math. Soc. Conf. Proc., AMS, Vol. 18, 1996, pp. 475–486.

[409] Z. Leszczyński and A. Skowroński, Tame triangular matrix algebras, *Colloq. Math.*, 86(2000), 259–303.

[410] Z. Leszczyński and A. Skowroński, Tame tensor products of algebras, *Colloq. Math.*, 98(2003), 125–145.

[411] Z. Leszczyński and A. Skowroński, Tame generalized canonical algebras, *J. Algebra,* 273(2004), 412–433.

[412] Z. Leszczyński, A. Skowroński and K. Yamagata, Derived tame locally hereditary algebras, Preprint, Toruń, 2007.

[413] Y. Lie and L. Peng, Elliptic Lie algebras and tubular algebras, *Advances Math.*, 196(2005), 487–530.

[414] S. Liu, The degrees of irreducible maps and the shapes of the Auslander–Reiten quivers, *J. London Math. Soc.*, 45(1992), 32–54.

[415] S. Liu, Semi-stable components of an Auslander–Reiten quiver, *J. London Math. Soc.*, 47(1993), 405–416.

[416] S. Liu, Tilted algebras and generalized standard Auslander–Reiten components, *Archiv Math. (Basel)*, 61(1993), 12–19.

[417] S. Liu, The connected components of the Auslander–Reiten quiver of a tilted algebra, *J. Algebra,* 161(1993), 505–523.

[418] S. Liu, Almost split sequences for non-regular modules, *Fund. Math.,* 143(1993), 183–190.

[419] S. Liu, Infinite radical in standard Auslander–Reiten components, *J. Algebra,* 166(1994), 245–254.

[420] S. Liu, On short cycles in a module category, *J. London Math. Soc.,* 51(1995), 62–74.

[421] S. Liu and R. Schultz, The existence of bounded infinite DTr-orbits, *Proc. Amer. Math. Soc.,* 122(1994), 1003–1005.

[422] Y. Liu and C. C. Xi, Constructions of stable equivalences of Morita type for finite dimensional algebras, I, *Trans. Amer. Math Soc.,* 358(2006), 2537–2560 (electronic).

[423] Y. Liu and C. C. Xi, Constructions of stable equivalences of Morita type for finite dimensional algebras, II, *Math. Z.,* 251(2005), 21–39.

[424] P. Malicki, On the composition factors of indecomposable modules in almost cyclic coherent Auslander-Reiten components, *J. Pure Appl. Algebra,* 207(2006), 469–490.

[425] P. Malicki and A. Skowroński, Almost cyclic coherent components of an Auslander-Reiten quiver, *J. Algebra,* 229(2000), 695–749.

[426] P. Malicki and A. Skowroński, Algebras with separating almost cyclic coherent Auslander-Reiten components, *J. Algebra,* 291(2005), 208–237.

[427] P. Malicki and A. Skowroński, Concealed generalized canonical algebras and standard stable tubes, *J. Math. Soc. Japan,* 59(2007), 521–539.

[428] P. Malicki and A. Skowroński, On the additive categories of generalized standard almost cyclic coherent Auslander-Reiten components, Preprint, Toruń, 2006.

[429] P. Malicki and A. Skowroński, Hochschild cohomologies of algebras with separating almost cyclic coherent Auslander-Reiten components, Preprint, Toruń, 2007.

[430] P. Malicki, A. Skowroński and B. Tomé, Indecomposable modules in coils, *Colloq. Math.,* 93(2002), 67–130.

[431] E. Marmolejo and C. M. Ringel, Modules of bounded length in Auslander–Reiten components, *Archiv Math. (Basel),* 50(1988), 128–133.

[432] R. Martinez-Villa and J. A. de la Peña, The universal cover of a quiver with relations, *J. Pure Appl. Algebra,* 142(1983), 397–408.

[433] R. Martinez-Villa and J. A. de la Peña, Automorphisms of representation-finite algebras, *Invent. Math.,* 72(1983), 359–362.

[434] R. Martinez-Villa and D. Zacharia, Auslander-Reiten sequences, locally-free sheaves and Chebysheff polynomials, *Compositio Math.,* 142(2006), 397–408.

[435] M. I. R. Martins and J. A. de la Peña, Comparing the simplicial and the Hochschild cohomologies of a finite dimensional algebra, *J. Pure Appl. Algebra,* 138(1999), 45–58.

[436] H. Meltzer, Auslander–Reiten components for concealed canonical algebras, *Colloq. Math.,* 71(1996), 183–202.

[437] H. Meltzer, Exceptional vector bundles, tilting sheaves and tilting complexes for weighted projective lines, *Memoirs Amer. Math. Soc.,* No. 808, 2004, pp. 1–138.

[438] H. Meltzer and A. Skowroński, Group algebras of finite representation type, *Math. Z.,* 182(1983), 129–148; Correction 187(1984), 563–569.

[439] H. Nakajima, Varieties associated with quivers, In: *Representation Theory of Algebras and Related Topics,* Canad. Math. Soc. Conf. Proc., AMS, Vol. 19, 1996, pp. 139–157.

[440] H. Nakajima, Quiver varieties and Kac-Moody algebras, *Duke Math. J.,* 91(1998), 515–560.

[441] L. A. Nazarova, Representations of quadruples, *Izv. Akad. Nauk SSSR*, 31(1967), 1361–1378 (in Russian).

[442] L. A. Nazarova, Representations of quivers of infinite type, *Izv. Akad. Nauk SSSR*, 37(1973), 752–791 (in Russian).

[443] L. A. Nazarova and A. V. Roiter, On a problem of I. M. Gelfand, *Funk. Anal. i Priložen.*, 7(1973), 54–69 (in Russian).

[444] L. A. Nazarova and A. V. Roiter, Categorical matrix problems and the Brauer-Thrall conjecture, Akad. Nauk Ukr. S.S.R., Inst. Matem, Kiev, 1973, Preprint No. 73.9, pp. 1–189 (in Russian).

[445] L. A. Nazarova and A. V. Roiter, Kategorielle Matrizen-Probleme und die Brauer-Thrall-Vermutung, *Mitt. Math. Sem. Giessen*, 115(1975), 1–153.

[446] L. A. Nazarova and A. G. Zavadskij, Partially ordered sets of tame type, In: *Matrix Problems*, Akad. Nauk Ukr. S.S.R., Inst. Matem, Kiev, 1977, 122–143 (in Russian).

[447] L. A. Nazarova and A. G. Zavadskij, On finiteness and a bound of numbers of parameters, Akad. Nauk Ukr. S.S.R., Inst. Matem., Kiev, 1981, Preprint No. 81.27, pp. 21–29 (in Russian).

[448] L. A. Nazarova and A. G. Zavadskij, Partially ordered sets of finite growth, *Funkc. Analiz i Priložen.*, 16(1982), 72–73 (in Russian).

[449] J. Nehring and A. Skowroński, Polynomial growth trivial extensions of simply connected algebras, *Fund. Math.*, 132(1989), 117–134.

[450] R. Nörenberg and A. Skowroński, Tame minimal non-polynomial growth strongly simply connected algebras, In: *Representation Theory of Algebras*, Canad. Math. Soc. Conf. Proc., AMS, Vol. 18, 1996, pp. 519–538.

[451] R. Nörenberg and A. Skowroński, Tame minimal non-polynomial growth simply connected algebras, *Colloq. Math.*, 73(1997), 301–330.

[452] A. Ostermann and A. Pott, Schwach positive ganze quadratische Formen, die eine aufrichtige, positive Wurzel mit einem Koefficient 6 besitzen, *J. Algebra*, 126(1989), 80–118.

[453] S. A. Ovsienko, Integral weakly positive forms, In: *Schur Matrix Problems and Quadratic Forms*, Inst. Mat. Akad. Nauk Ukr. S.S.R., Preprint 78.25, 1978, pp. 3–17 (in Russian)

[454] S. A. Ovsienko, A bound of roots of weakly positive forms, In: *Representations and Quadratic Forms*, Akad. Nauk Ukr. S.S.R., Inst. Matem., Kiev, 1979, pp. 106–123 (in Russian).

[455] J. A. de la Peña, On the representation type of one-point extensions of tame algebras, *Manuscr. Math.* 61(1988), 183–194.

[456] J. A. de la Peña, Algebras with hypercritical Tits form, In: *Topics in Algebra, Part I: Rings and Representations of Algebras*, (Eds: S. Balcerzyk, T. Józefiak, J. Krempa, D. Simson, W. Vogel), Banach Center Publications, Vol. 26, PWN Warszawa, 1990, pp. 353–369.

[457] J. A. de la Peña, On the dimension of module varieties of tame and wild algebras, *Comm. Algebra*, 19(1991), 1795–1807.

[458] J. A. de la Peña, Tame algebras with sincere directing modules, *J. Algebra,* 161(1993), 171–185.

[459] J. A. de la Peña,, Coxeter transformations and the representation theory of algebras, In: *Finite Dimensional Algebras and Related Topics*, NATO ASI Series, Series C: Mathematical and Physical Sciences, Vol. 424, Kluwer Academic Publishers, Dordrecht, 1994, pp. 223–253.

[460] J. A. de la Peña, On the corank of the Tits form of a tame algebra, *J. Pure Appl. Algebra,* 107(1996), 89–105.

[461] J. A. de la Peña, The Tits form of a tame algebra, In: *Representation Theory of Algebras and Related Topics*, Canad. Math. Soc. Conf. Proc., AMS, Vol. 19, 1996, pp. 159–183.

[462] J. A. de la Peña and D. Simson, Prinjective modules, reflection functors, quadratic forms and Auslander-Reiten sequences, *Trans. Amer. Math. Soc.*, 329(1992), 733–753.

[463] J. A. de la Peña and A. Skowroński, Characterizations of strongly simply connected polynomial growth algebras, *Archiv Math. (Basel)*, 65(1995), 391–398.

[464] J. A. de la Peña and A. Skowroński, Geometric and homological characterizations of polynomial growth strongly simply connected algebras, *Invent. Math.*, 126(1996), 287–296.

[465] J. A. de la Peña and A. Skowroński, Forbidden subcategories of non-polynomial growth algebras, *Canad. J. Math.*, 48(1996), 1018–1043.

[466] J. A. de la Peña and A. Skowroński, Substructures of non-polynomial growth algebras with weakly non-negative Tits form, In: *Algebras and Modules, II*, Canad. Math. Soc. Conf. Proc., AMS, Vol. 24, 1998 pp. 415–431.

[467] J. A. de la Peña and A. Skowroński, The Tits and Euler forms of a tame algebra, *Math. Ann.,* 315(1999), 37–59.

[468] J. A. de la Peña and A. Skowroński, Substructures of algebras with weakly non-negative Tits form, Preprint, Toruń, 2007.

[469] J. A. de la Peña and M. Takane, Spectral properties of Coxeter transformations and applications, *Archiv Math. (Basel)*, 55(1990), 120–134.

[470] J. A. de la Peña and M. Takane, On the number of terms in the middle of almost split sequences over tame algebras, *Trans. Amer. Math. Soc.*, 351(1999), 3857–3868.

[471] L. Peng and Y. Tan, Derived categories, tilted algebras and Drinfeld doubles, *J. Algebra*, 206(2003), 723–748.

[472] L. Peng and J. Xiao, On the number of DTr-orbits containing directing modules, *Proc. Amer. Math. Soc.*, 118(1993), 753–756.

[473] L. Peng and J. Xiao, Root categories and simple Lie algebras, *J. Algebra*, 198(1997), 19–56.

[474] L. Peng and J. Xiao, Triangulated categories and Kac-Moody Lie algebras, *Invent. Math.*, 140(2000), 563–603.

[475] R. S. Pierce, *Associative Algebras*, Springer-Verlag, New York, Heidelberg, Berlin, 1982.

[476] Z. Pogorzały, Invariance of Hochschild cohomology algebras and stable equivalence of Morita type, *J. Math. Soc. Japan,* 53(2001), 913–918.

[477] Z. Pogorzały, A new invariant of stable equivalences of Morita type, *Proc. Amer. Math. Soc.,* 131(2003), 343–349.

[478] Z. Pogorzały, On the Auslander-Reiten periodicity of self-injective algebras, *Bull. London Math. Soc.,* 36(2004), 156–168.

[479] Z. Pogorzały and A. Skowroński, On algebras whose indecomposable modules are multiplicity-free, *Proc. London. Math. Soc.,* 47(1983), 463–479.

[480] Z. Pogorzały and A. Skowroński, Selfinjective biserial standard algebras, *J. Algebra,* 138(1991), 491–504.

[481] Z. Pogorzały and A. Skowroński, Symmetric algebras stably equivalent to the trivial extensions of tubular algebras, *J. Algebra,* 181(1996), 95–111.

[482] M. Prest, Representation embeddings and the Ziegler spectrum, *J. Pure Appl. Algebra,* 113(1996), 315–323.

[483] M. Prest, Ziegler spectra of tame hereditary algebras, *J. Algebra,* 207(1998), 146–164.

[484] M. Prest, Topological and geometric aspects of the Ziegler spectrum, In: *Infinite Length Modules* (Eds: H. Krause and C. M. Ringel), Trends in Mathematics, Birkhäuser Verlag, Basel-Boston-Berlin, 2000, pp. 369–392.

[485] M. Reineke, Generic extensions and multiplicative bases of quantum groups at $q = 0$, *Representation Theory*, 5(2001), 147–163.

[486] M. Reineke, Quivers, desingularizations and canonical bases, In: *Studies in memory of Issai Schur*, Progress Math., 210, Birkhäuser, Boston, 2003, pp. 325–344.

[487] I. Reiten and C. Riedtmann, Skew group algebras in the representation theory of artin algebras, *J. Algebra*, 92(1985), 224–282.

[488] I. Reiten and C. M. Ringel, Infinite dimensional representations of canonical algebras, *Canad. J. Math.*, 58(2006), 180–224.

[489] I. Reiten and A. Skowroński, Sincere stable tubes, *J. Algebra*, 232(2000), 64–75.

[490] I. Reiten and A. Skowroński, Characterizations of algebras with small homological dimensions, *Adv. Math.*, 179(2003), 122–154.

[491] I. Reiten and A. Skowroński, Generalized double tilted algebras, *J. Math. Soc. Japan*, 56(2004), 269–288.

[492] I. Reiten, A. Skowroński and S. Smalø, Short chains and short cycles of modules, *Proc. Amer. Math. Soc.*, 117(1993), 343–354.

[493] I. Reiten and M. Van den Bergh, Grothendieck groups and tilting objects, *Algebras and Repesentation Theory*, 4(2001), 257–272.

[494] I. Reiten and M. Van den Bergh, Noetherian hereditary abelian categories satisfying Serre duality, *J. Amer. Math. Soc.*, 15(2002), 295–366.

[495] J. Rickard, Morita theory for derived categories, *J. London Math. Soc.*, 39(1989), 436–456.

[496] J. Rickard, Derived categories and stable equivalence, *J. Pure Appl. Algebra*, 61(1989), 303–317.

[497] J. Rickard, Some recent advances in modular representation theory, In: *Algebras and Modules I*, Canad. Math. Soc. Conf. Proc., AMS, Vol. 23, 1998, pp. 157–178.

[498] J. Rickard, Equivalences of derived categories for symmetric algebras, *J. Algebra*, 257(2002), 460–481.

[499] C. Riedtmann, Algebren, Darstellungsköcher, Überlagerungen und zurück, *Comment. Math. Helvetici*, 55(1980), 199–224.

[500] C. Riedtmann, Representation-finite selfinjective algebras of class $\mathbb{A}_n$, In: *Representation Theory* II, Lecture Notes in Math., No. 832, Springer-Verlag, Berlin, Heidelberg, New York, 1980, pp. 449–520.

[501] C. Riedtmann, Representation-finite selfinjective algebras of class $\mathbb{D}_n$, *Compositio Math.*, 49(1983), 231–282.

[502] C. Riedtmann, Degenerations for quivers with relations, *Ann. Sci. École Norm. Sup.*, 4(1986), 275–301.

[503] C. Riedtmann, Lie algebras and coverings, *Comment. Math. Helvetici*, 69(1994), 291–310.

[504] C. Riedtmann, Lie algebras generated by indecomposables, *J. Algebra*, 170(1994), 526–546.

[505] C. Riedtmann, Tame quivers, semi-invariants and complete intersections, *J. Algebra*, 279(2004), 362–382.

[506] C. Riedtmann and A. Schofield, On open orbits and their complements, *J. Algebra*, 130(1990), 388–411.

[507] C. Riedtmann and A. Schofield, On a simplicial complex associated with tilting modules, *Comment. Math. Helvetici*, 66(1991), 70–78.

[508] C. Riedtmann and G. Zwara, On the zero set of semi-invariants for quivers, *Ann. Sci. École Norm. Sup.*, 36(2003), 969–976.

[509] C. Riedtmann and G. Zwara, On the zero set of semi-invariants for tame quivers, *Comment. Math. Helvetici*, 79(2004), 350–361.

[510] C. Riedtmann and G. Zwara, The zero set of semi-invariants for extended Dynkin quivers, Preprint, Toruń, 2006.

[511] C. M. Ringel, The indecomposable representations of the dihedral 2-groups, *Math. Ann.*, 214(1975), 19–34.

[512] C. M. Ringel, Representation type of local algebras, In: *Representations of Algebras*, Lecture Notes in Math., No. 488, Springer-Verlag, Berlin, Heidelberg, New York, 1975, pp. 282–305.

[513] C. M. Ringel, Unions of chains of indecomposable modules, *Comm. Algebra*, 3(1975), 1121–1144.

[514] C. M. Ringel, Representations of K-species and bimodules, *J. Algebra*, 41(1976), 269–302.

[515] C. M. Ringel, Finite dimensional hereditary algebras of wild representation type, *Math. Z.*, 161(1978), 235–255.

[516] C. M. Ringel, Report on the Brauer-Thrall conjectures: Rojter's theorem and the theorem of Nazarova and Rojter, In: *Representation Theory I*, Lecture Notes in Math., No. 831, Springer-Verlag, Berlin, Heidelberg, New York, 1980, pp. 104–136.

[517] C. M. Ringel, Report on the Brauer-Thrall conjectures: Tame algebras, In: *Representation Theory I*, Lecture Notes in Math. No. 831, Springer-Verlag, Berlin, Heidelberg, New York, 1980, pp. 137–287.

[518] C. M. Ringel, The rational invariants of the tame quivers, *Invent. Math.*, 58(1980), 217–239.

[519] C. M. Ringel, Reflection functors for hereditary algebras, *J. London Math. Soc.*, 21(1980), 465–479.

[520] C. M. Ringel, Kawada's theorem, In: *Abelian Group Theory,* Lecture Notes in Math., No. 874, Springer-Verlag, Berlin, Heidelberg, New York, 1981, pp. 431–447.

[521] C. M. Ringel, Indecomposable representations of finite-dimensional algebras, In: *Proceedings Intern. Congress of Math.*, PWN - Polish Scientific Publishers, Warszawa, 1983, pp. 425–436.

[522] C. M. Ringel, Bricks in hereditary length categories, *Result. Math.,* 6(1983), 64–70.

[523] C. M. Ringel, Separating tubular series, In: *Séminaire d'Algèbre Paul Dubreil et Marie-Paul Malliavin,* Lecture Notes in Math., No. 1029, Springer-Verlag, Berlin, Heidelberg, New York, 1983, pp. 134–158.

[524] C. M. Ringel, Unzerlegbare Darstellungen endlich-dimensionaler Algebren, *Jber. Deutche Math. Verein,* 85(1983), 86–105.

[525] C. M. Ringel, *Tame Algebras and Integral Quadratic Forms,* Lecture Notes in Math., No. 1099, Springer-Verlag, Berlin, Heidelberg, New York, 1984, pp. 1–371.

[526] C. M. Ringel, Representation theory of finite-dimensional algebras, In: *Representations of Algebras,* London Math. Soc. Lecture Notes Series 116(1986), pp. 1–79.

[527] C. M. Ringel, The regular components of Auslander–Reiten quiver of a tilted algebra, *Chinese Ann. Math.*, 9B(1988), 1–18.

[528] C. M. Ringel, The canonical algebras, with an appendix by W. Crawley-Boevey, In: *Topics in Algebra, Part 1: Rings and Representations of Algebras,* (Eds: S. Balcerzyk, T. Józefiak, J. Krempa, D. Simson, W. Vogel), Banach Center Publications, Vol. 26, PWN - Polish Scientific Publishers, Warszawa, 1990, pp. 407–439.

[529] C. M. Ringel, Hall polynomials for the representation-finite hereditary algebras, *Advances Math.* 84(1990), 137–178.

[530] C. M. Ringel, Hall algebras and quantum groups, *Invent. Math.,* 101(1990), 583–591.

[531] C. M. Ringel, Recent advances in the representation theory of finite dimensional algebras, In: *Representation Theory of Finite Groups and Finite-Dimensional Algebras,* Progress in Mathematics 95, Birkhäuser-Verlag, Basel, 1991, pp. 137–178.

[532] C. M. Ringel, The category of modules with good filtrations over quasi-hereditary algebra has almost split sequences, *Math. Z.,* 208(1991), 209–223.

[533] C. M. Ringel, The spectral radius of the Coxeter transformations for a generalized Cartan matrix, *Math. Ann.*, 300(1994), 331–339.

[534] C. M. Ringel, Cones, In: *Representation Theory of Algebras*, Canad. Math. Soc. Conf. Proc., AMS, Vol. 18, 1996, pp. 583–586.

[535] C. M. Ringel, The Liu-Schultz example, In: *Representation Theory of Algebras*, Canad. Math. Soc. Conf. Proc., AMS, Vol. 18, 1996, pp. 587–600.

[536] C. M. Ringel, A construction of endofinite modules, In: *Advances in Algebra and Model Theory*, (Eds: M. Droste and R. Göbel), Algebra, Logic and Applications Series, Vol. 9, Gordon & Breach Science Publishers, Australia, 1997, pp. 387–399.

[537] C. M. Ringel, The repetitive algebra of a gentle algebra, *Bol. Soc. Mat. Mexicana*, 3(1997), 235–253.

[538] C. M. Ringel, The preprojective algebra of a tame quiver: the irreducible components of the module varieties, In: *Trends in Representation Theory of Finite Dimensional Algebras, Contemp. Math.*, 229(1998), 293–306.

[539] C. M. Ringel, The prepojective algebra of a quiver, In: *Algebras and Modules II*, Canad. Math. Soc. Conf. Proc., AMS, Vol. 24, 1998, pp. 467–480.

[540] C. M. Ringel, The Ziegler spectrum over a tame hereditary algebra, *Colloq. Math.*, 76(1998), 105–115.

[541] C. M. Ringel, Tame algebras are Wild, *Algebra Colloq.*, 6(1999), 473–490.

[542] C. M. Ringel, Infinite length modules. Some examples as introduction, In: *Infinite Length Modules* (Eds: H. Krause and C. M. Ringel), Trends in Mathematics, Birkhäuser Verlag, Basel-Boston-Berlin, 2000, pp. 1–73.

[543] C. M. Ringel, On generic modules for string algebras, *Bol. Soc. Mat. Mexicana*, 7(2001), 85–97.

[544] C. M. Ringel, Combinatorial representation theory. History and future, In: *Representations of Algebras*, Vol I, Proc. Conf. ICRA IX, Beijing 2000, (Eds: D. Happel and Y. B. Zhang), Beijing Normal University Press, 2002, pp. 122–144.

[545] C. M. Ringel, The Gabriel-Roiter measure, *Bull. Sci. Math.*, 129(2005), 726–748.

[546] C. M. Ringel, Foundation of the representation theory of Artin algebras, using the Gabriel-Roiter measure, In: *Trends in Representation Theory of Algebras and Related Topics, Contemp. Math.*, 406(2006), 105–135.

[547] C. M. Ringel, The theorem of Bo Chen and Hall polynomials, *Nagoya Math. J.*, 183(2006), 143–160.

[548] C. M. Ringel and K. W. Roggenkamp, Diagrammatic methods in representation theory of orders, *J. Algebra*, 60(1979), 11–42.

[549] C. M. Ringel and H. Tachikawa, QF-3 rings, *J. reine angew. Math.*, 272(1975), 49–72.

[550] C. M. Ringel and D. Vossieck, Hammocks, *Proc. London Math. Soc.*, 54(1987), 216–246.

[551] A. V. Roiter, Unboundedness of the dimension of the indecomposable representations of an algebra which has infinitely many indecomposable representations, *Izv. Akad. Nauk. SSSR. Ser. Mat.*, 32(1968), 1275–1282.

[552] A. V. Roiter, Matrix problems and representations of bisystems, *Zap. Nauchn. Sem. Leningrad. Otdel. Mat. Inst. Steklov. (LOMI)*, 28 (1972), pp. 130–143 (in Russian).

[553] A. V. Roiter, Matrix problems, In: *Proceedings Intern. Congress of Math.*, Helsinki, 1978, Vol. 1, Academia Scientiarum Fennica, 1980, pp. 319–322.

[554] A. V. Roiter, Matrix problems and representations of bocses, In: *Representations and Quadratic Forms*, Akad. Nauk Ukr. S.S.R., Inst. Matem., Kiev, 1979, 3–38 (in Russian).

[555] R. Rouquier, Representation dimensions of exterior algebras, *Invent, Math.*, 165(2006), 357–367.

[556] A. H. Schofield, *Representations of Rings over Skew Fields*, London Math. Soc. Lecture Notes Series, No. 92, Cambridge University Press, 1985.

[557] A. H. Schofield, Artin's problems for skew field extensions, *Math. Proc. Camb. Phil. Soc.*, 97(1985), 1–6.

[558] A. H. Schofield, Hereditary artinian rings of finite representation type and extensions of simple artinian rings, *Math. Proc. Camb. Phil. Soc.* 102(1987), 411–420.

[559] A. Schofield, Semi-invariants of quivers, *J. London Math. Soc.*, 43(1991), 385–395.

[560] A. Schofield, General representations of quivers, *Proc. London Math. Soc.*, 65(1992), 46–64.

[561] A. Schofield and M. Van den Bergh, Semi-invariants of quivers for arbitrary dimension vectors, *Indag. Math.*, (N.S.) 12(2001), 125–138.

[562] J. Schröer, On the infinite radical of a module category, *Proc. London Math. Soc.*, 81(2000), 651–674.

[563] J. Schröer, On the Krull-Gabriel dimension of an algebra, *Math. Z.*, 233(2000), 287–303.

[564] J. Schröer, Varieties of pairs of nilpotent matrices annihilating each other, *Comment. Math. Helvetici,* 79(2004), 396–426.

[565] V. V. Sergeichuk, Canonical matrices for linear matrix problems, *Linear Algebra Appl.,* 317(2000), 53–102.

[566] W. Sierpiński, *Elementary Theory of Numbers,* Warszawa, 1964.

[567] D. Simson, Functor categories in which every flat object is projective, *Bull. Polish Acad. Sci., Ser. Math.,* 22(1974), 375–80.

[568] D. Simson, On pure global dimension of locally finitely presented Grothendieck categories, *Fund. Math.* 96(1977), 91–116.

[569] D. Simson, Pure semisimple categories and rings of finite representation type, *J. Algebra,* 48(1977), 290–296; Corrigendum 67(1980), 254–256.

[570] D. Simson, Categories of representations of species, *J. Pure Appl. Algebra* 14(1979), 101–114.

[571] D. Simson, Partial Coxeter functors and right pure semisimple hereditary rings, *J. Algebra* 71(1981), 195-218.

[572] D. Simson, Vector space categories, right peak rings and their socle projective modules, *J. Algebra* 92(1985), 532–571.

[573] D. Simson, Socle reductions and socle projective modules, *J. Algebra,* 103(1986), 18–68.

[574] D. Simson, Representations of bounded stratified posets, coverings and socle projective modules, In: *Topics in Algebra, Part 1: Rings and Representations of Algebras,* (Eds: S. Balcerzyk, T. Józefiak, J. Krempa, D. Simson, W. Vogel), Banach Center Publications, Vol. 26, PWN - Polish Scientific Publishers, Warszawa, 1990, pp. 499–533.

[575] D. Simson, *Linear Representations of Partially Ordered Sets and Vector Space Categories,* Algebra, Logic and Applications, Vol. 4, Gordon & Breach Science Publishers, 1992.

[576] D. Simson, Right peak algebras of two-separate stratified posets, their Galois coverings and socle projective modules, *Comm. Algebra* 20 (1992), 3541–3591.

[577] D. Simson, On representation types of module subcategories and orders, *Bull. Pol. Acad. Sci., Ser. Math.,* 41(1993), 77–93.

[578] D. Simson, Triangles of modules and non-polynomial growth, *C. R. Acad. Sci. Paris,* Série I, 321(1995), 33–38.

[579] D. Simson, A reduction functor, tameness and Tits form for a class of orders, *J. Algebra,* 174(1995), 430–452.

[580] D. Simson, On bimodule matrix problems and artinian bipartite piecewise peak PI-rings of finite prinjective module type, *Math. J. Okayama Univ.,* 35(1995), 89–138.

[581] D. Simson, Representation embedding problems, categories of extensions and prinjective modules, In: *Representation Theory of Algebras*, Canad. Math. Soc. Conf. Proc., AMS, Vol. 18, 1996, pp. 601–639.

[582] D. Simson, Prinjective modules, propartite modules, representations of bocses and lattices over orders, *J. Math. Soc. Japan*, 49(1997), 31–68.

[583] D. Simson, A class of potential counter-examples to the pure semisimplicity conjecture, In: *Advances in Algebra and Model Theory*, (Eds: M. Droste and R. Göbel), Algebra, Logic and Applications Series, Vol. 9, Gordon & Breach Science Publishers, Australia, 1997, pp. 345–373.

[584] D. Simson, Representation types, Tits reduced quadratic forms and orbit problems for lattices over orders, *Contemp. Math.*, 229(1998), 307–342.

[585] D. Simson, An Artin problem for division ring extensions and the pure semisimplicity conjecture, II, *J. Algebra*, 227(2000), 670–705.

[586] D. Simson, A reduced Tits quadratic form and tameness of three-partite subamalgams of tiled orders, *Trans. Amer. Math. Soc.*, 352(2000), 4843–4875.

[587] D. Simson, Cohen-Macaulay modules over classical orders, In: *Interactions between Ring Theory and Representations of Algebras*, Lecture Notes in Pure and Appl. Math., Vol. 221, Marcel-Dekker, 2000, pp. 345–382.

[588] D. Simson, On coalgebras of tame comodule type, In: *Representations of Algebras*, Vol II, Proc. Conf. ICRA IX, Beijing 2000, (Eds: D. Happel and Y. B. Zhang), Beijing Normal University Press, 2002, pp. 450–486.

[589] D. Simson, Coalgebras, comodules, pseudocompact algebras and tame comodule type, *Colloq. Math.*, 90(2001), 101–150.

[590] D. Simson, An endomorphism algebra realisation problem and Kronecker embeddings for algebras of infinite representation type, *J. Pure Appl. Algebra*, 172(2002), 293–303.

[591] D. Simson, On large indecomposable modules, endo-wild representation type and right pure semisimple rings, *Algebra and Discr. Math.*, 2(2003), 93–118.

[592] D. Simson, Path coalgebras of quivers with relations and a tame-wild dichotomy problem for coalgebras, *Lecture Notes in Pure and Appl. Math.*, Marcel-Dekker, 236(2004), pp. 465–492.

[593] D. Simson, On Corner type Endo-Wild algebras, *J. Pure Appl. Algebra*, 202(2005), 118-132.

[594] D. Simson, Irreducible morphisms, the Gabriel quiver and colocalisations for coalgebras, *Intern. J. Math. Math. Sci.*, 72(2000), 1–10.

[595] D. Simson, Hom-computable coalgebras, a composition factors matrix and a bilinear form of an Euler coalgebra, *J. Algebra*, 315(2007), in press.

[596] D. Simson, Localising embeddings of comodule categories with applications to tame and Euler coalgebras, *J. Algebra*, 312(2007), 455–494.

[597] D. Simson, Representation-directed incidence coalgebras of intervally finite posets and the tame-wild dichotomy, *Comm. Algebra*, 35(2007), in press.

[598] D. Simson, Incidence coalgebras of intervally finite posets, their integral quadratic forms and comodule categories, Preprint, Toruń, 2006.

[599] D. Simson, Tame-wild dichotomy for coalgebras, Preprint, Toruń, 2006.

[600] D. Simson and A. Skowroński, Extensions of artinian rings by hereditary injective modules, In: *Representations of Algebras*, Lecture Notes in Math., No. 903, Springer-Verlag, Berlin, Heidelberg, New York, 1981, pp. 315–330.

[601] D. Simson and A. Skowroński, The Jacobson radical power series of module categories and the representation type, *Bol. Soc. Mat. Mexicana*, 5(1999), 223–236.

[602] D. Simson and A. Skowroński, Hereditary stable tubes in module categories, Preprint, Toruń, 2007.

[603] A. Skowroński, The representation type of group algebras, CISM Courses and Lectures No. 287, pp. 517–531, Springer-Verlag, Wien - New York, 1984.

[604] A. Skowroński, Tame triangular matrix algebras over Nakayama algebras, *J. London Math. Soc.*, 34(1986), 245–264.

[605] A. Skowroński, On tame triangular matrix algebras, *Bull. Polish Acad. Sci., Ser. Math.*, 34(1986), 517–523.

[606] A. Skowroński, Group algebras of polynomial growth, *Manuscr. Math.*, 59 (1987), 499–516.

[607] A. Skowroński, On algebras with finite strong global dimension, *Bull. Polish Acad. Sci., Ser. Math.*, 35(1987), 539–547.

[608] A. Skowroński, Generalization of Yamagata's theorem on trivial extensions, *Archiv Math. (Basel)*, 48(1987), 68–76.

[609] A. Skowroński, Selfinjective algebras of polynomial growth, *Math. Ann.*, 285 (1989), 177–199.

[610] A. Skowroński, Algebras of polynomial growth, In: *Topics in Algebra, Part 1: Rings and Representations of Algebras*, (Eds: S. Balcerzyk,

T. Józefiak, J. Krempa, D. Simson, W. Vogel), Banach Center Publications, Vol. 26, PWN - Polish Scientific Publishers, Warszawa, 1990, pp. 535–568.

[611] A. Skowroński, Simply connected algebras and Hochschild cohomologies, In: *Representations of Algebras*, Canad. Math. Soc. Conf. Proc., AMS, Vol. 14, 1993, pp. 431–447.

[612] A. Skowroński, Generalized standard Auslander–Reiten components without oriented cycles, *Osaka J. Math.*, 30(1993), 515–527.

[613] A. Skowroński, Regular Auslander–Reiten components containing directing modules, *Proc. Amer. Math. Soc.*, 120(1994), 19–26.

[614] A. Skowroński, Cycles in module categories, In: *Finite Dimensional Algebras and Related Topics*, NATO ASI Series, Series C: Mathematical and Physical Sciences, Vol. 424, Kluwer Academic Publishers, Dordrecht, 1994, pp. 309–345.

[615] A. Skowroński, Generalized standard Auslander–Reiten components, *J. Math. Soc. Japan*, 46(1994), 517–543.

[616] A. Skowroński, On the composition factors of periodic modules, *J. London Math. Soc.*, 49(1994), 477–492.

[617] A. Skowroński, Minimal representation-infinite artin algebras, *Math. Proc. Cambridge Phil. Soc.*, 116(1994), 229–243.

[618] A. Skowroński, Module categories over tame algebras, In: *Representation Theory of Algebras and Related Topics*, Canad. Math. Soc. Conf. Proc., AMS, Vol. 19, 1996, pp. 281–313.

[619] A. Skowroński, On omnipresent tubular families of modules, In: *Proceedings of the Seventh International Conference on Representation Theory of Algebras*, Canad. Math. Soc. Conf. Proc., AMS, Vol. 18, 1996, pp. 641–657.

[620] A. Skowroński, Simply connected algebras of polynomial growth, *Compositio Math.*, 109(1997), 99–133.

[621] A. Skowroński, Tame algebras with strongly simply connected Galois coverings, *Colloq. Math.*, 72(1997), 335–351.

[622] A. Skowroński, Tame quasi-tilted algebras, *J. Algebra*, 203(1998), 470–480.

[623] A. Skowroński, Tame module categories of finite dimensional algebras, In: *Trends in Ring Theory*, Canad. Math. Soc. Conf. Proc., AMS, Vol. 22, 1998, pp. 187–219.

[624] A. Skowroński, On the structure of periodic modules over tame algebras, *Proc. Amer. Math. Soc.*, 127(1999), 1941–1949.

[625] A. Skowroński, Generalized canonical algebras and standard stable tubes, *Colloq. Math.*, 90(2001), 77–93.

[626] A. Skowroński, Directing modules and double tilted algebras, *Bull. Polish Acad. Sci., Ser. Math.*, 50(2002), 77–81.

[627] A. Skowroński, Generically directed algebras, *Archiv Math. (Basel)*, 78(2002), 358–361.

[628] A. Skowroński, On artin algebras with almost all indecomposable modules of projective or injective dimension at most one, *Centr. Eur. J. Math.*, 1(2003), 108–122 (electronic).

[629] A. Skowroński, A construction of complex syzygy periodic modules over artin algebras, *Colloq. Math.*, 103(2005), 61–69.

[630] A. Skowroński, Selfinjective algebras: finite and tame type, In: *Trends in Representation Theory of Algebras and Related Topics, Contemp. Math.*, 406(2006), 169–238.

[631] A. Skowroński, Classification of selfinjective algebras of domestic type, Preprint, Toruń, 2007.

[632] A. Skowroński, Classification of selfinjective algebras of polynomial growth, Preprint, Toruń, 2007.

[633] A. Skowroński, Module categories of selfinjective algebras of polynomial growth, Preprint, Toruń, 2007.

[634] A. Skowroński, Derived categories and Euler forms, Preprint, Toruń, 2007.

[635] A. Skowroński and S. Smalø, Artin algebras with only preprojective or preinjective modules are of finite type, *Archiv Math. (Basel)*, 64(1995), 8–10.

[636] A. Skowroński, S. Smalø and D. Zacharia, On the finiteness of the global dimension of Artin rings, *J. Algebra*, 251(2002), 475–478.

[637] A. Skowroński and K. Yamagata, Socle deformations of self-injective algebras, *Proc. London Math. Soc.*, 72(1996), 545–566.

[638] A. Skowroński and K. Yamagata, Stable equivalence of selfinjective algebras of tilted type, *Archiv Math. (Basel)*, 70(1998), 341–350.

[639] A. Skowroński and K. Yamagata, Galois coverings of selfinjective algebras by repetitive algebras, *Trans. Amer. Math. Soc.*, 351(1999), 715–734.

[640] A. Skowroński and K. Yamagata, On selfinjective artin algebras having nonperiodic generalized standard Auslander-Reiten components, *Colloq. Math.*, 96(2003), 235–244.

[641] A. Skowroński and K. Yamagata, On invariability of selfinjective algebras of tilted type under stable equivalences, *Proc. Amer. Math. Soc.*, 132(2004), 659–667.

[642] A. Skowroński and K. Yamagata, Positive Galois coverings of selfinjective algebras, *Advances Math.*, 194(2005), 398–436.

[643] A. Skowroński and K. Yamagata, Stable equivalence of selfinjective algebras of Dynkin type, *Algebras and Representation Theory*, 9(2006), 33–45.

[644] A. Skowroński and K. Yamagata, A general form of non-Frobenius selfinjective algebras, *Colloq. Math.*, 105(2006), 135–141.

[645] A. Skowroński and K. Yamagata, Selfinjective artin algebras with all Auslander-Reiten components generalized standard, Preprint, Toruń, 2007.

[646] A. Skowroński and J. Waschbüsch, Representation-finite biserial algebras, *J. reine angew. Math.*, 345(1983), 172–181.

[647] A. Skowroński and M. Wenderlich, Artin algebras with directing indecomposable projective modules, *J. Algebra*, 165(1994), 507–530.

[648] A. Skowroński and J. Weyman, Semi-invariants of canonical algebras, *Manuscr. Math.*, 100(1999), 391–403.

[649] A. Skowroński and J. Weyman, The algebras of semi-invariants of quivers, *Transform. Groups*, 5(2000), 361–402.

[650] A. Skowroński and G. Zwara, Degenerations of modules with nondirecting indecomposable direct summands, *Canad. J. Math.*, 48(1996), 1091–1120.

[651] A. Skowroński and G. Zwara, On indecomposable modules without self extensions, *J. Algebra*, 195(1997), 151–169.

[652] A. Skowroński and G. Zwara, On the number of discrete indecomposable modules over tame algebras, *Colloq. Math.*, 73(1997), 93–114.

[653] A. Skowroński and G. Zwara, Degenerations of indecomposable modules and tame algebras, *Ann. Sci. École Norm. Sup.*, 31(1998), 153–180.

[654] A. Skowroński and G. Zwara, Degenerations for modules over blocks of group algebras, *Manuscr. Math.*, 97(1998), 143–154.

[655] A. Skowroński and G. Zwara, Degenerations in module varieties with finitely many orbits, In: *Trends in Representation Theory of Finite Dimensional Algebras, Contemp. Math.*, 229(1998), 343–356.

[656] A. Skowroński and G. Zwara, Degeneration-like order on the additive categories of generalized standard Auslander-Reiten components, *Archiv Math. (Basel)*, 74(2000), 11–21.

[657] A. Skowroński and G. Zwara, Derived equivalences of selfinjective algebras preserve singularities, *Manuscr. Math.*, 112(2003), 221–230.

[658] S. Smalø, The inductive step of the second Brauer–Thrall conjecture, *Canad. J. Math.*, 2(1980), 342–349.

[659] S. Smalø, Almost split sequences in categories of representations of quivers, *Proc. Amer. Math. Soc.*, 129(2001), 095–098.

[660] D. Smith, On generalized standard Auslander-Reiten components having only finitely many non-directing modules, *J. Algebra*, 279(2004), 493–513.

[661] N. Snashall and Ø. Solberg, Support varieties and Hochschild cohomology rings, *Proc. London. Math. Soc.*, 88(2004), 705–732.

[662] Ø. Solberg, Support varieties for modules and complexes, In: *Trends in Representation Theory of Algebras and Related Topics, Contemp. Math.*, 406(2006), 239–270.

[663] H. Strauss, On the perpendicular category of a partial tilting module, *J. Algebra*, 144(1991), 43–66.

[664] H. Tachikawa and T. Wakamatsu, Tilting functors and stable equivalence for self-injective algebras, *J. Algebra*, 109(1987), 138–165.

[665] L. Unger, The concealed algebras of the minimal wild hereditary algebras, *Bayr. Math. Schriften*, 31(1990), 145 154.

[666] L. Unger, The simplicial complex of tilting modules over quiver algebras, *Proc. London Math. Soc.*, 73(1996), 27–46.

[667] R. Vila-Freyer and W. W. Crawley-Boevey, The structure of biserial algebras, *J. London. Math. Soc.*, 57(1998), 41–54.

[668] D. Vossieck, Représentations de bifoncteurs et interprétation en termes de modules, *C. R. Acad. Sci. Paris*, Série I, 307(1988), 713–716.

[669] B. Wald and J. Waschbüsch, Tame biserial algebras, *J. Algebra*, 95(1985), 480–500.

[670] J. Waschbüsch, Symmetrische Algebren vom endlichen Modultyp, *J. reine angew. Math.*, 321(1981), 78–98.

[671] J. Waschbüsch, On selfinjective algebras of finite representation type, Monographs of Institute of Mathematics, Vol. 14, UNAM Mexico, 1983.

[672] C. E. Watts, Intrinsic characterisation of some additive functors, *Proc. Amer. Math. Soc.*, 11(1960), 5–8.

[673] P. Webb, The Auslander-Reiten quiver of a finite group, *Math. Z.*, 179(1982), pp. 79–121.

[674] M. Wenderlich, Krull dimension of strongly simply connected algebras, *Bull. Polish Acad. Sci., Ser. Math.*, 44(1996), 473–480.

[675] D. Woodcock, Some categorical remarks on the representation theory of coalgebras, *Comm. Algebra*, 25(1997), 2775–2794.

[676] C. C. Xi, On the finitistic dimension conjecture, I. Related to representation-finite algebras, *J. Pure Appl. Algebra*, 193(2004), 287–305; Erratum: 202(2005), 325–328.

[677] C. C. Xi, On the finitistic dimension conjecture, II. Related to finite global dimension, *Advances Math,* 201(2006), 116–142.

[678] J. Xiao and B. Zhu, Locally finite triangulated categories, *J. Algebra,* 290(2005), 473–490.

[679] K. Yamagata, Frobenius algebras, In: *Handbook of Algebra,* (Ed.: M. Hazewinkel), Vol. 1, North-Holland Elsevier, Amsterdam, 1996, pp. 841–887.

[680] T. Yoshi, On algebras of bounded representation type, *Osaka Math. J.,* 8(1956), 51–105.

[681] Y. Yoshino, *Cohen-Macaulay Modules over Cohen-Macaulay Rings,* London Math. Soc. Lecture Notes Series, Vol. 146, Cambridge University Press, 1990.

[682] A. G. Zavadskij, Sincere partially ordered sets of finite growth, Akad. Nauk Ukr. S.S.R., Inst. Matem., Kiev, 1981, Preprint 81.27, pp. 30–42 (in Russian).

[683] A. G. Zavadskij, Representations of partially ordered sets of finite growth, Kievskij Ordena Trudovovo Krasnovo Znameni Inžinerno-Stroitelnyi Institut (KISI), Dep. Ukr. NIINTI, 413 - Yk -D83, Kiev, 1983, pp. 1–76 (in Russian).

[684] A. G. Zavadskij, The structure of representations of partially ordered sets of finite growth, In: *Linear Algebra and Representation Theory,* Akad. Nauk Ukr. S.S.R., Inst. Matem., Kiev, 1983, pp. 55–67 (in Russian).

[685] A. G. Zavadskij, The Auslander-Reiten quivers for posets of finite growth, In: *Topics in Algebra, Part 1: Rings and Representations of Algebras,* (Eds: S. Balcerzyk, T. Józefiak, J. Krempa, D. Simson, W. Vogel), Banach Center Publications, Vol. 26, PWN - Polish Scientific Publishers, Warszawa, 1990, pp. 569–587.

[686] A. G. Zavadskij, Equipped posets of finite growth, In: *Representatios of Algebras and Related Topics, Fields Inst. Commun.,* 45(2005), 363–396.

[687] M. V. Zeldich, Sincere weakly positive unit quadratic forms, In: *Representations of Algebras,* Canad. Math. Soc. Conf. Proc., AMS, Vol. 14, 1993, pp. 453–461.

[688] Y. Zhang, The modules in any component of the *AR*-quiver of a wild hereditary algebra are uniquely determined by their composition factors, *Archiv Math. (Basel),* 53(1989), 250–251.

[689] Y. Zhang, The structure of stable components, *Canad. J. Math.,* 43(1991), 652–672.

[690] W. Zimmermann, Einige Charakterisierung der Ringe über denen reine Untermoduln direkte Summanden sind, *Bayer. Akad. Wiss. Math.-Natur.*, Abt. II(1973), 77–79.

[691] B. Zimmermann-Huisgen and W. Zimmermann, On the sparsity of representations of rings of pure global dimension zero, *Trans. Amer. Math. Soc.*, 320(1990), 695-711.

[692] B. Zimmermann-Huisgen, Homological domino effects and the first finitistic dimension conjecture, *Invent. Math.*, 108(1992), 369–383.

[693] B. Zimmermann-Huisgen, The finitistic dimension conjectures - a tale of 3.5 decades, In: *Abelian Groups and Modules*, Math. Appl., 343, Kluwer Acad. Publ., Dordrecht, 1995, pp. 501–517.

[694] G. Zwara, Degenerations for modules over representation-finite algebras, *Proc. Amer. Math. Soc.*, 127(1999), 1313–1322.

[695] G. Zwara, Immersions of module varieties, *Colloq. Math.*, 82(1999), 287–299.

[696] G. Zwara, Degenerations of finite dimensional modules are given by extensions, *Compositio Math.*, 121(2000), 205–218.

[697] G. Zwara, Smooth morphisms of module schemes, *Proc. London Math. Soc.*, 84(2002), 239–258.

[698] G. Zwara, Unibranch orbit closures in module varieties, *Ann. Sci. École Norm. Sup.*, 35(2002), 877–895.

[699] G. Zwara, An orbit closure for a representation of the Kronecker quiver with bad singularities, *Colloq. Math.*, 97(2003), 81–86.

[700] G. Zwara, Regularity of codimension one orbit closures in module varieties, *J. Algebra*, 283(2005), 821–848.

[701] G. Zwara, Orbit closures for representations of Dynkin quivers are regular in codimension two, *J. Math. Soc. Japan*, 57(2005), 859–880.

[702] G. Zwara, Singularities of orbit closures in module varieties and cones over rational normal curves, *J. London Math. Soc.*, 74(2006), 623–638.

[703] G. Zwara, Codimension two singularities for representations of extended Dynkin quivers, *Manuscr. Math.*, 2007, in press.

Index

451

List of symbols

Printed in the United States
by Baker & Taylor Publisher Services

Printed in the United States
by Baker & Taylor Publisher Services